Lifestyle in Heart Health and Disease

Lifestyle in Heart Health and Disease

Edited By

Ronald Ross Watson

Sherma Zibadi

ACADEMIC PRESS

An imprint of Elsevier

Academic Press is an imprint of Elsevier
125 London Wall, London EC2Y 5AS, United Kingdom
525 B Street, Suite 1800, San Diego, CA 92101-4495, United States
50 Hampshire Street, 5th Floor, Cambridge, MA 02139, United States
The Boulevard, Langford Lane, Kidlington, Oxford OX5 1GB, United Kingdom

Notices
Knowledge and best practice in this field are constantly changing. As new research and experience broaden our
understanding, changes in research methods, professional practices, or medical treatment may become necessary.

Practitioners and researchers must always rely on their own experience and knowledge in evaluating and
using any information, methods, compounds, or experiments described herein. In using such information or
methods they should be mindful of their own safety and the safety of others, including parties for whom they
have a professional responsibility.

To the fullest extent of the law, neither the Publisher nor the authors, contributors, or editors, assume any liability
for any injury and/or damage to persons or property as a matter of products liability, negligence or otherwise, or
from any use or operation of any methods, products, instructions, or ideas contained in the material herein.

Library of Congress Cataloging-in-Publication Data
A catalog record for this book is available from the Library of Congress

British Library Cataloguing-in-Publication Data
A catalogue record for this book is available from the British Library

ISBN 978-0-12-811279-3

For information on all Academic Press publications
visit our website at https://www.elsevier.com/books-and-journals

Working together
to grow libraries in
developing countries

www.elsevier.com • www.bookaid.org

Publisher: Mica Haley
Acquisition Editor: Stacy Masucci
Editorial Project Manager: Samuel Young
Production Project Manager: Anusha Sambamoorthy
Cover Designer: Greg Harris

Typeset by SPi Global, India

Contents

Part I
Overview and Mechanisms

1. Cardiovascular Diseases, Obesity, and Lifestyle Changes

John Kwagyan, Otelio S. Randall

2. Public Knowledge of Cardiovascular Risk Numbers: Contextual Factors Affecting Knowledge and Health Behavior, and the Impact of Public Health Campaigns

Jeffrey L. Kibler, Mindy Ma, Jacquelyn Hrzich, Roberta A. Roas

3. Extension of Peer Support from Diabetes Management to Cardiovascular Disease Prevention and Management in Primary Care and Community Settings in Anhui Province, China

Xuefeng Zhong, Bert Potemans, Chanuantong Tanasugarn, Edwin B. Fisher

12. Lifestyle and Heart Diseases in Choice Experiments

José M. Grisolía

13. Lost in Translation: What Does the Physical Activity and Health Evidence Actually Tell Us?

Darren E.R. Warburton, Shannon S.D. Bredin

14. Community-Based Maintenance Cardiac Rehabilitation

Sandra Mandic, Anna Rolleston, Garrick Hately, Stacey Reading

15. Determinants of Exercise Ventilatory Inefficiency in Heart Failure With Reduced or Preserved Ejection Fraction: Application of Classical and Emerging Integrative Physiology Concepts

Erik H. Van Iterson, Thomas P. Olson

Part III
Alcohol, Tobacco, and Other Drugs of Abuse

16. Relationships of Alcohol Consumption With Risks for Type 2 Diabetes Mellitus and Cardiovascular Disease in Men and Women

Ichiro Wakabayashi

17. Lifestyle Features and Heart Disease

Vijay Singh, Ronald Ross Watson

18. Prevention of Cardiovascular Disease Among People Living With HIV: A Tailored Smoking Cessation Program Treating Depression

Stephanie Wiebe, Lousie Balfour, Paul A. McPherson

19. Factors Associated With Tobacco Use Among Patients With MCC: Multidisciplinary Visions about the Lifestyle on Health and Cardiovascular Disease

Arise G. de Siqueira Galil, Arthur da Silva Gomes, Bárbara A.B.B. de Andrade, Mariana M. Gusmão, Tatiane da Silva Campos, Marcela M. de Melo, Eliane F.C. Banhato

Part IV
Social, Population, and Family Effects on the Heart and Arteries

Contributors

Numbers in parenthesis indicate the pages on which the authors' contributions begin.

Tolulope A. Adesiyun (99) Division of Cardiology, Department of Medicine, Johns Hopkins School of Medicine, Baltimore, MD, United States

Amelia Aranega (285) Department of Experimental Biology, University of Jaén, Jaén, Spain

Lousie Balfour (227) Department of Psychology, The Ottawa Hospital, University of Ottawa, Ottawa, ON, Canada

Eliane F.C. Banhato (233) IMEPEN Foundation, Faculty of Medicine, Federal University of Juiz de Fora; Department of Psychology, Higher Education Center of Juiz de Fora, Juiz de Fora, Minas Gerais, Brazil

Ayşegül Bayir (297) Faculty of Medicine, Emergency Department, Selçuk University, Konya, Turkey

Yihua Bei (87) Cardiac Regeneration and Ageing Lab, School of Life Science, Shanghai University, Shanghai, China

Khaldoun Ben Hamda (285) Cardiology Service, Fattouma Bourguiba Hospital, Monastir, Tunisia

Courtney Boen (291) University of Pennsylvania, Philadelphia, PA, United States

Shannon S.D. Bredin (175) Physical Activity Promotion and Chronic Disease Prevention Unit, University of British Columbia (UBC), Vancouver, BC, Canada

Cecilia Bukutu (275) Institutional Research and Program Development, Concordia University of Edmonton, Edmonton, AB, Canada

Jemni B.E. Chibani (285) Biochemistry and Molecular Biology Laboratory, Faculty of Pharmacy, University of Monastir, Monastir, Tunisia

Tatiane da Silva Campos (233) IMEPEN Foundation, Faculty of Medicine, Federal University of Juiz de Fora, Juiz de Fora, Minas Gerais; Faculty of Nursing, State University of Rio de Janeiro, Rio de Janeiro, Brazil

Arthur da Silva Gomes (233) IMEPEN Foundation, Faculty of Medicine, Federal University of Juiz de Fora, Juiz de Fora, Minas Gerais; Department of Biomedical Sciences, Federal University of Fluminense, Niterói, Rio de Janeiro, Brazil

Qiying Dai (87) Department of Cardiology, The First Affiliated Hospital of Nanjing Medical University, Nanjing, China; Metrowest Medical Center, Framingham, MA, United States

Houria Daimi (285) Department of Experimental Biology, University of Jaén, Jaén, Spain; Biochemistry and Molecular Biology Laboratory, Faculty of Pharmacy, University of Monastir, Monastir, Tunisia

Veronica D'Ambra (267) Department of Internal Medicine, Section on Cardiology, Wake Forest University School of Medicine, Winston-Salem, NC, United States

Bárbara A.B.B. de Andrade (233) IMEPEN Foundation, Faculty of Medicine; Institute of Human Sciences, Department of Psychology, Federal University of Juiz de Fora, Juiz de Fora, Minas Gerais, Brazil

Marcela M. de Melo (233) IMEPEN Foundation, Faculty of Medicine, Federal University of Juiz de Fora, Juiz de Fora, Minas Gerais, Brazil

Arise G. de Siqueira Galil (233) IMEPEN Foundation, Faculty of Medicine, Federal University of Juiz de Fora, Juiz de Fora, Minas Gerais, Brazil

Edwin B. Fisher (21) Peers for Progress and Department of Health Behavior, Gillings School of Global Public Health, University of North Carolina at Chapel Hill, Chapel Hill, NC, United States

Ricardo Fontes-Carvalho (153) Cardiology Department, Gaia Hospital Center, Vila Nova de Gaia; Department of Surgery and Physiology, University of Porto, Porto; Unidade de Investigação Cardiovascular (UnIC), Porto, Portugal

Diego Franco (285) Department of Experimental Biology, University of Jaén, Jaén, Spain

Siyi Fu (87) Cardiac Regeneration and Ageing Lab, School of Life Science, Shanghai University, Shanghai, China

Sandra Gilbertson (35) Mosaic Life Care, St. Joseph, MO, United States

Petro Gjini (267) Department of Internal Medicine, Section on Cardiology, Wake Forest University School of Medicine, Winston-Salem, NC, United States

Aaron Gluth (247) Division of Hospital Medicine, Department of Psychiatry and Behavioral Sciences, Emory University, Atlanta, GA, United States

Pedro Gonçalves-Teixeira (153) Cardiology Department, Gaia Hospital Center, Vila Nova de Gaia, Portugal; Department of Surgery and Physiology, University of Porto, Porto, Portugal

Barbara A. Graves (35) Capstone College of Nursing, University of Alabama, Tuscaloosa, AL, United States

Theresa L. Green (301) Queensland University of Technology; Royal Brisbane and Women's Hospital, Brisbane, QLD, Australia

José M. Grisolía (163) Department of Applied Economics, Universidad de Las Palmas de Gran Canaria, Las Palmas de GC, Spain

Mariana M. Gusmão (233) IMEPEN Foundation, Faculty of Medicine, Federal University of Juiz de Fora, Juiz de Fora, Minas Gerais, Brazil; Institute of Human Sciences, Department of Psychology, Federal University of Juiz de Fora, Juiz de Fora, Minas Gerais, Brazil

Amel Haj Khelil (285) Biochemistry and Molecular Biology Laboratory, Faculty of Pharmacy, University of Monastir, Monastir, Tunisia

Garrick Hately (187) University of Otago, Dunedin, New Zealand

Jacquelyn Hrzich (11) College of Psychology, Nova Southeastern University, Fort Lauderdale, FL, United States

Kathleen P. Ismond (275) Department of Pediatrics, University of Alberta, Edmonton, AB, Canada

Sergey Kachur (137) Department of Graduate Medical Education, Ocala Regional Medical Center, Ocala, FL, United States

Jeffrey L. Kibler (11) College of Psychology, Nova Southeastern University, Fort Lauderdale, FL, United States

John Kwagyan (3) Howard University College of Medicine, Washington, DC, United States

Carl J. Lavie (137) Department of Cardiovascular Diseases, John Ochsner Heart and Vascular Institute, Ochsner Clinical School-University of Queensland School of Medicine, New Orleans, LA, United States

Yongqin Li (87) Cardiac Regeneration and Ageing Lab, School of Life Science, Shanghai University, Shanghai, China

Mindy Ma (11) College of Psychology, Nova Southeastern University, Fort Lauderdale, FL, United States

Hayley V. MacDonald (115) Department of Kinesiology, The University of Alabama, Tuscaloosa, AL, United States

Sandra Mandic (187) University of Otago, Dunedin, New Zealand

Paul A. McPherson (227) Department of Psychology, The Ottawa Hospital, University of Ottawa, Ottawa, ON, Canada

A. Menotti (47) Association for Cardiac Research, Rome, Italy

Richard V. Milani (137) Department of Cardiovascular Diseases, John Ochsner Heart and Vascular Institute, Ochsner Clinical School-University of Queensland School of Medicine, New Orleans, LA, United States

Michael D. Morledge (137) Department of Cardiovascular Diseases, John Ochsner Heart and Vascular Institute, Ochsner Clinical School-University of Queensland School of Medicine, New Orleans, LA, United States

Lee Moylan (301) Royal Brisbane and Women's Hospital, Brisbane, QLD, Australia

James H. O'Keefe (137) St. Luke's Mid America Heart Institute, University of Missouri-Kansas City, Kansas City, MO, United States

Thomas P. Olson (65, 199) Department of Cardiovascular Medicine, Mayo Clinic, Rochester, MN, United States

Janine Pampellonne (301) Queensland University of Technology, Brisbane, QLD, Australia

Parham Parto (137) Department of Cardiovascular Diseases, John Ochsner Heart and Vascular Institute, Ochsner Clinical School-University of Queensland School of Medicine, New Orleans, LA, United States

Katha Patel (255) University of Arizona, Mel and Enid Zuckerman College of Public Health, Tucson, AZ, United States

Linda S. Pescatello (115) Department of Kinesiology, University of Connecticut, Storrs, CT, United States

Bert Potemans (21) Vårdcentral Rättvik, Rättvik, Sweden

P.E. Puddu (47) Department of Cardiovascular, Respiratory, Nephrological, Anesthesiological and Geriatric Sciences, Sapienza University of Rome, Rome, Italy

Otelio S. Randall (3) Howard University College of Medicine, Washington, DC, United States

Stacey Reading (187) Department of Exercise Sciences, University of Auckland, Auckland, New Zealand

Roberta A. Roas (11) College of Psychology, Nova Southeastern University, Fort Lauderdale, FL, United States

Anna Rolleston (187) Faculty of Medicine and Health Science, University of Auckland, Auckland, New Zealand

Stuart D. Russell (99) Division of Cardiology, Department of Medicine, Johns Hopkins School of Medicine, Baltimore, MD, United States

Vijay Singh (223) University of Arizona, Mel and Enid Zuckerman College of Public Health, Tucson, AZ, United States

Richard B. Stacey (267) Department of Internal Medicine, Section on Cardiology, Wake Forest University School of Medicine, Winston-Salem, NC, United States

Chanuantong Tanasugarn (21) Department of Health Education and Behavioral Science, Faculty of Public Health, Mahidol University, Bangkok, Thailand

Erik H. Van Iterson (65, 199) Department of Cardiovascular Medicine, Mayo Clinic, Rochester, MN, United States

Eduardo M. Vilela (153) Cardiology Department, Gaia Hospital Center, Vila Nova de Gaia, Portugal

Sunita Vohra (275) Department of Pediatrics, University of Alberta, Edmonton, AB, Canada

Ichiro Wakabayashi (213) Department of Environmental and Preventive Medicine, Hyogo College of Medicine, Nishinomiya, Japan

Darren E.R. Warburton (175) Physical Activity Promotion and Chronic Disease Prevention Unit, University of British Columbia (UBC), Vancouver, BC, Canada

Martha Ward (247) Department of Psychiatry and Behavioral Sciences; Department of General Medicine, Emory University, Atlanta, GA, United States

Ronald Ross Watson (223, 255) University of Arizona, Mel and Enid Zuckerman College of Public Health, Tucson, AZ, United States

DeJuan White (247) Department of Psychiatry and Behavioral Sciences; Department of General Medicine, Emory University, Atlanta, GA, United States

Stephanie Wiebe (227) Department of Psychology, The Ottawa Hospital, University of Ottawa, Ottawa, ON, Canada

Junjie Xiao (87) Cardiac Regeneration and Ageing Lab, School of Life Science, Shanghai University, Shanghai, China

Yang C. Yang (291) Lineberger Comprehensive Cancer Center, University of North Carolina at Chapel Hill, Chapel Hill, NC, United States

Xuefeng Zhong (21) Anhui Provincial Center for Disease Control and Prevention, Hefei, China

Acknowledgments

The work of Dr. Watson's editorial assistant, Bethany L. Stevens, in communicating with authors, editors, and working on the manuscripts was critical to the successful completion of the book. It is very much appreciated. Support for Ms. Stevens' and Dr. Watson' editing was graciously provided by the Natural Health Research Institute (http://www.naturalhealthresearch.org) and Southwest Scientific Editing and Consulting, LLC. The encouragement and support of Elwood Richard and Dr. Richard Sharpee was vital. Direction and guidance from Elsevier's staff was critical. Finally, the work of the librarian at the Arizona Health Sciences Library, Mari Stoddard, was vital and very helpful in identifying key researchers who participated in the book.

Part I

Overview and Mechanisms

Chapter 1

Cardiovascular Diseases, Obesity, and Lifestyle Changes

John Kwagyan, Otelio S. Randall

Howard University College of Medicine, Washington, DC, United States

OBESITY AND CARDIOVASCULAR DISEASES

Subjectively, obesity is defined as excess weight or excess body fat. Objectively, obesity is defined by a body mass index, $BMI \geq 30 \, kg/m^2$. Globally, more than 1.5 billion persons are overweight, and the prevalence of obesity has doubled since 1980 [1]. Current reports on the prevalence in the United States estimate that 35.5% of adult men and 35.8% of adult women are obese [2]. Disparities in prevalence among race have been noted with rates higher among minority populations. For the African American population, the age-adjusted rates are 49.5% compared with 39.1% in Hispanics and 34.3% in Caucasians [2]. In recent years, there has been a rise in obesity-specific cardiovascular disease (CVD) morbidity and mortality [3,4]. There is also the concern of dramatic increase of the prevalence and magnitude of obesity in children and adolescent [5,6]. Obesity has been linked with major cardiovascular disease risk factors including hypertension, diabetes mellitus, and dyslipidemia [7–10].

A report by Fontaine et al. [11] indicated a marked reduction in life expectancy due to obesity with significant differences observed between races. In the study, it was reported that the maximum years of life lost for the young adult obese African American man was estimated to be 20 years compared with 13 years for Caucasians. In another study, Fontaine et al. [12] reported the impact of obesity on the health-related quality of life and indicated more impairment in obese persons when compared with the general population. The pattern of the results indicated that, as weight increases, the quality of life related to the physical domains becomes adversely affected. In the general population, the prevalence of these risk factors increases progressively with increasing BMI [7,13–15]. Randall et al. [15] reported a major finding in a cohort of obese African Americans that suggests a threshold of between 30 and $35 \, kg/m^2$ for maximal appearance of CVD risk factors, after adjusting for age and sex. The implication of the results from the study is that preventive measures need to be implemented before reaching the threshold BMI, and there is the need to curtail the incidence of risk factors sooner rather than later. Fig. 1 provides a conceptual pathway for the development of various CV outcomes including diabetes, myocardial infarction, congestive heart failure, left ventricular hypertrophy, and stroke. The pathway clearly suggests the autonomous influence of obesity in the development of these risk factors and associated CV outcomes.

Coexistence of obesity related cardiovascular disease risk factors including hypertension, diabetes, and dyslipidemia is termed the metabolic syndrome [16–19]. Clinically, the metabolic syndrome is described according to the Third National Cholesterol Education Program Adult Treatment Panel (NCEP ATP III) when at least three of the following five features are present: abdominal obesity reflected by a large waistline and characterized with waist circumference > 102 cm (40.2 in.) in men and >88 cm (34.6 in) in women, low high-density lipoprotein (HDL) cholesterol level defined as HDL < 40 mg/dL in men or < 50 mg/dL in women, high serum triglycerides (TG) defined as TG ≥ 150 mg/dL, high blood pressure (BP) defined as ≥ 130/85 mmHg, and high fasting serum glucose (BS) defined as ≥ 110 mg/dL. The metabolic syndrome, an important marker for atherosclerosis—hardening and narrowing of the arteries—including premature coronary artery disease and cardiovascular mortality affects about 25%–35% of the US adults [18–20], and the age-specific rates increased rapidly. A person's risk for heart disease, diabetes, and stroke increases with the number of metabolic syndrome risk factors, and the risk of having metabolic syndrome is closely linked to obesity and a lack of physical activity. There is the report that, in the future, metabolic syndrome may overtake smoking as the leading risk factor for heart disease (https://www.nhlbi.nih.gov/health/health-topics/topics/ms). The prevalence in 50-year-old persons was >30% and that in persons aged 60 years and above was 40%. Epidemiological studies [18] show a highest prevalence of the metabolic syndrome in Hispanics and

Lifestyle in Heart Health and Disease. https://doi.org/10.1016/B978-0-12-811279-3.00001-X

lower in Caucasian and in African Americans. Haffner and Heinrich [20] explained that the lower prevalence of the metabolic syndrome among African Americans may be explained by the two separate lipid criteria defined by the NCEP (high triglycerides and low HDL cholesterol), which offset the higher rates of hypertension and glucose intolerance observed in this ethnic group. Among the African Americans, the metabolic syndrome has been linked with increased insulin levels, elevated blood pressure, dyslipidemia, and endothelial dysfunction [16–18]. In a sample of adolescents, the prevalence of the metabolic syndrome was estimated at 6.8% among overweight adolescents and 28.7% among obese adolescents [20]. Cherqaoui et al. [21] recently noted in a two-factorial analysis that even in the so-called metabolically healthy but obese, the state of the phenotype appears to be only transient. The finding of increased risk of death or cardiovascular events for obese individuals suggests that gaining excess weight is associated with risk that may accumulate over time, even before metabolic and cardiovascular signs become apparent in laboratory tests.

CORONARY HEART DISEASE

Heart disease remains the leading cause of death in the United States [22,23] and worldwide. Overall, heart disease accounts for about one in four deaths. Coronary heart disease (CHD) affects 15.5 million Americans with a prevalence of 6.2% and is responsible for one in seven deaths in the United States [2]. For the African American and Hispanic, however, the morbidity and mortality rate is disproportionally high [22,23]. From 2006 to 2010, the age-adjusted prevalence of coronary heart disease (CHD) in the United States declined overall [23] from 6.7% to 6.0%. However, while declines from 2006 to 2010 were observed across all racial/ethnic groups, the prevalence among Blacks rose in the same period. For Whites, the rates declined from 6.4% to 5.8% and for Hispanics from 6.9% to 6.1%. Among Blacks, the rate increases marginally [23] from 6.4% to 6.5%. It is reported that genetics, environment, lifestyle, and access to health care are the major contributing factors to the negative outcomes seen in African Americans [24]. Coronary heart disease is a multifactorial disease, and a

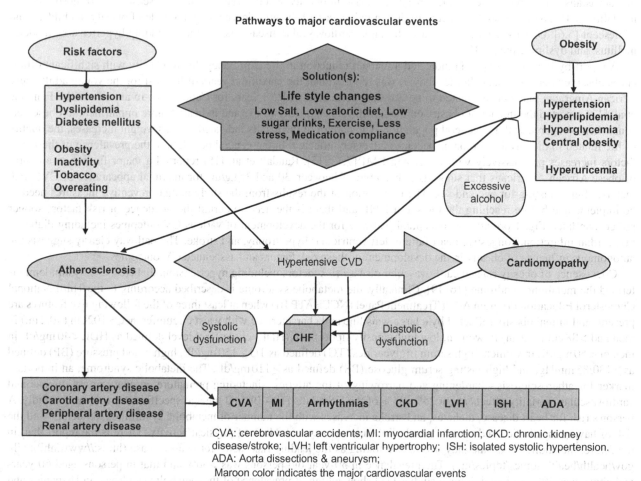

FIG. 1   Risk factors affecting the developmental pathway to major cardiovascular disease events.

Source: *Ethnicity & Disease © 2015. (Kwagyan et al.). Published by ISHIB. Printed with permission. All rights reserved.*

measure of the overall disease risk is necessary and has been derived from the Framingham Heart Study [22], a prospective epidemiological study of cardiovascular diseases. Elevated blood pressure and lipid levels are known to be associated with an increasing risk of cardiovascular disease (CVD) and are part of the Framingham score algorithm.

High serum cholesterol is a known risk factor and a treatment target in primary and secondary prevention of CHD. Results from the MESA study [25] observed that in a cohort free of CHD, there was no significant difference in the prevalence of dyslipidemia among African Americans, Hispanics, and Caucasians. The report showed that African Americans and the Hispanic group were 30% less likely than Caucasians to have their dyslipidemia under control. When these metrics were adjusted for socioeconomic status and health care access, there was attenuation of the disparity, implying that biological differences alone were not responsible for the differences in control.

Hypertension or high blood pressure is a common disease in which blood flows through the blood vessels (arteries) at higher than normal pressures. It is the major risk factor for atherosclerosis and by far the leading risk factor for the development of CVD and the leading cause of death and disability [26–28].

Over time, consistently high blood pressure weakens and damages the blood vessels, which can lead to severe clinical complications including but not limited to the development of heart failure, heart attack, chronic kidney disease, cognitive changes, eye damage, and stroke, to list a few (NHLBI site). The prevalence and incidence of hypertension and hypertension-related complications are reported to be disproportionately higher among African Americans than other ethnicities [29,30]. This has been linked with low social economic status, the lack of access to health care, and a higher non-compliance with prescribed medication usage.

BIOPHYSICAL PROPERTIES OF THE ARTERIAL SYSTEM

While high blood pressure is a major risk factor for cardiovascular events, the role of pulse pressure, an independent predictor of arterial stiffness, is modestly emphasized. Pulse pressure, a marker of the biophysical properties of the arterial system, has been demonstrated in observational and longitudinal studies to show a direct association with carotid atherosclerosis [31,32], left ventricular mass [31], white matter lesions [33], and heart failure [34]. A report by Randall et al. [35] discussed the effect of intensive blood pressure control on cardiovascular remodeling in hypertensive patients with nephrosclerosis and showed that, after covariate adjustment, a higher pulse pressure is associated with increased risk of CV outcome (relative risk $= 1.28$ and CI $= 1.11–1.47$) and development of LVH (relative risk $= 1.26$ and CI $= 1.04–1.54$). Although an association between increased BMI and elevated blood pressure has been established [36,37], limited data exist on the relationship between BMI and PP. Martin and colleagues [38] examined the relationship between BMI and PP in older adults with isolated systolic hypertension and reported PP to increase with BMI values greater than $30 \, \text{kg/m}^2$, while pulse pressure decreased with increasing BMI values less than or equal to $30 \, \text{kg/m}^2$. Randall et al. [39] and Kwagyan et al. [40] showed a closer significant correlation between pulse pressure and BMI than between systolic blood pressure and BMI. These studies further reported that diet and exercise program averaging 10% reductions in BMI, translated into a significant improvement in pulse pressure, stroke volume, a measure of the amount of blood the heart pumps per beat, and arterial compliance, the ratio of stroke volume to pulse pressure. In a recent study, Kwagyan et al. [40] showed in multivariate models in obese African Americans that pulse pressure is associated positively with increasing BMI ($P < .01$), systolic blood pressure ($P = .01$), diabetes mellitus ($P = .04$), and age above 50 years ($P = .01$) and negatively with heart rate ($P = .04$). There was also a strong correlation between pulse pressure and systolic blood pressure ($r = 0.67$ and $P < .01$) and no correlation between pulse pressure and diastolic blood pressure ($r = 0.02$ and $P = .83$). In multiple regression analysis, pulse pressure correlated positively with systolic blood pressure ($P < .01$) and BMI ($P < .01$) and negatively with heart rate ($P = .02$). Fig. 2 shows the relationship of pulse pressure with number of metabolic syndrome components.

As expected, the figure depicts a positive relationship between the number of components and pulse pressure. The results indicate that CVD risk factors including hypertension, dyslipidemia, and diabetes that are also features of the components of the metabolic syndrome constellation impact negatively on arterial vascular properties, as reflected in increased pulse pressure, suggesting that the metabolic syndrome is detrimental to vascular health.

DISTRIBUTION AND CORRELATES OF MAJOR CVD RISK FACTORS IN OBESE AFRICAN AMERICANS

Randall et al. [41] examined the distribution and correlates of major CVD risk factors in 515 obese African Americans, with average age $\pm$ SD was 48.3 ± 9.9 years and BMI of these individuals was $42.9 \pm 6.8 \, \text{kg/m}^2$ (see Table 1). Of these individuals, 27% had dyslipidemia, 56.9% had hypertension, and 24.1% had diabetes mellitus. These rates are higher than those found in the African American population by the third NHANES survey. The study noted that, after adjusting for age and

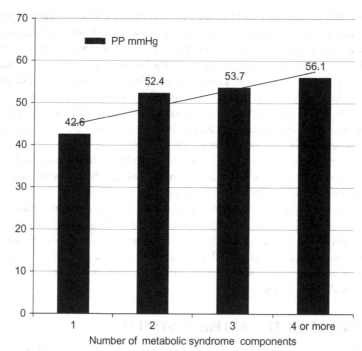

FIG. 2 Relationships of pulse pressure (PP mmHg) to number of metabolic syndrome components.

TABLE 1 Prevalence Rates of the CVD Risk Factors of Hypertension, Dyslipidemia, and Diabetes Mellitus Across BMI Categories

| Risk Factors | BMI Category | | | | | |
	30–34.9 (N=86)	35–39.9 (N=125)	40–44.9 (N=110)	45–49.9 (N=92)	50+ (N=105)	Total (515)
No risk factor	26	22	29	29	27	26
Dyslipidemia	26	34	29	26	19	27
Hypertension	63	62	53	48	63	57
Diabetes mellitus	20	29	22	20	30	24

Data are percent.

sex, there were no significant differences in the prevalence rates of these risk factors according to increasing BMI, suggesting a threshold of between 30 and 34.99 kg/m^2 for maximal appearance of these risk factors. Ninety-nine percent met the metabolic syndrome criteria for waist circumference, 40% met the criteria for BP, and 37% for HDL. Thirty-nine percent had exactly two features of the metabolic syndrome, and 36% were diagnosed with it. The proportion of participants who met the MS criteria for HDL and those diagnosed with it increases significantly across BMI categories. In particular, a logistic regression analysis showed that in these obese individuals, a 5 kg/m^2 rise in BMI increased the risk of the MS by 44%.

AMBULATORY BLOOD PRESSURE MONITORING

Several studies support the prognostic significance of 24 h ambulatory blood pressure monitoring [24–32]. This offers a convenient means of obtaining a large number of repeated BP measurements on each subject and thereby increases the statistical power of detecting significant hemodynamic changes. The diurnal variation of ABP provides a measure of the percentage of BP readings that are elevated, the extent of BP fall during sleep, and thus the overall BP load. Although a nocturnal fall in BP is usual, various groups of individuals including Blacks, diabetics, and subjects who have developed left ventricular hypertrophy (LVH) or renal damage have been noted to have little or no nocturnal fall [24,42–44]. Both observational and longitudinal studies alike have shown that target organ damage accompanying hypertension is closely related to 24 h ABP. Individuals with a lack of a significant nocturnal fall in BP has been linked to more serious target organ damage and higher incidence of cardiovascular complications than those with a nocturnal fall [45–48]. Kwagyan et al. [49] reported in a cohort of obese African Americans that 44% were nondippers in any BP component, with dipping defined as those who exhibit a reduction in average BP by at least 10% from the day to night. The proportions of dippers in SBP were 17%, DBP 54%, and MBP 40%. Ninety-nine percent of SBP dippers were also DBP dippers. Consistent with other studies, the report showed that increasing BMI correlated negatively with DBP dipping ($P=.01$). Systolic BP dipping correlated positively with HTN ($P=.02$) and HR ($P=.06$) and negatively with DM ($P=.02$). The higher percentage of nondipping in these obese African Americans is alarming. A lifestyle change in these obese populations is necessary to improve dipping status and CV health. Evaluation of nocturnal fall in DBP may be of clinical relevance, and more research is needed to establish its prognostic value. The high rate of nondipping in this obese population reinforces the need to reduce BMI and improve hemodynamic and lipid profiles through therapeutic lifestyle changes. This is likely to shift nondippers to dipper status and reduce the potential risk for target organ damage.

DIET AND PHYSICAL ACTIVITY INTERVENTION

The modifiable environmental factors that have been clearly shown to correlate with blood pressure levels and CVD risk are salt intake, physical activity, and excess body weight [39]. A key element of prevention is therefore to identify those individuals most likely to have cardiovascular events and intervention strategies for reduction of associated risk factors. Preventive measures, such as physical activity and nutritious low-salt and low-fat diets, particularly diets low in cholesterol and saturated fats, have been investigated and established to reduce the risk of hypertension and CHD. Kwagyan et al. [49] examined the effect of participating in a diet-exercise program on cardiovascular risk factors and CHD risk prediction among obese African Americans. Variables included in the estimate of CHD risk prediction were age, systolic BP, diastolic BP, total cholesterol, HDL, cigarette smoking, and diabetes. Table 2 showed the average percent changes in anthropometric, hemodynamic, and lipid measures after 6-month diet and exercise intervention from the study.

TABLE 2 Anthropometric, Hemodynamic, and Lipid Profiles at Baseline and 6 Months ($N=129$)

	Baseline	Six months	Percent Change	P-value
Variable				
BMI (kg/m^2)	42.5±1.8	37.9±1.5	−10.8	<.01
Waist circumference (cm)	118.4±4.2	110.8±3.7	−6.4	.02
Systolic BP (mmHg)	131.9±2.2	124.2±1.9	−5.8	.02
Diastolic BP (mmHg)	79.8±1.6	76.2±1.3	−3.6	.05
Pulse pressure (mmHg)	52.3±1.4	48.0±1.2	−8.2	<.01
Heart rate (bpm)	82.7±1.9	76.6±2.1	−7.4	<.01
Glucose (mg/dL)	111.0±8.0	101.4±11.6	−5.7	.24
Total cholesterol (mg/dL)	224.2±8.6	212.6±8.6	−5.2	.10
LDL (mg/dL)	137.7±7.8	126.8±6.3	−7.9	.01
HDL (mg/dL)	56.4±2.3	61.8±3.2	+9.6	<.01
Smokers (number (%))	1(3.4)	1(3.4)	0	1
10-year CHD risk				
Women	6%	4%	−33.3	<.01
Men	16%	13%	−18.5	<.01

The results show an 11% decrease in BMI, an 8% decrease in LDL, and a 10% increase in HDL. The LDL/HDL ratio decreased from 2.3 to 2.1, while the total cholesterol/HDL ratio decreased from 4.4 to 3.4. The initial 10-year CHD risk, as per the Framingham risk calculator, was 6% in women and 16% in men. The 10-year CHD risk prediction decreased on average from 6% to 4% for women and 16% to 13% for men. This was accomplished by the improvements in BMI (kg/m^2), waist circumference, blood pressure, LDL, and HDL. The results clearly highlight the importance of continuous and persistence adherence to lifestyle changes to risk reduction and cardiovascular health (see Fig. 1). Implementing similar programs that are culturally sensitive across minority groups could significantly improve cardiovascular health and CHD risk for an at-risk population at lower costs than long-term pharmacological treatment.

CONCLUSION

In cardiovascular research, the primary objective is to identify and attempt to prevent or reduce the onset of risk factors that are predictive of cardiovascular disease morbidity and mortality. Obesity, and the related condition, the metabolic syndrome, are noted to be risk factors for the development of CVD [50]. The prevalence of the major CVD risk factors is disproportionately higher among minority populations, in general and particularly among obese individuals. Obesity is a multifactorial condition, with some individuals suffering from an endocrine disease or from genetic propensity. Many individuals, however, have developed the condition from poor eating habits and less or no physical activity and have become an epidemic in minority populations. Management of obesity, metabolic syndrome, and CVD risks consists primarily of modification or reversal of the root causes and direct therapy of the risk factors. Moreover, the high prevalence and increasing incidence of obesity, the metabolic syndrome, and associated cardiovascular risk factors in adults and children emphasize the need to focus on obesity reduction in this population. There is the report that, in the future, metabolic syndrome may overtake smoking as the leading risk factor for heart disease (NHLBI site).

The first strategy involves weight reduction and increased physical activity, both of which can improve all components of the CHD risk profile and components of the metabolic syndrome. Lifestyle changes are the most important target for decreasing these risk factors. As we have discussed and others have shown that this can be achieved through a successful diet and exercise program and likely at less cost in the long-term than through pharmacological means [39,40]. Education and awareness are of the utmost importance in ensuring success with the diet an exercise program [51]. Understanding why they are overweight, the consequences of excess weight and the associated deliberating complications should be emphasized in educational programs.

In conclusion, we emphasize that with the rise in prevalence and incidence of obesity and associated conditions worldwide and particularly the high onset of the obesity and the metabolic syndrome among children and the adolescent, earlier intervention through lifestyle modification to stop this epidemic will have potential benefit to the society now and the future. This includes regularly engaging in physical activity; consumption of low-salt, low-fat diet; consumption of fruits and vegetables; and refraining from high sugary foods and drinks. Other healthy lifestyle changes include managing stress, having a good night sleep, quitting smoking, and adhering to prescription medication.

REFERENCES

[1] Basu S, McKee M, Stuckler D. Relationship of soft drink consumption to global overweight, obesity, and diabetes: a cross national analysis of 75 countries. Am J Public Health 2013;103(11):2071–7.

[2] Flegal KM, Carroll MD, Kit BK, Ogden CL. Prevalence of obesity and trends in the distribution of body mass index among US adults, 1999–2010. JAMA 2012;307(5):491–7.

[3] Yang Q, Cogswell ME, Flanders WD, Hong Y, Zhang Z, Loustalot F, et al. Trends in cardiovascular health metrics and associations with all-cause and CVD mortality among US adults. JAMA 2012;307(12):1273–83.

[4] World Health Organization. Obesity: preventing and managing the global epidemic. Report of a WHO consultation on obesity. Geneva: World Health Organization; 1998.

[5] Weiss R, Dziura J, Burgert TS, Tamborlane WV, Taksali SE, Yeckel CW, et al. Obesity and the metabolic syndrome in children and adolescents. N Engl J Med 2004;350(23):2362–74.

[6] National Institutes of Health, National Heart, Lung, and Blood Institute. Clinical guidelines on the identification, evaluation, and treatment of overweight and obesity in adults. The Evidence Report; 1998; NHLB: NIH Pub No. 98-4083.

[7] Denke MA, Sempos CT, Grundy SM. Excess body weight. An under recognized contributor to high blood cholesterol levels in white American men. Arch Intern Med 1993;153:1093–103.

[8] Kucamarski RJ, Flegal KM, Campbell SM, Johnson CL. Increasing prevalence of over-weight among US adults, the National Health and Nutrition Examination Surveys, 1960–1991. JAMA 1994;272:205–11.

[9]   Burt VL, Whelton P, Roccella EJ, Brown C, Cutler JA, Higgins M, et al. Prevalence of hypertension in the U.S. adult population: results from the third National Health and Nutrition Examination Survey, 1988–1991. Hypertension 1995;25:305–13.

[10] Ford ES, Williamson DF, Liu S. Weight change and diabetes incidence: findings from a national cohort of US adults. Am J Epidemiol 1997;146:214–22.

[11] Fontaine KR, Redden DT, Wang C, Westfall AO, Allison DB. Years of lost due to obesity. JAMA 2003;289:187–93.

[12] Fontaine KR, Cheskin LJ, Barofsky I. Health related quality of life in obese persons seeking treatment. J Fam Pract 1996;43:265–79.

[13] He Q, Ding ZF, Fong DY, Kalberg J. Blood pressure is associated with body mass index in both normal and obese children. Hypertension 2000;36:165–70.

[14] Masuo K, Mikami H, Ogahara T, Tuck ML. Weight gain-induced blood pressure elevation. Hypertension 2000;35:1135–40.

[15] Randall OS, Retta TM, Kwagyan J, Gordeuk VR, Xu S, Maqbool AR, et al. Obese African Americans: the prevalence of dyslipidemia, hypertension, and diabetes mellitus. Ethn Dis 2004;14(3):384–8.

[16] Ford ES, Giles WH, Dietz WH. Prevalence of the metabolic syndrome among US adults. JAMA 2002;287:356–9.

[17] Cook S, Weitzman M, Auinger P, Nguyen M, Dietz WH. Prevalence of a metabolic syndrome phenotype in adolescents: findings from the third National Health and Nutrition Examination Survey, 1988–1994. Arch Pediatr Adolesc Med 2003;157(8):821.

[18] Ford ES. Prevalence of the metabolic syndrome defined by the International Diabetes Federation among adults in the US. Diabetes Care 2005;28(11):2745–9.

[19] NHLBI 2016. Retrieved from https://www.nhlbi.nih.gov/health/health-topics/topics/ms.

[20] Haffner S, Heinrich T. Epidemic obesity and the metabolic syndrome. Circulation 2003;108:1541–5.

[21] Cherqaoui R, Kassim TA, Kwagyan J, Freeman C, Nunlee-Bland G, Ketete M, et al. The metabolically healthy but obese phenotype in African Americans. J Clin Hypertens 2012;14(2):92–6.

[22] Cook S, Weitzman M, Auinger P, Nguyen M, Dietz WH. Prevalence of a metabolic syndrome phenotype in adolescents: findings from the Third National Health and Nutrition Examination Survey, 1988–1994. Arch Pediatr Adolesc Med 2003;157:821–7.

[23] Sidney S, Rosamond WD, Howard VJ, Luepker RV. The "Heart Disease and Stroke Statistics—2013 Update" and the need for a national cardiovascular surveillance system. Circulation 2013;127(1):21–3.

[24] Prevalence of coronary heart disease—United States 2006–2010. CDC Weekly 2011;60(40):1377–81. Available at: http://www.cdc.gov/mmwr/preview/mmwrhtml/mm6040a1.htm.

[25] Leigh JA, Alvarez M, Rodriguez CJ. Ethnic minorities and coronary heart disease: an update and future directions. Curr Atheroscler Rep 2016;18(2):1.

[26] Kumar J. Epidemiology of hypertension. Clin Queries Nephrol 2013;2(2):56–61.

[27] Padwal RS, Bienek A, McAlister FA, Campbell NR, Outcomes Research Task Force of the Canadian Hypertension Education Program. Epidemiology of hypertension in Canada: an update. Can J Cardiol 2016;32(5):687–94.

[28] Wang AYM. Epidemiology of hypertension in chronic kidney disease in core concepts in hypertension in kidney disease. New York: Springer; 2016. p. 1–13.

[29] Diaz KM, Veerabhadrappa P, Brown MD, Whited MC, Dubbert PM, Hickson DA. Prevalence, determinants, and clinical significance of masked hypertension in a population-based sample of African Americans: the Jackson Heart Study. Am J Hypertens 2015;28(7):900–8.

[30] Williams SF, Nicholas SB, Vaziri ND, Norris KC. African Americans, hypertension and the renin angiotensin system. World J Cardiol 2014;6(9):878.

[31] Franklin SS, Sutton-Tyrell K, Belle S, Steve H, Weber M, Michael H, et al. The importance of pulsatile components of hypertension in predicting carotid stenosis in older adults. J Hypertens 1997;15:1143–50.

[32] Suurkula M, Agewall S, Fagerberg B, et al, for the Risk Intervention Study (RIS) Group. Ultrasound evaluation of atherosclerotic manifestations in the carotid artery in high-risk hypertension patients. Arterioscler Thromb 1994;14:1297–304.

[33] Krüger B, Döhler B, Opelz G, Mundt H, Krämer BK, Süsal C. High pulse pressure and outcome in kidney transplantation: results from the collaborative transplant study. J Hypertens 2016;34:e20.

[34] Pannier B, Brunel P, El Aroussy W. Pulse pressure and echocardiographic findings in essential hypertension. J Hypertens 1989;7:127–32.

[35] Randall O, Kwagyan J, Retta T, et al. Effect of intensive blood pressure control on cardiovascular remodeling in hypertensive patients with nephrosclerosis. Int J Nephrol 2013;2013:120167.

[36] Liao D, Cooper L, Toole J. The prevalence and severity of white lesions, their relationship with age, ethnicity, gender and cardiovascular disease risk factors: ARIC study. Neuroepidemiology 1997;16:149–62.

[37] Ferrannini E. The hemodynamics of obesity: a theoretical analysis. J Hypertens 1992;10:1417–23.

[38] Martins D, Tareen N, Pan D, Norris K. The relationship between body mass index and pulse pressure in older adults with isolated systolic hypertension. Am J Hypertens 2002;15:538–43.

[39] Randall OS, Kwagyan J, Huang Z, Xu S, Ketete M, Maqbool AR. Effect of diet and exercise on pulse pressure and cardiac function in morbid obesity: analysis of 24-hour ambulatory blood pressure. J Clin Hypertens 2005;7(8):455–63.

[40] Kwagyan J, Tabe CE, Xu S, Maqbool AR, Gordeuk VR, Randall OS. The impact of body mass index on pulse pressure in obesity. J Hypertens 2005;23(3):619–24.

[41] Randall OS, Feseha HB, Kachi I, Xu S, Ketete M, Kwagyan J, et al. Response of lipoprotein(a) levels to therapeutic life-style change in obese African-Americans. Atherosclerosis 2004;172:155–60.

[42] Verdecchia P, Porcellati C, Schillaci G, Borgioni C, Ciucci A, Battiselli M, et al. Ambulatory blood pressure. An independent predictor of prognosis in essential hypertension. Hypertension 1994;34:793–801.

[43] Verdecchia P, Schillaci G, Borgioni C. Ambulatory pulse pressure: a potent predictor of total cardiovascular risk in hypertension. Hypertension 1998;32:983–8.

[44] Verdecchia P, Schillaci G, Reboldi G, et al. Different prognostic impact of 24-hour mean blood pressure and pulse pressure on stroke and coronary artery disease in essential hypertension. Circulation 2001;103:2579–84.

[45] Devereax RB, Pickering TG. Relationship between the level, pattern and variability of ambulatory blood pressure and target organ damage in hypertension. J Hypertens 1991;8:S34–8.

[46] Redon J, Campos C, Narciso ML, Rodicio JL, et al. Prognostic value of ABPM in refractory hypertension. Hypertension 1998;31:712–8.

[47] Mancia G, Parati G. Ambulatory blood pressure monitoring and organ damage. Hypertension 2000;36:894–900.

[48] Staessen JA, Thijs L, Fagard R. Predicting cardiovascular risk using conventional versus ambulatory blood pressures in older patients with systolic hypertension. JAMA 1999;282:539–46.

[49] Kwagyan J, Retta TM, Ketete M, Bettencourt CN, Maqbool AR, Xu S, et al. Obesity and cardiovascular diseases in a high-risk population: evidence-based approach to CHD risk reduction. Ethn Dis 2015;25(2):208.

[50] RVega GL. Obesity, the metabolic syndrome, and cardiovascular disease. Am Heart J 2001;142:1108–16.

[51] Randall OS, Randall D. Menu for life: African Americas get healthy, eat well, loss weight and live beautifully. New York, NY: Random House Inc.; 2003 22.

FURTHER READING

[1] Masuo K, Mikami H, Ogahara T, Tuck ML. Weight gain-induced blood pressure elevation. Hypertension 2000;35:1135–40.

Chapter 2

Public Knowledge of Cardiovascular Risk Numbers: Contextual Factors Affecting Knowledge and Health Behavior, and the Impact of Public Health Campaigns

Jeffrey L. Kibler, Mindy Ma, Jacquelyn Hrzich, Roberta A. Roas
College of Psychology, Nova Southeastern University, Fort Lauderdale, FL, United States

INTRODUCTION

Cardiovascular disease (CVD) is the leading cause of mortality globally, accounting for 31% of all deaths in 2013 [1]. In the United States, although the CVD mortality rate has declined in the past few decades, it remains the leading cause of death, with a rate of approximately 800,000 per year in 2013 [1,2]. Despite the considerable progress made in the prevention, detection, and management of CVD, heart disease and stroke remain important health concerns impacting 84.6 million Americans [1,3]. The annual expenditure in CVD-related direct and indirect costs total more than $316.6 billion [1]. Although genetics and demographic characteristics may predispose individuals to CVD, a significant proportion of the variance may be attributable to social, cultural, economic, and behavioral factors [4,5].

The relationships between lifestyle and CVD risks have been well established [6–8]. Recognizing the importance of enhancing the public's knowledge and management of modifiable risk factors, the American Heart Association (AHA) has developed a strategic plan to improve heart health by 20% and CVD-related mortality by 20% for all Americans by 2020 [6]. The AHA's 2020 impact goals aim to target all stages of prevention throughout the life span by promoting core factors that define ideal cardiovascular health. Known as the seven metrics [6], ideal cardiovascular health consists of the presence of four favorable cardiovascular health factors (normal blood pressure (BP), cholesterol/lipid levels, body mass index (BMI), and glucose level) and three health behaviors (healthy diet, physical activity, and abstention from smoking).

Conceptual Significance of the Health Belief Model

The health belief model (HBM) is a theoretical framework that has been widely used to predict whether individuals are likely to engage in risk reduction/disease prevention [9]. This model provides for a contextual understanding of variables that impact behaviors related to cardiovascular health. According to this model, motivations to initiate and maintain health-protecting behaviors are influenced by perception variables, which include beliefs surrounding personal susceptibility to disease, seriousness of the disease, benefits of taking action, and barriers to behavioral change. The greater the perceived risk, the greater the likelihood that individuals will change their behaviors to decrease risk. In a study that examined knowledge of heart disease in women [10], the perceptions of susceptibility and seriousness of heart disease, knowledge of risk factors, and general health motivation accounted for 76% of the variance in cardiovascular preventive behaviors. An intervention study aimed at increasing the knowledge of risks associated with uncontrolled diabetes, personal susceptibility to the disorder, and beliefs about the benefits of adherence to treatment revealed a significant decrease in fasting blood glucose in the treatment group relative to the control group [11]. These findings suggest knowledge and individual perceptions identified by the HBM are associated with the seven key CVD metrics highlighted in the AHA recommendations [6].

Research has addressed ways in which knowledge of personal and target levels of health indicators may interact with HBM variables [12,13] to predict engagement in preventive behaviors [14]. In order to be aware of personal risks, individuals must possess knowledge of their own and target levels of the CVD risk factors [12]. For example, knowledge of target

BP levels has been associated with improvements in diet [15] and rates of BP control [16]. Additionally, increased physical activity was observed among women with awareness of optimal blood glucose levels [15].

Additional factors, such as race/ethnicity, age, and level of education, may influence knowledge and perceptions according to the HBM. Health-related perceptions vary across ethnic and racial groups and appear to be influenced by internal and external stimuli. African Americans and Hispanic Americans are less likely to perceive themselves as obese compared with Caucasian Americans [17,18]. Cultural factors related to race/ethnicity may play an important role in determining body image and weight parameters that are desirable and acceptable. Relative to Caucasian women, obese African American women are more likely to report their weight as normal and to overestimate their height and underestimate their weight [18]. This leads to inaccurate estimates of BMI and distorted perceptions of susceptibility to disorders that are comorbid with obesity, such as diabetes and cardiovascular disease [18]. Obesity rates are higher, and trends reflect a faster rise in obesity, among African American and Hispanic populations; thus, greater health disparities in comorbid illnesses may emerge than are observed currently if effective preventative efforts are not initiated [17,19]. With regard to education, a study with African American women revealed a positive relationship between years of education and likelihood of perceiving heart disease as a serious health risk [20]. Further, women 50 years of age and older were more likely than their younger counterparts to list heart disease as a serious health threat [20].

Purpose of the Present Chapter

In the present chapter, we review the recent literature on adults' knowledge for personal and target levels of four key cardiovascular variables (BP, lipids, BMI, and glucose) and relationships of knowledge to health behaviors; searches of PsycInfo and Medline were utilized to identify relevant studies from the past 12 years (2005–17). We address some of the potential determinants of knowledge that have been suggested (i.e., race/ethnicity and level of risk). The current literature on the effectiveness of community interventions is then presented. Finally, we present our conclusions and future directions for each of these areas of research.

KNOWLEDGE OF TARGET NUMBERS AND PERSONAL LEVELS FOR CVD RISK INDICATORS

The research on knowledge and awareness of key cardiovascular indicators has largely focused on four variables (BP, lipid levels, BMI, and glucose levels), and AHA has identified these variables as key elements of national public health initiatives [6]. The AHA recommendations incorporate an "ideal health" perspective, articulating that in order to meet the definition of ideal cardiovascular health, an individual would achieve the ideal levels for each of these four factors, along with maintaining a healthy diet, engaging in physical activity, and abstaining from smoking.

Blood Pressure

High BP has been recognized as a major independent modifiable risk factor for cardiovascular health [21]. It can be argued that knowledge of healthy/optimal BP levels, along with awareness of one's own personal BP levels, is one of the first important steps in preventing hypertension in the community. Although awareness of hypertension as a risk factor for CVD is relatively high, with estimates from 66% to 84% of individuals being aware [22,23], research on knowledge of actual cutoffs for normal and clinically high BP reflects variability in estimates of awareness. The levels of awareness/knowledge of healthy BP levels reported in the studies reviewed here ranged widely, from 37% observed by Mochari et al. [24] to 91% reported by Wright-Nunes et al. [25], but a cluster of estimates was observed around 50% [15,26–29]. A methodological consideration in interpreting the results from these studies is that the definition of an optimal BP is defined slightly differently across studies. It is notable that the Wright-Nunes et al. study of 401 patients with chronic kidney disease [25] defined the accurate optimal BP as less than 130 systolic and 80 diastolic. In a study of 449 pharmacy patients [30], only 40% of participants reported that a healthy BP is in the range less than 140/90; only 35% of the participants who reported healthy BP readings in this range (13% of the total sample) identified a BP under 120/80 as the optimal range. Relative to awareness of healthy BP levels, fewer studies assessed awareness of one's own BP, but the available estimates reflected a similarly wide range from 42% observed by Slark et al. [28] to 82% reported by Chun et al. [31]. These estimates were more evenly distributed across this range with no apparent cluster of awareness levels across studies. In terms of accuracy in categorically estimating hypertension status, Huerta et al. [32] reported data from a large representative community sample in Spain ($N=1556$) that indicated underestimation of hypertension status; only 20% of participants reported hypertension, whereas measured BP data reflected that 35% of participants would meet the criteria for hypertension.

Most of the studies on BP awareness/knowledge did not assess whether race or ethnic group differences were a demographic variable that accounted for variability in rates. However, a few studies that examined demographics did identify significant differences by race/ethnicity [12,15,24]. Ma and Ma [12] reported significantly greater awareness of target and personal BP levels for Caucasians relative to African Americans. In another study [24], Caucasians were more likely than Hispanics to know healthy BP levels; no significant group differences were observed for African Americans in this study. The Mosca et al. study [15] revealed significantly greater awareness of optimal BP levels for Caucasians relative to both African American and Hispanic participants. In all, these studies reflect racial/ethnic group differences, with minority groups tending to report lower levels of awareness for BP.

A subset of the studies examining awareness of BP also examined relationships of awareness to health behaviors/outcomes [12,15,27]. In the Ma and Ma study [12], awareness of personal BP levels was associated with physical activity, and awareness of both personal and optimal levels was associated with healthier diet for Caucasians only (this relationship was not observed for African Americans). Mosca et al. [15] also reported that awareness of healthy BP levels was associated with improvements in diet, and the effect in this study was not affected by racial/ethnic group. In a study by Prugger et al. [27], the lack of awareness of optimal BP levels was associated with more obesity and higher cholesterol. In general, these studies support the notion that awareness of BP levels is related to healthier behaviors/outcomes. There is not yet sufficient information to evaluate whether there is racial/ethnic group moderation of these relationships.

Lipids

Awareness of healthy/optimal lipid levels and one's own personal lipid levels is generally lower than the awareness for BP. This is likely due in large part to the added complexity of having different types of lipid readings (i.e., total cholesterol (TC), low-density lipoprotein (LDL) levels, high-density lipoprotein (HDL) levels, and triglycerides). The range of awareness/knowledge of healthy lipid levels for the studies reviewed here ranged from 4% awareness of TC among a large sample of Korean women [33] to 50% overall awareness of healthy TC for a sample of Caucasian and African American men and women in the Southern United States [12]. However, there were only a limited number of studies that reported on awareness of healthy lipid levels [12,15,26,29,33], and the findings clustered around 40%–50% awareness for TC. The two studies that reported awareness of healthy LDL and HDL [15,34] showed lower levels of knowledge for these indexes (2%–37%).

There were a greater number of studies that reported awareness of participants' own lipid levels, and these results ranged more widely than the awareness for healthy levels. In the Choi study [33] of Korean women, 5% awareness of participant's own personal TC was observed, whereas in a very large study of women who work as health professionals [35], 84% of participants reported their own TC level. In all, these findings for personal lipid levels did not produce a clear cluster, with six studies reporting personal TC awareness on the low end (5%–29%) and five others observing levels on the high end (43%–84%). As with awareness of healthy lipid levels, the available data [34–36] suggest that awareness of personal levels of LDL and HDL tend to be lower overall (7%–28%) compared with awareness for TC. Studies generally support the accuracy of participants' knowledge of their own lipid level numbers, but their interpretation of these numbers may not be consistent with clinical guidelines; Huerta et al. [32] observed that participants significantly underestimated hyperlipidemia as defined by elevated TC or triglycerides, and Huang et al. [35] observed an underestimate of hyperlipidemia by 6% based on TC. Also, those being treated for CVD or who are identified as being high risk do not necessarily have better awareness. The Cheng et al. [34] study indicated that a greater number of risk factors and/or history of CVD did not predict better awareness.

Several studies examined whether race/ethnicity was associated with knowledge/awareness of lipid levels. In a large study of 467 inner-city men and women, Kaplan et al. [37] observed that Caucasians had significantly greater awareness of personal TC level (25%) than African American (5%) and Hispanic participants (14%). Likewise, in a large sample of 1008 women, Mosca et al. [15] found that Caucasian participants had significantly greater awareness of healthy LDL and HDL levels compared with African American and Hispanic participants. Ma and Ma [12] also reported significantly greater awareness of both personal TC and healthy TC among Caucasians relative to African American participants. In the Mochari et al. [24] study, Caucasians were significantly more likely to know healthy HDL levels than Hispanics. In another study [34], Caucasians displayed significantly greater awareness of personal TC levels than a group of unspecified nonwhite participants. There were two studies, however, that did not observe race/ethnic differences with respect to awareness of lipid levels [29,35]. Although these studies are mixed overall, in terms of observing significant race/ethnic differences, the majority of studies do reflect group differences. Moreover, there is a clear pattern identified in these studies, with Caucasians demonstrating greater awareness. This reinforces the need for additional research on the mechanisms underlying awareness/the lack of awareness and better programs to enhance awareness of lipid levels for minority groups.

A few of the studies investigating lipid awareness examined whether this knowledge is associated with health behaviors/outcomes. In the large study of 39,876 female health professionals, Huang et al. [35] observed a significant association of personal lipid level awareness with less obesity, more physical activity, and lower smoking prevalence. In the Ma and Ma [12] study of Caucasian and African American men and women, awareness of healthy TC levels was only associated with healthier diet among Caucasian participants. However, in a large study of 1008 women, Mosca et al. [15] did not observe significant associations between awareness of healthy LDL/HDL levels and physical activity or diet. This pool of studies is too limited to draw firm conclusions, but it is interesting to note that the lack of findings in the Mosca et al. [15] study involved relationships with LDL and HDL. Although speculative without further research, these preliminary results suggest the possibility that knowledge/awareness of LDL and HDL translate less easily into health behaviors due to the more limited understanding of these indexes in the general community.

Body Mass/Obesity

The evaluation of knowledge/awareness of body weight-related information has been conducted using two general constructs: (1) the awareness of specific BMI levels and (2) the awareness of weight status/category (i.e., normal, overweight, and obese). With regard to BMI study, participants' awareness of their own level has been found to range widely, from 8% [38] to 66% [39]. However, most studies have observed low levels of awareness for personal BMI, with a cluster of studies reporting estimates from 8% to 18% [12,38,40,41]. A similar pattern was observed for knowledge/awareness of the healthy/clinical cutoffs for BMI; knowledge of accurate BMI cutoffs ranged from 8% in a study of 657 obese British men and women [41] to 62% in a study of 53 renal transplant recipients [39]. As with awareness of personal BMI, these estimates for knowledge of BMI cutoffs tended to be low, with several studies observing rates in the 8%–22% range [12,38,40,41].

When knowledge/awareness was interpreted from the standpoint of weight status/category, rather than BMI, higher estimates of weight-related knowledge were evident. With regard to personal weight status, awareness ranged from 56% in a study of 266 British men who were professional drivers [42] to 89% in a study of 970 obese adults [13]. Only three studies provided data utilizing this construct of weight status/category; in the third study of 182 obese men and women, 60% accurately knew their own weight status [43]. In the study of male drivers [42], it was notable that far fewer obese participants (19%) were accurate in their assessment of weight status relative to the accurate awareness among normal/healthy weight participants (82%) and overweight participants (74%). This finding was consistent with another study of 253 mothers [44] that examined the participants' accuracy in identifying their own personal body size using pictorial silhouettes—29% of the mothers in the obese group were accurate, compared with 63% of the overweight participants and 67% of the healthy weight participants. This pattern suggests that obese individuals are less aware or knowledgeable about their weight in relation to guidelines for these categorical descriptions.

Only two of the studies reviewed here tested racial/ethnic differences in weight-related knowledge/awareness. In the large study of 970 obese men and women [13], in which high rates of awareness were reported, Caucasians had higher levels of awareness (95%) compared with African American (84%) and Hispanic participants (86%). Ma and Ma [12] found no racial/ethnic differences in awareness of personal BMI but observed significantly higher knowledge of the healthy/normal cutoff for BMI among Caucasians (28%) relative to African Americans (14%). The data from this limited number of studies suggest the possibility that, similar to the findings for BP and lipids, Caucasians may have higher levels of awareness/knowledge with regard to BMI and/or weight status. Further study of this issue is needed with regard to weight, given the limited number of investigations. In addition, the Ma and Ma study [12] was the only one reviewed here that examined the relationship of weight-related awareness to behavior—knowledge of personal BMI was significantly related to greater physical activity in both the Caucasian and African American groups.

Blood Glucose

Knowledge of personal and healthy blood glucose levels has generally been found to be low, with a range for knowledge/awareness of personal blood glucose ranging from 25% [45] to 40% [46]. The range for awareness of healthy blood glucose levels was wider, with estimates from 19% [12] to 53% [47]. The index for blood glucose varies across studies, and only one of the studies presently reviewed reported on awareness of HbA1c levels [45]. It is notable, therefore, that use of HbA1c produced the lowest estimate of awareness for personal blood glucose levels. This is likely due to the more complex nature of indexing blood glucose regulation over an extended period, and/or the relative infrequency of providers introducing this index to patients. This idea was underscored by Ezenwaka et al. [48] in a study of 89 type 2 diabetes patients—none of the participants in this study reported knowing about HbA1c as a measure. Although Iqbal et al. [49] did not measure awareness of actual blood glucose levels, their study indicated higher awareness of the term HbA1c (41% of participants were familiar with the term).

Only one study reported race/ethnic differences in awareness of blood glucose levels—in the Ma and Ma study [12], Caucasian participants had significantly higher awareness of personal blood glucose levels (34%) relative to African American participants (23%). This study did not observe any difference between the two groups in awareness of healthy blood glucose levels. Two additional studies that assessed race/ethnic differences in awareness of healthy blood glucose levels [15,24] indicated no significant group differences. Therefore, the limited data suggest that there may not be race/ethnic differences in awareness of healthy blood glucose levels, and more research is needed to substantiate the observed differences related to personal blood glucose.

Both of the studies that examined awareness of blood glucose levels in relation to health behaviors/outcomes observed significant associations [12,15]. Mosca et al. [15] identified a significant relationship between awareness of healthy blood glucose levels and greater physical activity. Ma and Ma [12] observed a significant association between awareness of personal blood glucose and healthy diet. Based on the data from this limited number of studies, there is preliminary support for the idea that knowledge/awareness for blood glucose may translate into better health behaviors and CVD prevention efforts.

INTERVENTIONS TARGETING AWARENESS OF CVD RISK FACTORS

Knowledge of healthy cutoffs for cardiovascular risk factors and personal levels has been associated with preventative action and positive lifestyle changes [15,50,51]. Additionally, interventions that promoted awareness have led to increases in individuals seeking medical attention, enhanced compliance with medical treatments, and improved lipid levels [52,53]. Conversely, the lack of awareness regarding cardiovascular risk factors may contribute to ethnic/racial disparities in CVD [12]. In this section, we review the range of knowledge interventions conducted across different settings and summarize their effectiveness. Based on the findings, we also make recommendations for elements that appear to be important to consider when developing future knowledge interventions.

Effectiveness of Interventions

Community-programs. Interventions conducted in the community have been the most common approach to enhancing knowledge of CVD risk. Health fair promotion interventions, the *know your numbers* campaigns, and other community-based interventions have provided community members with educational resources, while some interventions extend access to BP and cholesterol screenings. Landy et al. [54] found that their health fair program effectively identified large numbers of individuals with abnormal cholesterol levels (58% had LDL cholesterol over 129 mg/dL; 24% had HDL cholesterol below 40 mg/dL). Their follow-up screening indicated that 67% of participants reduced their LDL levels by 16 mg/dL and 61% increased HDL levels by 5 mg/dL [54]. Studies of "know your numbers" (KYN) programs have shown success with enhancing knowledge. A KYN program in Australia [50] registered close to 60,000 individuals who received feedback about their BP, referral to their doctor if BP was high, educational materials about CVD risks including a self-assessment survey, and guidance regarding reducing risk; this program was shown to increase participants' knowledge of their BP by 9% ($P < .05$). Further, all respondents who were followed longitudinally reported at least one health promotion action following the campaign, such as increasing exercise or eating healthier [50]. This suggests the possibility that enhanced knowledge may have contributed to the healthier behaviors in this Australian cohort.

Public health campaigns may benefit from the inclusion of a low-cost prime prior to the intervention to increase the audience's level of personal involvement. Ma et al. [51] found that participants who received a health knowledge questionnaire as a priming stimulus reported greater awareness of information from a KYN campaign, in comparison with the control group who did not complete the questionnaire. The primed individuals also reported higher knowledge of their own and normal ranges for cholesterol, glucose, and BMI levels, as well as engaging in positive cardiovascular health choices [51]. This suggests that utilizing low-cost primes may maximize the effectiveness of public health campaigns.

Two other community programs, the *CVD prevention* campaign and the *REACH* campaign, consisted of culturally tailored educational programs targeting African American and Hispanic populations. Altman et al. [55] found that the *CVD prevention* campaign yielded significant improvements in accurate knowledge of personal CVD risk levels following a 4-month bilingual educational program—accurate estimates of hypertension increased by 48% and obesity increased by 49%. The *REACH* campaign [56] was found to be effective in improving knowledge of the relationships between blood glucose and healthy eating and physical activity for individuals with diabetes. This intervention also resulted in a significant decrease in A1C levels compared with usual care control groups. Overall, the evidence across studies of community programs supports their effectiveness for enhancing knowledge of CVD risk and reducing risk as associated with BP and lipids.

The *Mosque campaign* was a 3-year campaign that utilized culturally sensitive interventions, including language adaptation and private medical consultations, to target the specialized needs of Turkish-immigrant women in Austria experiencing barriers to health-care access. This campaign effectively reduced levels of unawareness about personal cholesterol levels from 57% to 32%, glucose levels from 50% to 25%, and BP from 41% to 18% within this population [57]. This demonstrates the need for culturally sensitive programs to target minority groups in order to ensure equality of access to campaigns and health-care services.

Media. The mass media provides a highly cost-effective method to inform a large audience about relevant health information [43,58]. For example, movie theater ads apply to a uniquely captive and attentive audience, which may capture the attention of the audience. Peddecord et al. [59] found that advertisements with animation and sound had higher recall rates (80%). As an example of interventions that integrate mass media methods, a hypertension campaign utilized multiple advertising methods, including television, radio, billboard, and other print ads, to promote healthy behaviors [60]. This intervention was found to be effective in raising awareness and increasing positive health monitoring behavior—the percentage of participants who knew their BP levels increased from 45% to 52%, and the percentage of participants who checked their BP levels within the past 2 months rose from 34% to 40% [60]. Petrella et al. [61] examined the effects of a social marketing media campaign on two similar Canadian cities and found that different methods of media had varying levels of effectiveness—television had the highest reported recall for both test and control groups (58.8% in test group and 48.8% in the control group), followed by radio (28.6% test and 19.5% control), billboard (25.9% test and 16.0% control), and bus board (14.6% test and 10.8% control). Overall, they found an increased awareness of personal hypertension rates from 33% at baseline to 41% following the campaign. However, improvements in this knowledge did not translate into lasting effects on health behaviors. Results demonstrated an increase in physician visits and control of BP levels (61.3%) immediately following the campaign, but not at the 6-month follow-up (68.3%), which suggests a need for continued implementation of this message [61]. The combination of multiple media modalities has been found to be most effective for increasing awareness for large target audiences [58,60]. Peterson et al. [58] found that the exposure to the combination of two types of television advertisements and a billboard had the highest impact on individual's awareness and behavior change in comparison with any of the advertisements individually. Further, this social marketing campaign was found to reach 76.5% of the population and resulted in 45.2% of participants reporting a healthy lifestyle change [58]. Another mass media campaign utilized television programs, radio segments, and various printed advertisements (billboards, bus boards, posters, and print ads) to reach almost 2000 individuals of varying demographics and communities within a 3-month period due to the multitude of exposure methods [60]. Although the research is limited to date, mass media interventions appear to hold promise for increasing knowledge about numbers such as BP and monitoring of BP levels.

Other Potential Strategies and Venues for Addressing Awareness of Risk Numbers

Population level programs, such as mass media campaigns, can be effective, but may not be sufficient in reaching underserved subgroups [62]. The evidence on knowledge/awareness of risk numbers suggest that greater efforts are needed when targeting minority populations for CVD knowledge interventions. Utilizing indigenous members and leaders from a community has shown to be beneficial in raising awareness due to higher levels of acceptance and trust. Environmental barriers, such as crime or issues with cost/availability of fresh, healthy foods in minority neighborhoods, need to be taken into consideration when developing intervention strategies [56].

Faith-based programs may represent a fruitful area for additional research on knowledge interventions. Most Americans attend a church or religious institution, making this a potential setting for interventions [63]. Many religious organizations are already involved in health promotion and outreach practices as part of their mission. Therefore, they are generally open to health promotion programs to keep their members and community healthy [63]. The study by Ma et al. [51] illustrated the potential for engagement and enhancing knowledge in a faith-based setting—potential volunteers were approached in church foyers after services and asked if they would complete a brief survey on cardiovascular health. In the next phase of the study, KYN brochures were placed in the church foyers and were available to all church attendees for 2 months. The brochures were designed to educate about the importance of knowing one's own cardiovascular health indicators (i.e., blood cholesterol, BP, blood glucose, and BMI). No participants had knowledge of the KYN campaign prior to the distribution of the educational brochures for the current study. After 2 months of the KYN campaign, a post-KYN campaign survey was conducted, and those who completed each phase of the study showed enhanced knowledge of cholesterol, blood glucose, and BMI levels [51].

Health-care providers also represent a pivotal element of awareness. Providers may often be educating patients for the first time about health risk information. However, the significance of following up with patients after a first educational encounter to assess retention of information and to reinforce positive health behaviors is also underscored by the findings related to knowledge levels reviewed in this chapter. Approximately 88% of respondents report seeing a health-care professional regularly, which indicates a large potential for educational opportunities [15]. The lower levels of knowledge we have noted for target levels of lipids, blood glucose, and BMI across all racial/ethnic groups [12,15] are more striking than for BP and may be attributable to the lack of exposure to information about these risks and screening measures. In the Mosca et al. study [15], respondents reported much higher rates of BP checks (96% in past 2 years) in comparison with cholesterol screenings (80% in past 5 years). Routine measurement of BP at each medical visit likely contributes to this heightened awareness of personal and target levels. Physician recommendation is one of the most important cues to screening [64]. Mosca et al. [15] identified that the most frequently cited reason why women have not discussed their risk of CVD with their physician was that health-care professionals did not bring it up (38%). These variations demonstrate the importance of the physician's role in educating and exposing patients to screenings, especially with minority patients; health-care providers and friends/relatives were shown to have a more significant impact on motivating preventative action in nonwhites than whites [15]. In addition, Hispanic patients view physicians as powerful, respectable, authority figures and tend to listen to their suggestions [64]. The *COACH* and home visit programs have utilized community health workers to provide individualized interventions in African American communities; these workers often share the same ethnicity, community, and socioeconomic backgrounds as their patients, which can foster trust through a shared perspective [65]. The *COACH* program utilized community health workers to deliver educational and behavioral counseling on health behavior modification, and found greater improvements in lipids and BP relative to usual care [65].

Mosca et al. [15] identified top motivations for taking preventative action as wanting to improve health (95%), feel better (92%), live longer (90%), and avoid taking medications (69%), and 67% did it for their family. These factors should be emphasized as motivators for raising awareness and sustaining healthy behaviors.

Recommendations

Although trends in cardiovascular risks have decreased over the past two decades, disparities remain between sociodemographic groups; further, non-Latino white and educated populations have evidenced greater improvements in health risks over time [66]. Prevention programs should focus on bridging this gap through utilization of culturally tailored interventions. While programs to further increase awareness of CVD risks across the general population are greatly needed, the lower levels of awareness in minority groups suggest that it is crucial to continuously evaluate the focus and approach to delivering knowledge interventions to target minority groups that may fall through the cracks of larger-scale programs [67]. Having community members and leaders assist in tailoring intervention strategies to be more culturally specific may help to enhance intervention accessibility, quality, and outcomes [62]. Given the issues related to trust, motivation to enhance health for one's family, and need for community engagement [15,65], it is recommended that minority programs include extended social supports, including friends, relatives, church members, and other community members, to raise awareness of CVD risks and likelihood of sustaining preventative behaviors [15].

Faith-based programs could be especially beneficial for reaching the large segment of the population who participate in services or other activities related to their faith. Such programs may also enhance engagement with "hard to reach" members of minority populations that experience mistrust of traditional health care [63]. Faith-based programs often utilize leaders or volunteers from the congregation to learn relevant health information and pass on this knowledge to peers in order to facilitate culturally effective interventions from a trustworthy source [63].

CONCLUSIONS

From the presently reviewed research, we conclude that approximately half of adults in the general community are aware of the cutoffs for healthy BP. Likewise, awareness of personal BP levels appears to be in the range of 40%–80%, but more research is needed to provide greater confidence in these estimates. Awareness of healthy TC levels is estimated to be 40%–50%, with awareness of LDL and HDL levels being 30% or lower; awareness of healthy lipid levels is also an area where more research is needed. A greater number of studies are reported on awareness of personal lipid levels, but these estimates varied widely (even for TC), with no apparent central cluster. The available data indicate that awareness of personal and target BMI is low (10%–20%) and that awareness of blood glucose (20%–50% for healthy levels and 25%–40% for personal levels) is somewhat higher than levels for BMI. For studies that evaluated racial/ethnic group differences in awareness, minority groups tended to report lower levels of awareness. In general, the available data also suggested that greater levels of awareness are associated with better health behaviors.

The HBM has been applied across a broad range of contexts and populations and provides a useful theoretical framework for predicting engagement in CVD risk reduction [9]. While perception variables are directly linked to both knowledge and behavior [6,14–16], additional factors such as race/ethnicity, education, and age also appear to influence knowledge/ perceptions and motivation for risk reduction [17–20]. Consideration of these elements is indispensable to inform the development of efficacious interventions and tailor them to the needs of specific groups.

Interventions to enhance awareness of cardiovascular risk numbers, including community and mass media campaigns, hold promise for enhancing knowledge. However, some forums for research on enhancing knowledge have been understudied—unique elements of faith-based or worksite approaches may be helpful to incorporate into future intervention efforts for knowledge/awareness. Given the lower rates of awareness for racial and ethnic minority groups, strong arguments can be made for the advantages of grass roots approaches that utilize members of minority communities or faith-based organizations to better engage with peers [63,65]. Health-care professionals have a significant responsibility in the prevention of CVD due to their great influence on patient care. Increasing patient's awareness of risk factors and knowledge of personal levels is a primary step in facilitating this process.

In all, it is concerning that knowledge/awareness is not higher than the figures reported in studies of the key CVD risk variables. There are strong conceptual models for understanding the key roles of knowledge, along with the perceptions of risk and disease, in preventing CVD risk. Nonetheless, the available studies on knowledge campaigns and programs are encouraging in terms of the potential to enhance knowledge/awareness. Considerations of the potential for more intensive programs to provide even stronger results in the future or for interventions that utilize innovative methods and/or venues to enhance awareness among underserved populations serve as the bases for enthusiasm in this area.

REFERENCES

[1]   Writing Group Members, et al. Heart disease and stroke statistics-2016 update: a report from the American Heart Association. Circulation 2016;133(4):e38–360.

[2]   National Heart, Lung, and Blood Institute. 2012 NHLBI morbidity and mortality chart book; 2012 (cited 2016 9/29). p. 1–98. Available from https://www.nhlbi.nih.gov/files/docs/research/2012_ChartBook.pdf.

[3]   Viera AJ, et al. High blood pressure knowledge among primary care patients with known hypertension: a North Carolina Family Medicine Research Network (NC-FM-RN) study. J Am Board Fam Med 2008;21(4):300–8.

[4]   Centers for Disease Control and Prevention. Heart disease risk factors; 2016 (cited 2016 9/30).

[5]   Sharma S, et al. Racial, ethnic and socioeconomic disparities in the clustering of cardiovascular disease risk factors. Ethn Dis 2004;14(1):43–8.

[6]   Lloyd-Jones DM, et al. Defining and setting national goals for cardiovascular health promotion and disease reduction: the American Heart Association's strategic Impact Goal through 2020 and beyond. Circulation 2010;121(4):586–613.

[7]   Ahmed HM, Blaha MJ, Blumenthal RS. Modifiable lifestyle risks, cardiovascular disease, and all-cause mortality. Int J Cardiol 2014;e199–200.

[8]   Pischke CR, et al. Long-term effects of lifestyle changes on well-being and cardiac variables among coronary heart disease patients. Health Psychol 2008;27(5):584–92.

[9]   Rosenstock IM. Encyclopedia of psychology. Washington, DC: American Psychological Association; 2000. p. 78–80.

[10] Ali NS. Prediction of coronary heart disease preventive behaviors in women: a test of the health belief model. Women Health 2002;35(1):83–96.

[11] Sharifirad G, et al. The effectiveness of nutritional education on the knowledge of diabetic patients using the health belief model. J Res Med Sci 2009;14(1):1–6.

[12] Ma M, Ma A. Racial/ethnic differences in knowledge of personal and target levels of cardiovascular health indicators. J Community Health 2015;40(5):1024–30.

[13] Sivalingam SK, et al. Ethnic differences in the self-recognition of obesity and obesity-related comorbidities: a cross-sectional analysis. J Gen Intern Med 2011;26(6):616–20.

[14] Mathieu RAt, et al. Physical activity participation, health perceptions, and cardiovascular disease mortality in a multiethnic population: the Dallas Heart Study. Am Heart J 2012;163(6):1037–40.

[15] Mosca L, et al. National study of women's awareness, preventive action, and barriers to cardiovascular health. Circulation 2006;113(4):525–34.

[16] Knight EL, et al. Predictors of uncontrolled hypertension in ambulatory patients. Hypertension 2001;38(4):809–14.

[17] Dorsey RR, Eberhardt MS, Ogden CL. Racial/ethnic differences in weight perception. Obesity 2009;17(4):790–5.

[18] Johnson WD, et al. Ethnic differences in self-reported and measured obesity. Obesity 2009;17(3):571–7.

[19] Dubowitz T, et al. Neighborhood socioeconomic status and fruit and vegetable intake among whites, blacks, and Mexican Americans in the United States. Am J Clin Nutr 2008;87(6):1883–91.

[20] Sadler GR, et al. African-American women's perceptions of their most serious health problems. J Natl Med Assoc 2005;97(1):31–40.

[21] Ayala C, et al. Prevalence of self-reported high blood pressure awareness, advice received from health professionals, and actions taken to reduce high blood pressure among US adults—Healthstyles 2002. J Clin Hypertens (Greenwich) 2005;7(9):513–9.

[22] Winham DM, Jones KM. Knowledge of young African American adults about heart disease: a cross-sectional survey. BMC Public Health 2011;11:248.

[23] Qvist I, et al. Self-reported knowledge and awareness about blood pressure and hypertension: a cross-sectional study of a random sample of men and women aged 60–74 years. Clin Epidemiol 2014;6:81–7.

[24] Mochari-Greenberger H, et al. Knowledge, preventive action, and barriers to cardiovascular disease prevention by race and ethnicity in women: an American Heart Association national survey. J Womens Health (2002) 2010;19(7):1243–9.

[25] Wright-Nunes JA, et al. Patient knowledge of blood pressure target is associated with improved blood pressure control in chronic kidney disease. Patient Educ Couns 2012;88(2):184–8.

[26] Cheng S, et al. Knowledge of blood pressure levels and targets in patients with coronary artery disease in the USA. J Hum Hypertens 2005;19(10):769–74.

[27] Prugger C, et al. Blood pressure control and knowledge of target blood pressure in coronary patients across Europe: results from the EUROASPIRE III survey. J Hypertens 2011;29(8):1641–8.

[28] Slark J, et al. Knowledge of blood pressure in a U.K. general public population. J Hum Hypertens 2014;28(8):500–3.

[29] Mochari H, et al. Cardiovascular disease knowledge, medication adherence, and barriers to preventive action in a minority population. Prev Cardiol 2007;10(4):190–5.

[30] Lam JY, Guirguis LM. Patients' blood pressure knowledge, perceptions and monitoring practices in community pharmacies. Pharm Pract 2010;8(3):187.

[31] Chun H, Kim IH, Min KD. Accuracy of self-reported hypertension, diabetes, and hypercholesterolemia: analysis of a representative sample of Korean older adults. Osong Public Health Res Perspect 2016;7(2):108–15.

[32] Huerta JM, et al. Accuracy of self-reported diabetes, hypertension and hyperlipidemia in the adult Spanish population. DINO study findings. Rev Esp Cardiol 2009;62(2):143–52.

[33] Choi EJ, et al. Middle-aged women's awareness of cholesterol as a risk factor: results from a national survey of Korean Middle-aged Women's Health Awareness (KomWHA) study. Int J Nurs Stud 2010;47(4):452–60.

[34] Cheng S, et al. Knowledge of cholesterol levels and targets in patients with coronary artery disease. Prev Cardiol 2005;8(1):11–7.

[35] Huang PY, et al. Awareness, accuracy, and predictive validity of self-reported cholesterol in women. J Gen Intern Med 2007;22(5):606–13.

[36] Neufingerl N, Cobain MR, Newson RS. Web-based self-assessment health tools: who are the users and what is the impact of missing input information? J Med Internet Res 2014;16(9):163–75.

[37] Kaplan RC, et al. Differences by age and race/ethnicity in knowledge about hypercholesterolemia. Cardiol Rev 2006;14(1):1–6.

[38] Cardozo ER, et al. Knowledge of obesity and its impact on reproductive health outcomes among urban women. J Commun Health Publ Health Promot Dis Prev 2013;38(2):261–7.

[39] Mucha K, et al. Weight gain in renal transplant recipients in a Polish single centre. Ann Transplant 2015;20:16–20.

[40] Post RE, et al. Patient understanding of body mass index (BMI) in primary care practices: a two-state practice-based research (PBR) collaboration. J Am Board Fam Med 2015;28(4):475–80.

[41] Johnson F, et al. Do weight perceptions among obese adults in Great Britain match clinical definitions? Analysis of cross-sectional surveys from 2007 and 2012. BMJ Open 2014;4(11).

[42] DeVille-Almond J, et al. Awareness of obesity and diabetes: a survey of a subset of British male drivers. Am J Mens Health 2011;5(1):30–7.

[43] Morley B, et al. Impact of a mass media campaign linking abdominal obesity and cancer: a natural exposure evaluation. Health Educ Res 2009;24(6):1069–79.

[44] Paul TK, et al. Size misperception among overweight and obese families. J Gen Intern Med 2015;30(1):43–50.

[45] Heisler M, et al. The relationship between knowledge of recent HbA1c values and diabetes care understanding and self-management. Diabetes Care 2005;28(4):816–22.

[46] Mozumdar A, Liguori G. Statewide awareness study on personal risks of cardiovascular disease in women: a Go Red North Dakota study. Womens Health (Lond) 2010;6(1):37–50.

[47] Li C, et al. Knowledge of blood sugar control standard brings the higher attainment rate of HbA1c. Zhong Nan Da Xue Xue Bao Yi Xue Ban = J Central South Univ Med Sci 2013;38(8):773–8.

[48] Ezenwaka CE, et al. Is diabetes patients' knowledge of laboratory tests for monitoring blood glucose levels associated with better glycaemic control? Arch Physiol Biochem 2014;120(2):86–90.

[49] Iqbal N, et al. Improving patients' knowledge on the relationship between HbA1c and mean plasma glucose improves glycaemic control among persons with poorly controlled diabetes. Ann Clin Biochem 2008;45:504–7.

[50] Cadilhac DA, et al. The Know Your Numbers (KYN) program 2008 to 2010: impact on knowledge and health promotion behavior among participants. Int J Stroke 2015;10(1):110–6.

[51] Ma M, et al. The effects of priming on a public health campaign targeting cardiovascular risks. Prev Sci 2011;12(3):333–8.

[52] Grover SA, et al. Patient knowledge of coronary risk profile improves the effectiveness of dyslipidemia therapy: the CHECK-UP study: a randomized controlled trial. Arch Intern Med 2007;167(21):2296–303.

[53] Kilkenny MF, et al. Feasibility of a pilot programme to increase awareness of blood pressure as an important risk factor for stroke in Australia. Int J Stroke 2010;5(5):344–50.

[54] Landy DC, et al. Increasing access to cholesterol screening in rural communities catalyzes cardiovascular disease prevention. J Rural Health 2013;29(4):360–7.

[55] Altman R, de Ybarra JN, Villablanca AC. Community-based cardiovascular disease prevention to reduce cardiometabolic risk in Latina women: a pilot program. J Women's Health 2014;23(4):350–7.

[56] Two Feathers J, et al. Racial and ethnic approaches to community health (REACH) detroit partnership: improving diabetes-related outcomes among African American and Latino adults. Am J Public Health 2005;95(9):1552–60.

[57] Bader A, Musshauser D, Sahin F, Bezirkan H, Hochleitner M. The mosque campaign: a cardiovascular prevention program for female Turkish immigrants. Wien Klin Wochenschr 2006;118:217–23.

[58] Peterson M, Chandlee M, Abraham A. Cost-effectiveness analysis of a statewide media campaign to promote adolescent physical activity. Health Promot Pract 2008;9(4):426–33.

[59] Peddecord KM, et al. Can movie theater advertisements promote health behaviors? Evaluation of a flu vaccination pilot campaign. J Health Commun 2008;13(6):596–613.

[60] Oto MA, et al. Impact of a mass media campaign to increase public awareness of hypertension. Turk Kardiyol Dern Ars 2011;39(5):355–64.

[61] Petrella RJ, Speechley M, Kleinstiver PW, Ruddy T. Impact of a social marketing media campaign on public awareness of hypertension. Am J Hypertens 2005;18(2):270–5.

[62] Pearson TA, et al. American Heart Association Guide for Improving Cardiovascular Health at the Community Level, 2013 update: a scientific statement for public health practitioners, healthcare providers, and health policy makers. Circulation 2013;127(16):1730–53.

[63] Campbell MK, et al. Church-based health promotion interventions: evidence and lessons learned. Annu Rev Public Health 2007;28:213–34.

[64] Austin LT, et al. Breast and cervical cancer screening in Hispanic women: a literature review using the health belief model. Womens Health Issues 2002;12(3):122–8.

[65] Allen JK, et al. Community Outreach and Cardiovascular Health (COACH) Trial: a randomized, controlled trial of nurse practitioner/community health worker cardiovascular disease risk reduction in urban community health centers. Circ Cardiovasc Qual Outcomes 2011;4(6):595–602.

[66] Chatterji P, Joo H, Lahiri K. Racial/ethnic- and education-related disparities in the control of risk factors for cardiovascular disease among individuals with diabetes. Diabetes Care 2012;35(2):305–12.

[67] American Heart Association Cardiovascular Disease. Bridging the gap: CVD and health equity. 2011; Available from https://www.heart.org/idc/groups/heart-public/@wcm/@adv/documents/downloadable/ucm_301731.pdf.

Chapter 3

Extension of Peer Support from Diabetes Management to Cardiovascular Disease Prevention and Management in Primary Care and Community Settings in Anhui Province, China*

Xuefeng Zhong*, Bert Potemans†, Chanuantong Tanasugarn‡, Edwin B. Fisher§

*Anhui Provincial Center for Disease Control and Prevention, Hefei, China, †Vårdcentral Rättvik, Rättvik, Sweden, ‡Department of Health Education and Behavioral Science, Faculty of Public Health, Mahidol University, Bangkok, Thailand, §Peers for Progress and Department of Health Behavior, Gillings School of Global Public Health, University of North Carolina at Chapel Hill, Chapel Hill, NC, United States

China has experienced a rapid epidemiological transition from acute to chronic or noncommunicable disease over the past 40 years, and chronic conditions have become the major causes of disease burden in recent decades. Along with the rest of the world, this transition has seen not only increased life expectancy but also increased disability through the burdens and complications of noncommunicable diseases [1]. From 1990 to 2013, the life expectancy for females in China increased from 70.21 to 79.99 years and for males from 66.01 to 73.53 [1].

The changes in lifestyle in China, linked to its rapid socioeconomic development and urbanization, are occurring at an unprecedented rate. As part of these trends, the prevalence of diabetes mellitus in China has increased markedly during the past four decades [2]. The most recent 2010 national survey estimated the prevalence among Chinese adults at 11.6% or 113.9 million adults and the prevalence of prediabetes at 50.1%, 493.4 million [3]. Two-thirds (67%) of Chinese adults with diabetes have complications [4], and only 11.5% exhibit satisfactory glycemic control [5].

Cardiovascular disease (CVD) also is a growing burden in China. Based on a report of the Chinese Ministry of Public Health [6], a 2007 review noted an increased prevalence of CVD morbidity, including hypertension, heart disease, and cerebrovascular disease, between 1993 and 2003 from 31.4 to 50.0% [7]. Similarly, a 2013 study of the prevalence of stroke in China [8] found an age-standardized prevalence of 1.11%.

Hence, the burden of chronic diseases and the associated preventable premature morbidity and mortality in China will continue to have a very substantial impact on the health-care costs, the labor force, and the country's economy over the coming years [9,10].

Posing a challenge to self-management, the Chinese health system, as do many others [11], tends to emphasize clinical care but places little emphasis on patient education. Traditional diabetes self-management programs envision support provided on a one-to-one basis. This support, however, is often labor-intensive so that it is too costly in time and money to be provided by professionals. Therefore, the need to identify effective and evidence-based, affordable, and scalable approaches to prevent and control chronic diseases, such as diabetes and CVD, is very urgent [12].

There is now substantial evidence that peer support approaches can contribute significantly to improved diabetes self-management behavior and blood sugar control [5–7], but this has not been adopted widely in China. An exception is a

---

* Portions of this chapter are drawn from Zhong X, Wang Z, Fisher EB, Tanasugarn C. Peer support for diabetes management in primary care and community settings in Anhui Province, China. Ann Fam Med 2015;13(Suppl 1):S50–58. The sections on extending peer support to cardiovascular disease prevention report work that has also been previously reported in Zhong X, Potemans B, Zhang L, Oldenburg B. Getting a grip on NCDs in China: an evaluation of the implementation of the Dutch-China cardiovascular prevention program. Int J Behav Med 2015;22(3):393–403.

recent report of telephone peer support provided by a diabetes specialty service for patients of primary care physicians in Hong Kong [8]. Peer support approaches are especially promising for rapidly developing countries, such as China, which are facing rapid increases in the rates of diabetes and a very serious shortage in health resources to address diabetes and related complications.

Given the prominence of both diabetes and CVD, peer support to address them is of great interest. This chapter reports on the development of peer support programs for diabetes management in Anhui Province, China, and their extension then to CVD.

BACKGROUND ON PEER SUPPORT

A 2014 review of peer support interventions [13] documented contributions to basic health needs (e.g., reducing childhood undernutrition), to primary care and health promotion, and to chronic disease management. A review conducted by Peers for Progress, a program dedicated to promoting research on and improved practices of peer support worldwide [14], included peer support interventions from 14 different countries [15]. Papers addressed a wide variety of prevention and health objectives entailing the kinds of sustained behavior change central to chronic disease management (in contrast to relatively isolated acts such as cancer screening). A total of 65 papers addressed the prevention or management of a variety of health conditions including drug, alcohol, and tobacco addiction (3 papers); cardiovascular disease (10); diabetes (9); HIV/AIDS (6); other chronic diseases (12); maternal and child health (17); and mental health (8).

Across all 65 papers, 54 (83.1%) reported significant impacts of peer support, 40 (61.5%) reported between-group differences, and another 14 (21.5%) reported significant within-group changes. When limited to papers reporting controlled designs and using objective or standardized measures, 36 of 43 or 83.7% reported significant effects of peer support. Among an additional 24 reviews of peer support, 19 provided quantifiable findings with a median of 64.5% of studies reviewed reported significant effects of peer support.

Focusing on peer support for diabetes, an extension of the original review through 2015 included 30 papers of which 17 (56.7%) reported significant, between-group differences favoring peer support. Among the 30 diabetes studies, 19 reported pre- and postintervention values of HbA1c among those receiving peer support [16–33], and another 4 reported average changes from pre- to postintervention [34–37]. Across all 23, the average change score was a decrease of 0.76 ($p < 0.001$). Among the 19 reporting pre- and postintervention values, repeated measures analysis controlled for the number of participants in the peer support condition and the duration from pre- to postintervention/follow-up appointment. The average HbA1c declined from 8.42% to 7.63% ($p = 0.004$) with an average of 140 participants in peer support condition (range = 14–781) followed an average of 12.21 months (range = 4–24). Neither the number of participants nor the duration of intervention and follow-up significantly influenced the change in HbA1c over time.

Based in part on the project in Anhui described here and starting in 2009, Peers for Progress has collaborated with the leaders of the Chinese Diabetes Society and colleagues at Southeast University and Zhongda Hospital in Nanjing to train over 450 program managers, clinicians, and diabetes educators to develop and implement peer support. Over 30 programs have been developed, including a demonstration project of the Beijing Diabetes Prevention and Treatment Association that engaged 50 hospitals and community health centers and over 5000 individuals with diabetes.

With the general growth of evidence for peer support, developing interest in China, and interest in peer support implemented through primary care [38–40], this dissemination and implementation project was administered through the Provincial Center for Disease Control (CDC) in Anhui Province [41]. It examined a peer support intervention for type 2 diabetes in cooperating primary care community health services centers (CHSCs). Following models for implementation and dissemination research [42–44], it included initial formative evaluation of acceptability and feasibility, implementation and advantages, disadvantages and barriers, reach and recruitment, effectiveness within an evaluation subsample, and program sustainability and adoption.

An important context of the peer support project was the community chronic disease management system implemented across China through population-wide primary care provided by community health service centers (CHSCs) and their community health service stations (CHSSs). Through this initiative, the government provides ¥25/person/year (in 2009–10, approximately, US$3.50, and increased to ¥50/person/year in 2017) to CHSCs to deliver a defined package of 11 basic public health services. For diabetes, these include yearly clinical assessment; quarterly blood sugar assessment; and health education addressing healthy diet, physical activity, medication adherence, and routine home follow-up visits. Traditionally, the CHSCs provide health education mainly through leaflets and brochures for patients with diabetes or through diabetes education information in community locations. As in most areas of China, these activities are coordinated through local CDC.

Within the context of the community chronic disease management system, the peer leader-support program (PLSP) was implemented in three cities in Anhui. One city in each of the northern, southern, and middle regions of Anhui was selected.

A district was then selected within each city and, within each district, one community (≈20,000 residents) randomly selected. These were Yang Guang Community of Tonglin, Da Qin Community of Bangbu, and He Ye-di Community of Hefei. Within each of these, two subcommunities were chosen with similar characteristics in terms of demographics, customs, and social norms. The two chosen subcommunities, each served by its own CHSS, were then randomly assigned to intervention and control conditions.

FORMATIVE EVALUATION AND LESSONS FOR IMPLEMENTATION

Formative evaluation addressed the feasibility, adaptability, and acceptability of the program and its key features relative to community and organization policy. It was also intended to engage and empower local communities as part of program development. Leaders of each of six CHSSs, two in each community that had agreed to be in the study, along with the leader of the district health bureau and the leader of the neighborhood committee associated with each of the six CHSSs participated in focus groups and individual interviews.

As detailed in Table 1, the representatives from local community neighborhood committees indicated that PLSP would be acceptable and feasible for their neighborhoods. Responses of CHSC directors, staff, and patients were also positive and included expressions of desiring "more training and direction" along with some concern among staff that "the program may bring a large work burden for us." Accordingly, health authorities in the three cities agreed to provide policy, technical, and modest financial support to the PLSP. With this support from health authorities and endorsement by community and CHSC leaders of the PLSP and a systematic study of its effectiveness, implementation proceeded without substantial change to initial program plans.

RECRUITMENT

Participants were identified through the community chronic disease management system in each CHSC as having type 2 diabetes diagnosed at hospitals, based on WHO criteria [45]. They also were at least 15 years of age and had resided in a project community for at least 1 year. Those unable to participate due to physical and/or mental disabilities were excluded.

After contact by CHSS staff, those willing to participate attended their CHSS to provide demographic, self-report, and biophysical data. Baseline data were gathered between June and December of 2009 with follow-up between October and December of 2010.

PEER LEADER SELECTION AND TRAINING

Across the 3 PLSP communities, 19 peer leaders were recruited by CHSC staff. They were selected according to the following criteria: type 2 diabetes for at least 1 year, willingness to participate as a volunteer, and clinician's rating of general adherence to both medication and behavioral management. Additional criteria were interest in helping others, positive and sociable personality, time available, understanding the importance of patient confidentiality, good relationships with community residents, and leadership in their communities. Further selection was based on willingness to liaise with CHSC staff in response to unanticipated problems, to commit to the project schedule, to adhere to the responsibilities of peer leaders and program policies, to attend 3 days' training, and to contact group members frequently.

Peer leaders were retired adults with a mean duration of diabetes of 9.3 years. Although some were nonprofessional, a number had pertinent prior work experience such as in teaching or nursing. Sixteen of 19 were male (84.2%).

The Anhui CDC research team provided 3 days' training for the peer leaders, including an introduction to the PLSP, basic skills, and introduction to self-management as well as how to promote it. Emphasis was on key functions of peer support [46,47] promoted by Peers for Progress: (1) assisting and encouraging daily management; (2) social and emotional support; (3) linkage with community resources and primary care, largely built into close ties to the CHSSs; and (4) ongoing support. Peer leaders were told how they could seek support from their CHSS staff, CDC professionals, and tertiary hospital specialists if needed. They also met with each other every 2 weeks, sharing their experience and exchanging lessons. A peer leader handbook included materials for use with participants. CHSC staff also received training in their roles supporting and working with the PLSP.

PLSP MEETINGS AND ACTIVITIES

Nineteen "peer support groups" were set up among CHSSs randomized to the PLSP condition. Each group consisted of 10–15 participants. Twelve biweekly education meetings over 6 months were coled by peer leaders with CHSC staff's

TABLE 1 Observations From Participants, Peer Leaders, and Clinical Staff and Administrators Regarding Acceptability, Feasibility, Implementation, and Emergent Features of Peer Leader-Support Program (PLSP)

Category	Observation or Quotation (Type of Respondent)
Acceptability and feasibility	"We will provide meeting rooms or other instrumental support if this project is implemented in our community" (representative of community neighborhood committee) Community residents "are familiar with each other and like group activities such as mah-jongg, dancing, walking, and chatting" (representative of community neighborhood committee) "Peer leaders are like the bridge to link our CHSC with patients in their served neighborhood" (community health center director) "This is an innovative program from which we could learn new knowledge and skills. Also, it may help us to use this approach for other health issues and diseases" (community health center director) "We hope the Anhui CDC will give us more training and direction during project implementation" (community health center director) "This project can help us to manage our registered patients with diabetes" (community health center station staff member) "We don't have many intervention services from our CHSC" (patient) "If the community organized a 'diabetes peer support group,' we would be keen to attend this group because this is beneficial for our disease and health" (patient)
Facilitating factors of the CHSCs	CHSC director indicated that peer leaders can help CHSC staff implement primary care more effectively, e.g., through group monitoring and care with peer leaders helping to inform patients to attend. Instead of telephoning patients individually, peer leaders can contact them in groups and through their neighborhoods CHSC staff suggested (i) ongoing network from the provincial level to community levels to facilitate program implementation, (ii) requiring that diabetes patients receive diabetes self-management education, and (iii) improving PLSP outcomes through implementation over a longer time
Barriers of CHSCs	Limited time and human resources, system coordination, and external locus; "we worry a little bit if the program may bring a large work burden for us" (community health center station staff member) Although required to provide public health services, CHSCs tended to focus on fee-producing clinical services, while public health services depended largely on local government
Facilitating factors of peer leaders	*Instrumental support* for peer leaders through collaboration with CHCS staff: "we contacted with CHSC staff and professionals more frequently. I have more close ties with CHSC staff. If we want them to help us, they always can do that" *Confidence:* Training, discussions, and participation in the PLSP had increased knowledge and skills for diabetes self-management and confidence in leading group meetings/activities *Sense of importance:* PLSP meetings and activities viewed as best ways for those with diabetes to help each other. Sense of importance, pride from meeting needs, providing help, being valuable for others and their communities
Barriers of peer leaders	Need for modest financial support for materials (e.g., table tennis balls), transportation, refreshments, and limited knowledge and skills about promoting diabetes management, "I would like to have more time training" (peer leader) "I would like to have a chance to learn from others. Could the CHSC organize us to visit another community to learn from them?" (peer leader)
Facilitating factors of participants	Most participants expressed positive views of group meetings and activities and valued the openness of their implementation, such as by being told in advance of topics to be discussed
Participant barriers	Attending group meetings/activities was discouraged by weather, family matters, inconvenient or nonpreferred meeting times, or planned topics of little interest
Practical, specific support	"…it is difficult to get up in the morning to do exercise. My peer leader had a phone call in the morning for weeks and waited for me under my building to get me doing morning exercise. Then, I can get up every morning to do exercise with group" (participant) "I often forgot the date to get blood sugar checked that is free provided by CHSC. My peer leader always give me a call to mention to me that next day I need to go to CHSC to get routine check. If I still forgot on check day, he will call me again to ask me to come to CHSC for check" (participant)
Emergence of emotional support	"Peer support group is like our 'second home' in which you can say what you want to say, without worry about misunderstanding and discrimination" (participant) "We can easily talk with peer leaders and peer group members when we felt unhappy, and we didn't want to talk with our children because we didn't want to burden them" (participant)
Sustainability	CHSC leaders also saw peer support as an innovative strategy to needs other than diabetes management; "the peer support approach is a good method to help manage chronic disease, not only for diabetes. I am preparing to set up peer groups to target hypertension patients" (CHSC director)

involvement titrated to peer leaders' needs. Meetings lasted from 1½ to 2 h and covered a range of topics, e.g., diet; physical activity; medications; foot care; stress management and depression; barriers to self-management; and obtaining resources and supports from the community, family, friends, or the health system. Content and order of presentation could be changed according to leaders' judgment of needs of the group. For efficiency, groups were often combined, resulting in more than 30 participants and providing limited opportunity for discussion. Accordingly, peer leaders also led 12 biweekly discussion meetings over 6 months. These reviewed the topics of the education meetings and included sharing experiences and modeling self-management practices.

PEERS LINK THE RESIDENTIAL AND CLINICAL AND PROMOTE INFORMAL MUTUAL SUPPORT

An important feature of the settings in Anhui was their integration of primary care with communities. Individuals in a particular housing site receive their care through a clinical team assigned to that site. Fig. 1, for example, shows the director of a participating CHSC pointing to a sign in the entrance of a residence indicating the names and telephone contacts of the clinical team serving that building. Peer leaders receive care through the same team as the others with whom they live. Thus, as in Fig. 2, peer leaders, their neighbors, and their shared clinical service team are connected so as to facilitate coherence of clinical care, peer support, and community activity.

An important result of the integration of clinical teams and peers within neighborhoods was the facilitation of mutual and informal support extending the influence of the regular group meetings. Peer leaders led or encouraged informal activities (e.g., walking, fishing, dancing, and tai chi groups) among group members. Within Chinese culture, these kinds of informal activities and groups are popular, especially among older or retired adults. The visitor to China (or to sections of cities around the world with large ethnic Chinese populations) will have seen numbers of folks in parks or other gathering places doing tai chi or other exercises together in mornings or evenings. The activity may entail mah-jongg, chess, or simply discussing neighborhood gossip. With the encouragement of the peer support program, this kind of naturally occurring gathering took on the added dimension of providing an opportunity for those with diabetes to engage in activities together, with then the added attention to improving health together. As described in further detail below, group participants mentioned that they liked to talk about worries or concerns about their health in these groups that became like a "second home." And, again, because peer leaders and participants lived within the same subcommunities, these kinds of casual interactions and activities were greatly facilitated among them.

VARIABLE IMPLEMENTATION ACROSS CHSCS

Table 2 includes, for each subcommunity, the average frequency of educational and discussion meetings and the numbers of participants enrolled and participating. With only one CHSC staff member assigned to manage the program and

FIG. 1 Community health center director pointing to sign in residence entry way that indicates the names and contact information for clinical team that serves residents of the building.

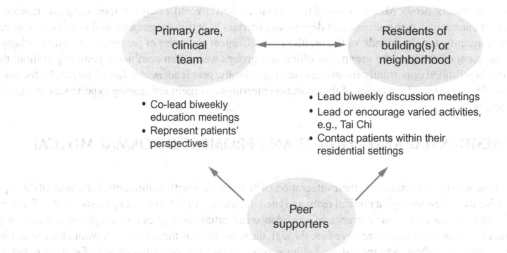

FIG. 2 Relationships among peer leaders, clinical teams, and community residents and varied components of peer leader-support program.

TABLE 2 Average Frequency of Key Meetings and Activities and Numbers of Participants in Each Community

Category	Subcommunity 1	Subcommunity 2	Subcommunity 3
Number of biweekly educational meetings coled by CHSC staff and peer leaders (12 by protocol)	8	7	4
Number of biweekly discussion meetings by peer leaders (12 by protocol)	12	9	6
Number of enrollees	123	120	122
Actual number participants (attended ≥6 meetings)	111	128	40

colead group meetings, implementation in community 3 did not achieve protocol objectives. Only three peer leaders were recruited, and only three peer groups of 10–15 participants were organized. As a result, most participants did not have the opportunity to attend group meeting and activities. Accordingly, this community and its participants were not included in the evaluations of program effectiveness.

EVALUATION OF EFFECTIVENESS

With the elimination of community 3 from analyses, effectiveness evaluations were completed with 137 PLSP and 94 control participants from communities 1 and 2 who completed both baseline and follow-up evaluations. Because of some differences among communities and by condition at baseline, age, education, and income were included as control variables in all statistical evaluations. Additionally, because women were underrepresented in those completing follow-up (41% in comparison with 55.3% of those not completing, $p = .002$), sex also was added as a control variable in statistical evaluations in addition to age, education, and income.

Table 3 includes baseline and follow-up measures of knowledge, attitudes, self-management behaviors, and clinical outcomes for the PLSP and control groups in communities 1 and 2, controlling for sex, age, education, income, and community. Significant interactions between time (baseline vs follow-up) and condition indicated greater favorable changes for PLSP participants for fasting glucose, 2 h postprandial glucose, BMI, SBP, DBP, knowledge, and self-efficacy. For fasting glucose, for example, PLSP participants showed a reduction from 7.68 to 6.76 mmol/L (138.38–121.80 mg/dL), while those in the control communities showed a slight increase from 6.38 to 6.66 mmol/L (114.96–20.00 mg/dL). The difference between these two patterns was highly significant ($p = 0.000$). For mean BMI, there was a modest decrease in PLSP from 24.26 to 23.66 but an increase in the controls from 23.51 to 23.97 (p for difference between patterns = 0.000). These values should be understood with recognition that, for groups of Chinese origin, BMI values of 24·0 (for men) and 23·0 kg/m [2] (women) have been recommended as cutoffs for obesity [49].

TABLE 3 Means (Standard Deviations) of Participants on Study Variables in PLSP Intervention and Control Communities at Baseline and Follow-up and p Levels for Tests of Changes Over Time and Interactions Between Condition (PLSP vs Control) and Time (Changes from Baseline to Follow-up)[a]

| Variable | PLSP (n = 135) | | Control (n = 94) | | p Levels | |
	Baseline	Follow-up	Baseline	Follow-up	Baseline Versus Follow-up	Interaction: Condition X Time
Knowledge	6.49 (2.53)	7.33 (2.82)	7.88 (2.04)	6.39 (3.24)	.121	.000
Attitudes	27.15 (6.03)	34.08 (4.87)	27.61 (4.96)	33.29 (4.52)	.593	.372
Self-efficacy	36.05 (4.27)	35.21 (3.95)	36.67 (5.52)	32.90 (4.04)	.407	.002
Social support	32.46 (8.02)	31.71 (7.15)	30.37 (3.65)	30.51 (6.51)	.410	.136
Diet	18.92 (1.96)	20.00 (2.36)	18.95 (1.54)	19.71 (2.11)	.747	.395
Physical activity	10.22 (3.19)	10.32 (3.04)	9.66 (2.65)	10.13 (2.88)	.164	.482
Glucose monitoring	3.86 (1.56)	4.13 (1.67)	5.12 (1.30)	4.86 (1.38)	.348	.152
Medication adherence	3.93 (1.86)	4.04 (1.70)	4.18 (1.45)	4.39 (1.20)	.047	.638
BMI (kg/m²)	24.26 (2.88)	23.66 (2.80)	23.51 (2.50)	23.97 (2.49)	.401	.000
SBP	135.92 (14.91)	127.95 (11.29)	130.44 (11.80)	130.76 (14.29)	.838	.000
DBP	82.51 (8.72)	79.08 (7.33)	79.04 (8.67)	78.63 (8.58)	.841	.020
Fasting glucose (mmol/L)	7.68 (2.13)	6.76 (1.80)	6.38 (1.48)	6.66 (1.63)	.744	.000
Fasting glucose (mg/dL)[b]	138.38	121.80	114.96	120.00		
Two hour postprand glucose (mmol/L)	11.75 (3.54)	10.68 (3.65)	10.36 (2.05)	10.51 (3.39)	.765	.022
Two hour post prand glucose (mg/dL)[b]	211.71	192.43	186.67	189.37		

[a]*GLM repeated measures analyses controlling for sex, age, education, family income, and community.*
[b]*Glucose levels in milligrams per deciliter are conversions from means expressed as mmol/L.*
Description of self-report measures
Knowledge: Total of 12 items, 4, glucose; 3, diabetic complications; 2, diet; and 3, insulin. Correct responses received one point and incorrect answers or "don't know/unsure" received zero. Possible range of knowledge scores was 0–12. High scores indicate greater knowledge.
Attitudes toward self-management: Nine items scored 1–5 ("strongly disagree," ... "strongly agree") with the range of 9–45. High scores indicate more positive attitudes.
Self-efficacy for self-management: Nine items scored 1–5 ("strongly disagree," ... "strongly agree") with the range of 9–45. High scores indicate greater self-efficacy.
Perceived social support for self-management: Nine items scored 1–5 ("never," ... "always") with the range of 9–45. High scores indicate greater perceived support.
Self-management: Nine items scored 1–5 resulting in subscores for diet, four items, maximum score=20; physical activity, two items, maximum score=10; glucose monitoring, two items, maximum=10; and medication adherence, one item, maximum=5 [48]. High scores indicate greater self-management practices.

QUALITATIVE EVALUATION: ADVANTAGES, DISADVANTAGES, AND BARRIERS TO IMPLEMENTATION

Qualitative evaluation with CHSC staff, peer supporters, and patients at the conclusion of PLSP identified barriers to and facilitators of implementation, detailed in Table 1. Barriers ranged from staff and financial resources within the CHSSs to inconvenient meeting times.

Most peer leaders felt that the PLSP benefited both diabetes patients and themselves, viewed the PLSP as important, were happy with their work, and said they would like to remain peer leaders. Facilitating factors included assistance peer leaders reported from CHSC staff, skills staff reported learning through the program, and one CHSC director's opinion that the PLSP improved the provision of care. Staff also advocated greater duration of the PLSP. Participants noted no negative aspects of the PLSP but described the value of both practical, specific assistance and emotional support it provided.

CULTURAL INFLUENCES ON SUPPORT WITHIN THE PLSP

Group activities are traditional in Chinese culture, especially for older adults. This and the fact that peers and participants lived in the same housing complexes or neighborhoods may have enhanced participation in the informal groups based on shared interests, e.g., jogging groups. Also of note, all peer leaders and most participants were older adults (average age 63 ± 9.41) and retired (62.8%) providing them time to attend group meetings and activities. Practical, specific support from the PLSP may have contributed to participants' increased self-efficacy [50].

As noted above, 84% of peer leaders were men. Although retired, most women invited to become peer leaders felt that tasks of housework and looking after grandchildren prevented their commitment to group meetings/activities. Also, most females in the generation sampled have lower education levels than males.

EMOTIONAL SUPPORT

Emotional support emerged as a benefit of the PLSP. With cultural emphases on harmony and the interests of the family, relatives are obligated to support each other. That relatives are obligated to support each other, however, confers on individuals a reluctance to share concerns in order not to put a burden on their families [51]. The PLSP apparently provided a setting in which individuals could discuss feelings with those outside their family and, thereby, without fear of obligating others.

Several patients commented on the value of emotional support through the PLSP.

When I knew I had diabetes, I felt upset and that my life is meaningless. When I joined the peer support program, I found that many people have the same illness as me, but they lived very well. Some are more than 80 years old. This encouraged me. I also can live long and healthy if I can control my blood sugar and managed my life suitably.

(61-year-old female, Da Qing community)

I feel that the 'peer group' is like my second family. The people in the group are so kind. Normally, I did not like to talk about my illness with other people, but in here, I can talk about my disease. We also discussed diet, activities, medicine, and blood sugar control.

(66-year-old female, He Yedi community)

The observations noting the value of emotional support are consistent with other reports of benefits of peer support in diabetes management [27,34,36,52–55]. A study of telephone peer support in Hong Kong [34], for example, examined the addition of peer support to comprehensive care that included coordination through a registry; initial appraisals and patient education; and ongoing, quarterly reports and algorithm-based recommendations for care to patients and clinicians [56]. In terms of clinical outcomes, the peer support did not add incremental effectiveness—apparently, the comprehensive primary care was quite effective in this setting. Consonant with the present study's identification of the importance of emotional support, however, the Hong Kong peer support substantially reduced emotional distress among the 20% who were initially above cut points for a measure of depression, anxiety, and stress [57]. That is, the high-quality clinical care appeared sufficient for clinical outcomes, but the peer support added incrementally in the area of emotional distress.

As is commonly observed worldwide [58], emotional distress is a common predictor of hospitalization among those with chronic diseases. This was so in the Hong Kong sample in which 20% who were high on distress and assigned to the control group with comprehensive primary care but no peer support had an especially high rate of hospitalization. Among those with high distress who received peer support, however, the rate of hospitalization was equivalent to that among those with low distress [34].

PEER LEADER ROLES

Several aspects of the roles of peer leaders deserve attention. Coleading educational meetings with CHSS staff might be thought to compromise the "peerness" of the peer leaders. They reported, however, that they valued not having to worry about making mistakes, but being able to focus on their roles as peers and the activities they promoted. This reflected a view of their role as distinct from and complementary to health professionals, not as "junior doctors or nurses."

In their distinct roles, peer leaders reported good working relationships with CHSS staff, providing feedback to CHSC staff about group members' needs, questions raised, or needs for specialty referrals. Participants valued the peer leaders as "our" representative who could express "our" needs and requirements to CHSCs and tertiary hospital specialists. In this, peer leaders may offer a valued counterforce to the otherwise hierarchical nature of medical care as it has evolved not only in China but also in many health-care systems. The PLSP also embodied community participation strategies needed for chronic disease prevention and control in the daily lives of individuals [59,60]. This was also valued by CHSC directors, one of whom noted that peer leaders may serve as "the bridge to link our CHSC with patients in their served neighborhood." This reflects broader discussion of the contributions of peer support to primary care [38–40].

IMPORTANCE OF ORGANIZATIONAL SUPPORT

Variability in implementation underscores the importance of organizational support to program implementation, issues just beginning to be studied in peer support. [40] The low level of implementation in community 3 also reflects recent transitions of health care in China. Hospitals formerly supported by state-owned enterprises have had to find new funding as those enterprises have become private and, often, discontinued hospital support. Many hospitals have become CHSCs such as the CHSC in community 3. At the time of the study, it had not yet obtained sufficient new funding and so focused on clinical revenues. Additionally, with its history as a hospital, it lacked public health professionals and experience. Reflecting these influences, the CHSC in community 3 did not assign many staff to the PLSP. Fortunately, these transitions of former hospitals have now been generally completed, and the associated limitations are less common.

SUSTAINABILITY, ADOPTION, AND EXTENSION TO CARDIOVASCULAR DISEASE

Based on the results of the PLSP for diabetes, the Anhui Provincial Health Bureau has expanded the program to other communities in the province and to CVD prevention and management. This has been facilitated by the policies of the national government noted above that encourage 11 primary care and preventive packages including the management of hypertension as well as diabetes and severe mental illness along with, e.g., health education, care for older adults, and electronic health records. National policies also encourage primary care CHSCs to serve as "health gatekeepers" and to manage referrals to specialist care and hospitals [61].

The decision to "scale up" the PLSP as a strategy for risk factors for other noncommunicable diseases led to a Dutch-China cardiovascular prevention program that was conducted in Anhui province during 2010–13. The program included population screening with a self-report risk measure, individual consultation and assessment for those who screened as at high risk, and classes to review risk profiles and ways to reduce them. In its original plan, the three components of this program would be delivered by primary care professionals from local CHSCs.

Adaptation of CVD Prevention Program

The cardiovascular prevention program was undertaken in two phases: a pilot phase in the Heyedi community in Shushan district of Hefei City and a wider implementation phase, in which the program was implemented in a further seven communities [62]. The strategy of using a self-administered questionnaire to identify people at high risk of cardiometabolic diseases in the community and the curricula for the lifestyle education and intervention sessions and a cardiovascular risk-factor management online website were all refined following the pilot phase. To prepare for implementation, the staff of the seven additional CHSCs received intensive training in all aspects of program delivery over 2 days.

The focus of the program was to help residents at high risk for CVD to reduce and manage those risks through primary care and community settings. The program focused on common risk factors such as smoking, inactivity, salty diet, and low levels of vegetable and fruit intake. Local community residents 35 years of age or older were invited by CHSC staff to complete a self-report measure of CVD risk based on measures developed in a variety of global settings [63,64]. The measure includes age (0 points for those <35 years up to 3 points for those ≥55 years), BMI (0 for those <24 up to 2 points for those ≥28), smoking (2 points for current), personal history of diabetes or hypertension (2 points), and family history of diabetes or CVD (1 point). High risk was based then on a score of 5 or greater.

Those who were characterized as high risk on the self-report measure were then invited to visit their CHSC for an individual follow-up consultation that reviewed their risk status and added measures of waist circumference, blood pressure, and blood glucose and an optional cholesterol measure. Confirmatory evaluations were scheduled for those with clinical measures reaching criteria, and individuals were then encouraged to participate in the prevention program and, as appropriate, offered further clinical care for high blood pressure, diabetes, etc.

Those at confirmed high risk were invited to attend a health education class. The content of this class included "what is cardiovascular disease" and "what are common high-risk factors of CVD (inactivity, unhealthy diet, smoking, excess alcohol consumption, family, or personal history)." Participants then identified their own risk factors based on this information. The core of the intervention was a personalized assessment and action plan that followed from this.

LIFESTYLE INTERVENTION AND PEER-LED GROUP ACTIVITIES

All participants who attended the health education class and completed a personalized action plan volunteered to attend lifestyle intervention groups composed of others with a similar action plan to reduce the same high risk. Lifestyle intervention groups addressed physical activity, smoking cessation, reducing salt intake, and eating a healthier diet by increasing vegetable and fruit intake. Most participants in the group focusing on reducing salt intake were women who were responsible for meal preparation in their families. In contrast, most participants in the group focusing on healthy diet and increasing fruit and vegetable consumption were men who tended to eat more meat and fewer vegetables and fruits. Peer leaders were actively involved in leading these lifestyle intervention group activities.

In summary, the intervention following the personalized assessment centered around the following:

(1) What risk factors do you have?
(2) What risk factors are you confident that you can change in the next 3 months or 6 months?
(3) Which activity groups do you want to join to help you change those risk factors?

Roles and Contributions of Peer Group Leader

Peer group leaders were recommended by CHSC staff or by group members. They received 1.5 days of training that covered (a) the purpose and overall strategy of the program, (b) the basic knowledge of CVD prevention, (c) the roles and responsibilities of group leaders, (d) the communication skills, and (e) the skills for helping people set and follow behavior change plans. The general role of peer leaders was to organize group meetings at least once a week and remind members to attend these activities. These meetings were very casual. Their main purpose was to assess how many times participants attended activities and to discuss what prevented them from attending. This led to discussions with peer leaders about how to address barriers to attendance and to following through with intended changes.

As with the PLSP for diabetes, peer leaders also provided important linkage to CHSCs and their staff. For example, they reported how many participants attended group activities to CHSC staff who entered these data in a cardiovascular risk-factor management online website that tracked individual attendance and participation. The group leader also advanced to CHSC staff their participants' requests for more professional support if they needed it. For example, members of smoking cessation groups asked for CHSC staff assistance with quit-smoking counseling and medication.

The peer group leaders also coordinated group activities on a weekly or daily basis. These included physical activity groups varying according to member interests, such as a Taiji group, a jogging group, or a dancing group. Healthy diet and cooking groups including groups focusing on salt reduction and cooking less salty dishes or shopping for vegetables and fruit.

EVALUATION LIFESTYLE EDUCATION AND INTERVENTION SESSIONS

From 10,642 self-report measures distributed by the seven CHSCs, 9067 individuals completed the measures, of whom 2720 (30%) had a score indicating high risk. Of these, 926 (34%) received an individual follow-up consultation, and 457 attended lifestyle classes led by CHSC professionals. These classes were organized and formed by CHSC staff, and each class included 15–20 participants. According to their personal action plans, individuals joined lifestyle intervention groups addressing smoking cessation, healthy diet, physical activity, etc. In addition to being organized around personal action plans (smoking cessation, etc.), participants were organized into groups based on the neighborhood in which they lived so that they would be participating with neighbors whom they would also see in daily activities outside the groups. Table 4 shows that the majority of males (64.8%) chose the quit-smoking intervention group, consistent with the sharp difference in smoking rates among males (as high as 66%) [65] and females. Among females, on the other hand, 53.6% chose exercise groups.

TABLE 4 Choice of Lifestyle Intervention Group

Life-style Intervention Group	Percentage N (%)	Female N (%)	Male N (%)
Cooking	63 (13.8)	41 (24.1)	22 (7.7)
Exercise	167 (36.5)	91 (53.6)	76 (26.5)
Quit smoking	218 (47.7)	32 (18.8)	186 (64.8)
No choice	9 (2.0)	6 (3.5)	3 (1.0)
Total	457 (100)	170 (37.2)	287 (62.8)

Program Acceptability by Health Professionals and Program Participants

Health professionals from all seven CHSCs returned questionnaires asking them about their feedback on the program. They noted a number of positive aspects including the value of the program in detecting and raising awareness of CVD risk and improving lifestyles. They also noted that the program was simple, convenient and practical to use, and very innovative in China. Negative features of the program and surrounding it included insufficient funding to cover expenses and CHSC efforts, difficulty in following up individuals during the different stages of the intervention, and difficulty in generating awareness. Respondents also indicated that they felt that the program should be scaled up and extended to not only the initial communities but also the whole city. In summary, while the respondents from the CHSCs were pleased with the design and concept, some felt that the level of funding to the CHSCs to implement the program and the execution of some stages were problematic.

Participants from the program ($N = 15$) also gave quite positive feedback concerning the benefits of the program, including comments such as the following:

- It is good because all these activities are good for my health. For me, I have begun to know more about how to do exercise, how to do diet, which things are harmful for health, and which things are better for my health.
- This program has taught me about my high-risk factors, and the doctors then taught us about a healthy lifestyle in the lifestyle class. Peer leaders organized activities.
- I have changed some unhealthy lifestyle by attending this program. Now, I go for a walk every morning and after dinner every day. I now also have a plan to quit smoking.
- I now take my hypertension pills more regularly, compared with before … as a result of knowing my "risk score" and attending intervention group.
- Although I previously felt my health is good and I had no disease, I have now been diagnosed with diabetes through this program.

One participant placed the program in the context of general developments in China:
- Now, our general lives have begun to improve in China, so we can begin to pay more attention to our own health. This program has let me know how to live a more healthy life.

Many individuals also indicated that they would attend the program in the future and/or introduce it to others:

- I want to introduce my friends to this program in the future.
- I hope that I can attend again if there is a similar program in our community in the future.
- I would willingly recommend to others.
- I hope this program continues to exist in the future.

REFLECTIONS, CONCLUSIONS, AND FUTURE DIRECTIONS

The projects on peer support in diabetes and CVD prevention have had a substantial impact on similar practices throughout Anhui Province and across China. The Anhui CDC has integrated these approaches into the prevention and control of noncommunicable diseases across 24 of the 105 counties in the province. Nationally, the Chinese CDC launched an initiative to create "national noncommunicable disease (NCD) demonstration counties or districts." Patient self-management groups are one of the seven key components of these demonstrations, conducted at the time of writing in 265 counties across China.

Standardization and Adaptability

Variability in some peer leader activities (e.g., encouraging informal activities not specified in the protocol) raises issues concerning standardization and achievement of treatment fidelity. The programs have addressed this through standardization not around specific activities but around the key functions of peer support described above (assistance in daily management, social and emotional support, linkage to clinical care and community resources, and ongoing availability of support) [14,47]. This reflects an emphasis of standardization by functions that activities and interventions pursue rather than by specific procedures or materials [66–68]. In the context of these programs in Anhui, this not only provided sufficient focus and specification of what the peer support should entail to guide program development but also provided flexibility for CHSSs to tailor the interventions to their own needs, resources, and the communities they serve.

REFERENCES

[1] DALYs GBD, HALE Collaborators, Murray CJ, et al. Global, regional, and national disability-adjusted life years (DALYs) for 306 diseases and injuries and healthy life expectancy (HALE) for 188 countries, 1990-2013: quantifying the epidemiological transition. Lancet 2015;386(10009):2145–91.

[2] Yang W, Lu J, Weng J, et al. Prevalence of diabetes among men and women in China. N Engl J Med 2010;362(12):1090–101.

[3] Xu Y, Wang L, He J, et al. Prevalence and control of diabetes in Chinese adults. JAMA 2013;310(9):948–59.

[4] Pan X. Strive to prevent and control diabetes in China. Chin Med J 1995;108(2):83.

[5] Pan C. Study on exploring diabetes present characteristics information in China. Section Endocrinal Foreign Med Sci 2002;22(3):135–8.

[6] Chinese Ministry of Public Health. Annual Statistical Reports of Death, Injury and Cause of Death in China, 1993–2005.

[7] Wang Y, Mi J, Shan XY, Wang QJ, Ge KY. Is China facing an obesity epidemic and the consequences? The trends in obesity and chronic disease in China. Int J Obes (Lond) 2007;31(1):177–88.

[8] Wang W, Jiang B, Sun H, et al. Prevalence, incidence, and mortality of stroke in China: results from a nationwide population-based survey of 480 687 adults. Circulation 2017;135(8):759–71.

[9] Yang G, Wang Y, Zeng Y, et al. Rapid health transition in China, 1990–2010: findings from the Global Burden of Disease Study 2010. Lancet 2013;381(9882):1987–2015.

[10] World Health Organization. Regional Office for the Western Pacific. Country health information profiles; 2011.

[11] Carpenter D, Fisher E, Greene S. Shortcomings in public and private insurance coverage of diabetes self-management education and support. Popul Health Manag 2011;15:144–8.

[12] Langenbrunner JC, Marquez PV, Wang S. Toward a healthy and harmonious life in China: stemming the rising tide of non-communicable diseases. Washington, DC: The World Bank; 2011.

[13] Perry HB, Zulliger R, Rogers MM. Community health workers in low-, middle-, and high-income countries: an overview of their history, recent evolution, and current effectiveness. Annu Rev Public Health 2014;35:399–421.

[14] Boothroyd RI, Fisher EB. Peers for progress: promoting peer support for health around the world. Fam Pract 2010;27(Suppl 1):i62–8.

[15] Fisher EB, Boothroyd RI, Elstad EE, et al. Peer Support of Complex Health Behaviors in Prevention and Disease Management with Special Reference to Diabetes: Systematic Reviews. Clin Diabetes Endocrinol. In Press.

[16] Babamoto KS, Sey KA, Camilleri AJ, Karlan VJ, Catalasan J, Morisky DE. Improving diabetes care and health measures among hispanics using community health workers: results from a randomized controlled trial. Health Educ Behav 2009;36(1):113–26.

[17] Greenhalgh T, Campbell-Richards D, Vijayaraghavan S, et al. New models of self-management education for minority ethnic groups: pilot randomized trial of a story-sharing intervention. J Health Serv Res Policy 2011;16(1):28–36.

[18] Heisler M, Vijan S, Makki F, Piette JD. Diabetes control with reciprocal peer support versus nurse care management: a randomized trial. Ann Intern Med 2010;153(8):507–15.

[19] Sacco WP, Malone JI, Morrison AD, Friedman A, Wells K. Effect of a brief, regular telephone intervention by paraprofessionals for type 2 diabetes. J Behav Med 2009;32(4):349–59.

[20] Smith SM, Paul G, Kelly A, Whitford DL, O'Shea E, O'Dowd T. Peer support for patients with type 2 diabetes: cluster randomised controlled trial. BMJ 2011;342:d715.

[21] Ruggiero L, Riley BB, Hernandez R, et al. Medical assistant coaching to support diabetes self-care among low-income racial/ethnic minority populations: randomized controlled trial. West J Nurs Res 2014;36(9):1052–73.

[22] Palmas W, March D, Darakjy S, et al. Community health worker interventions to improve glycemic control in people with diabetes: a systematic review and meta-analysis. J Gen Intern Med 2015;30(7):1004–12.

[23] Collinsworth AW, Vulimiri M, Schmidt KL, Snead CA. Effectiveness of a community health worker-led diabetes self-management education program and implications for CHW involvement in care coordination strategies. Diabetes Educ 2013;39(6):792–9.

[24] Rothschild SK, Martin MA, Swider SM, et al. Mexican American trial of community health workers: a randomized controlled trial of a community health worker intervention for Mexican Americans with type 2 diabetes mellitus. Am J Public Health 2014;104(8):1540–8.

[25] Islam NS, Wyatt LC, Patel SD, et al. Evaluation of a community health worker pilot intervention to improve diabetes management in Bangladeshi immigrants with type 2 diabetes in New York City. Diabetes Educ 2013;39(4):478–93.

[26] Prezio EA, Balasubramanian BA, Shuval K, Cheng D, Kendzor DE, Culica D. Evaluation of quality improvement performance in the Community Diabetes Education (CoDE) program for uninsured Mexican Americans: results of a randomized controlled trial. Am J Med Qual: Off J Am Coll Med Qual 2014;29(2):124–34.

[27] Gagliardino JJ, Arrechea V, Assad D, et al. Type 2 diabetes patients educated by other patients perform at least as well as patients trained by professionals. Diabetes Metab Res Rev 2013;29:152–60.

[28] Ayala GX, Ibarra L, Cherrington AL, et al. Puentes hacia una mejor vida (Bridges to a Better Life): outcome of a diabetes control peer support intervention. Ann Fam Med 2015;13(Suppl 1):S9–S17.

[29] Micikas M, Foster J, Weis A, et al. A community health worker intervention for diabetes self-management among the Tz'utujil Maya of Guatemala. Health Promot Pract 2015;16(4):601–8.

[30] Simmons D, Prevost AT, Bunn C, et al. Impact of community based peer support in type 2 diabetes: a cluster randomised controlled trial of individual and/or group approaches. PLoS ONE 2015;10(3). e0120277.

[31] Willard-Grace R, Chen EH, Hessler D, et al. Health coaching by medical assistants to improve control of diabetes, hypertension, and hyperlipidemia in low-income patients: a randomized controlled trial. Ann Fam Med 2015;13(2):130–8.

[32] Ryabov I. Cost-effectiveness of community health workers in controlling diabetes epidemic on the U.S.-Mexico border. Public Health 2014;128(7):636–42.

[33] Assah FK, Atanga EN, Enoru S, Sobngwi E, Mbanya JC. Community-based peer support significantly improves metabolic control in people with type 2 diabetes in Yaounde, Cameroon. Diabet Med 2015;32(7):886–9.

[34] Chan JC, Sui Y, Oldenburg B, et al. Effects of telephone-based peer support in patients with type 2 diabetes mellitus receiving integrated care: a randomized clinical trial. JAMA Intern Med 2014;174(6):972–81.

[35] Shaya FT, Chirikov VV, Howard D, et al. Effect of social networks intervention in type 2 diabetes: a partial randomised study. J Epidemiol Community Health 2014;68(4):326–32.

[36] Thom DH, Ghorob A, Hessler D, De Vore D, Chen E, Bodenheimer TA. Impact of peer health coaching on glycemic control in low-income patients with diabetes: a randomized controlled trial. Ann Fam Med 2013;11(2):137–44.

[37] Tang TS, Funnell MM, Sinco B, Spencer MS, Heisler M. Peer-Led, empowerment-based approach to self-management efforts in diabetes (PLEASED): a randomized controlled trial in an African American community. Ann Fam Med 2015;13(Suppl 1):S27–35.

[38] Collinsworth A, Vulimiri M, Snead C, Walton J. Community health workers in primary care practice: redesigning health care delivery systems to extend and improve diabetes care in underserved populations. Health Promot Pract 2014;15(2 Suppl):51S–61S.

[39] Findley S, Matos S, Hicks A, Chang J, Reich D. Community health worker integration into the health care team accomplishes the triple aim in a patient-centered medical home: a Bronx tale. J Ambul Care Manage 2014;37(1):82–91.

[40] Mayer MK, Urlaub DM, Guzman-Corrales LM, Kowitt SD, Shea CM, Fisher EB. "They're doing something that actually no one else can do": a qualitative study of peer support and primary care integration. J Ambul Care Manage. in press.

[41] Zhong X, Wang Z, Fisher EB, Tanasugarn C. Peer support for diabetes management in primary care and community settings in Anhui Province, China. Ann Fam Med 2015;13(Suppl 1):S50–8.

[42] Glasgow RE. RE-AIMing research for application: ways to improve evidence for family medicine. J Am Board Fam Med 2006;19(1):11–9.

[43] Glasgow RE, Nelson CC, Strycker LA, King DK. Using RE-AIM metrics to evaluate diabetes self-management support interventions. Am J Prev Med 2006;30(1):67–73.

[44] Feldstein AC, Glasgow RE. A practical, robust implementation and sustainability model (PRISM) for integrating research findings into practice. Jt Comm J Qual Patient Saf/Jt Comm Resour 2008;34(4):228–43.

[45] World Health Organization. Definition, diagnosis, and classification of diabetes mellitus and its complications: report of a WHO consultation. Geneva: World Health Organization; 1999.

[46] Fisher EB, Brownson CA, O'Toole ML. Ongoing follow up and support for chronic disease management in the Robert Wood Johnson Foundation Diabetes Initiative. Diabetes Educ 2007;33(Suppl 6):201S–7S.

[47] Fisher EB, Earp JA, Maman S, Zolotor A. Cross-cultural and international adaptation of peer support for diabetes management. Fam Pract 2010;27(Suppl 1):i6–i16.

[48] Zhang L, Liu Z, Wang Z, Yi J, Wang Q. Diabetes health education and effectiveness evaluation. Chin J Prev Contr Chron Non-Commun Dis 1995;3(4):170–2.

[49] Zeng Q, He Y, Dong S, et al. Optimal cut-off values of BMI, waist circumference and waist: height ratio for defining obesity in Chinese adults. Br J Nutr 2014;112(10):1735–44.

[50] Bandura A. Self-efficacy mechanism in human agency. Am Psychol 1982;37:122–47.

[51] Kim HS, Sherman DK, Taylor SE. Culture and social support. Am Psychol 2008;63(6):518–26.

[52] Fisher EB, Boothroyd RI, Coufal MM, et al. Peer support for self-management of diabetes improved outcomes in international settings. Health Aff (Millwood) 2012;31(1):130–9.

[53] Moskowitz D, Thom DH, Hessler D, Ghorob A, Bodenheimer T. Peer coaching to improve diabetes self-management: which patients benefit most? J Gen Intern Med 2013;28(7):938–42.

[54] Perez-Escamilla R, Damio G, Chhabra J, et al. Impact of a community health workers-led structured program on blood glucose control among Latinos with type 2 diabetes: the DIALBEST trial. Diabetes Care 2015;38(2):197–205.

[55] Kowitt SD, Urlaub D, Guzman-Corrales L, et al. Emotional support for diabetes management: an international cross-cultural study. Diabetes Educ 2015;41(3):291–300.

[56] Chan JC, So WY, Yeung CY, et al. Effects of structured versus usual care on renal endpoint in type 2 diabetes: the SURE study: a randomized multicenter translational study. Diabetes Care 2009;32(6):977–82.

[57] Henry JD, Crawford JR. The short-form version of the Depression Anxiety Stress Scales (DASS-21): construct validity and normative data in a large non-clinical sample. Br J Clin Psychol 2005;44(Pt 2):227–39.

[58] Suzuki T, Shiga T, Kuwahara K, et al. Impact of clustered depression and anxiety on mortality and rehospitalization in patients with heart failure. J Cardiol 2014;.

[59] Fisher EB, Brownson CA, O'Toole ML, Shetty G, Anwuri VV, Glasgow RE. Ecologic approaches to self-management: the case of diabetes. Am J Public Health 2005;95(9):1523–35.

[60] Marrero DG, Ard J, Delamater AM, et al. Twenty-first century behavioral medicine: a context for empowering clinicians and patients with diabetes: a consensus report. Diabetes Care 2013;36(2):463–70.

[61] Yip WCM, Hsiao WC, Chen W, Hu S, Ma J, Maynard A. Early appraisal of China's huge and complex health-care reforms. Lancet 2012;379(9818):833–42.

[62] Zhong X, Potemans B, Zhang L, Oldenburg B. Getting a grip on NCDs in China: an evaluation of the implementation of the Dutch-China cardiovascular prevention program. Int J Behav Med 2015;22(3):393–403.

[63] Conroy RM, Pyorala K, Fitzgerald AP, et al. Estimation of ten-year risk of fatal cardiovascular disease in Europe: the SCORE project. Eur Heart J 2003;24(11):987–1003.

[64] Lindstrom J, Tuomilehto J. The diabetes risk score: a practical tool to predict type 2 diabetes risk. Diabetes Care 2003;26(3):725–31.

[65] Yang GH, Ma JM, Liu N, Zhou LN. Smoking and passive smoking in Chinese, 2002. Zhonghua Liu Xing Bing Xue Za Zhi 2005;26(2):77–83.

[66] Hawe P, Shiell A, Riley T. Complex interventions: how "out of control" can a randomised controlled trial be? BMJ 2004;328(7455):1561–3.

[67] Aro A, Smith J, Dekker J. Contextual evidence in clinical medicine and health promotion. Eur J Pub Health 2008;18(6):548.

[68] Fisher EB, Ballesteros J, Bhushan N, et al. Key features of peer support in chronic disease prevention and management. Health Aff (Millwood) 2015;34(9):1523–30.

Chapter 4

Heart Health and Children

Sandra Gilbertson*, Barbara A. Graves[†]
*Mosaic Life Care, St. Joseph, MO, United States, [†]Capstone College of Nursing, University of Alabama, Tuscaloosa, AL, United States

INTRODUCTION

According to the American Heart Association (AHA), improvements in the current status of health among US children in the areas of tobacco smoke exposure, diet, physical activity levels, serum lipid and blood sugar levels, and blood pressure have the potential to decrease the future proportion of American adults affected by cardiovascular (CV) disease (CVD) [1]. As the leading cause of adult death in the United States [2], CVD affects approximately one of every three American adults and has an annual economic burden of over \$316.6 billion [1]. Although heart attack, stroke, and peripheral artery disease are medical conditions of adulthood, the processes leading up to these conditions begin at an early age.

Atherosclerotic lesions, including fatty streaks and fibrous plaques, begin to develop in childhood as lipids accumulate in response to nonmodifiable factors (family history, familial hypercholesterolemia, and history of high-risk illness) and modifiable factors (tobacco smoke exposure, poor diet, inactivity, excess body weight, dyslipidemia, hypertension, and elevated blood sugar) [3–5]. Early studies, laboratory and autopsy data, and clinical evidence produced in landmark studies [6–8] formed the foundation for the current understanding of how higher levels of risk in childhood influence the onset and severity of clinical CVD in adulthood [9]. Evidence that is more recent suggested that slowing the progression of childhood atherosclerotic lesions through lifestyle modification can prevent or delay the development of CVD in adulthood, thereby extending life expectancy and improving quality of life [10,11].

The AHA [1] has identified health behaviors and measures that contribute to optimal CV health in children and adults. According to the AHA, ideal CV health is the absence of clinical manifestations of CVD together with the presence of the following seven metrics, termed "Life's Simple 7":

- Not smoking
- Having a healthy diet pattern
- Getting enough physical activity
- Maintaining a healthy body weight
- Maintaining healthy cholesterol levels
- Maintaining healthy blood pressure levels
- Maintaining healthy blood sugar levels

Shay et al. [12] compared the AHA standards with National Health and Nutrition Examination Survey (NHANES) data from 2005 to 2010 and determined that none of the 4673 participants in the 6-year cross-sectional study (representing approximately 33 million nonpregnant, noninstitutionalized US adolescents) displayed ideal health in the seven CV behaviors, and only 50% exhibited five or more [12]. The high prevalence of nonideal health behaviors and factors in US adolescents serve as a reminder of the need to intervene in childhood to prevent or slow the development of CVD in adulthood. Hong [13] and Le et al. [14] agreed that atherosclerotic lesions in childhood increase at a rate that corresponds to the number and severity of risk factors. Although some risk factors cannot be changed (family history, ethnicity, age, and gender), up to 80% of adult CVD may be preventable through optimal health behaviors started early in life [12]. Health-promoting behaviors include avoiding tobacco smoke, consuming a healthy diet, and engaging in adequate amounts of physical activity, behaviors that contribute to healthy body weight, cholesterol, blood pressure, and serum glucose. Guiding children to develop patterns of healthy behaviors early in life is a prevention strategy that will lead to improvements in the prevalence of modifiable CV risk factors throughout the life span.

Tobacco Smoke Exposure

Tobacco smoke causes multiple human diseases, including CVD, which is the main cause of death from smoking [15]. Tobacco smoke contributes to arterial inflammation, increased clotting tendencies, heart rate elevations, hypertension, hyperlipidemia, and suppression of the protective effect of high-density lipoprotein cholesterol (HDL-C) [16]. The dangers of smoke are not limited to smokers, but are shared with those who are passively exposed to second- or thirdhand tobacco smoke by inhalation, skin absorption, or ingestion [17–19]. According to the American Lung Association, the 36.5 million adult smokers in the United States expose more than 24 million children to secondhand smoke [20].

Diet

A healthy diet in childhood and adolescence consists of a portion-controlled variety of foods from the groups found in the Dietary Guidelines for Americans [21]. The guidelines list items to be included in a healthy diet (vegetables, fruits, grains, fat-free or low-fat dairy, lean protein, and healthy oils) and those to be limited (added sugar, saturated fat, and sodium).

The consumption of vegetables and fruits leads to a reduction in CV risk by improving lipid levels, lowering blood pressure, and improving the intake of antioxidant micronutrients. In one longitudinal cohort study with data drawn from the Coronary Artery Risk Development in Young Adults (CARDIA) study, researchers found a significant inverse relationship between vegetable and fruit consumption and the prevalence of CVD later in life [22]. Additionally, a diet rich in vegetables and fruits causes a feeling of satiety that helps discourage unhealthy snacking behaviors and overindulgences in foods that are high in fat, salt, and sugar. Despite published recommendations, 93% of US children and adolescents do not eat enough vegetables, and 60% do not eat enough fruit [23].

Any food made from wheat, rice, oats, corn, or another cereal is a grain product. Whole grains are those products that contain the entire grain, whereas refined grains have been ground, a process that removes the bran and the germ and along with them B vitamins, iron, and dietary fiber. In a systematic review and meta-analysis of 45 studies on the benefits of dietary whole grains, Aune et al. concluded that the consumption of adequate amounts of whole grains improves CV health by lowering cholesterol levels, blood pressure, blood sugar, and body weight [24]. Despite the documented benefits of whole grain intake, 2011–12 NHANES data revealed that less than 1% of US children and adolescents consume the recommended daily amount [25].

The role of fat-free or low-fat dairy in reducing CV risk is unclear. In a review of the literature, Rice et al. [26] proposed that the components of milk, including milk fat, vitamin D, calcium, magnesium, potassium, and whey proteins, play a role in improving serum lipid levels and reducing blood pressure, blood sugar, and body mass index. Full-fat and whole milk products not only provide a nutritional benefit similar to that of low-fat and fat-free milk products but also contain saturated fat, cholesterol, and unnecessary calories. The 2010 Dietary Guidelines for Americans recommends the consumption of three servings of low-fat or fat-free dairy per day, but according to 2007–10 NHANES data, the average American over the age of 2 years consumes only 1.9 servings per day [27].

The nutrients in low-fat sources of protein help to lower cholesterol levels, improve blood pressure and body weight, enhance the sense of satiety, and reduce the overall intake of carbohydrates, especially refined carbohydrates [28]. Unlike other food groups, the majority of American children consume far *more* protein than they actually need for good health [29] but are choosing protein sources high in saturated fat (red meat and whole milk products) rather than low-fat alternatives (poultry, fish, and nuts) [30].

Dietary fats are critical for proper brain and hormone development and to fuel other body functions. Fat is a good source of energy and causes a feeling of fullness but has twice the number of calories per gram as carbohydrates or protein. The four types of fats, saturated, trans, monounsaturated, and polyunsaturated, have different chemical structures and affect the body in different ways. The "bad" fats, including saturated and trans fats, are generally solid at room temperature, while the "good" fats, mono- and polyunsaturated fats, are more liquid. For children and adolescents, dessert and snack foods, whole milk products, fatty meats (bacon, hotdogs, and fatter cuts of red meat), and fast foods are significant sources of dietary saturated and trans fat that contribute to obesity, raise blood cholesterol levels, and increase the risk of developing heart disease [31]. Polyunsaturated and monounsaturated fats are the fats found in some fish, avocados, olives, walnuts, and liquid vegetable oils such as olive, corn, safflower, soybean, canola, and sunflower. These fats, when consumed in moderation, contribute to improvements in lipid metabolism leading to lower levels of LDL cholesterol; a reduction in oxidative damage, inflammation, and endothelial dysfunction; improvements in insulin sensitivity; and reductions in blood pressure, thrombosis formation, and abdominal fat [32–34]. The AHA recommends that children and adolescents over the age of 3 years consume between 25% and 35% of their total daily calories in the form of fat with only 10% of calories coming from saturated fat and less than 1% from trans fat. Data from NHANES and calculations from the National Cancer Institute demonstrated that the average daily intake of saturated and trans fat among 2- to 18-year-olds in the United States was 433 cal/day (equivalent to 47 g), mainly from pizza, desserts, whole milk products, and fatty meats, an amount 2–4 times greater than what is recommended [35,36].

In addition to the promotion of vegetables and fruits, whole grains, fat-free or low-fat dairy, lean protein, and healthy oils, the AHA recommends limits on the intake of added sugar, especially in the form of sugar-sweetened beverages, to promote CV health [37,38]. The most recent guidelines suggest the following:

- Children under the age of two should not consume any added sugars, including sugar-sweetened beverages.
- Children age 2 years through adolescence should limit their intake of added sugar to less than six teaspoons per day.
- Children and teens should consume no more than 8 oz of sugar-sweetened beverages per week.

An average 20 oz soda contains 15–18 teaspoons (68–75 g) of sugar, which is equivalent to approximately 240 nonnutritive or "empty" calories. Consumption of these beverages by children has increased dramatically over the past 50 years and is associated with higher caloric intake, higher body weight, reduced intake of milk, and increased risk of medical problems, including hypertension, dyslipidemia, metabolic disease, fatty liver disease, and dental caries [39,40]. A recent analysis of NHANES data from 1999 to 2010 revealed that sugar-sweetened beverages account for 155 nonnutritive calories per day per average American youth and have displaced adequate milk consumption in many girls and boys [41,42]. The reduction or elimination of sugar-sweetened beverages can lead to improvements in health by reducing caloric intake, promoting healthy weight and improving the associated comorbidities in children.

In addition to limits on intake of dietary sugar, the AHA recommends that children and adolescents limit the amount of sodium consumed. As in adults, excessive sodium intake in children and teens causes water retention in the circulatory system. Over time, the increased volume of fluid leads to increases in the workload of the heart and higher pressures in the vessels and later stiffening of the vessels, enlargement of the chambers of the heart, heart attacks, and strokes [43]. The 2010 Dietary Guidelines for Americans recommends that children eat less than 2300 mg of sodium per day, but according to NHANES 2011–12 survey data results, the average sodium intake for US children between 6 and 10 years of age is 3051 mg/day; for children between 11 and 13 years, 3117 mg/day; and for adolescents, 3565 mg/day [44].

Physical Activity

The integrated guidelines for cardiovascular health and risk reduction in children and adolescents recommend that children engage in at least 60 min of age-appropriate moderate to vigorous physical activity on most days to maintain optimal health and prevent the development of CV risk factors [45]. Evidence supports the assumption that physical activity reduces adiposity, improves muscle strength, lowers blood pressure, improves lipid levels, and may even be more effective against obesity than dietary interventions [46,47]. Walker et al. [48] contributed that vigorous physical activity is associated with a reduction in CVD risk regardless of the participant's BMI [48]. Despite the overwhelming evidence in support of regular physical activity, the Family Nutrition and Physical Activity survey found that only 16% of US children achieve the recommended daily amount of exercise [49].

For children and adolescents, limiting physical inactivity in the form of "screen time" to less than 2 h per day is as important as engaging in adequate physical activity. The Campaign for a Commercial-Free Childhood reports that 29% of babies under the age of one watch TV for an average of 90 min/day (and 23% have a television in their room) and 8- to 18-year-olds consume an average of more than 7 h of screen time per day [50]. The American Academy of Pediatrics (AAP) Council on Communications and Media stated that excessive engagement in "screen time" results in a reduction in physical activity and an increase in unhealthy eating due to snacking and exposure to marketing strategies that target children with the promotion of foods and beverages that are high in fat, salt, and sugar [51]. The result is an increase in rates of obesity, hypertension, dyslipidemia, insulin resistance, and type 2 diabetes. Lambiase conducted a review of literature and concluded that limits on sedentary behavior led to increases in physical activity and decreases in caloric intake, body fat, and BMI percentiles [52].

Body Weight

The prevalence of childhood obesity has doubled in the last 30 years with current rates of childhood overweight and obesity at 32%–33% for 2- to 19-year-olds [53]. Excess weight in childhood has immediate health consequences including gastrointestinal, musculoskeletal, orthopedic, and respiratory problems and leads to increases in the heart disease risk factors of prediabetes and type 2 diabetes, hypertension, and elevated levels of serum cholesterol [45,54]. In a systematic review of the literature, Singh et al. reported an increased risk for overweight or obese youth to become overweight or obese adults, with the likelihood of persistence directly related to the extent of the BMI elevation [55]. The AAP estimated that up to 80% of obese children would develop into obese adults [56]. Fontaine et al. added that obese children may have life spans that are up to 13 years shorter than their average weight peers [57]. Despite these facts, research indicates that parents often do not consider excess body weight as problematic in their own children, nor do they consider excess body weight a health concern in childhood [58–64].

Cholesterol

Cholesterol is a type of fat that circulates in the bloodstream and helps the body build cell walls and produce hormones. Although the cholesterol needed by the body is primarily produced by the liver, additional cholesterol enters the bloodstream through the digestion of cholesterol-containing food products such as full-fat dairy and fatter cuts of meat. Cholesterol is carried in the blood by low-density lipoproteins (LDL) and high-density lipoproteins (HDL). When LDL or "bad cholesterol" levels become elevated, plaques form within the blood vessels that lead to vessel narrowing, blood pressure elevations, and other CVD antecedents. HDL or "good cholesterol" carries cholesterol out of the vessels contributing to improvements in CV health. The AAP suggests that acceptable levels of total cholesterol in children and adolescents are less than 170 mg/dL, with LDL less than 110 and HDL greater than 35 mg/dL [65].

Controversy exists regarding cholesterol screening in childhood. In 2011, the National Heart, Lung, and Blood Institute (NHLBI) published the expert panel on integrated guidelines for cardiovascular health and risk reduction in children and adolescents, a summary of over 1000 published documents on CVD as it relates to children and adolescents, and recommended cholesterol screening by primary care providers for all children once between the ages of 9 and 11 years and again between 18 and 21 years [45]. The guidelines were designed to identify those children and adolescents with undiagnosed familial hypercholesterolemia and to identify individuals with high cholesterol levels early so that corrective lifestyle interventions could be implemented. However, since the original publication, the NHLBI determined that the yield of universal screenings did not support the cost and risk [66]. Although the AAP has not withdrawn its support of universal screenings, the current recommendations of the NHLBI are to screen those children and adolescents who demonstrate the following:

- A positive family history of high cholesterol or premature CVD
- An unknown CV family history
- CV risk factors such as overweight, obesity, hypertension, cigarette smoking, or diabetes

For children and adolescents with elevated cholesterol panels, treatment focuses on lifestyle changes in the areas of nutrition and physical activity. For those with significant elevations or associated risk factors (family history of heart disease and diabetes), pharmacological intervention is considered.

Blood Pressure

High blood pressure is a condition in which blood flows through the blood vessels with more force than normal and over time can cause health problems such as aneurysm, heart attack, heart failure, peripheral artery disease, and stroke [67]. Elevated blood pressure readings in children and adolescents are often asymptomatic and only discovered on routine screenings in primary care settings. Unlike adult measurement, there is not one single diastolic or systolic blood pressure reading to identify hypertension in children and adolescents. Instead, hypertension in childhood is identified as systolic and/or diastolic blood pressure that is >95th percentile for age, gender, and height until the age of 12 years when the adult level of 120/80 mmHg serves as the boundary [68]. Using these standards, Hansen et al. estimated the prevalence of hypertension in individuals between 3 and 18 years of age to be 3.6% and another 3.4% in a "prehypertensive" state, with a strong correlation between hypertension and obesity and an increase in incidence as children age [69].

The cause of blood pressure elevations may be primary (essential) or secondary to identifiable causes such as renal anomalies, coarctation of the aorta, bronchopulmonary dysplasia, Wilms' tumor, neuroblastoma, endocrine issues, and certain medications [70]. A variety of factors contribute to the development of primary hypertension including race, heredity, poor diet, inadequate exercise, smoking cigarettes, and obesity. Management of hypertension and prehypertension in children and adolescents involves treatment of identifiable underlying cause; lifestyle modifications in the areas of diet, exercise, and cigarette smoking; weight reduction; and medications when necessary.

Blood Sugar

One of the many reasons children and adolescents need to consume a healthy diet is to reduce the risk of developing diabetes. Foods and beverages that contain carbohydrates (grains, fruits, and vegetables) are broken down by the body into sugars that the body can use for energy or store as fat for potential later use. Some carbohydrates, like whole grains, certain fruits and vegetables, and other high-fiber foods, take longer to digest, which helps maintain stability in the blood sugar level. Refined grains, potatoes, and foods high in added sugar are digested quickly and cause blood sugar level to spike more rapidly. If blood sugar levels undergo repeated quick spikes, the body eventually begins to struggle to maintain stable levels and over time may develop diabetes. Type 1 diabetes develops suddenly and occurs when the body does not make enough insulin to use sugar in the blood properly. Type 2 diabetes, on the other hand, develops over time in response to lifestyle choices including those that lead to obesity (poor diet, inadequate physical activity, and excessive sedentary time). Diabetes causes plaques to develop within blood vessels at a faster than normal rate and increases the risk of developing heart disease, kidney damage, peripheral vascular disease, high blood pressure, and eye damage [71]. Because the development of type 2 diabetes is largely preventable, education regarding healthy eating and active living is crucial to reduce the personal, social, and economic burden of the disease.

PREVENTION, SCREENING, AND TREATMENT OF CVD IN CHILDREN AND ADOLESCENTS

The AHA's scientific statement, Cardiovascular Health Promotion in Children: Challenges and Opportunities for 2020 and Beyond [72], recommends prevention of CVD throughout the life span at three different levels. Primordial prevention refers to efforts that prevent the development of CV risk factors, generally through the promotion of healthy lifestyle behaviors. Primary prevention efforts seek to identify individuals at risk and prevent them from developing disease. Secondary prevention interventions are aimed at halting the progression of disease or preventing complications in those who already have CV disease. Individuals interested in promoting CV health in childhood and adolescence actively seek to develop, implement, and evaluate interventions that discourage the development of risk factors by

- promoting healthy lifestyle choices (primordial prevention);
- promptly identifying risk factors and acting to reduce them through early detection, counseling, and education (primary prevention);
- treating those already affected by CVD through medication therapy, changes in diet and exercise patterns, blood pressure control, lipid management, and regulation of blood sugar (secondary prevention) [72].

Prevention strategies are appropriate first-line therapies to reduce the lifetime risk of CVD when they inform individuals and families about CV risk factors, engage the public in learning about healthy lifestyle behaviors, and focus on health education reinforced over time in the areas of tobacco smoke exposure, nutrition, and physical activity [73–80].

Prevention: Smoke Exposure

There is no safe level of tobacco smoke exposure for children and adolescents. Regular counseling aimed at parents and adolescents regarding the detrimental health effects of smoking (including the addictive properties of tobacco and the risks of secondary smoke exposure) should be provided at clinic visits, through the educational system, in public health messaging, and in the community. Individual, organizational, and community efforts to develop and support smoke-free environments protect the public from the dangers of cigarette smoke, motivate smokers to quit, and reduce the number of people who initiate the habit of smoking [81]. Additional efforts include individual, organizational, and community support of the reduction of smoking images in the media [82] and taxation of tobacco products [83].

Prevention: Nutrition

Good nutrition beginning at birth has immeasurable health benefits, including the potential to reduce the burden of CVD by preventing the development of obesity, dyslipidemia, hypertension, and prediabetes/diabetes. The 2015–20 edition of Dietary Guidelines for Americans provides recommendations for good nutrition across the life span to prevent chronic diseases including CVD [21]. The guidelines, available online, encourage a reduction in the amount of sugar, saturated fat, and sodium in the diets of US children and teens. Furthermore, the guidelines recommend eating a variety of nutritious foods such as vegetables, fruits, whole grains, low-fat and fat-free dairy, and lean meats. Encouraging parents and their families to make small changes in each of these areas over time will lead to long-term health benefits. Strategies for prevention in the area of nutrition include recommending exclusive breastfeeding for the first 6 months of life, promoting water as the beverage of choice for thirst, limiting fat intake by decreasing the frequency of eating meals outside of the home, and reducing sodium by offering more vegetables and fruits, and fewer processed, preserved, and packaged foods. Community-wide prevention strategies include using the guidelines to influence school meal programs, restaurants, and media.

Prevention: Physical Activity and Inactivity

Several long-term studies demonstrated that daily physical activity decreases the risk of developing CV risk factors and CVD. The STRIP study revealed that lower levels of physical activity correlated with an increase in subclinical atherosclerosis, obesity, dyslipidemia, and hypertension [84–86]. The cardiovascular risk in young Finns study showed that low levels of physical activity corresponded with accelerated atherosclerosis over 27 years of follow-up [87]. Based on these and similar studies and studies documenting the benefits of physical activity in adulthood, the US Department of Health and Human Services [88] and the AAP [89] recommend the following:

- All children over the age of 5 years should participate in physical activity for at least 60 min/day with at least 3 days including vigorous activity.
- Children younger than 5 years should be given unlimited active playtime.
- Leisure screen time should be limited to less than 2 h/day except in children under age 2 years who should have no screen time.

These concepts, incorporated into patient education at well-child visits and through advocacy in schools, organizations, and throughout the community, have the potential to provide cardioprotective effects throughout the life span [90].

Screening and Treatment in Primary Care

Screenings in pediatric primary care are common. Health-care providers screen for anemia, lead poisoning, diminished hearing and vision, and many other conditions. Screenings for CV risk factors help identify conditions that increase the chance of developing CVD then intervene with cost-effective measures to reduce or eliminate risk. Although there are no long-term data available to support the precept that screening children for CV risk factors will reduce the prevalence of clinical disease, the AAP [91], the AHA [1], and the National Heart, Lung, and Blood Institute (NHLBI) [45] support the use of such screenings. All three organizations recommend routinely screening children at well-child visits for high-risk cardiac conditions (diabetes and kidney disease, history of heart transplant, and Kawasaki disease with coronary aneurysms), moderate-risk conditions (chronic inflammatory disease, human immunodeficiency virus infection, and nephrotic syndrome), smoke exposure, poor diet, inactivity, overweight and obesity, dyslipidemia (when appropriate), hypertension, elevated fasting blood sugar (when appropriate), and family history of CVD. Treatment targets the identified risk factors and may include lifestyle change, structured counseling or group support, weight loss, smoking cessation, and medication therapy.

Health-care providers interested in promoting heart-healthy lifestyles across the lifestyle should participate in using simple, clear, and consistent good health messaging that can be incorporated into clinic visits, public health campaigns, organizational philosophies, school programs, day care centers, and community events. To aid in the communication of simple, consistent key messaging that informs and educates parents, families, and other caregivers about activities that promote optimal CV health, the AAP endorses the use of clinical guidelines focused on the "5210" plan [92]. The "5210" plan was developed as a collaborative effort between the Maine Chapter of the AAP and the Maine Center for Public Health and encourages all children to eat five servings of fruits or vegetables per day, limit nonschool-related "screen time" to less than 2 h/day, engage in at least 1 h of physical activity per day, and eliminate sugar-sweetened beverages in favor of more water and low-fat milk. The APP [91] further recommends children and teens be strongly discouraged from cigarette smoking, and that tobacco smoke be eliminated from places where children live, learn, and play. Parents not meeting the recommendations of the 5210 plan are encouraged to set small, simple, and concrete goals for their families and to schedule regular health-care visits to evaluate progress. Recommendations such as these are evidence-based, safe for virtually all children, and effective in providing reinforcement for those already meeting the guidelines and in promoting early intervention for those who are not [10,93–95]. Community-wide and statewide programs, such as Maine's *5210 Let's Go!* are demonstrating significant improvements in parental awareness of positive lifestyle behaviors by using a consistent format to promote healthy eating and regular physical activity throughout childhood.

Primary care providers working with children and adolescents to promote ideal CV health must actively participate in strategies to assess and document the CV health of children and teens in primary care and provide individualized plans of care to prevent, reduce, or eliminate CV risk factors. These strategies include the following:

- Routine physical assessment including measurements of body mass index, blood pressure, and laboratory values with tracking over time.
- Appropriate collection of past medical and family history.
- Collection of data related to tobacco use or secondhand smoke exposure, dietary intake, and physical activity.
- Education directed at parents about the health status of their children and teens. In children, goals are more likely to be met when parents are motivated to encourage behavior changes [74]. Parental motivation has been shown to be positively influenced by health-care providers when they provide results of child screening tests along with lifestyle counseling, but research has revealed that these activities are not routinely implemented in primary care settings [96,97].
- Counseling the parent and child or adolescent about current health status and health-promoting behaviors. Rogers et al. documented the impact of the 5210 Let's Go program in 12 communities in Maine and concluded that the multisetting community-based program had a strong positive impact on childhood health behaviors [98]. In a smaller study by Gilbertson and Graves, the authors used the 5210 Let's Go material in a 10 min teaching intervention to educate parents of 9- to 11-year-olds about their child's current CV health status, healthy lifestyle choices, and modifiable CVD risk factors [80]. After incorporating the evidence-based primordial and primary prevention techniques endorsed by the AAP, caregiver responses indicated significant increases in awareness of adequate fruit and vegetable intake and physical activity levels. Furthermore, responses indicated caregivers were likely to consider encouraging health behavior change in each area of the 5210 plan with 92%–100% positive responses for likelihood of encouraging change.
- Setting specific, measurable, attainable, relevant goals with appropriate time frames for achievement. In a study by Shilts and Townsend, a goal-setting intervention had a positive effect on dietary habits and physical activity of adolescent study participants [99]. A review by Bodenheimer and Handley supported the idea that goal setting in health care is effective when goals are specific and developed collaboratively between the patient/family and the clinician [100].
- Referrals when indicated (dietician, exercise specialist, cardiologist, and child obesity expert).
- Follow-up with support and encouragement.

In addition to office-based health promotion, interested primary care providers engage in community activities that influence policies and environments to discourage tobacco smoke exposure, and encourage healthy eating and active living. Examples include support for smoke-free communities, taxation of tobacco products and sugar-sweetened beverages, removal of high-fat/high-sugar snacks from schools, supporting changes in school meal options, promoting breastfeeding, building playgrounds, and supporting physical activity classes throughout the school years.

THE ROLE OF CAREGIVERS AND FAMILIES

According to Murray Bowen [101], family systems theory is a theory of human behavior that defines the family unit as a complex social system in which members interact to influence each other's behavior. Family members interconnect, making it appropriate to view the system as a whole rather than as individual elements. Any change in one individual within a family is likely to influence the entire system and may even lead to change in other members. Many interventions designed to promote behavior change in children are directed at the parent-child unit, although it may be more beneficial to focus on the family as a whole.

In a review, Vedanthan et al. [102] stated that parents and other adult caregivers have a dominant effect on the physical and social elements that impact the CV health of the family members. Beginning at birth, parents make choices about feeding breastmilk or formula, then decide when and how to introduce solid foods. After children begin to move on their own, caregiver attention influences levels of physical activity and inactivity, as well as continuing to influence food choices and behaviors. Caregivers make decisions related to the quality of foods brought to the home, the number of meals eaten outside of the home, the amount and quality of snacking (including the use of sugary beverages), and the social interactions during mealtimes (the use of electronics at the dinner table, television viewing while eating, and the presence or absence of conversation). Additionally, caregivers decide what, when, where, how, and how much family members consume and often participate in activities that place tight control on the intake of children and teens (e.g., "cleaning" the plate, bribery to eat, and rewards with unhealthy foods for consuming healthy foods). Regarding physical activity, children are naturally active and curious but become sedentary when freedom to explore becomes restricted or when less active behaviors are encouraged (television viewing, playing video games, riding in a car rather than walking, etc.) [100]. Furthermore, it is well established that caregivers who engage in unhealthy eating behaviors and maintain relatively inactive lifestyles influence family members to do the same [103,104] and that children of overweight parents are more likely to become overweight than their peers of healthy weight parents [103,105–108].

Parenting style has been implicated as a contributing factor in the development of CV disease. There are four major parenting styles: authoritative, neglectful, permissive, and authoritarian. Each style has different characteristics and produces different effects on the children. According to Vedanthan et al. [102], authoritative parenting is considered the style most conducive to the development of optimal CV health. This style is characterized by respect for the child, high expectations for achievement, understanding, and support. Generally considered the most effective style of parenting, authoritative parents are more likely to discuss healthy behaviors with their children, negotiate, communicate, and praise behaviors that lead to good health. Children of authoritative parents are inclined to eat more fruits and tend to exhibit healthy behaviors into adolescence compared with peers of other parenting styles. They are less likely to be overweight, more likely to be physically active, less likely to engage in excessive screen time, and less likely to smoke [104,109–114].

Vendanthan et al. [102] continue to describe the other three parenting styles and their effect on child CV risk factor development. The second style of parenting is permissive. The permissive parent caters to the child's emotional needs without expecting the child to have any responsibility or self-control. They are more indulgent and impose fewer restrictions on sedentary activity, screen time, and consumption of energy dense and sugary foods. The third style, neglectful parenting, shows low levels of sensitivity toward the child and places minimal demands on the child's self-control. These parents may seem disinterested as they rarely provide effective discipline or establish boundaries. Children of neglectful parents are more likely to be obese, have a higher risk of CV disease, and are more likely to smoke and use illicit drugs. The fourth style, authoritarian parenting, demands maturity from the child without warmth or understanding and compromises the child's sense of self-regulation. Authoritarian parents maintain tight control, and engage in coercive feeding practices (clean plate club, forcing vegetables, and strict restriction of "junk food"). Children of permissive, neglectful, or authoritarian parents are less likely to use internal cues to regulate food intake, thereby having an increased risk of developing risk factors for CV disease.

To help children and teens achieve and maintain ideal CV health, it is incumbent upon parents to develop parenting skills that foster healthy lifestyle behaviors within the family unit. The parenting qualities of nurturance, structure and behavioral control have been shown to positively influence child health behaviors, whereas overprotection and coercion are negatively associated [115]. Parents who focus on good health are more successful at influencing child health behaviors than parents who focus on thinness or physical appearance [116]. Successful parenting factors related to eating behaviors proposed by Scaglioni et al. include appropriate role modeling, engaging in scheduled family meal times, suggesting healthy food choices and portion sizes, encouraging self-regulation, and encouraging social interaction during meals [117]. The AAP further recommends that parents foster appropriate levels of physical activity in children and adolescents by finding activities that are fun and developmentally appropriate, providing access to sports leagues or dance/fitness programs of interest, providing active toys, ensuring safety in all environments, limiting screen time (and monitoring content, including advertising), being a good role model, and engaging in activities with the child or teen [118].

Parents can also help children and teens achieve or maintain optimal CV health by stressing the importance of avoiding tobacco smoke and maintaining a healthy body weight. To be successful, parents must make provisions for routine health care including wellness examinations and be advocates for their children by broaching the topic of CV health promotion in childhood with educated questions about body mass index, blood pressure, lipid, and glucose levels. Highly engaged parents may also promote heart-healthy attitudes in schools and throughout the community and campaign for legislative changes that promote heart health (e.g., smoke-free communities and taxation of tobacco products and sugary drinks).

CONCLUSION

The problem of the early development of CVD risk factors is negatively affecting the nation's health and placing a tremendous burden on the economy. This review highlighted the importance of optimizing long-term heart health beginning in childhood by limiting exposure to tobacco smoke, improving diet and exercise habits, improving child weight, and monitoring child blood pressure, cholesterol, and glucose levels. Activities that prevent (primordial prevention) or reduce (primary prevention) the development of modifiable CVD risk factors are centered in the family unit, and those that identify and treat them (secondary prevention) are supported by primary care providers. Families, health-care systems, and the community all have significant roles to play in CVD prevention and risk reduction in children.

REFERENCES

[1] Mozaffarian D, Benjamin E, Go AS, et al. Heart disease and stroke statistics—2016 update. Circulation 2015;134(20):E1. https://doi.org/10.1161/CIR0000000000000350.

[2] Leading causes of death. Centers for Disease Control and Prevention, National Center for Health Statistics web site. CDC.org. https://www.cdc.gov/nchs/fastats/leading-causes-of-death.htm; 2016 [accessed 28.12.16].

[3] Juonala M, Magnussen CG, Venn A, et al. Parental smoking in childhood and brachial artery flow-mediated dilatation in young adults: the cardiovascular risk in young Finns study and the childhood determinants of adult health study. Arterioscler Thromb Vasc Biol 2012;32(4):1024–31. https://doi.org/10.1161/ATVBAHA.111.243261.

[4] McBride PE, Kavey RW. Lipid screening and treatment recommendations for children and adolescents. Pediatr Ann 2012;41(7):1–10. https://doi.org/10.3928/00904481-20120625-08.

[5] Peinemann F, Moebus S, Dragano N, et al. On behalf of the Heinz Nixdorf Recall Study Investigative Group. Second hand smoke exposure and coronary artery calcification among nonsmoking participants of a population-based cohort. Environ Health Perspect 2011;119(11):1556–61. https://doi.org/10.1289/ehp.1003347.

[6] Berenson GS, Srinivasan SR, Bao W, Newman WP, Tracy RE, Wattigney WA. Association between multiple cardiovascular risk factors and atherosclerosis in children and young adults. The Bogalusa Heart Study. N Engl J Med 1998;338(23):1650–6. https://doi.org/10.1161/CIRCULATIONAHA.107.184595.

[7] Davis PH, Dawson JD, Riley WA, Lauer RM. Carotid intimal-medial thickness is related to cardiovascular risk factors measured from childhood through middle age: The Muscatine Study. Circulation 2001;104(23):2815–9. https://doi.org/10.1161/hc4601.099486.

[8] McGill HC, McMahan CA, Malcom GT, Oalmann MC, Strong JP, for the PDAY Research Group. Effects of serum lipoproteins and smoking on atherosclerosis in young men and women. Arterioscler Thromb Vasc Biol 1997;17:95–106. https://doi.org/10.1161/01.ATV.17.1.95.

[9] Shah AS, Dolan LM, Gao Z, Kimball TR, Urbina EM. Clustering of risk factors: a simple method of detecting cardiovascular disease in youth. Pediatrics 2011;127:e312–8. https://doi.org/10.1542/peds.2010-1125.

[10] McGill HC, McMahan CA, Gidding SS. Preventing heart disease in the 21st century: implications of the pathobiological determinants of atherosclerosis in youth (PDAY) study. Circulation 2008;111:1216–27. https://doi.org/10.1161/CIRCULATIONAHA.107.717033.

[11] Coronary Artery Risk Development in Young Adults Study. National Institute of Health. Updated July 2014. https://www.nhlbi.nih.gov/research/resources/obesity/population/cardia.htm.

[12] Shay CM, Ning H, Daniels SR, Rooks CR, Gidding SS, Lloyd-Jones DM. Status of cardiovascular health in US adolescence: prevalence estimates from the National Health and Nutrition Examination Surveys (NHANES) 2005–2010. Circulation 2013;127:1369–76. https://doi.org/10.1161/CIRCULATIONAHA.113.001559.

[13] Hong YM. Atherosclerotic cardiovascular disease beginning in childhood. Korean Circ J 2010;40(1):1–9. https://doi.org/10.4070/kcj.2010.40.1.1.

[14] Le J, Zhang D, Menees S, Chen J, Raghuveer G. "Vascular age" is advanced in children with atherosclerosis-promoting risk factors. Circ Cardiovasc Imaging 2010;3:8–14. https://doi.org/10.1161/CIRCIMAGING.109.880070.

[15] US Department of Health and Human Services, Public Health Service. Office of the Surgeon General. The Health Consequences of Smoking – 50 Years of Progress. A Report of the Surgeon General. Executive Summary. Rockville, MD. https://www.surgeongeneralgov/library/reports/50-years-of-progress/exec-summary.pdf; 2014 [accessed 28.12.16].

[16] Campbell SC, Moffatt RJ, Stamford BA. Smoking and smoking cessation—the relationship between cardiovascular disease and lipoprotein metabolism: a review. Atherosclerosis 2008;201:225–35. https://doi.org/10.1016/j.atherosclerosis.2008.04.046.

[17] Martins-Green M, Adhami N, Frankos M, et al. Cigarette smoke toxins deposited on surfaces: implications for human health. PLoS ONE 2014;9(1):e86391. https://doi.org/10.1371/journal.pone.0086391.

[18] Prignot JJ. Recent contributions of air and biomarkers to the control of second hand smoke (SHS): a review. Int J Environ Res Public Health 2011;8(3):648–82. https://doi.org/10.3390/ijerph8030648.

[19] Barnoya J, Glantz SA. Cardiovascular effects of secondhand smoke: nearly as large as smoking. Circulation 2005;111:2684–98. https://doi.org/10.1161/CIRCULATIONAHA.104.492215.

[20] Health effects of secondhand smoke. American Lung Association web site. http://www.lung.org/stop-smoking/smoking-facts/health-effects-of-secondhand-smoke.html?referrer=https://www.google.com/; 2016.

[21] US Department of Health and Human Services and US Department of Agriculture. 2015–2020 Dietary Guidelines for Americans. 8th ed. http://health.gov/dietaryguidelines/2015/; December 2015.

[22] Miedema MD, Petrone A, Shikany JM, et al. The association of fruit and vegetable consumption during early adulthood with the prevalence of coronary artery calcium after 20 years of follow-up: The CARDIA Study. Circulation 2015;135(1). https://doi.org/10.1161/CIRCULATIONAHA.114.012562.

[23] Children eating more fruit, but fruit and vegetable intake still too low. CDC Newsroom. Atlanta, GA. https://www.cdc.gov/media/releases/2014/p0805-fruits-vegetables.html; August 5, 2014 [accessed 28.12.16].

[24] Aune D, Keum N, Giovannucci E, et al. Whole grain consumption and risk of cardiovascular disease, cancer, and all cause and cause specific mortality: systematic review and dose response meta-analysis of prospective studies. BMJ 2016;353. https://doi.org/10.1136/bmj.i2716.

[25] Albertson AM, Reicks M, Joshi N, Gugger CK. Whole grain consumption trends and associations with body weight measures in the United States: results from the cross sectional national Health and Nutrition Examination Survey 2001–2012. Nutr J 2016;15:8. https://doi.org/10.1186/s12937-016-0126-4.

[26] Rice BH, Cifelli CJ, Pikosky MA, Miller GD. Dairy components and risk factors for cardiometabolic syndrome: recent evidence and opportunities for future research. Adv Nutr 2011;2(5):396–407. https://doi.org/10.3945/an.111.000646.

[27] Quann EE, Fulgoni VL, Auestad N. Consuming the daily recommended amounts of dairy products would reduce the prevalence of inadequate micronutrients intakes in the United States: diet modeling study based on NHANES 2007-2010. Nutr J 2015;14:90. https://doi.org/10.1186/s12937-015-0057-5.

[28] Why is it important to make lean or low-fat choices from the protein foods group? US Department of Agriculture. Choose My Plate web site. https://www.choosemyplate.gov/protein-food-nutrients-health; 2016 [accessed 28.12.16].

[29] Daniel CR, Cross AJ, Koebnick C, Sinha R. Trends in meat consumption in the USA. Public Health Nutr 2011;14(4):575–83. https://doi.org/10.1017/S1368980010002077.

[30] Bernstein AM, Sun Q, Hu FB, Stampfer MJ, Manson JE, Willett WC. Major dietary protein sources and risk of coronary heart disease in women. Circulation 2010;122(9):876–83. https://doi.org/10.1161/CIRCULATIONAHA.109.915165.

[31] Saturated Fats. American Heart Association web site. Last Updated 10/12/2016 [accessed 28.12.16].

[32] Schwingshackl L, Strasser B, Hoffman G. Effects of monounsaturated fatty acids on cardiovascular risk factors; a systematic review and meta-analysis. Ann Nutr Metab 2011;59:2–4. 176–186.

[33] Covas MI. Olive oil and the cardiovascular system. Pharmacol Res 2007;55(93):175–86. https://doi.org/10.1016/j.phrs.2007.01.010.

[34] Covas MI, Konstantinidou V, Fito M. Olive oil and cardiovascular health. J Cardiovasc Pharmacol 2009;54(6):477–82. https://doi.org/1097?FJC.0b013e3181c5e7fd.

[35] Sources of Energy among the US Population, 2005–2006. Epidemiology and Genomics Research Program website. National Cancer Institute. http://epi.grants.cancer.gov/diet/foodsources/energy/; Update April 22, 2016 [accessed 10.12.16].

[36] Reedy J, Krebs-Smith SM. Dietary sources of energy, solid fats, and added sugars among children and adolescents in the United States. J Am Diet Assoc 2010;110(10):1477–84.

[37] Vos MB, Kaar JL, Welsch JA, et al. Added sugars and cardiovascular disease risk in children: a scientific statement from the American Heart Association. Circulation 2016;134:1–18. https://doi.org/10.1161/CIR.0000000000000439.

[38] Yang Q, Zhang Z, Gregg EW, Flanders D, Merritt R, Hu FB. Added sugar intake and cardiovascular diseases mortality among US adults. JAMA Intern Med 2014;174(4):516–24. https://doi.org/10.1001/jamainternmed.2013.13563.

[39] Friedman R. Study synopses: sugar-sweetened beverages (SSBs) and health risks. New Haven, CT: Yale Rudd Center for Food Policy and Obesity. http://www.yaleruddcenter.org/resources/upload/docs/what/policy/SSBtaxes/SSBStudiesHealth.pdf; Updated 11/21/2013 [accessed 28.12.16].

[40] Kosova EC, Auinger P, Bremer AA. The relationship between sugar-sweetened beverage intake and cardiometabolic markers in young children. J Acad Nutr Diet 2013;113:219–27.

[41] Ervin RB, Ogden CL. Trends in intake of energy and macronutrients in children and adolescents from 1999–2000 through 2009–2010. National Center for Health Statistics Data Brief. CDC NCHS Web site. http://www.cdc.gov/nchs/data/databriefs/db113.htm; Updated February 21, 2013 [accessed 10.12.16].

[42] Kit BK, Fakhouri TH, Park S, Nielsen SJ, Ogden CL. Trends in sugar-sweetened beverage consumption among youth and adults in the United States: 1999–2010. Am J Clin Nutr 2013. https://doi.org/10.3945/ajcn.112.057943.

[43] Yang Q, Liu T, Kuklina EV, et al. Sodium and potassium intake and mortality among US adults: prospective data from the Third National Health and Nutrition Examination Survey. Arch Intern Med 2011;171(13):1183–91. https://doi.org/10.1001/archinternmed.2011.257.

[44] High sodium intake in children and adolescents: Cause for concern. Centers for Disease Control and Prevention Web site. https://www.cdc.gov/salt/pdfs/children_sodium.pdf [accessed 28.12.16].

[45] US Department of Health and Human Services. National Institutes of Health Web site. Expert panel on integrated guidelines for cardiovascular health and risk reduction in children and adolescents. http://www.nhlbi.nih.gov/files/docs/guidelines/peds_guidelines_full.pdf; October 2012 [accessed 10.12.16].

[46] Doyle-Baker PK, Venner AA, Lyone ME, Fung T. Impact of a combined diet and progressive exercise intervention for overweight and obese children. Appl Physiol Nutr Metab 2011;36:515–25. https://doi.org/10.1139/H11-042.

[47] Guntin B. Diet vs. exercise for the prevention of pediatric obesity: the role of exercise. Int J Obes 2011;35(1):29–32. https://doi.org/10.1038/ijo.2010.140.

[48] Walker DJ, MacIntosh A, Kozyrskyj A, Becker A, McGavock J. The associations between cardiovascular risk factors, physical activity, and arterial stiffness in youth. J Phys Act Health 2013;10(2):198–204.

[49] The state of family nutrition and physical activity. Are we making any progress? American Dietetic Association Foundation web site. http://www.eatright.org/foundation.org; 2011 [accessed 10.12.16].

[50] Campaign for a Commercial Free Childhood. Selected Research on Screen Time and Children. www.screenfree.org/wp-content/uploads/2014/01/screentimefs.pdf [accessed 10.12.16].

[51] Strasburger VC, The Council on Communications and Media. Children, adolescents, obesity, and the media. Pediatrics 2011;128(1):201–8. https://doi.org/10.1542/peds.2011-1066.

[52] Lambiase M. Treating pediatric overweight through reductions in sedentary behavior: a review of the literature. J Pediatr Health Care 2009;23(1):29–36. https://doi.org/10.1016/j.pedhc.2008.04.005.

[53] Ogden CL, Carroll MD, Kit BK, Flegal KM. Prevalence of childhood and adult obesity in the United States, 2011-2012. JAMA 2014;311(8):806–14.

[54] Freedman DS, Mei Z, Srinivasan SR, Bergenson GS, Dietz WH. Cardiovascular risk factors and excess adiposity among overweight children and adolescents: The Bogalusa Heart Study. J Pediatr 2007;150(1):12–7. https://doi.org/10.1016/j.peds.2006.08.042.

[55] Singh AS, Mulder C, Twisk JWR, van Mechelen W, Chinapaw JM. Tracking of childhood overweight into adulthood: a systematic review of the literature. Obes Rev 2008;9:474–88. https://doi.org/10.1111/j.1467-789X.2008.00475.x.

[56] American Academy of Pediatrics. Policy statement: prevention of pediatric overweight and obesity. Pediatrics 2003;112(2):424–30.

[57] Fontaine KR, Redden DT, Wang C. Years of life lost to obesity. J Am Med Assoc 2010;289(2):187–93. https://doi.org/10.1001/jama.289.2.187.

[58] DeLaO A, Jordan KC, Ortiz K, et al. Do parents accurately perceive their child's weight status? J Pediatr Health Care 2009;23(4):216–21. https://doi.org/10.1016/j.pedhc.2007.12.014.

[59] He M, Evans A. Are parents aware that their children are overweight or obese? Do they care? Can Fam Physician 2007;53:1493–9.

[60] Jones AR, Parkinson KN, Drewett RF, Hyland RM, Pearce MS, Adamson AJ, et al. Parental perceptions of weight status in children: the Gateshead Millennium Study. Int J Obes 2011;35:953–62. https://doi.org/10.1038/ijo.2011.106.

[61] Towns N, D'Auria J. Parental perceptions of their child's overweight: an integrative review of the literature. J Pediatr Nurs 2009;24(2):115–30. https://doi.org/10.1016/jpedn.2008.02.032.

[62] Warschburger P, Kroeller K. Maternal perception of weight status and health risks associated with obesity in children. Pediatrics 2009;124(1):e60–8. https://doi.org/10.1542/peds.2008-1845.

[63] Hansen AR, Duncan DT, Tarasenko YN, Yan F, Zhang J. Generational shift in parental perceptions of overweight among school aged children. Pediatrics 2014;134(3):481–8.

[64] Walusimbi F. Accurate parental perception as a milestone in managing childhood obesity. Dayton, OH: Wright State University; Core Scholar Libraries; 2015. http://corescholar.libraries.wright.edu/cgi/viewcontent.cgi?article; .

[65] Cholesterol levels in children and adolescents. Healthychildren.org Web site. https://www.healthychildren.org/English/healthy-living/nutrition/Pages/Cholesterol-Levels-in-Children-and-Adolescents.aspx [accessed 28.12.16].

[66] Lipid disorders in children: screening. US Preventive Services Task Force Web site. https://www.uspreventiveservicestaskforce.org/Page/Document/RecommendationStatementFinal/lipid-disorders-in-children-screening; 2013 [accessed 28.12.16].

[67] Description of high blood pressure. National Heart, Lung, and Blood Institute Web site. https://www.nhlbi.nih.gov/health/health-topics/topics/hbp; Updated September 10, 2015 [accessed 10.12.16].

[68] National High Blood Pressure Education Working Group on High Blood Pressure in Children and Adolescents. The fourth report on the diagnosis, evaluation, and treatment of high blood pressure in children and adolescents. NIH Publication No. 05-5267. http://www.nhlbi.nih.gov/health/prof/heart/hbp/hbp_ped.pdf; 2005 [accessed 10.12.16].

[69] Hansen ML, Gunn PW, Kaelber DC. Under diagnosis of hypertension in children and adolescents. JAMA 2007;298:874–9. https://doi.org/10.1001/jama.298.8.874.

[70] Rodriguez-Cruz, E. Pediatric Hypertension. Medscape. http://emedicine.medscape.com/article/889877-overview#a2; Update July 2015 [accessed 10.12.16].

[71] Dokken BB. The pathophysiology of CVD and diabetes: beyond blood pressure and lipids. Diabetes Spectr 2008;21(3):160–5. https://doi.org/10.2337/diaspect.21.3.160.

[72] Mohamed TN, Afonso LC, Ramappa P, Hari P. Primary and secondary prevention of coronary artery disease. Medscape. http://emedicine.medscape.com/article/164214-overview#a1; 2015 [accessed 28.12.16].

[73] Rhee KE, DeLago CW, Arscott-Mills T, Mehta SD, Davis RK. Factors associated with parental readiness to make changes for overweight children. Pediatrics 2005;116(1):94–101. https://doi.org/10.1542/peds.2004-2479.

[74] Moore LC, Harris CV, Bradlyn AS. Exploring the relationship between parental concern and the management of childhood obesity. Matern Child Health J 2012;16:902–8. https://doi.org/10.107/s10995-11-0813-x.

[75] Vaczy E, Seaman B, Peterson-Sweeny K, Hondorf C. Passport to health: an innovative tool to enhance healthy lifestyle choices. J Pediatr Health Care 2011;25(1):31–7. https://doi.org/10.1016/j.pedhc.2010.04.006.

[76] West F, Sanders MR, Cleghorn GJ, Davies PSW. Randomized clinical trial of a family-based lifestyle intervention for childhood obesity involving parents as the exclusive agents of change. Behav Res Ther 2010;48(12):1170–9. https://doi.org/10.1016/j.brat.2010.08.008.

[77] Barlow SE, Dietz WH. Obesity evaluation and treatment: Expert Committee recommendations. Pediatrics 1998;102(3):e29. https://doi.org/10.1542/peds.102.3.e29. 1–11.

[78] Barlow SE, The Expert Committee. Expert Committee recommendations regarding the prevention, assessment, and treatment of child and adolescent overweight and obesity: summary report. Pediatrics 2007;120(4):s164–92. https://doi.org/10.1542/peds.2007-2329C.

[79] Kelishadi R, Malekahmadi M, Hashemipour M, et al. Can a trial of motivational lifestyle counseling be effective for controlling childhood obesity and the associated cardiometabolic risk factors? Pediatr Neonatol 2012;53(2):90–7. https://doi.org/10.1016/j.pedneo.2012.01.005.

[80] Gilbertson SL, Graves BA. Motivating parents to promote cardiovascular health in children. J Cardiovasc Nurs 2013;30(1):E8–18. https://doi.org/10.1097/JCN0000000000000126.

[81] Smokefree Environments. American Lung Association web site. www.lung.org/our-initiatives/tobacco/smokefree-environments/ [accessed 28.12.16].

[82] Smoking in the movies. Centers for Disease Control and Prevention web site. https://www.cdc.gov/tobacco/data_statistics/fact_sheets/youth_data/movies/; 2016 [accessed 28.12.16].

[83] Campaign for Tobacco-Free Kids. Raising cigarette taxes reduces smoking, especially among kids (and the cigarette companies know it). https://www.tobaccofreekids.org/research/factsheet/pdf/0146.pdf; 2016 [accessed 28.12.16].

[84] Pahkala K, Heinonen OJ, Simell O, et al. Association of physical activity with vascular endothelial function and intima-medial thickness. Circulation 2011;124:1956.

[85] Pahkala K, Laitinen TT, Heinonen OJ, et al. Association of fitness with vascular intima-media thickness and elasticity in adolescents. Pediatrics 2013;132:e77.

[86] Pahkala K, Heinonen OJ, Langstrom H, et al. Clustered metabolic risk and leisure-time physical activity in adolescents: effect of dose? Br J Sports Med 2012;46:131.

[87] Juonala M, Viikari JS, Kahonen M, et al. Life-time risk factors and progression of carotid atherosclerosis in young adults: The Cardiovascular Risk in Young Finns study. Eur Heart J 2010;31:1745.

[88] 2008 Physical Activity Guidelines for Americans. US Department of Health and Human Services web site. www.health.gov/paguidelines; 2008 [accessed 17.12.16].

[89] Daniels SR, Hassink SG. The role of the pediatrician in primary prevention of obesity. Pediatrics 2015;136:1.

[90] Varghese T, Shultz WM, McCue AA, et al. Physical activity in the prevention of coronary heart disease: implications for the clinician. Heart 2016;102:904–9. https://doi.org/10.1136/heartjnl-2015-308773.

[91]   Recommendations for Preventive Pediatric Health Care. Bright Futures. American Academy of Pediatrics. https://www.aap.org/en-us/professional-resources/practice-transformation/managing-patients/Pages/Periodicity-Schedule.aspx; Updated 10/20/15 [accessed 09.01.17].

[92]   American Academy of Pediatrics. Pediatric obesity decision support chart 5210 clinical guidelines. Elk Grove, IL: American Academy of Pediatrics; 2008.

[93]   Hayman LL, Meininger JC, Daniels SR, et al. Primary prevention of cardiovascular disease in nursing practice: focus on children and youth: a scientific statement from the American Heart Association Committee on atherosclerosis, hypertension, and obesity in Youth of the Council on Cardiovascular Disease in the Young, Council on Cardiovascular Nursing, Council on Epidemiology and Prevention, and Council on Nutrition, Physical Activity, and Metabolism. Circulation 2007;116:344–57. https://doi.org/10.1161/CIRCULATIONAHA.107.184595.

[94]   Joanna Briggs Institute. Effective dietary intervention for managing overweight and obesity in childhood. Best Practice 2007;11(1):1–4. http://connect.jbiconnectplus.org/ViewSourceFile.aspx?0=4344; [accessed 07.04.13].

[95]   Morrison JA, Glueck CJ, Wang P. Childhood risk factors predict cardiovascular disease, impaired fasting glucose plus type 2 diabetes mellitus, and high blood pressure 26 years later at a mean age of 38 years: The Princeton-Lipid Research Clinics follow-up study. Metabolism 2012;61(4):531–41. https://doi.org/10.1016/j.metabol.2011.08.010.

[96]   Kubik MY, Story M, Davey C, Dudovitz B, Zuehlke EU. Providing obesity prevention counseling on children during a primary care clinic visit: results from a pilot study. J Am Diet Assoc 2008;108(11):1902–6. https://doi.org/10.1016/j.jada.2008.08.017.

[97]   Cook S, Weitzman M, Auinger P, Barlow SE. Screening and counseling associated with obesity diagnosis in a national survey of ambulatory pediatric visits. Pediatrics 2005;116(1):112–6. https://doi.org/10.1542/peds.2004-1517.

[98]   Rogers VW, Hart PH, Motyka E, Rines EN, Vine J, Deatrick DA. Impact of Let's Go! 5-2-1-0: a community-based, multisetting childhood obesity prevention program. J Pediatr Psychol 2013;38(9):1010–20. https://doi.org/10.1093/jpepsy/jst057.

[99]   Shilts MK, Townsend MS. A goal setting intervention positively impacts adolescents dietary behaviors and physical activity self-efficacy. J Youth Develop 2012;7(4):93–108.

[100]   Bodenheimer T, Handley MA. Goal-setting for behavior change in primary care: an exploration and status report. Patient Educ Couns 2009;76(2):174–80.

[101]   Theory. The Bowen Center for the Study of the Family. https://www.thebowencenter.org/theory/ [accessed 28.12.16].

[102]   Vendanthan R, Bansilal S, Soto AV, et al. Family-based approaches to cardiovascular health promotion. J Am Coll Cardiol 2016;67(14):1725–37. https://doi.org/10.1016/j.jacc.2016.01.036.

[103]   Johannsen DL, Johannsen NM, Specker BL. Influence of parents' eating behaviors and child feeding practices on children's weight status. Obesity 2006;14(3):431–9. https://doi.org/10.1038/oby.2006.57.

[104]   Savage JS, Orlet Fishe J, Birch LL. Parental influence on eating behavior. J Law Med Ethics 2007;35(1):22–34. https://doi.org/10.1111/j.1748-720X.2007.00111.x.

[105]   Agras WS, Hammer LD, McNicholas F, Kraemer HC. Risk factors for childhood overweight: a prospective study from birth to 9.5 years. J Pediatr 2004;145(1):20–5. https://doi.org/10.1016/j.peds.2004.030023.

[106]   Agras WS, Kraemer HC, Berkowitz RI, Hammer LD. Influence of early feeding style on adiposity at 6 years of age. J Pediatr 1990;116(5):805–9. https://doi.org/10.1016/S0022-3476(05)82677-0.

[107]   Fuemeler BF, Lovelady CA, Zucker NL, Ostby T. Parental obesity moderates the relationship between childhood appetitive traits and weight. Obesity 2013;21(4):815–23. https://doi.org/10.1002/oby.20144.

[108]   Tzou IL, Chu NF. Parental influence on childhood obesity: a review. Health 2012;4(12A):1464–70. https://doi.org/10.4236/health.2012.412A211.

[109]   Rhee KE, Lumeng JC, Appugliese DP, et al. Parenting styles and overweight status in first grade. Pediatrics 2006;117(6):2047–54.

[110]   Johnson SL, Birch LL. Parents' and children's adiposity and eating style. Pediatrics 1994;94:653–61.

[111]   Berge JM, Wall M, Loth K, et al. Parenting style as predictor of adolescent weigh and weight related behaviors. J Adolesc Health 2010;46:331–8.

[112]   Schmitz KH, Lytle LA, Phillips GA, et al. Psychosocial correlates of physical activity and sedentary leisure habits in young adolescents: The Teens Eating for Energy and Nutrition at School study. Prev Med 2002;34:266–78.

[113]   O'Byrne KK, Haddock CK, Poston WS. Parenting style and adolescent smoking. J Adolesc Health 2002;30:418–25.

[114]   Fuemeler BF, Yang C, Costanzo P, et al. Parenting styles and body mass index trajectories from adolescence to adulthood. Health Psychol 2012;31:441–9.

[115]   Phillips N, Sioen I, Michels N, Sleddens E, DeHenauw S. The influence of parenting style on health related behavior of children: findings from the ChiBS study. Int J Behav Nutr Phys Act 2014;11(95).

[116]   Gonea M, Crows S. Parents as key players in the prevention and treatment of weight related problems. Nutr Rev 2004;62(1):39–50.

[117]   Scaglioni S, Arizza C, Vecchi F, Tedeschi S. Determinants of children's eating behavior. Am J Clin Nutr 2011;94(6). https://doi.org/10.3945/ajcn.110.001685.

[118]   American Academy of Pediatrics. 11 ways to encourage your child to be physically active. Available at www.healthychildren.org; Updated 11/18/15 [accessed 02.01.17].

FURTHER READING

[1]   Rhee K. Childhood overweight and the relationship between parent behaviors, parenting style, and family functioning. Ann Am Acad Pol Soc Sci 2008;615:11–37.

Chapter 5

Lifestyle Factors and the Impact on Lifetime Incidence and Mortality of Coronary Heart Disease

P.E. Puddu*, A. Menotti[†]

*Department of Cardiovascular, Respiratory, Nephrological, Anesthesiological and Geriatric Sciences, Sapienza University of Rome, Rome, Italy,
[†]Association for Cardiac Research, Rome, Italy

HISTORICAL INTRODUCTION

The first identification of coronary heart disease (CHD) risk factors coincided with the beginning of cardiovascular epidemiology as a research discipline in the second half of the past century. They were biophysical or biochemical measurements, but since the beginning, smoking habits—a typical behavioral, lifestyle risk factor—were considered. Only later on, other lifestyle factors attracted the attention of investigators, and among them, two were the major ones, physical activity and eating-dietary habits. Studies on physical activity started a little later, as well as those on eating-dietary habits, due to the complexity and uncertainties of their measurement, although hypotheses and preliminary evidence on their possible role were already available. We will confine our attention to smoking, physical activity, and dietary habits in order to simplify the approach and also to recruit sufficiently representative information when these habits are considered concomitly.

During the first decades of cardiovascular epidemiology, not only most studies on physical activity and eating-dietary habits but also some on cigarette smoking were conducted in an independent way, that is, without the concomitant consideration of other classical risk factors, such as age, blood pressure, or serum cholesterol.

Cigarette smoking measurement was usually based on the questionnaire of various types. In rare studies, validation was provided by biomarkers essay such as the measurement of thiocyanate in blood—a generic residual of combustion—or of cotinine and other nicotine metabolites in blood or urine. Summary of acquired evidence was provided by large monographs mainly produced by the US Department of Health, Education, and Welfare [1–4]. Among the single studies, a pivotal role was played by the Doll and Peto investigation on British doctors started in 1951 that reached the half century of follow-up documented in a publication of 2004 [5]. Evidence of the role of cigarette smoking as a major risk factor for CHD events was always clear, including the role of passive smoking.

Studies on physical activity were based, in the majority of cases, on retrospective activity questionnaires; activity recall questionnaires and diaries self-compiled or by interview; and much more rarely by oxygen uptake, heart rate recording, direct time and motion observation, and food consumption. Frequently, information gathered by those approaches was converted into metabolic equivalent task (MET in kcal). Studies on physical activity of the first decades of cardiovascular epidemiology were summarized between 1978 and 2008 by at least five major contributions from several sources [6–11]. Several studies were based on working physical activity, but the more recent ones were oriented toward leisure physical activity since physical engagement at work was sharply declining during the second part of the last century. In the first contribution [6], 20 studies were reported on the relationship of physical activity with CHD; in the last one [11], findings of 36 studies were reported in a meta-analysis showing that physically-active people were protected from CHD events when compared with sedentary people. Recent contributions tended to indicate that even moderate leisure time physical activity represents a protective behavior [12].

The possible role of eating-dietary habits was more complex to demonstrate, and studies were carried out with some delay. Several measurement techniques were employed, and among them, there were (a) the weighed record method, time-consuming and very expensive, validated by food composition tables or by chemical analysis of food composites; (b) the dietary history method; (c) the food frequency method; and more recently; (d) the videotel system. Techniques based on questionnaire were the most commonly used, and validation versus food composition tables was always needed. In this area, there were sporadic observations based on the so-called geographic pathology made in the first part of the 20th century. Then, apart from a few other scattered and incomplete studies in the second part of the 1900s [13–19], the first systematic population investigation of diet versus CHD was made within the Seven Countries Study of cardiovascular diseases started in 1957 and still running. In the 1970 monograph [20], strong ecological relationships were shown between average intake of saturated fat and population CHD incidence and mortality during a 5-year follow-up, a finding confirmed in subsequent analyses for longer follow-up [21–23]. Initial focus on nutrients and fats in particular was bound to the notion that serum cholesterol was a major risk factor for CHD and that among nutrients the one more strongly influencing serum cholesterol level was saturated fat [24]. Only later, the interest shifted toward food groups, either alone or combined into dietary patterns [25–27], with the identification of the so-called Mediterranean diet [28] roughly characterized by a predominance of plant versus animal food and by its protective effects on CHD and other conditions. During the last decade of 1900 and the beginning of the years 2000, there was an explosion of population studies dealing with food groups and dietary patterns, many of them based on the concept of Mediterranean diet. A systematic review of these concepts was made by several investigators [29–33]. Moreover, along the second half of the year 1900s, a few studies based on FAO nutritional data and WHO mortality data of several countries provided gross evidence of the relationship of nutrients and food groups versus CHD mortality [34–39].

The brief summary presented above was deliberately confined to observational population studies with the exclusion of intervention trials and consensus statements or guidelines. Interest for the above major lifestyle behaviors expanded, during the last 20 years, and findings were more and more frequently presented in a combined fashion. However, an exhaustive review of all these elements is far beyond our scopes. Instead, we intended to concentrate on observational investigations since a major purpose was to address the problem of lifestyle projected toward lifetime end points and survival, an effort that might be undertaken only on these research pieces. A synoptic table was composed to facilitate a synthetic overview of all these evidences (Table 1).

SHORT-TERM FOLLOW-UP STUDIES

Whether health behavior risk factors (smoking, drinking, physical activity, and body weight) for mortality vary by age and gender were investigated among 6109 adults, 45–74 years old in the National Health and Nutrition Examination Survey (NHANES I) 1971–75 were traced during the 1982–84 NHANES I [40]. The idea of this public survey, whose database is now freely available, was at ascertaining whether the declined mortality rates for older US adults observed since 1960 were and to which extent related to prevention and health promotion in improving the health and survival of older adults. For middle-aged men (45–54 years old) and for older men (65–74 years old), both smoking and nonrecreational physical activity were predictors of survival time. Additionally, for older men, drinking and low body mass index were associated with shorter survival time. Among women, there was less consistency of associations across age groups. As with men, nonrecreational physical activity and low body mass index were associated with shorter survival among older women. Therefore, health behaviors appeared to be associated with survival in older adults and in middle-aged adults, although the specific behavioral risk factors may vary by age and gender. No specific relationships were determined specifically with CHD-related outcomes.

A prospective cohort study of 81,722 US women in the Nurses' Health Study from June 1984 to June 2010 investigated the degree to which adherence to a healthy lifestyle may lower the risk of sudden coronary death (SCD), defined as death occurring within 1 h after symptom onset without the evidence of circulatory collapse, among women [41]. Lifestyle factors were assessed via questionnaires every 2–4 years. A low-risk lifestyle was defined as not smoking, body mass index of less than 25, exercise duration of 30 min/day or longer, and top 40% of the alternate Mediterranean diet score, which emphasizes high intake of vegetables, fruits, nuts, legumes, whole grains, and fish and moderate intake of alcohol. There were 321 cases of SCD during 26 years of follow-up. Women had 72 years at the time of the SCD event, thus confirming a relatively late occurrence of this outcome [59,60] also among women. All four low-risk lifestyle factors were significantly and independently associated with a lower risk of SCD. The absolute risks of SCD were 22 cases/100,000 person-years among women with zero low-risk factors, 17 cases/100,000 person-years with one low-risk factor, 18 cases/100,000 person-years with two low-risk factors, 13 cases/100,000 person-years with three low-risk factors, and 16 cases/100,000 person-years with four low-risk factors. Compared with women with zero low-risk factors, the multivariable relative risk of SCD was 0.54 (95% confidence interval CI, 0.34–0.86) for women with one low-risk factor, 0.41 (95% CI, 0.25–0.65) for 2 low-risk factors, 0.33 (95% CI, 0.20–0.54) for 3 low-risk factors, and 0.08 (95% CI, 0.03–0.23) for four low-risk factors. The proportion of SCD attributable to smoking, inactivity, overweight, and poor diet was 81% (95% CI, 52%–93%). Among women without clinically diagnosed CHD, the percentage of population attributable risk for CHD was 79% (95% CI, 40%–93%). Thus, the study clearly showed that adherence to a low-risk lifestyle is associated with a low risk of SCD.

TABLE 1 Observational Population Studies Whereby Behavioral Characteristics Were Concomitantly Investigated In Relation to All-Cause or Cause-Specific Outcomes to Conclude That Healthy Versus Unhealthy Lifestyle is Advantageous for Prolonging Life in Humans

Study	Country	Year	N	Age Range	Gender	Factors					Outcomes	Years of F-up	RR	CHD Assessment
						S	D	P	B	MD				
Short-term studies														
NHANES I [40]	The United States	1994	6109	45–74	F, M	S	D	P	B	MD	Survival time	10	0.73	N
Nurses' [41]	The United States	2011	81,722	30–55	F	S	D	P	B	MD	SD<1 h	26	0.15	Y
Nurses' [42]	The United States	2008	77,782	34–59	F	S	D	P	B	MD	Mortality	24		N
											-All-cause		4.31 (3.51–5.31)[a]	
											-CVD		8.17 (4.96–13.47)[a]	
											-Cancer		3.26 (2.45–4.34)[a]	
Widepop [43]	The United Kingdom	2010	4886	>18	F, M	S	D	P		MD	Mortality	20	3.49 (2.31–5.26)[a,b]	N
Shanghai [44]	PRC	2010	71,243	40–70	F	S[c]		P	B[d]	MD	Mortality	9		N
											-All-cause		0.57 (0.44–0.74)	
											-CVD		0.29 (0.16–0.54)	
											-Cancer		0.76 (0.54–1.06)	
China origin [45]	SGP	2011	50,466	45–74	F, M	S	D	P	B[e]	MD	CVD mortality	15	0.24 (0.17–0.34)	Y
											CHD mortality	15	0.23 (0.14–0.37)	
											Stroke mortality	15	0.25 (0.14–0.46)	

Continued

TABLE 1 Observational Population Studies Whereby Behavioral Characteristics Were Concomitantly Investigated In Relation to All-Cause or Cause-Specific Outcomes to Conclude That Healthy Versus Unhealthy Lifestyle is Advantageous for Prolonging Life in Humans—cont'd

Study	Country	Year	N	Age Range	Gender	Factors					Outcomes	Years of F-up	RR	CHD Assessment
						S	D	P	B	MD				
Severance [46]	ROK	2012	59,941	30–84	F, M	S	D	P	B	MD	Mortality	10		N
											-All-cause		2.00[a,f]	
											-Noncancer		1.92[a]	
											-Cancer		2.04[a]	
Karolinska [47]	S	2013	2193	>60	F	S	D	P	B[g]	MD	Mortality	11	0.25 (0.15–0.44)	N
											-CVD		0.44 (0.26–0.75)	
		2039		> 60	M	S	D	P	B[g]	MD	Mortality	11	0.35 (0.23–0.54)	N
											-CVD		0.39 (0.25–0.61)	
AARP [48]	The United States	2013	170,672	51–71	F	S		P	B[h]	MD	Mortality	13	0.27 (0.25–0.29)	N
Lombardy [49]	I	2015	1693	40–74	F, M	S		P		MD	Mortality	17	0.27	N
ARIC++ [50]	The United States	2016	55,685	44–80	F, M	S		P	B	MD	CHD events	19	0.54 (0.47–0.63)	Y
MetaAnal [51]	D	2012	531,804[i]		F, M	S	D	P	B	MD	Mortality	13	0.34 (0.27–0.42)	N
Lifetime studies														
IRA-SCS [52]	I	2014	1564	45–64	M	S		P		MD	Mortality	40		Y
											-All-cause		0.83 (0.60–0.96)	
											-CVD		0.43 (0.35–0.61)	
											-CHD		0.37 (0.18–0.74)	

Study	Country	Year	N	Age	Sex	S	D	P	j	MD	Outcome	FU (yr)	RR (95% CI)	
IRA-SCS [53]	I	2015	1677	40–59	M	S		P		MD	Incidence	50		Y
											-CHD		0.37 (0.19–0.47)	
											-Cancer		0.39 (0.24–0.52)	
IRA-SCS [54,55]	I	2015	1712	40–59	M	S		P		MD	Mortality	50		Y
											-All-cause		0.85 (0.65–0.92)	
											-CVD		0.85 (0.63–0.91)	
											-CHD		0.45 (0.27–0.72)	
											-Cancer		0.60 (0.42–0.83)	

Studies in very elderly people

Study	Country	Year	N	Age	Sex	S	D	P	j	MD	Outcome	FU (yr)	RR (95% CI)	
Barcelona [56]	E	1995	1219	> 65	F, M	S	D	P	j		Mortality[k]	5		N
											-Smokers		3.0 (1.78–5.04)	
											-Drinkers		0.38 (0.17–0.86)	
											-Sedentary		1.53 (1.08–2.15)	
SENECA [57]	B, DK, I, NL, P, E	2002	1281	70–75	F, M	S		P		MD	Mortality	10	3–4[a]	N
HALE [58]	B, DK, I, NL, P, E, F	2004	2339	70–90	F, M	S	D	P		MD	Mortality	10	0.27 (0.25–0.29)	Y
											-All-cause		0.35 (0.28–0.44)	
											-CVD		0.33 (0.22–0.47)	

Continued

TABLE 1 Observational Population Studies Whereby Behavioral Characteristics Were Concomitantly Investigated In Relation to All-Cause or Cause-Specific Outcomes to Conclude That Healthy Versus Unhealthy Lifestyle is Advantageous for Prolonging Life in Humans—cont'd

Study	Country	Year	N	Age Range	Gender	Factors	Outcomes	Years of F-up	RR	CHD Assessment
							-CHD		0.27 (0.14–0.53)	
							-Cancer		0.31 (0.19–0.50)	

Studies, see text and references to disclose the acronym significances; countries, defined by automobile definitions; year, year of publication; N, population size; age range, minimum and maximum when the study started (in years); gender, F=female and M=male; factors, S=smoking habits, D=alcohol drinking habits, P=physical activity, B=body weight, and MD=Mediterranean or an healthy diet; outcomes, cause-specific death (or incident events) considered and when not specified mortality means all-cause mortality; years of F-up, years of follow-up either rounded or averaged; RR, relative risks (±95% confidence intervals) either presented or computed by comparing the respective outcome effects of healthy versus unhealthy or unhealthy versus healthy ([a]) lifestyles (in general four or five positive factors versus zero or the reverse); CHD assessment, whether CHD cause-specific deaths or events were considered, N=no and Y=yes CHD, coronary heart disease; CVD, cardiovascular disease.

[a]*Unhealthy versus healthy lifestyles (four or five poor factors vs zero poor factors).*

[b]*Death was seen an average of 12 years earlier in those with all poor lifestyle factors.*

[c]*Spouse smoking habits in never-smoking women who never drink.*

[d]*In addition to B, there were also waist-to-hip measurements.*

[e]*In addition to B, there was also a quantitation of usual sleep.*

[f]*Attributable risk of all-cause mortality was 26.5% in F and 44.5% in M.*

[g]*Factors adjusted for body mass index and educational level.*

[h]*B was replaced by abdominal leanness.*

[i]*There were 21 Studies selected in 18 cohorts, and 15 were included whereby an healthy lifestyle was defined as the presence of three of five positive factors.*

[j]*Hours of daily sleep instead of B.*

[k]*Adjusted for age, educational level, and perceived health status and factors measured at baseline, respectively, versus nonsmokers, abstainers, and more active people.*

Nurses' Health Study was previously used to evaluate prospectively the impact of combinations of lifestyle factors on mortality in 77,782 middle-aged (34–59 years) women free from cardiovascular disease (CVD) and cancer in 1980 [42]. Relative risk of mortality during 24 years of follow-up in relation to five lifestyle factors (cigarette smoking, being overweight, taking little moderate-to-vigorous physical activity, no light-to-moderate alcohol intake, and low diet quality score) was the main outcome variable, and again, no specific CHD-related assessments were performed. There were 8882 deaths, including 1790 from CVD and 4527 from cancer (all types). Each lifestyle factor independently and significantly predicted mortality. Relative risks for five compared with zero lifestyle risk factors were 3.26 (95% CI, 2.45–4.34) for cancer mortality, 8.17 (4.96–13.47) for cardiovascular mortality, and 4.31 (3.51–5.31) for all-cause mortality. A total of 28% (25–31%) of deaths during follow-up could be attributed to smoking and 55% (47–62%) to the combination of smoking, being overweight, the lack of physical activity, and a low diet quality. Additionally, considering alcohol intake did not substantially change this estimate. Importantly, these results indicated that, among middle-aged women at least, adherence to lifestyle guidelines is associated with markedly lower mortality of not only the most incident causes, cardiovascular in particular, as the most related to these five factors but also cancer and all-cause mortality. These were the first indications about the importance to maximize efforts to eradicate cigarette smoking and to stimulate regular physical activity and a healthy diet.

The individual and combined influence of physical activity, diet, smoking, and alcohol consumption were examined prospectively on total and cause-specific mortality among 4886 individuals at least 18 years old from the United Kingdom-wide population in 1984–85 [43]. A health behavior score was calculated, allocating one point for each poor behavior: smoking, fruits and vegetables consumed less than three times daily, less than 2 h of physical activity per week, and weekly consumption of more than 14 units of alcohol (in women) and more than 21 units (in men) (range of points, 0–4). During a mean follow-up period of 20 years, 1080 participants died, 431 from CVD, 318 from cancer, and 331 from other causes. Adjusted hazard ratios and 95% CI for total mortality associated with one, two, three, and four poor health behaviors compared with those with none were 1.85 (95% CI, 1.28–2.68), 2.23 (95% CI, 1.55–3.20), 2.76 (95% CI, 1.91–3.99), and 3.49 (95% CI, 2.31–5.26), respectively (P value for trend <.001). The effect of combined health behaviors was strongest for other deaths and weakest for cancer mortality. Those with four compared with those with no poor health behaviors had an all-cause mortality risk equivalent to being 12 years older. Thus, the combined effect of poor health behaviors on mortality was substantial, in this investigation both in men and women, indicating that modest, but sustained, improvements to diet and lifestyle could have significant public health benefits.

The overall impact of lifestyle-related factors beyond that of active cigarette smoking and alcohol consumption on all-cause and cause-specific mortality was again investigated in women from China using data from the Shanghai Women's Health Study, an ongoing population-based prospective cohort study including 71,243 women aged 40–70 years enrolled during 1996–2000 who never smoked or drank alcohol regularly [44]. A healthy lifestyle score was created on the basis of five lifestyle-related factors shown to be independently associated with mortality outcomes (normal weight, lower waist-hip ratio, daily exercise, never exposed to spouse's smoking, and higher daily fruit and vegetable intake). The score ranged from zero (least healthy) to five (most healthy) points. During an average follow-up of 9 years, 2860 deaths occurred, including 775 from CVD and 1351 from cancer. Adjusted hazard ratios for mortality decreased progressively with an increasing number of healthy lifestyle factors. Compared with women with a score of zero, hazard ratios (95% CI) for women with four to five factors were 0.57 (0.44–0.74) for total mortality, 0.29 (0.16–0.54) for CVD mortality, and 0.76 (0.54–1.06) for cancer mortality. The inverse association between the healthy lifestyle score and mortality was seen consistently regardless of chronic disease status at baseline. The population attributable risks for not having four to five healthy lifestyle factors were 33% for total deaths, 59% for CVD deaths, and 19% for cancer deaths. Thus, in Chinese women too, a healthier lifestyle pattern was associated with reductions in total and cause-specific mortality among lifetime nonsmoking and nondrinking women, supporting the importance of overall lifestyle modification in disease prevention.

In another Chinese study dealing with both sexes, the association of six combined lifestyle factors (dietary pattern, physical activity, alcohol intake, usual sleep, smoking status, and body mass index) with CVD mortality was assessed among 50,466 (44,056 without a history of diabetes mellitus, CVD, or cancer and 6410 with diabetes mellitus or history of clinical CVD) Singapore inhabitants of Chinese origin who were 45–74 years of age during enrollment in 1993–98 and followed up through 2009 [45]. Each lifestyle factor was independently associated with CVD mortality. When combined, there was a strong, monotonic decrease in age- and sex-standardized CVD mortality rates with an increasing number of protective lifestyle factors. Relative to participants with no protective lifestyle factors, the hazard ratios of CVD mortality for one, two, three, four, and five to six protective lifestyle factors were 0.60 (95% CI, 0.45–0.84), 0.50 (95% CI, 0.38–0.67), 0.40 (95% CI, 0.30–0.53), 0.32 (95% CI, 0.24–0.43), and 0.24 (95% CI, 0.17–0.34), respectively, among those without a history of diabetes mellitus, CVD, or cancer (P for trend <.0001). A parallel graded inverse association was observed in participants with a history of CVD or diabetes mellitus at baseline. Results were consistent for CHD and cerebrovascular disease mortality. Therefore, the study clearly indicated that an increasing number of protective lifestyle factors are associated with a marked decreased risk of CHD, cerebrovascular, and overall CVD mortality in Chinese men and women.

In a study population including 59,941 Koreans, 30–84 years of age, who had visited the Severance Health Promotion Center between 1994 and 2003, Cox regression models were fitted to establish the association between combined lifestyle factors (current smoker, heavy daily alcohol use, overweight or obese weight, physical inactivity, and unhealthy diet) and mortality outcomes [46]. During 10.3 years of follow-up, there were 2398 all-cause deaths. Individual and combined lifestyle factors were found to be associated with the risk of mortality. Compared with those having none or only one risk factor, in men with a combination of four lifestyle factors, the relative risk for cancer mortality was 2.04-fold, for noncancer mortality 1.92-fold, and for all-cause mortality 2.00-fold. In women, the relative risk was 2.00-fold for cancer mortality, 2.17-fold for noncancer mortality, and 2.09-fold for all-cause mortality. The population attributable risks for all-cause mortality for the four risk factors combined was 44.5% for men and 26.5% for women. This study suggests that having a high (unhealthy) lifestyle score, in contrast to a low (healthy) score, can substantially increase the risk of death by any cause, cancer, and noncancer in Korean men and women.

Loef and Walach [51] performed a systematic review and meta-analysis of the combined effects of healthy lifestyle behaviors on all-cause mortality. Prospective studies were selected if they reported the combined effects of at least three of five lifestyle factors (obesity, alcohol consumption, smoking, diet, and physical activity). The mean effect sizes that certain numbers of combined lifestyle factors have on mortality were compared with the group with the least number of healthy lifestyle factors by meta-analysis. Sensitivity analyses were also conducted to explore the robustness of the results. Up to February 2012 from searched Medline, Embase, Global Health, and Somed, there were 21 studies (18 cohorts) that met the inclusion criteria of which 15 were included in the meta-analysis that comprised 531,804 people with a mean follow-up of 13.24 years. The relative risks for all-cause mortality decreased proportionate to a higher number of healthy lifestyle factors for all-cause mortality. A combination of at least four healthy lifestyle factors was associated with a reduction of the all-cause mortality risk by 66% (95% CI, 58%–73%). This was the largest evidence accumulated thus far to indicate that adherence to a healthy lifestyle is associated with a lower risk of mortality.

More recently, Carlsson et al. [47] investigated on seven modifiable lifestyle factors to assess how they might predict reduced risk for CVD and all-cause mortality regardless of body mass index, educational level, and gender. It was a population-based prospective cohort study of representative 60-year-old women ($n=2193$) and men ($n=2039$). The following factors related to a healthy lifestyle were assessed using a questionnaire: nonsmoking, alcohol intake of 0.6–30 g/day, moderate physical activity at least once a week, low intake of processed meats, weekly intake of fish, daily intake of fruit, and daily intake of vegetables. These factors were combined to produce a total score of healthy lifestyle factors (0–7) and classified into four groups: unhealthy (0–2 lifestyle factors), intermediate (3), healthy (4–5), and very healthy (6–7). Over a follow-up of 11 years, there were 375 incident CVD and 427 all-cause deaths. Very healthy women and men exhibited a decreased risk for incident CVD compared with unhealthy individuals, with HR (and 95% CI) adjusted for educational level and BMI of 0.44 (0.26–0.75) and 0.39 (0.25–0.61), respectively. The corresponding HR for all-cause mortality for very healthy women and men were 0.25 (0.15–0.44) and 0.35 (0.23–0.54), respectively. Thus, with seven healthy lifestyle factors, it was possible to identify men and women with substantially lower relative risks of incident CVD and death, regardless of BMI and educational level.

In a cohort of 170,672 American women and men aged 51–71 years at baseline in 1996–97 and followed up through 2009, the individual and joint impact of four low-risk lifestyle factors were investigated [48]: abdominal leanness (waist circumference, 88 cm in women and 102 cm in men), recommended physical activity level (30 min or more of moderate exercise at least five times per week or 20 min or more of vigorous exercise at least three times per week), long-term nonsmoking (never smoker or quit smoking more than 10 years ago), and healthy diet (Mediterranean diet score within the upper two sex-specific quintiles). During 2,126,089 person-years of follow-up, 20,903 participants died. In multivariate Cox models, statistically significant decreased risks of mortality were observed for the low-risk factors abdominal leanness (RR, 0.86; 95% CI, 0.83–0.89), physical activity (RR, 0.86; 95% CI, 0.84–0.89), nonsmoking (RR, 0.43; 95% CI, 0.42–0.45), and healthy diet (RR, 0.86; 95% CI, 0.83–0.88). The larger the number of low-risk lifestyle factors, the lower was the mortality risk. The RR comparing adherence to all with none of the factors was 0.27 (95% CI, 0.25–0.29). It was estimated that 33% (95% CI, 30–35%) of deaths in this cohort were premature and could have been avoided if all study participants had adhered to all low-risk factors.

The specific association of the Mediterranean diet, smoking habits, and physical activity with all-cause mortality in an Italian population was studied by Prinelli et al. [49]. A total of 1693 subjects aged 40–74 who enrolled in the study in 1991–95 were asked about dietary and other lifestyle information at baseline. Adherence to the Mediterranean diet was evaluated by the Mediterranean dietary score. A healthy lifestyle score was computed by assigning 1 point each for a medium or high adherence to the Mediterranean dietary score, nonsmoking, and physical activity. The final sample included 974 subjects with complete data and without chronic disease at baseline. During a median of 17.4 years of follow-up, 193 people died. Subjects with high adherence to the Mediterranean dietary score (HR, 0.62; 95% CI, 0.43–0.89), nonsmokers (HR, 0.71; 95% CI, 0.51–0.98), and physically active subjects (HR, 0.55; 95% CI, 0.36–0.82) were at low risk of death. Each point increase in the Mediterranean dietary score was associated with a significant 5% reduction of death risk. Subjects with one, two, or three healthy lifestyle behaviors had a significantly 39, 56, and 73% reduced risk of death, respectively. A high adherence to Mediterranean diet, nonsmoking, and physical activity was strongly associated with a reduced risk of all-cause mortality in healthy subjects after relatively long-term follow-up. This reduction was even stronger when the healthy lifestyle behaviors were combined.

Lastly, using a polygenic score of DNA sequence polymorphisms, genetic risk for CHD was assessed in three prospective cohorts—7814 participants in the Atherosclerosis Risk in Communities (ARIC) study, 21,222 in the Women's Genome Health Study (WGHS), and 22,389 in the Malmö Diet and Cancer Study (MDCS)—and in 4260 participants in the cross-sectional BioImage Study for whom genotype and covariate data were available. Adherence to a healthy lifestyle among the participants was also evaluated using a scoring system consisting of four factors: no current smoking, no obesity, regular physical activity, and a healthy diet. The extent to which increased genetic risk can be offset by a healthy lifestyle was thus ascertained for the first time [50]. The relative risk of incident CHD events was 91% higher among participants at high genetic risk (top quintile of polygenic scores) than among those at low genetic risk (bottom quintile of polygenic scores) (HR, 1.91; 95% CI, 1.75–2.09). A favorable lifestyle (defined as at least three of the four healthy lifestyle factors) was associated with a substantially lower risk of coronary events than an unfavorable lifestyle (defined as no or only one healthy lifestyle factor), regardless of the genetic risk category. Among participants at high genetic risk, a favorable lifestyle was associated with a 46% lower relative risk of CHD events than an unfavorable lifestyle (HR, 0.54; 95% CI, 0.47–0.63). This finding corresponded to a reduction in the standardized 10-year incidence of coronary events from 10.7% for an unfavorable lifestyle to 5.1% for a favorable lifestyle in ARIC, from 4.6% to 2.0% in WGHS, and from 8.2% to 5.3% in MDCS. In the BioImage Study, a favorable lifestyle was associated with significantly less coronary artery calcification within each genetic risk category. Across the four considered studies involving 55,685 participants, genetic and lifestyle factors were independently associated with susceptibility to CHD. Among participants at high genetic risk, a favorable lifestyle was associated with a nearly 50% lower relative risk of CHD than was an unfavorable lifestyle.

LIFETIME EFFECTS OF LIFESTYLE AND EFFECTS IN VERY ELDERLY PEOPLE

Lifetime relations of lifestyle parameters have been a rarely investigated issue, mostly confined to the impact of isolated elements such as smoking habits to mortality effects on male British doctors during 50 years in the classic study by Doll et al. [5]. We have investigated instead quite extensively these relations within the Italian Rural Areas of the Seven Countries Study [52–55] since we were able to also assess the concomitant role of combined parameters (smoking habits, physical activity at work, and eating habits as derived from factor analysis) and therefore their multiplicative effect as seen above in relatively short-term investigations performed by other investigators [41–51]. Apart from exploring the relationships of these three lifestyle factors with long-term mortality, we dissected their role according to death causes, whether from all causes, CHD, CVD, or cancer, a particularly important clarification since, although cardiovascular deaths do impact enormously on all-cause mortality [61], there might be a need to consider the competition of risks when the follow-up is lifetime long and the role of most classic parameters, both cardiovascular and noncardiovascular, should be probably reevaluated under this perspective [62,63].

In the first investigation, out of a cohort of 1564 men aged 45–64 and examined in 1965, there were 693 men who died in 20 years and 1441 in 40 years [52]. In Cox proportional hazard models, men smoking cigarettes (vs never smokers), those having a sedentary activity (vs the very active), and those following the diet score 1, indexing an unhealthy diet (vs men with a diet close to the healthy Mediterranean style), had highly significant HR in relations with 20- and 40-year mortality from all causes, CHD, CVD, and cancer. HRs for CHD mortality in 40 years were 1.37 (95% CI, 1.01–1.84) for smokers, 1.89 (95% CI, 1.35–2.66) for sedentary people, and 1.68 (95% CI, 1.24–2.28) for men with unhealthy diet. Comparable but frequently smaller HR were found for all-cause, CVD, and cancers deaths.

Importantly, combination of three unhealthy risk factors versus their absence was associated [52] with 4.8-year life loss in the 20-year follow-up and 10.7-year life loss in the 40-year follow-up, implicitly confirming the results of the much younger women and men of the UK study during 20 years of follow-up [43]. Table 1, on the other hand, points to the global effects of healthy lifestyles versus poor ones in terms of decreasing RR. Our results show that 40-year long lifestyle behavior linked to physical activity and healthy smoking and eating habits is strongly associated with mortality and survival in middle-aged men (RR between 0.37 and 0.83).

In the second investigation, 1677 heart-disease-free men aged 40–59 years were followed up during 50 years for lifetime incidence of heart disease up to the age of 90 years [53]. They were classified as CHD and heart diseases of uncertain etiology (HDUE). As in the previous study [52], baseline cigarette smoking habits (nonsmokers and ex-smokers, moderate smokers, and heavy smokers), physical activity (sedentary, moderate, and vigorous), and eating habits (non-Mediterranean diet, prudent diet, and Mediterranean diet) were related to incidence of heart disease. Incidence of CHD and HDUE up to the age of 90 years was 28.8% and 17.7%, respectively. Univariate and multivariate analyses showed strong association of behavioral characteristics with CHD incidence, but not with HDUE incidence. Cox proportional hazard rates for CHD were 1.45 (95% CI, 1.11–1.90) for heavy smokers versus nonsmokers, 0.67 (95% CI, 0.50–0.89) for vigorous activity versus sedentary habits, and 0.62 (95% CI, 0.47–0.83) for Mediterranean diet versus non-Mediterranean diet. Combining CHD cases with HDUE cases made the predictive picture similar to that of CHD. When some basic risk factors were added to the model, results remained substantially unaltered, despite the existence of some correlations of behaviors with risk factors. RR when comparing people with all three positive lifestyle factors versus those who had none showed a substantial lifetime advantage also for CHD incidence (RR 0.37) among initially middle-aged men. Again, behavioral factors including cigarette smoking, physical activity, and diet are strong predictors of lifetime incidence of common heart diseases even adding other traditional risk factors.

Exploiting the available findings, a general score was constructed, assigning the value of 1 for each bad behavior, the value of 2 for each intermediate behavior, and the value of 3 for each good behavior. Adding up these values, a general score was obtained with a range from 3 (the worst) to 9 (the best). CHD incidence in 50 years was of 45.5% for score 3 and 21.4% for score 9, a relative risk for the latter of 0.47. Creating three numerically almost equivalent groups (scores 3, 4, 5, and 6; score 7; and scores 8 and 9), the relative risk between the extremes was 0.84, representing the advantage for the best behaviors.

Finally, the third study aimed at relating the major causes of death with lifestyle habits in an almost extinct male middle-aged population: a 40–59 aged male population of 1712 subjects was examined and followed up for 50 years [54,55]. Baseline smoking habits, working physical activity, and dietary habits were related to 50-year mortality subdivided into 12 simple and 3 composite causes of death by Cox proportional hazard models. Duration of survival was related to the same characteristics by a multiple linear regression model. Death rate in 50 years was of 97.5%. Out of 12 simple groups of the causes of death, 6 were related to smoking habits, 3 to physical activity, and 4 to dietary habits. Among composite groups of the causes of death, HRs (and their 95% CI) of never smokers versus smokers were 0.68 (0.57–0.81) for major CVD, 0.65 (0.52–0.81) for all cancers, and 0.72 (0.64–0.81) for all-cause deaths. HRs of vigorous physical activity at work versus sedentary physical activity were 0.63 (0.49–0.80) for major CVD, 1.01 (0.72–1.41) for all cancers, and 0.76 (0.64–0.90) for all-cause deaths. HRs of Mediterranean diet versus non-Mediterranean diet were 0.68 (0.54–0.86) for major CVD, 0.54 (0.40–0.73) for all cancers, and 0.67 (0.57–0.78) for all-cause deaths. Table 1 shows individual RR when comparing people with all three positive lifestyle factors with those who had none for all four outcomes, showing a substantial lifetime advantage (RR between 0.45 and 0.85) among initially middle-aged men. Expectancy of life was 12 years longer for men with the three best behaviors than for those with the three worst behaviors. These results clearly indicate that some lifestyle habits are strongly related to lifetime mortality, and a measurable and large survival duration difference exists between optimal and worst lifestyles.

An expansion of the above analyses [52–55] was represented by the estimate of CHD incidence (both fatal and nonfatal cases) in 50 years of follow-up including the relationship with entry levels of the three lifestyle behaviors, presented here for the first time. In a denominator of 1677 middle-aged men from the Italian Rural Areas of the Seven Countries Study, there were 493 CHD cases including a small subgroup of HDUE cases in whom angina pectoris was diagnosed and were thus considered much comparable with the typical CHD incident events [64,65]. CHD incidence was 29.4%. Figs. 1–3 show Kaplan-Meier curves for smoking habits (never smokers and ex-smokers vs smokers less than 20 cigarettes per day vs smokers 20 or more cigarettes per day, chi^2 of log rank test 5.6, $P=.0617$), physical activity (vigorous activity vs moderate activity vs sedentary activity, chi^2 of log rank test 9.6, $P=.0081$), and dietary habits (Mediterranean diet vs prudent diet vs non-Mediterranean diet, chi^2 of log rank test 26.1, $P<.0001$). These results represent the first lifetime evidence that baseline lifestyles impact on incident CHD cases, both fatal and nonfatal. A tremendously important fact is connected with what was illustrated above, in terms of overall mortality (from all causes, including CVD and cancer) and improved survival duration when the selected lifestyles are optimal [52–55].

A slightly different way of concentrating on what of life remains, yet with the great limitation of accepting the effects of natural selection for those at high risk of premature death, was to investigate groups of old individuals and follow them up to the end of life. There were a few studies that investigated the effects of lifestyle habits on these very elderly individuals [56–58].

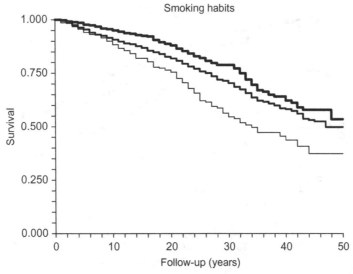

FIG. 1 Italian Rural Areas of the Seven Countries Study. Survival from CHD incidence during 50 years among 1677 CHD-free men aged 40–59 years at entry as a function of smoking habits. *Thick line*, never smokers and ex-smokers; *intermediate line*, smokers less than 20 cigarettes per day; *thin line*, smokers 20 or more cigarettes per day.

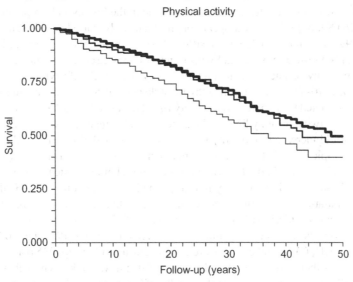

FIG. 2 Italian Rural Areas of the Seven Countries Study. Survival from CHD incidence during 50 years among 1677 CHD-free men aged 40–59 years at entry as a function of physical activity. *Thick line*, vigorous activity; *intermediate line*, moderate activity; *thin line*, sedentary activity.

To assess the risk of mortality associated with different health behaviors in a Mediterranean elderly population, the mortality experience of a cohort of 1219 noninstitutionalized men and women aged 65 years or over, who had participated in the 1986 health interview survey of Barcelona, was investigated [56]. At baseline, self-reported information on smoking, alcohol consumption, physical activity, and hours of daily sleep was collected through face-to-face home interviews. After 5 years of follow-up, a total of 224 (18.4%) participants had died (from all causes). After adjusting for age, level of education, and perceived health status, current smokers at baseline had a relative risk of dying during the follow-up of 3.0 (95% CI, 1.78–5.04) compared with nonsmokers. Those reporting moderate alcohol consumption had a relative risk of dying of 0.38 (95% CI, 0.17–0.86) compared with abstainers. Finally, individuals who reported a sedentary lifestyle had a risk of dying of 1.53 (95% CI, 1.08–2.15) compared with the more active. Thus, adverse effects on survival of smoking and of being sedentary extend later into life. Because of the high prevalence of a sedentary lifestyle among elderly people, results suggest that promoting physical activity may have an important role in enhancing survival in this age group.

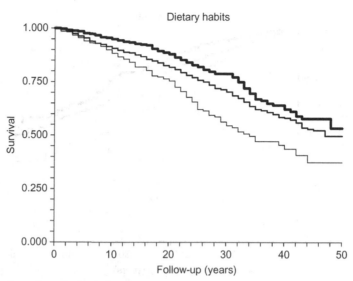

FIG. 3 Italian Rural Areas of the Seven Countries Study. Survival from CHD incidence during 50 years among 1677 CHD-free men aged 40–59 years at entry as a function of dietary habits. *Thick line*, Mediterranean diet; *intermediate line*, prudent diet, *thin line*: non-Mediterranean diet.

The single and combined effects of three healthy lifestyle behaviors—nonsmoking, being physically active, and having a high-quality diet—on survival were investigated among older people in the Survey in Europe on Nutrition and the Elderly: a Concerned Action (SENECA) Study [57]. This European longitudinal study started with baseline measurements in 1988–89 and lasted until 30 April 1999. The study population consisted of 631 men and 650 women aged 70–75 years from Belgium, Denmark, Italy, the Netherlands, Portugal, Spain, and Switzerland. A lifestyle score was calculated by adding the scores of the lifestyle factors smoking habits, physical activity, and dietary quality. The single lifestyle factors and the lifestyle score were related to mortality. Even at ages 70–75 years, the unhealthy lifestyle behaviors smoking, having a low-quality diet, and being physically inactive were singly related to an increased mortality risk (HR ranged from 1.2 to 2.1). The risk of death was further increased for all combinations of two unhealthy lifestyle behaviors. Finally, men and women with all three unhealthy lifestyle behaviors had a three- to fourfold increase in mortality risk. These results underline, also in the elderly, the importance of a healthy lifestyle, including multiple lifestyle factors and the maintenance of it with advancing age.

To investigate the single and combined effect of Mediterranean diet, being physically active, moderate alcohol use, and nonsmoking on all-cause and cause-specific mortality in European elderly individuals, the Healthy Ageing: a Longitudinal study in Europe (HALE) population was started [58], comprising individuals enrolled in the SENECA study, and the Finland, Italy, and the Netherlands' Elderly (FINE) studies include 1507 apparently healthy men and 832 women, aged 70–90 years in 11 European countries. This cohort study was conducted between 1988 and 2000, and in a 10-year follow-up mortality from all causes, CHD, CVD, and cancer were coded. During follow-up, 935 participants died: 371 from CVD, 233 from cancer, and 145 from other causes; for 186, the cause of death was unknown. Adhering to a Mediterranean diet (HR, 0.77; 95% CI, 0.68–0.88), moderate alcohol use (HR, 0.78; 95% CI, 0.67–0.91), physical activity (HR, 0.63; 95% CI, 0.55–0.72), and nonsmoking (HR, 0.65; 95% CI, 0.57–0.75) were associated with a lower risk of all-cause mortality (HRs controlled for age, sex, years of education, body mass index, study, and other factors). Similar results were observed for mortality from CHD, CVD, and cancer. Table 1 shows individual RR when comparing people with all four positive lifestyle factors versus those who had none for all four outcomes, showing a substantial advantage (RR between 0.27 and 0.35) also in very elderly women and men.

The combination of four low-risk factors lowered the all-cause mortality rate to 0.35 (95% CI, 0.28–0.44) [58]. In total, the lack of adherence to this low-risk pattern was associated with a population attributable risk of 60% of all deaths, 64% of deaths from CHD, 61% from CVD, and 60% from cancer. Therefore, among individuals aged 70–90 years, adherence to a Mediterranean diet and healthful lifestyle was associated with more than 50% lower rate of all-cause and cause-specific mortality during successive 10 years.

CONCLUSIONS

The literature is rich of reports documenting the separate role of the individual (between three and seven) behavioral habits on health, and this is particularly true, during the last few years, for eating habits (including healthier diets and moderate alcohol consumption) with the identification of subgroups of populations, almost in every country, that follow dietary habits resembling the so-called Mediterranean diet [28] and have reduced incidence and/or mortality from CHD, cancer, and all causes of death as described in this review and synthetically summarized in Table 1 with RR comparing healthier with unhealthier lifestyles from 0.14 to 0.75 depending on the outcome type considered and likely the duration of follow-up but in general representing a great advantage at long-term (around 15–20 years) with several (4–12) years of life saved. Reports dealing with all behaviors considered together are less frequent but still numerous [1–27]. There are impressive aspects in these findings. First, the contributions come from largely different countries and cultures, such as the United States, China, Singapore, Korea, and a dozen of European countries. Second, the results are rather similar across the various studies, and none contradicts the others. It seems that some universal rules apply everywhere when smoking habits, physical activity, and dietary habits are concerned in relation with health and disease and expectancy of life. Other parameters such as hours of daily sleep [56] might contribute positively to a lifelong healthy lifestyle, also applicable to elderly individuals of both genders [56–58], and might collaborate to improve survival and reducing not only CHD but also CVD and cancer mortality [52–58,61–65].

The Global Burden of Disease, Injuries, and Risk Factor study 2013 (GBD 2013) has just published [66] the results whereby risk factor quantification, particularly of modifiable risk factors, was essayed to help identifying emerging threats to population health and opportunities for prevention. Attributable deaths, years of life lost, years lived with disability, and disability-adjusted life-years (DALYs) have been estimated for 79 risks or clusters of risks using the GBD 2010 methods for 188 countries for the period of 1990–2013. All risks combined accounted for 57.2% of deaths and 41.6% of DALYs [66]. Risks quantified accounted for 87.9% of CVD DALYs, ranging to a low of 0% for neonatal disorders and neglected tropical diseases and malaria. In terms of global DALYs in 2013, six risks or clusters of risks each caused more than 5% of DALYs: dietary risks (low fruit, high sodium, low whole grains, low vegetables, and low nuts and seeds) accounting for 11.3 million deaths and 241.4 million DALYs, high systolic blood pressure for 10.4 million deaths and 208.1 million DALYs, child and maternal malnutrition for 1.7 million deaths and 176.9 million DALYs, tobacco smoke for 6.1 million deaths and 143.5 million DALYs, air pollution for 5.5 million deaths and 141.5 million DALYs, and high BMI for 4.4 million deaths and 134.0 million DALYs [66]. Risk factor patterns varied across regions and countries and with time. In women, high systolic blood pressure was the leading risk in most of Central and Eastern Europe and South and East Asia, whereas in men, high systolic blood pressure or tobacco use was the leading risks in nearly all high-income countries, in north Africa and the Middle East, Europe, and Asia [66,67].

Notwithstanding the large efforts produced to elaborate the abovementioned statistics, it is necessary to point out that when specific causes rather than large composite causes are selected, there are really few results on which to rely. Indeed, most investigators used raw mortality data as found in official death certificates and are probably aware of their limited value in terms of accuracy. This is why we reported with some emphasis our results on CHD and other specific causes of death [52–55,62,63] since each single event was evaluated there on the basis of multiple information and predetermined criteria anticipating methods that later were defined as verbal necroscopy.

Therefore, in line with what was previously stressed [68,69], it may be concluded that in view of the prominence of behavioral risk factors, behavioral and social science research on interventions for these risks should be strengthened which may be facilitated by the fact that many prevention and primary care policy options are available now to act on key risks. Acting effectively on smoking habits, physical activity, body weight, and dietary habits should be a good starting point for cardiovascular (and CHD in particular) and overall disease-burden prevention and accordingly for life prolongation.

REFERENCES

[1] US Dept of Health, Education and Welfare. Surgeon general of the US. Smoke and health. PHS publ N. 1103. Rockville, MD: US Dept of Health, Education and Welfare; 1964.

[2] Hammond EC. Smoking in relation to death rates of one million men and women. In: Haenszel W, editor. Epidemiological approaches to the study of cancer and other chronic diseases. National Cancer Institute Monograph 19. Rockville, MD: US Dept of Health, Education and Welfare; 1966. p. 127–204.

[3] Surgeon General of the US. The health consequence of involuntary smoking. A report of the 4. Surgeon General. DHHS Publ. no (PHS) 87-8398. Washington, DC: US Dept of Health and Human Services; 1986.

[4] Surgeon General of the US. The Health Benefits of Smoking Cessation. Rockville, MD: US Dept of HHS-PHS-CDC Centre for Chronic Diseases Prevention and Health Promotion; 1990.5.

[5] Doll R, Peto R, Boreham J, Sutherland I. Mortality in relation to smoking: 50 years' observations on male British doctors. Br Med J 2004;328:1519–27.

[6] Commission of the European Communities: Joint EEC/WHO Workshop on physical activity in primary prevention of ischaemic heart disease. Luxembourg 22–24 February 1977. Distributed 1978.

[7] Lange Andersen K, Masironi R, Rutenfranz J, Seliger V, Degré S, Trygg K. Habitual physical activity and health. WHO Regional Publication European Series No. 6. Copenhagen: World Health Organization, Regional office for Europe; 1978. p. 1–103.

[8] Public Health Reports 1985;100:N.2 (whole issue).

[9] US Department of Health and Human Services: Physical Activity and Health, A Report of the Surgeon General. Atlanta, GA. US Department of Health and Human Services, Centers for Disease Control and Prevention. National Centers for Chronic Disease Prevention and Health Promotion; 1996.

[10] Berlin JA, Colditz GA. A meta-analysis of physical activity in the prevention of coronary heart disease. Am J Epidemiol 1990;132:612–28.

[11] Sofi F, Capalbo A, Cesari F, Abbate R, Gensini GF. Physical activity during leisure time and primary prevention of coronary heart disease: an updated meta-analysis of cohort studies. Eur J Cardiovasc Prev Rehabil 2008;15:247–57.

[12] Carnethon MR. Physical activity and cardiovascular disease: how much is enough? Am J Lifestyle Med 2009;3:44S–9S.

[13] Bronte-Stewart B, Keys A, Brock JF. Serum-cholesterol, diet, and coronary heart-disease; an inter-racial survey in the Cape Peninsula Serum-cholesterol, diet, and coronary heart-disease; an inter-racial survey in the Cape Peninsula. Lancet 1955;269:1103–8.

[14] Grone JI, Tjiong BKK, Koster M, Verdonck G, Pierloot R, Willebrands AF. The influence of nutrition and way of life on physical condition, serum cholesterol content and incidence of atherosclerosis and coronary thrombosis in Trappist and Benedictine monks. Ned Tijdschr Geneeskd 1961;105:222–33.

[15] Cohen AM, Bavly S, Poznanski R. Change of diet of Yemenite Jews in relation to diabetes and ischaemic heart-disease. Lancet 1961;2:1399–401.

[16] Stuart KL, Schneckloth RE, Lewis LA, Moore FE, Corcoran AC. Diet, serum cholesterol, protein, blood haemoglobin, and glycosuria in a West Indian community (St. Kitts, W.I.). With observations on ischaemic heart disease. Br Med J 1962;2:1283–8.

[17] Kushi LH, Lew RA, Stare FJ, Ellison CR, El Lozy M, Bourke G, et al. Diet and 20-year mortality from coronary heart disease. The Ireland-Boston Diet-Heart Study. N Engl J Med 1985;312:811–8.

[18] Hu FB, Stampfer MJ, Manson JE, Rimm E, Colditz GA, Rosner BA, et al. Dietary fat intake and the risk of coronary heart disease in women. N Engl J Med 1997;337:1491–9.

[19] Gordon T, Kagan A, Garcia-Palmieri M, Kannel WB, Zukel WJ, Tillotson J, et al. Diet and its relation to coronary heart disease and death in three populations. Circulation 1981;63:500–15.

[20] Blackburn H, Menotti A, Buzina R, Mohacek I, Karvonen MJ, Punsar S, et al. Coronary heart disease in seven countries. Keys A., editor. Circulation 1970;41(1):1–211.

[21] Aravanis C, Blackburn H, Buzina R, Djordjevic BS, Dontas AS, Fidanza F, et al. In: Keys A, editor. Seven countries study. A multivariate analysis of death and coronary heart disease. Cambridge, MA: Harvard Univ. Press; 1980. p. 1–381.

[22] Keys A, Menotti A, Karvonen MJ, Aravanis C, Blackburn H, Buzina R, et al. The diet and 15-year death rate in the Seven Countries Study. Am J Epidemiol 1986;124:903–15.

[23] Kromhout D, Menotti A, Bloemberg B, Aravanis C, Blackburn H, Buzina R, et al. Dietary saturated and trans fatty acids and cholesterol and 25-year mortality from coronary heart disease: the Seven Countries Study. Prev Med 1995;24:308–15.

[24] Keys A, Anderson JT, Grande F. Prediction of serum-cholesterol responses of man to changes in fats in the diet. Lancet 1957;273:959–66.

[25] Huijbregts P, Feskens E, Rasanen L, Fidanza F, Nissinen A, Menotti A, et al. Dietary pattern and 20-year mortality in elderly men in Finland, Italy and the Netherlands: longitudinal cohort study. Br Med J 1997;315:13–7.

[26] Menotti A, Kromhout D, Blackburn H, Fidanza F, Buzina R, Nissinen A. Food intake patterns and 25-year mortality from coronary heart disease: cross-cultural correlations in the Seven Countries Study. Eur J Epidemiol 1999;15:507–15.

[27] Fidanza F, Alberti A, Lanti M, Menotti A. Mediterranean Adequacy Index: correlation with 25-year mortality from coronary heart disease in the Seven Countries Study. Nutr Metab Cardiovasc Dis 2004;14:254–8.

[28] Menotti A, Puddu PE. How the Seven Countries Study did contribute to the definition and development of the Mediterranean diet concept: a 50-year journey. Nutr Metab Cardiovasc Dis 2015;25:2445–52.

[29] Kushi LH, Lenart EB, Willett WC. Health implications of Mediterranean diets in light of contemporary knowledge. 1. Plant foods and dairy products. Am J Clin Nutr 1995;61(6 Suppl.):1407S–15S.

[30] Kushi LH, Lenart EB, Willett WC. Health implications of Mediterranean diets in light of contemporary knowledge. 2. Meat, wine, fats, and oils. Am J Clin Nutr 1995;61(6 Suppl.):1416S–27S.

[31] Sofi F, Macchi C, Abbate R, Gensini GF, Casini A. Mediterranean diet and health status: an updated meta-analysis and a proposal for a literature-based adherence score. Public Health Nutr 2014;17:2769–82.

[32] Grosso G, Marventano S, Yang J, Micek A, Pajak A, Scalfi L, et al. A comprehensive meta-analysis on evidence of Mediterranean diet and cardiovascular disease: are individual components equal? Crit Rev Food Sci Nutr 2015;57(15):3218–32. https://doi.org/10.1080/10408398.2015.11070210.

[33] Liyanage T, Ninomiya T, Wang A, Neal B, Jun M, Wong MG, et al. Effects of the Mediterranean diet on cardiovascular outcomes—a systematic review and meta-analysis. PLoS One 2016;11. e0159252.

[34] Yerushalmy J, Hilleboe HE. Fat in the diet and mortality from heart disease. N Y J Med 1957;57:2343–54.

[35] Jolliffe N, Archer M. Statistical associations between international coronary heart disease death rates and certain environmental factors. J Chronic Dis 1959;9:636–52.

[36] Masironi F. Dietary factors and coronary heart disease. Bull WHO 1970;42:103–14.

[37] Stamler J, Stamler R, Shekelle RB. Regional differences in prevalence, incidence and mortality from atherosclerotic coronary heart disease. In: de Haas JH, Hemker HC, Snellen HA, editors. Ischaemic heart disease. Leiden: Leiden University Press; 1970.

[38] Armstrong BK, Mann H, Adelstein AM, Eskin F. Commodity consumption and ischaemic heart disease mortality with special reference to dietary practices. J Chronic Dis 1975;28:455–69.

[39] Artaud-Wild SM, Connor SL, Seaton G, Connor WE. Differences in coronary mortality can be explained by differences in cholesterol and saturated fat intake in 40 countries, but not in France and Finland. A paradox. Circulation 1993;88:2771–9.

[40] Davis MA, Neuhaus JM, Moritz DJ, Lein D, Barclay JD, Murphy SP. Health behaviors and survival among middle-aged and older men and women in the NHANES I Epidemiologic Follow-up Study. Prev Med 1994;23:369–76.

[41] Chiuve SE, Fung TT, Rexrode KM, Spiegelman D, Manson JE, Stampfer MJ, et al. Adherence to a low-risk, healthy lifestyle and risk of sudden cardiac death among women. JAMA 2011;306:62–9.

[42] van Dam RM, Li T, Spiegelman D, Franco OH, Hu FB. Combined impact of lifestyle factors on mortality: prospective cohort study in US women. Br Med J 2008;337:a1440.

[43] Kvaavik E, Batty GD, Ursin G, Huxley R, Gale CR. Influence of individual and combined health behaviors on total and cause-specific mortality in men and women: the United Kingdom health and lifestyle survey. Arch Intern Med 2010;170:711–8.

[44] Nechuta SJ, Shu XO, Li HL, Yang G, Xiang YB, Cai H, et al. Combined impact of lifestyle-related factors on total and cause-specific mortality among Chinese women: prospective cohort study. 2010;7(9).

[45] Odegaard AO, Koh WP, Gross MD, Yuan JM, Pereira MA. Combined lifestyle factors and cardiovascular disease mortality in Chinese men and women: the Singapore Chinese health study. Circulation 2011;124:2847–54.

[46] Yun JE, Won S, Kimm H, Jee SH. Effects of a combined lifestyle score on 10-year mortality in Korean men and women: a prospective cohort study. BMC Public Health 2012;12:673.

[47] Carlsson AC, Wändell PE, Gigante Leander K, Hellenius ML, de Faire U. Seven modifiable lifestyle factors predict reduced risk for ischemic car-diovascular disease and all-cause mortality regardless of body mass index: a cohort study. Int J Cardiol 2013;30:946–52.

[48] Behrens G, Fischer B, Kohler S, Park Y, Hollenbeck AR, Leitzmann MF. Healthy lifestyle behaviors and decreased risk of mortality in a large pro-spective study of U.S. women and men. Eur J Epidemiol 2013;28:361–72.

[49] Prinelli F, Yannakoulia M, Anastasiou CA, Adorni F, Di Santo SG, Musicco M, et al. Mediterranean diet and other lifestyle factors in relation to 20-year all-cause mortality: a cohort study in an Italian population. Br J Nutr 2015;113:1001–11.

[50] Khera AV, Emdin CA, Drake I, Natarajan P, Bick AG, Cook NR, et al. Genetic risk, adherence to a healthy lifestyle, and coronary disease. N Engl J Med 2016;375:2349–58.

[51] Loef M, Walach H. The combined effects of healthy lifestyle behaviors on all cause mortality: a systematic review and meta-analysis. Prev Med 2012;55:163–70.

[52] Menotti A, Puddu PE, Lanti M, Maiani G, Catasta G, Alberti Fidanza A. Lifestyle habits and mortality from all and specific causes of death: 40-year follow-up in the Italian Rural Areas of the Seven Countries Study. J Nutr Health Aging 2014;18:314–21.

[53] Menotti A, Puddu PE, Maiani G, Catasta G. Lifestyle behaviour and lifetime incidence of heart diseases. Int J Cardiol 2015;201:293–9.

[54] Puddu PE, Menotti A. The impact of basic lifestyle behaviour on health: how to lower the risk of coronary heart disease, other cardiovascular dis-eases, cancer and all-cause mortality. Lifestyle adaptation: a global approach. E-J Cardiol Pract 2015;13:32.

[55] Menotti A, Puddu PE, Maiani G, Catasta G. Cardiovascular and other causes of death as a function of lifestyle habits in a quasi extinct middle-aged male population. A 50-year follow-up study. Int J Cardiol 2016;201:173–8.

[56] Ruigómez A, Alonso J, Antó JM. Relationship of health behaviors to five-year mortality in an elderly cohort. Age Ageing 1995;24:113–9.

[57] Haveman-Nies A, de Groot L, Burema J, Cruz JA, Osler M, van Staveren WA. Dietary quality and lifestyle factors in relation to 10-year mortality in older Europeans: the SENECA study. Am J Epidemiol 2002;156:962–8.

[58] Knoops KT, De Groot LC, Kromhout D, Perrin AE, Moreiras-Varela O, Menotti A, et al. Mediterranean diet, lifestyle factors, and 10-year mortality in elderly European men and women: the HALE project. JAMA 2004;292:1433–9.

[59] Menotti A, Puddu PE. Lifetime prediction of coronary heart disease and heart disease of uncertain etiology in a 50-year follow-up population study. Int J Cardiol 2015;196:55–60.

[60] Puddu PE, Menotti A. Natural history of coronary heart disease and heart disease of uncertain etiology: findings from a 50-year population study. Int J Cardiol 2015;197:260–4.

[61] Puddu PE, Menotti A, Tolonen H, Nedeljkovic S, Kafatos A. Determinants of 40-year all-cause mortality in the European cohorts of the Seven Countries Study. Eur J Epidemiol 2011;26:595–608.

[62] Puddu PE, Piras P, Menotti A. Competing risks and lifetime coronary heart disease incidence during 50 years of follow-up. Int J Cardiol 2016;219:79–83.

[63] Puddu PE, Piras P, Menotti A. Lifetime competing risks between coronary heart disease mortality and other causes of death during 50 years of follow-up. Int J Cardiol 2017;228:359–63.

[64] Menotti A, Puddu PE, Lanti M, Kromhout D, Tolonen H, Parapid B, et al. Epidemiology of typical coronary heart disease versus heart disease of uncertain etiology (atypical) fatalities and their relationships with classic coronary risk factors. Int J Cardiol 2013;168:3963–7.

[65] Puddu PE, Terradura Vagnarelli O, Mancini M, Zanchetti A, Menotti A. Typical and atypical coronary heart disease deaths and their different rela-tionships with risk factors. The Gubbio residential cohort Study. Int J Cardiol 2014;173:300–4.

[66] GBD 2013 Risk Factors Collaborators. Global, regional, and national comparative risk assessment of 79 behavioural, environmental and occupa-tional, and metabolic risks or clusters of risks in 188 countries, 1990–2013: a systematic analysis for the Global Burden of Disease Study 2013. Lancet. 2015;386:2287–323.

[67] Puddu PE, Menotti A. Coronary heart disease differences in Eastern versus Western Europe: a demanding situation. Int J Cardiol 2016; 217:S60–3.

[68] World Health Organization. The World Health Report 2002. Reducing risks—promoting healthy life. Geneva: World Health Organization; 2002.

[69] Ezzati M, Van der Hoorn S, Lopez AD, Danaei G, Rodgers A, Mathers CD, et al. Comparative quantification of mortality and burden of disease at-tributable to selected risk factors. In: Lopez AD, Mathers CD, Ezzati M, Jamison DT, Murray CJL, editors. Global burden of disease and risk factors. Washington, DC: World Bank; 2006 [chapter 4].

Part II

Exercise and Physical Activity

Chapter 6

Expanding the Clinical Classification of Heart Failure: Inclusion of Cardiac Function During Exercise

Erik H. Van Iterson, Thomas P. Olson
Department of Cardiovascular Medicine, Mayo Clinic, Rochester, MN, United States

INTRODUCTION

Chronic heart failure (HF) is a multi-organ syndrome of global epidemic proportions burdening men and women of all racial, ethnic, and socioeconomic groups [1–7]. In developed countries such as the United States, it has been estimated that >5.5 million adults demonstrate signs and symptoms of HF, which is a number expected to increase to >8 million by the year 2030 [7]. Importantly, on an annual basis, HF directly accounts for >65,000 deaths while contributing to >36% of all cardiovascular-related mortalities [7]. Because the lifetime risk for developing HF is >1 in 5 for both men and women >40 years of age with an incidence >10 per 1000 in adults >65 years of age [7], there is a high-priority need to better understand the underpinning integrated pathophysiology of this phenotype in clinical and research settings where signs and symptoms are commensurate with the true severity of this syndrome.

Although accumulating evidence to date suggests care for HF patients has improved over the past 30 years, controversial efficacy of advanced therapeutic technologies (e.g., cardiac resynchronization therapy) accompanied by broad use of variably effective traditional *class I* (Table 1) indicated therapies has placed a major strain on economic and health-care systems [1,5–13]. Recent estimates suggest HF accounts for >$30.7 billion annually equating to >68% of direct medical costs, which is purported to grow to >$69 billion by 2030 [7]. A large component of these costs is attributed to marked rates of HF-related (re)hospitalizations [1,7–9]. While in certain instances declines in HF mortality have been credited to effectiveness of routine implementation of class I therapies supported by level A standard of evidence, particularly in patients with reduced ejection fraction, others have demonstrated that HF-related mortality and hospitalization rates do not invariably decrease using traditional approaches in either reduced or preserved ejection fraction patients [5–8,14–18]. Moreover, although now hypothesized to derive from separate pathophysiology, there are no proven class I indicated therapies supported by level A evidence specifically for the management of HF patients with preserved ejection fraction [5–8,14–18]. Because of the high level of "nonresponders" to which standard clinical guidelines do not appear adequately applicable [2,3,5–7,10,17,19–22], we should strongly consider multifaceted alternative approaches for phenotyping patients in an effort to advance our understanding and ability to objectively quantify HF severity in clinical and research settings.

Consistent clinical and research observations suggest HF signs and symptoms are worsened during physical activity (e.g., exercise stress) while resolving to more favorable levels shortly after resting. In this context, many patients continue to report persistent exertional dyspnea and fatigue (i.e., exercise intolerance) despite receiving "optimal" therapy [3,5,6,16–18,22–41]. By contrast, a separate line of cardio-centric evidence emphasizes the importance of direct links between impaired exercise cardiac pumping capability and increased likelihood of HF-related hospitalizations and mortality [37–39,41–61]. However, there is paucity of evidence to suggest similar prognostic value associated with traditional resting measures of cardiac, circulatory, and pulmonary function for the clinical classification of HF [34–37,40,52,55,57–60,62–66]. Therefore, because exercise intolerance is a hallmark feature universal to HF, and debilitating signs and symptoms are most severe even during modest levels of physical activity (e.g., activities of daily living), this chapter is aimed at highlighting the appreciable role that both routine and advanced measures of cardiac pumping capability during exercise play in delivering high resolution and readily interpretable integrative physiological data used in the clinical classification of patients with HF.

Lifestyle in Heart Health and Disease. https://doi.org/10.1016/B978-0-12-811279-3.00006-9

TABLE 1 Traditional Heart Failure Classification Models

Severity	NYHA Functional Class	ACC/AHA Stages of Heart Failure	Weber-Janicki Class (Peak $\dot{V}O_2$ mL/kg/min)	Quality	Recommendation of Classes	Evidence Levels
	I — No limitation of physical activity. Ordinary physical activity does not cause undue fatigue, palpitation, or dyspnea	A — At high risk for developing heart failure. No identified structural or functional abnormality; no signs or symptoms	A — >20		I — Evidence and/or general agreement that a given treatment or procedure is beneficial, useful, and effective	A — Data derived from multiple randomized clinical trials or meta-analyses
	II — Slight limitation of physical activity. Comfortable at rest but ordinary physical activity results in fatigue, palpitation, or dyspnea	B — Developed structural heart disease that is strongly associated with the development of heart failure but without signs or symptoms	B — 16–20		IIa — Weight of evidence/opinion is in favor of usefulness/efficacy	B — Data derived from a single randomized clinical trial or large nonrandomized studies
	III — Marked limitation of physical activity. Comfortable at rest, but less than ordinary activity results in fatigue, palpitation, or dyspnea	C — Symptomatic heart failure associated with underlying structural heart disease	C — 10–16		IIb — Usefulness/efficacy is less well established by evidence/opinion	C — Consensus of opinion of the experts and/or small studies, retrospective studies, and registries
	IV — Unable to carry on any physical activity without discomfort. Symptoms at rest. If any physical activity is undertaken, discomfort is increased	D — Advanced structural heart disease and marked symptoms of heart failure at rest despite maximal medical therapy	D — 6–10		III — Evidence or general agreement that the given treatment or procedure is not useful/effective and in some cases may be harmful	
			E — <6			

MOVING BEYOND THE CLINICAL CLASSIFICATION OF HEART FAILURE AT REST: DOES HEART FAILURE WITH REDUCED OR PRESERVED EJECTION FRACTION MATTER?

Complicating the pathophysiological understanding of HF and precise role that abnormal exercise cardiac pumping capability plays in the clinical classification of patients are separate lines of study aimed at elucidating the mechanistic basis of HF with reduced (HFrEF) versus preserved (HFpEF) ejection fraction. While a detailed discussion of the exact clinical and pathophysiological factors that are suggested to demarcate the forms of HF is beyond the scope this chapter, it is noteworthy that exercise intolerance coupled with poor prognosis commonly aligns with the magnitude of impaired exercise cardiac pumping capability in all patients [5,6,16,17,37–39,41–54,57–61,67].

With acknowledgment that exercise intolerance has been defined using a spectrum of cardiac, circulatory, and pulmonary indexes across various clinical and research settings, it is worthwhile reiterating that classical clinical indicators (e.g., resting left ventricular ejection fraction (LVEF) percentage) do not traditionally correlate with severity of exercise intolerance in either HF group nor do those metrics invariably translate to prognosis [5,6,16,17,34–41,52,55,58,59,62–66,68,69]. By contrast, increased morbidity and mortality consistently and strongly relate with severity of exercise cardiac hemodynamic dysfunction in HF [18,37–39,41–57,60,67,70].

Traditional Classifying Models

First proposed in 1928, the construct of the New York Heart Association (NYHA) functional classification system has been used liberally and independently or with consideration from other outcome variables as the basis to assess syndrome severity and prognosis in HF [2,71–75].

Heart failure classification using NYHA criteria (increasing severity from I to IV, Table 1) is broadly based on subjective symptoms encountered at rest and during physical exertion [2,3,75]. Although NYHA functional class status is commonly typified within the context of activities of daily living, it is indeed generalizable as more than merely a marker of gross physiological function since it renders stratifications that are based on nonspecific or not well-identifiable mechanistic origins, which are susceptible to both physiological and nonphysiological factors (e.g., psychologic, social, and environmental). Therefore, interpretation of changes in NYHA class as a primary means of conceptualizing physiological-based causes of HF can in many instances lack real-world precision and clarity as this classifying system can describe a number of changes demonstrating multicollinearity such as functional capacity, symptoms, and/or quality of life [5,6,17,18,71–77].

Nevertheless, despite the continued use of the NYHA class system concurrent with recommended stage guidelines from the American College of Cardiology (ACC) and American Heart Association (AHA) joint taskforce for the clinical assessment of HF (based on structure and function of the myocardium, with increasing HF severity from stages A–D, Table 1) [2,6,71–76], there is paucity of evidence to suggest these systems independently, or together, consistently and reliably align with both prognosis and magnitude of objective signs of exercise intolerance [5–7,17,18,57,78,79]. This lack of robust association between classifying systems and objective assessment of exercise intolerance is particularly impactful to the clinical decision-making process as there are accumulating discussions directly attributing reduced exercise cardiac pumping capability as a major predictor of HF prognosis [37–39,41–56,59,60,67,70].

While acknowledging that NYHA class and ACC/AHA stage systems remain important components in the clinical classification of HF [2,6,71–76], related studies have used cardiopulmonary exercise testing (CPET) to develop a valuable adjunct classifying model. For example, the Weber-Janicki classification system (Table 1) is based on basic noninvasive exercise gas exchange and ventilatory responses and can be used to estimate integrative cardiac, pulmonary, and circulatory function determined via thresholds of pulmonary oxygen uptake ($\dot{V}O_2$) (increasing HF severity from stage A to D, Table 1) [36,37,80]. This system emphasizes the importance of an objective exercise-based clinical classifying system for HF patients [36,37,80]. However, because stratification thresholds used in the Weber-Janicki class system are in many respects based on physiological assumptions of direct relationships between adjustments in gas exchange and respiration with those of cardiac and circulatory hemodynamics, this system may not be without its own limitations as those physiological relationships are not mutually exclusive across patients.

What may be particularly problematic in relying on exercise $\dot{V}O_2$ as an independent stratifying index in determining clinical status is the understanding that, in addition to its assumed linear relationship with cardiac output ($\dot{Q}$), $\dot{V}O_2$ may demonstrate variable degrees of sensitivity to noncardiac and nonphysiological factors that may not be an intrinsic consequence of HF (e.g., aging, motivation, and sex) [81–85,36]. Examples of the lack of sensitivity exercise $\dot{V}O_2$ has in determining HF severity include notable work from groups such as Wilson et al. [68,69,81,86], Mancini et al. [60], and others [35,87,88]. Those related lines of evidence demonstrate peak $\dot{V}O_2$ responses do not unequivocally predict survival or other HF-related events beyond diagnostic information provided by exercise cardiac hemodynamics. Peak $\dot{V}O_2$ also does not routinely correlate with severity of cardiac dysfunction linked to exercise performance. While exact reasons for the inhomogeneous association between peak $\dot{V}O_2$ and degree of impaired cardiac pumping capability associated with exercise intolerance remains unclear, it has been proposed in both HFrEF and HFpEF that sensitivity of $\dot{V}O_2$ to pathological changes in skeletal muscle structure and function related to deconditioning decreased oxidative capacity, and fiber-type shifts favoring glycolytic pathways are likely candidate mechanisms [35,44,45,57,60,68,69,83,81,86–97].

Thus, to optimize the interpretive value of assessing cardiac pumping capability for the clinical classification of HF, it is desirable to implement "gold" standard testing methods such as invasive direct Fick or indicator-dilution techniques. Alternatively, because of greater ease of use and clinical and research availability compared with invasive methods, other indirect but robust breathing techniques involving blood-soluble inert gas transfer across the alveolar-capillary membrane barrier, such as those involving acetylene (C_2H_2), nitrous oxide (N_2O), or carbon dioxide (CO_2) gases, should be considered as efficacious surrogates to assess cardiac pumping capability during exercise in HF [98–113].

MEASUREMENT TECHNIQUES FOR CARDIAC PUMPING CAPABILITY DURING EXERCISE IN HEART FAILURE

In order to fully appreciate the clinical value of objectively quantifying exercise cardiac pumping capability, we briefly present a discussion on the science and real-world applicability of well-studied invasive and noninvasive techniques for directly or indirectly quantifying cardiac hemodynamics during exercise for the clinical classification of patients with HF.

Invasive Approaches

The first proper model used for the measurement of $\dot{Q}$ in humans was articulated by the renowned physiologist, mathematician, and physicist Adolph Fick, from which the terminology "Fick equation" or "direct Fick model" was coined. For this "gold" standard measurement of $\dot{Q}$ to work, one must be able to quantify (1) the change in concentration of a known substance (e.g., O_2) from the time it leaves the pump (e.g., the heart) and circulates through a closed system (e.g., systemic vasculature) to the time it returns to the pump and (2) the relative change in concentration of that substance deriving from differences in uptake (e.g., $\dot{V}O_2$) versus extraction (e.g., arterial-mixed venous O_2 difference = A-V O_2) as a function of time [100,114–120]. Therefore, calculation of $\dot{Q}$ relies on similarly accurate measurements $\dot{V}O_2$ and A-V O_2 during a quasi-steady-state condition.

The direct Fick technique used for cardiac hemodynamic measurement requires invasive placement of two catheters in order to quantify A-V O_2. Ideally, one catheter is used to attain a sample of true mixed venous O_2 content from a pulmonary artery, whereas the other is used at a peripheral arterial location (e.g., femoral, brachial, or radial) for measurement of arterial O_2 content. By contrast, to quantify $\dot{V}O_2$, open-circuit breath-by-breath spirometry systems must be utilized. Thus, in assuming that these technical considerations are met and with modest variability occurring in A-V O_2, direct Fick determination of $\dot{Q}$ can be reliably and accurately measured with consistency between 90% and 95% of the actual value [113,119–121].

In contrast to the direct Fick technique, but while still based on the physiological tenets of the Fick equation and also involving invasive procedures, indicator-dilution techniques involve the introduction of an indicator (e.g., indocyanine dye or cold saline) at a point in the circulation (e.g., right atrium) where uniform mixing of the indicator in blood can occur. Assuming there is complete mixing of the indictor solution within blood, analysis of the flow of blood occurring at a point downstream can be used to obtain a time dilution curve. In using this indicator time dilution curve with minimal mathematical modeling, $\dot{Q}$ can be determined by calculating the integrated area under the curve, which is inversely proportional to the average volume flow of blood such as [112,122–127]

$$\text{Dye dilution} = (I \times 60) / A$$

where I is the total amount of indicator injected (mg), 60 is the conversion factor (s/min), and A is the area under the dye-dilution curve (mg s/L), or by using one of many $\dot{Q}$ via thermodilution calculation approaches such as

$$\text{Thermodilution} = \left((\text{TB} - \text{TI})(10)(60)(1.08)(0.825) \right) / \left(\int_0^t Tb(t)\,dt \right)$$

where TB and TI are the initial temperature of the blood and injectate, respectively (°C); 10 is injectate volume (mL); t is time in seconds; 60 converts mL/s to mL/min; 1.08 is the ratio of the products of specific heats and gravities of 5% dextrose in water and blood; 0.825 is constant factor used to correct injectate warming; and $\int_0^t Tbdt$ is the concentration time integral [116,127–133].

The efficacy of the dye-dilution technique for measurement of $\dot{Q}$ has been tested against the direct Fick approach while demonstrating a coefficient of variation of 5%–10% accompanied by a strong correlation coefficient ($r \sim 0.95$ at rest and during exercise) [122–125,127,134]. By contrast, others suggest the thermodilution technique in relation to direct Fick is less robust versus its dye-dilution counterpart. For example, thermodilution may overestimate $\dot{Q}$ upwards of 39%–43% during rest and exercise compared with direct Fick measurements [112,126], which in such cases may be due to the presence of anatomical factors such as mitral and/or tricuspid valve prolapse with regurgitation leading to improper circulation of indicator. However, others demonstrate thermodilution to be a suitable technique for measuring $\dot{Q}$ at rest and/or during exercise in patient populations [104,116,127,132–136], whereas others report dye dilution does not demonstrate strong agreement with direct Fick [122]. Nevertheless, whether direct Fick, dye dilution, or thermodilution techniques used for the measurement of $\dot{Q}$ perfectly align across test and patient settings, it is clinically relevant and of great practical importance that these cardiac hemodynamic assessment methods require highly invasive procedures that increase patient risk (e.g., pneumothorax, infection, arrhythmia, and bleeding), are resource-dependent (e.g., related to the need for expert personnel and costs), values may be influenced by patient factors (e.g., mitral valve or pericardial disease, on anticoagulation therapy, etc.), and orders for these techniques are not routinely available and always necessary when considering involved elements in clinical and research settings [130,137,138]. Therefore, despite being considered the standard for measuring cardiac hemodynamics, there are alternative noninvasive methods proposed as effective options for assessing stroke volume (SV) and $\dot{Q}$, particularly during exercise in patient populations.

Noninvasive Approaches

While keeping in mind a primary intent of this chapter is to highlight the clinical value and utility of being able to routinely include measurements of cardiac hemodynamics during CPET in HF patients, this condensed discussion will not cover the broad spectrum of all noninvasive techniques that plausibly could be used to assess exercise cardiac hemodynamics. For example, we will not comment on the usefulness of cardiac magnetic resonance imaging (cMRI) as we are not aware of cases where CPET would routinely accompany this technique, nor will we overview methods involving arterial pulse wave contour modeling or impedance cardiography as the literature on these approaches appears less decided on the validity and reliability of these methods even when applied during resting states [139–145]. By contrast, this discussion will be focused on the strengths and limitations of three specific approaches that have garnered a considerable amount of scientific study to date while also demonstrating translation to real-world institutional settings where CPET in HF patients is performed. The following approaches used for noninvasive assessment of exercise cardiac hemodynamics will be discussed in no particular order of importance: echocardiography, inert gas uptake, and oxygen pulse.

Doppler Echocardiography

The most well-recognized noninvasive technique for evaluating cardiac hemodynamics is Doppler echocardiography (ECHO), which is currently used across a spectrum of institutional settings at moderate cost and negligible risk to patients [146,147]. Measurement of SV and $\dot{Q}$ using various ECHO approaches has been shown to be reproducible during exercise, demonstrating in some instances a coefficient of variation of 8.5 and 8.1% for SV and $\dot{Q}$, respectively, in healthy individuals [148]. Correlation coefficients > 0.90 and low inter- and intraobserver bias have also been reported between SV and $\dot{Q}$ assessed using ECHO and invasive techniques (direct Fick or thermodilution) in HF/cardiac patients at rest [149–155]. Despite those and other encouraging observations of reproducibility and validation studies across the health spectrum [148,153–160], ECHO can be influenced by body morphology (e.g., obesity related to chest wall size and adiposity), selection of image acquisition protocol, sonographer skill level, and reader interpretation. Importantly, those and other potential confounding influences are enhanced when ECHO is performed during exercise. Thus, routine efficacy of ECHO, particularly for exercise cardiac hemodynamics in the clinical setting, is suggested to require an expert level of skill by both sonographer and reader [153,159,161–166].

Inert Gas Uptake

Pulmonary inert gas uptake for the measurement of cardiac hemodynamics has been observed to reliably and accurately quantify cardiac hemodynamics consistent with invasive approaches at rest and exercise across a spectrum of populations with or without concurrent cardiac and/or pulmonary disease [48,89,98–107,109,115,167–173]. For example, while performing this technique using the C_2H_2 rebreathe approach at rest, Hoeper et al. [104] demonstrated acceptable limits of agreement for $\dot{Q}$ equal to -0.23 ± 1.14 L/min compared with direct Fick measurements in patients with pulmonary hypertension. Johnson et al. [100] and Liu et al. [109] demonstrated in healthy adults during exercise that $\dot{Q}$ measured via C_2H_2 open-circuit washin ($R^2 > 0.88$) or rebreathe ($r = 0.88$) techniques, respectively, strongly relate with $\dot{Q}$ via direct Fick.

Efficacious application of the inert gas uptake technique is in large-part based on the assumption of rapidly occurring homogeneous inert gas mixing within lungs and the principle of free diffusion of inert gas demonstrating high blood solubility (i.e., Bunsen solubility coefficient in blood) across the alveolar-capillary membrane barrier. Occurrence of these events, in theory, leads to inert gas uptake by passing pulmonary arterial blood commensurate with the rate at which blood is expelled from the LV [48,89,98–107,109,115,167–173]. By meeting these assumptions, available to the clinician or scientist when utilizing the inert gas uptake technique are unique gas compounds, which include the frequently used perfusion sensitive C_2H_2 or N_2O gases and the lesser used CO_2 gas.

With respect to C_2H_2, N_2O, and CO_2 gases, although a high Bunsen solubility coefficient in blood is a necessary chemical property for inert gas breathing to be used for assessment of cardiac hemodynamics, it is equally essential that inert gases do not demonstrate strong affinity or "competitive inhibitor" properties for O_2 binding sites on iron-rich hemoglobin (Hb) (i.e., high-affinity binding to Hb would increase the likelihood of inert gas recirculation, thereby falsely lessoning $\dot{Q}$), nor is it ideal for inert gases to be native physiological gases whereby post hoc breathing corrections are deemed necessary in order to calculate $\dot{Q}$ (e.g., necessary for CO_2 breathing techniques) [103,106,167–170,174].

Despite the proposed strengths, which include accuracy, low user dependence, and reliability of inert gas breathing for cardiac hemodynamic measurement, it has been suggested that because a central principle supporting the effectiveness of inert gas breathing is the supposition that C_2H_2, N_2O, or CO_2 gases are not diffusion limited, this technique may demonstrate variability in patients with overt diffusion limitations such as might be the case in interstitial lung disease or advanced pulmonary hypertension [29,98,101,104,175,176]. While there is merit for this concern, it is noted that in the absence of severe ventilation-perfusion ($\dot{V}_A/\dot{Q}_c$) inhomogeneity not due to pathology directly affecting the alveolar-capillary membrane barrier (e.g., R to L shunt and chronic obstructive pulmonary disease) [98,105,177], it is reasonable to suggest that methods used to evaluate the integrity of the alveolar-capillary membrane, such as lung diffusing capacity for carbon monoxide (DL_{CO}), may not accurately capture the true ability of a markedly stronger inert gas (Bunsen solubility coefficient in blood of CO is 0.0189 compared with 0.740 [167,170,174,178]) to diffuse across this barrier while demonstrating robust sensitivity to pulmonary capillary blood flow.

In this context, we highlight once more the compelling observations of Hoeper et al. [104] in patients with pulmonary hypertension whereby measurements of $\dot{Q}$ via C_2H_2 rebreathe aligned closely with those taken with direct Fick (despite being measured at rest where there is minimal possibility for collateral recruitment of long time-constant lung regions, which in contrast to what might be expected to occur during exercise). Likewise, work from the groups of Saur et al. [98,105] or McCLure et al. [173] demonstrated the presence of pulmonary or cardiac disease across a spectrum of patient cases did not influence measures of $\dot{Q}$ via N_2O (lower Bunsen solubility coefficient in blood = 0.405 versus C_2H_2 [170]) rebreathe compared with cMRI or thermodilution. Additional tests using the moderate blood-soluble CO_2 gas (~ 0.470, Bunsen solubility coefficient in blood lying between that of C_2H_2 and N_2O [179]) via rebreathe maneuver similarly suggest inert gas breathing compared with invasive techniques (e.g., dye dilution) may be used to assess cardiac hemodynamics at rest and during exercise in patients with HF [103].

Oxygen Pulse

Last of the well-studied methods for assessing exercise cardiac hemodynamics is the indirect noninvasive index termed oxygen pulse (O_2 pulse). Proposed as a direct surrogate of SV, O_2 pulse, the quotient of $\dot{V}O_2$ and heart rate (HR), should theoretically hold true based on meeting assumptions of the Fick equation [180,181]. In particular, it is an essential requirement that invariability of A-V O_2 be demonstrated, which permits independently occurring linearity between changes in $\dot{Q}$ and $\dot{V}O_2$ assuming a quasi-steady state of exercise. Thus, given a normal chronotropic response to exercise, the uptake of O_2 on a per beat basis can be computed, which is suggested to be commensurate with the volume of blood leaving the heart on a per beat basis and, hence, SV [180,181].

While O_2 pulse has been studied extensively, an eloquent description of the physiological rationale for the efficacy of O_2 pulse as a surrogate for SV is provided by Whipp et al. [181] during studies of healthy adults during CPET. Drawing on the fundamental relationships described in the Fick equation, Whipp et al. [181] demonstrated a correlation coefficient of 0.85 and coefficient of determination of 0.73 between O_2 pulse and SV measured via direct Fick. Similarly, in seeking to extend the work of Whipp et al. [181], Taylor et al. [180] demonstrated robust relationships ($r = 0.91$ and $P < 0.01$) between O_2 pulse and SV measured via direct Fick at peak exercise in HF patients with or without secondary pulmonary hypertension.

However, what may be particularly intriguing with regard to the comparable observations between Whipp et al. [181] and Taylor et al. [180], which is a potential limitation of O_2 pulse as a surrogate for SV in HF, is that β-blocking agents are a common HF therapy. This is relevant since β-blocking agents attenuate the rise in HR, which may artificially inflate O_2 pulse beyond SV. Nevertheless, although not being reported, visually estimated back extrapolation of the model goodness-of-fit line illustrated in Taylor et al. [180] for the relationship between O_2 pulse and SV does not suggest a y-intercept indicating a clear overexaggeration of O_2 pulse relative to SV, nor does the reported relationship appear as downward sloping curvilinearity with increasing O_2 pulse values.

Finally, despite appearing absent and/or noninfluential in the observations of Taylor et al. [180], it should also be acknowledged that in addition to the potential influences of β-blocker therapy in limiting the interpretation of O_2 pulse as a marker of SV in HF, limitations discussed above that are associated with either direct Fick or inert gas breathing technique have the potential of influencing the efficacy of O_2 pulse during exercise (e.g., A-V O_2 variability, assessed at a nonsteady state, and severe $\dot{V}_A/\dot{Q}_c$ mismatch).

HEART FAILURE CLASSIFICATION: BASIC AND ADVANCED MEASURES OF CARDIAC PUMPING CAPABILITY DURING EXERCISE

Although the two most common reasons that a HF patient presents to a specialist, clinic, or emergency department include unexplained *exertional* dyspnea and fatigue, paradoxically, it is not a standard practice for diagnostic tests of cardiac and circulatory function to be performed beyond resting evaluation. Despite being routinely evaluated at rest, LVEF (LV blood volume ejected ÷ LV end-diastolic volume) demonstrates no clear relationship with exercise capacity, exercise cardiac and circulatory function, or symptom severity in HF [35,45,50,52,58–60,62,63,67,70,182–186]. As it might be implied, this classical patient evaluation approach is limiting to the clinician in their efforts to identify a specific mechanistic cause of dyspnea based on signs and symptoms as neither peak nor reserve cardiac pumping capability can be quantified using resting measurements. This is a critical limitation of traditional clinical patient evaluation models because it is well recognized that HF signs and symptoms are most severe and recognizable during physical activity. Therefore, peak and reserve cardiac pumping capability measurements are suggested to play key roles in phenotyping the clinical severity of patients [18,44–49,57,59,60,67,187–190].

Consequently, but while in contrast to routine resting clinical evaluation models, it has been proposed that both direct (e.g., direct Fick) and indirect (e.g., inert gas breathing) methods of measuring exercise SV and $\dot{Q}$ more consistently linearize with exercise capacity and prognosis in HF [34,44–48,58,60,66,83,89,190,62,191,192,52,193,194]. Moreover, and in a related manner that strengthens this line of evidence, the topic of quantifying cardiac hemodynamics while directly accounting for the circulatory hemodynamic impact of cardiac function has gained traction. Quantifying the "hydraulic power" or "hydraulic work" capability of the heart (i.e., the product of flow or volume and pressure), described in simplest terms as cardiac power (CP = $\dot{Q}$ · blood pressure) or stroke work (SW = SV · blood pressure), reinforces the argument for the significance of assessing advanced cardiac hemodynamics during exercise as a robust model of cardiac pumping capability for the clinical classification of HF patients [44–49,57,59,60,67,187–190].

The Traditional Cardiac Pumping Capability Model in Heart Failure

Assessment of the pumping or working capability of the heart has traditionally been accomplished using measurements of SV and $\dot{Q}$. These indexes have historically proven useful for describing cardiac health because SV and $\dot{Q}$ are directly related to myocardial contractility across the spectrum of metabolic demand and/or disease severity [34,44–48,58,60,66,83,89,190,62,191,192,52,193,194]. While acknowledging that neither end-systolic volume nor LVEF can be equivalent to 0L or 100%, respectively, theoretically, the heart can only expel a maximal volume of blood equal to that, which is returned from the circulation. As such, SV and $\dot{Q}$ have concurrently been used as gross measures of cardiovascular function to describe integrated cardiac and circulatory hemodynamics [100,195–201].

Stroke volume is the amount of blood ejected from the right ventricle (RV) or LV toward the pulmonary or systemic circulation with each cardiac cycle, respectively [100,195–201]. The volume of blood is the absolute difference between ventricular filling (end-diastolic volume) and ventricular emptying (end-systolic volume), with the volume of blood entering the pulmonary and systemic circulations ideally demonstrating equivalence in the absence of shunting [100,195–201]. In contrast to SV, $\dot{Q}$ is the pumping capability of the heart every minute calculated as the product of HR and SV or using the Fick equation previously described as the quotient of $\dot{V}O_2$ and A-V O_2 [100]. Thus, assuming there is limited variation in A-V O_2, the maximal value of $\dot{Q}$ describes the integrated functional capacity of the cardiac system while also representing metabolic rate [181,202–204].

In this context, there are indeed important roles that abnormal changes in SV and $\dot{Q}$ play in contributing to exercise limitations and, hence, prognosis in HF. For example, as illustrated in Fig. 1, within the same study of carefully identified patients with HFrEF or HFpEF (*unpublished data from our laboratory; met respective European Society of Cardiology criteria for diagnosis of HFrEF or HFpEF* [2]) during CPET with SV index (SV_I) and cardiac index ($\dot{Q}_I$) measured via direct Fick (absolute responses presented in Table 2), these metrics lie in parallel with peak $\dot{V}O_2$ while similarly predicting exercise capacity across both patient groups. By contrast, when setting resting LVEF as an independent predictor of peak $\dot{V}O_2$, consistent with observations and interpretations of others [35,45,50,52,58–60,62,63,67,70,182–186], Fig. 2A illustrates that there is no discernable relationship between resting LVEF and exercise capacity across HF patients. Alternatively, in the same cohort, while using SV_I or $\dot{Q}_I$ as independent predictors of $\dot{V}O_2$, consistent with Fig. 1, these basic cardiac pumping responses at peak exercise indeed demonstrate moderate-to-large predictive strength for exercise capacity across the LVEF spectrum (Fig. 2). Moreover, and perhaps equally intriguing, it would appear that variance related to inotropic influences on cardiac hemodynamics accounts for the majority of the strength of relationships between peak SV_I or $\dot{Q}_I$ with peak $\dot{V}O_2$ since peak HR (Fig. 2B) did not relate with peak $\dot{V}O_2$ to any extent across patients. Finally, to further support the observations of others and our point concerning the variable effectiveness of utilizing NYHA class status as a marker of functional capacity and prognosis [205–209], Fig. 2 consistently demonstrates a scattering of HF patient NYHA class across the full range of resting LVEF and peak $\dot{V}O_2$, SV_I, $\dot{Q}_I$, and HR suggesting little translation of NYHA class to objective physiological responses during exercise.

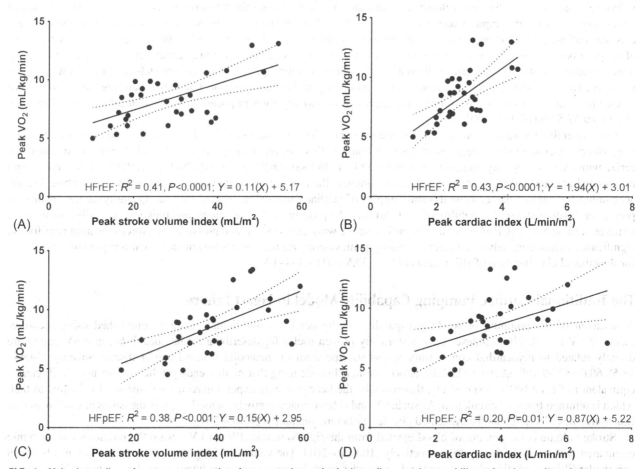

FIG. 1 Univariate ordinary least-squares regressions between peak exercise basic cardiac pumping capability and peak exercise pulmonary oxygen uptake ($\dot{V}O_2$). *Solid line* is the model goodness-of-fit line for the regression, whereas the *upper and lower dotted lines* are 95% confidence limits of the model goodness-of-fit line. (A and B) Heart failure patients with reduced ejection fraction (HFrEF). $N=32$; NYHA class II/III, 13/19, respectively; mean $\pm$ SD, LVEF, $22 \pm 7\%$; age, 55 ± 10 years; height, 174 ± 8 cm; BMI, 28 ± 4 kg/m^2; percent of predicted peak $\dot{V}O_2$, $31 \pm 7\%$. (C and D) Heart failure patients with preserved ejection fraction (HFpEF). $N=30$; NYHA class II/III, 7/23, respectively; mean $\pm$ SD, LVEF, $60 \pm 8\%$; age, 70 ± 10 years; height, 170 ± 10 cm; BMI, 34 ± 7 kg/m^2; percent of predicted peak $\dot{V}O_2$, $48 \pm 14\%$. All patients met respective European Society of Cardiology criteria for diagnosis of HFrEF or HFpEF [2].

TABLE 2 Rest, Peak Exercise, and Change Responses

	All	HFrEF	HFpEF	P-value
	N = 62	N = 32	N = 30	
Rest				
$\dot{V}O_2$, mL/kg/min	2.7 ± 0.6	2.7 ± 0.6	2.6 ± 0.6	.95
HR, bpm	68 ± 11	69 ± 69	67 ± 11	.49
MAP, mm Hg	88 ± 12	82 ± 11	94 ± 11	<.01
SVR_I, dynes s/cm^5/m^2	3139 ± 950	3250 ± 1082	3035 ± 813	.97
SV_I, mL/min	32 ± 8	28 ± 8	35 ± 7	.14
$\dot{Q}_I$, L/min/m^2	2.1 ± 0.5	1.9 ± 0.4	2.3 ± 0.5	.01
SW_I, J/m^2	0.30 ± 0.11	0.25 ± 0.11	0.36 ± 0.08	<.01
CP_I, W/m^2	0.38 ± 0.12	0.32 ± 0.10	0.43 ± 0.11	<.01
AVO_2, mL/dL	5.61 ± 1.24	6.11 ± 1.20	5.1 ± 1.1	.32
DO_2, mL/min/m^2	328 ± 77	313 ± 87	344 ± 61	.06
Peak exercise				
$\dot{V}O_2$, mL/kg/min	8.4 ± 2.3*	8.3 ± 2.2*	8.5 ± 2.4*	.87
HR, bpm	101 ± 19*	104 ± 22*	99 ± 15*	.95
MAP, mm Hg	101 ± 18*	92 ± 17*	111 ± 14*	<.01
SVR_I, dynes s/cm^5/m^2	2095 ± 590*	2188 ± 674*	1998 ± 480*	.78
SV_I, mL/min	33 ± 12	28 ± 11	38 ± 10	.04
$\dot{Q}_I$, L/min/m^2	3.3 ± 1.2*	2.8 ± 0.7*	3.8 ± 1.3*	<.01
SW_I, J/m^2	0.32 ± 0.16	0.24 ± 0.13	0.40 ± 0.14	<.01
CP_I, W/m^2	0.62 ± 0.32*	0.46 ± 0.24*	0.78 ± 0.32*	<.01
AVO_2, mL/dL	11.8 ± 2.8*	12.9 ± 2.31*	10.5 ± 2.7*	.02
DO_2, mL/min/m^2	537 ± 181*	474 ± 132*	604 ± 202*	<.01
Percent change				
$\dot{V}O_2$	228 ± 114	211 ± 79	246 ± 141	.25
HR	52 ± 31	53 ± 38	50 ± 20	.41
MAP	15 ± 14	11 ± 13	19 ± 14	.13
SVR_I	−30 ± 17	−28 ± 16	−31 ± 18	.47
SV_I	6 ± 34	1 ± 32	11 ± 34	.30
$\dot{Q}_I$	54 ± 39	44 ± 26	65 ± 47	.13
SW_I	5 ± 36	−1 ± 33	12 ± 38	.40
CP_I	62 ± 56	46 ± 43	78 ± 62	.02
AVO_2	113 ± 44	115 ± 33	112 ± 55	.44
DO_2	64 ± 41	53 ± 29	75 ± 49	.18

Data are mean ± SD. Peak exercise workload = 39 ± 11 versus 37 ± 13 W, HFrEF and HFpEF, respectively; P = .26. Relative percent (%) change, rest to peak exercise. Oxygen uptake ($\dot{V}O_2$); Heart rate (HR); Mean arterial pressure (MAP); Systemic vascular resistance index (SVR_I); Stroke volume index (SV_I); Cardiac index (Q_I); Stroke work index (SW_I); Cardiac power index (CP_I); Arterial-mixed venous O_2 content difference (AVO_2); O_2 delivery index (DO_2). Between and/ or within group differences for rest and peak exercise were tested using mixed models repeated measures ANCOVA tests, setting group-by-time interactions in models, which were significant for all dependent variables at an alpha set at 0.05. Whereas age and sex were set as covariates. Table P-values reflect post-hoc Tukey-Kramer corrections. In contrast, between group differences for %change were tested using ANCOVA models setting group as the main effect, and setting age and sex as covariates. Overall models were significant for MAP and CP_I with table P-values reflecting main effect differences. P-values in the table for the remaining dependent variables reflect that of the overall model fit.

*Different compared with rest, P < 0.05.

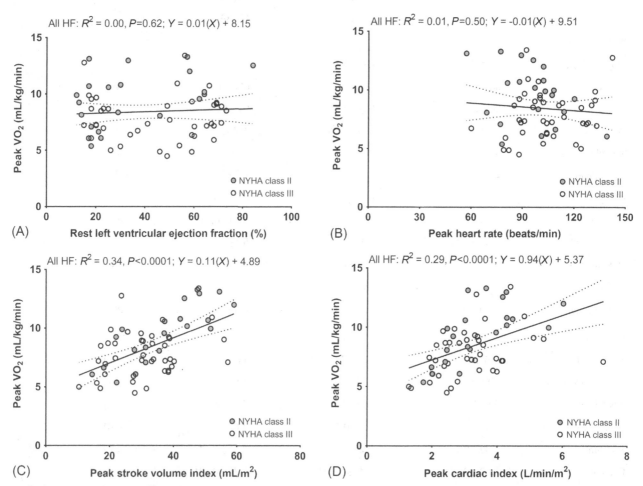

FIG. 2   Univariate ordinary least-squares regressions between resting cardiac function, peak exercise heart rate, or peak exercise basic cardiac pumping capability and peak exercise pulmonary oxygen uptake ($\dot{V}O_2$) across heart failure patients with reduced (HFrEF) or preserved ejection fraction (HFpEF) ($N=62$; NYHA class II/III, 20/42, respectively; mean $\pm$ SD, LVEF, $42 \pm 22\%$; age, 63 ± 13 years; height, 172 ± 9 cm; BMI, 31 ± 7 kg/m^2; percent of predicted peak $\dot{V}O_2$, $40 \pm 15\%$). *Solid line* is the model goodness-of-fit line for the regression, whereas the *upper and lower dotted lines* are 95% confidence limits of the model goodness-of-fit line.

The Advanced Cardiac Pumping Capability Model in Heart Failure

Cardiac power and SW effectively describe the hydraulic pumping capability of the heart as a function of time or a per beat basis, respectively. In applying these advanced models to describe cardiac hemodynamics, it is noteworthy that changes in systemic arterial pressure are accounted for in series with the magnitude of flow or volume exiting the heart, which can be directly interpreted as the perfusion pressure-generating capability of the heart. Therefore, because HF patients are predominately afterload dependent while demonstrating augmented systemic vascular resistance, the ability to articulate the relationship between SV or $\dot{Q}$ and adjustments in systemic arterial pressure is of great clinical relevance. This is particularly important in HF as this syndrome is further associated with peripheral perfusion impairments, which is a known contributor to exercise intolerance [69,86,210–214,191].

Accounting for CP using breath-by-breath open-circuit spirometry for $\dot{V}O_2$ accompanied by invasive techniques including right heart and arterial catheters to attain right atrial pressure (RAP), mean arterial pressure (MAP), and $\dot{Q}$ (via direct Fick), CP can be calculated in units of watts (W) as $\left(\text{MAP} - \text{RAP}\right) \times \dot{Q} \times 2.22 \times 10^{-3}$ [51]. Taking an additional measurement of mean pulmonary capillary wedge pressure (PCWP), SW can be calculated in units of joules (J) as, $\left(\text{MAP} - \text{PCWP}\right) \times \dot{Q} \div \text{HR} \times 0.133$ [51]. By replacing $\dot{Q}$ with $\dot{Q}_I$ in both CP and SW computations, one can also ascertain CP index (CP$_I$) and SW index (SW$_I$), respectively, which may be a necessary correction in HF patients due to the propensity for obesity in these individuals.

While making reference to absolute or percent change (rest to peak exercise) cardiovascular responses to CPET demonstrated by HFrEF and HFpEF (see basic demographics in caption of Fig. 1) in Table 2, Fig. 3 illustrates how invasively derived measures of SW$_I$ and CP$_I$ appear to drive exercise capacity within and across patients. These data indicate that

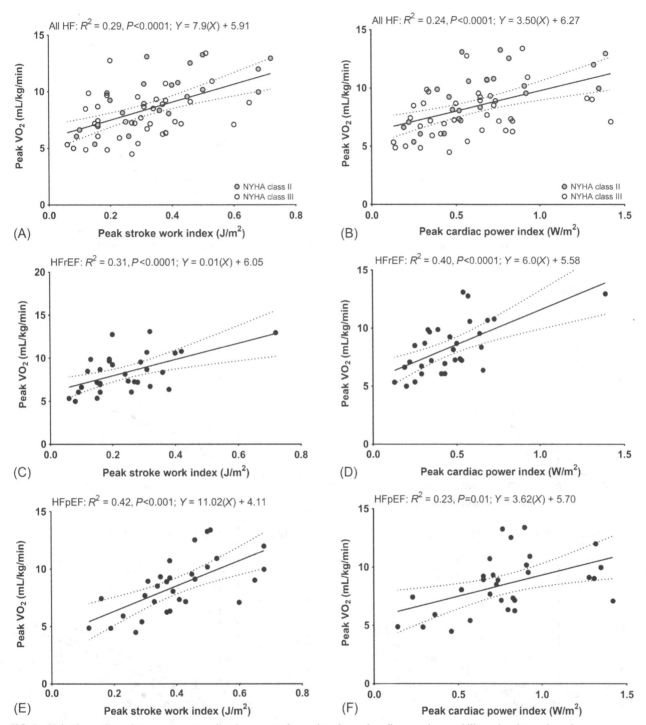

FIG. 3 Univariate ordinary least-squares regressions between peak exercise advanced cardiac pumping capability and peak exercise pulmonary oxygen uptake ($\dot{V}O_2$). Solid line is the model goodness-of-fit line for the regression, whereas the upper and lower dotted lines are 95% confidence limits of the model goodness-of-fit line. (A and B) Across heart failure patients with reduced or preserved ejection fraction (see caption in Fig. 2 for basic demographics). (C and D) Heart failure patients with reduced ejection fraction (HFrEF) (see caption in Fig. 1 for basic demographics). (E and F) Heart failure patients with preserved ejection fraction (HFpEF) (see caption in Fig. 1 for basic demographics).

both SW_I and CP_I are moderately to strongly related with the positive increase in exercise capacity across HF patients and when stratified by type. This is consistent with studies of HF survival using both pharmacological and exercise stress tests, which confirm SW_I and CP_I as robust prognostic indicators [44–48,51,60,89,215–218,190]. Likewise, commensurate with positive associations depicted in Fig. 3, Fig. 4 shows that cardiac pumping reserve (i.e., maximally achieved percent increase during stress from rest) strongly relates with the increase in reserve $\dot{V}O_2$ when aligned with $\dot{Q}$ and CP across patients.

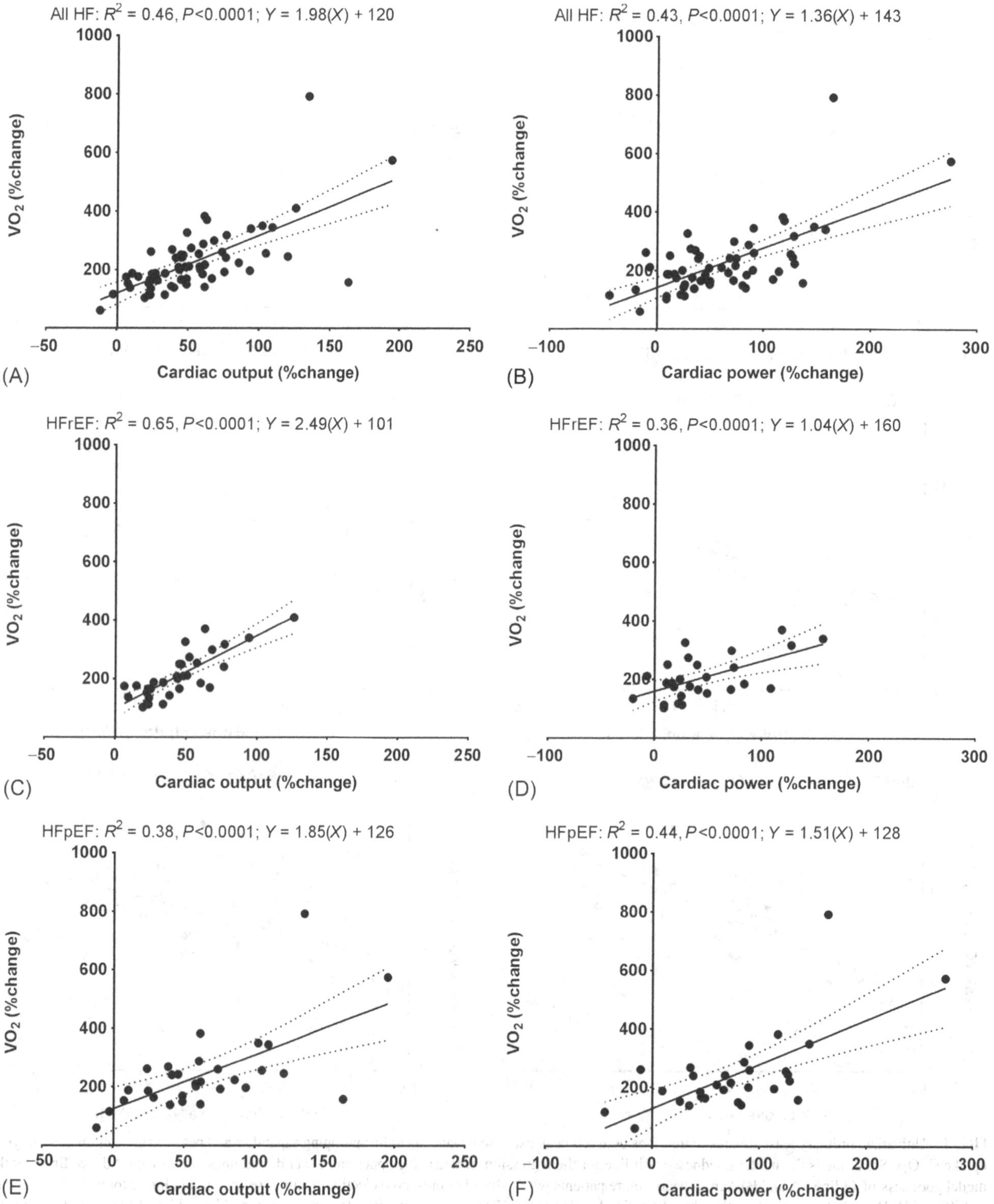

FIG. 4 Univariate ordinary least-squares regressions between percent change (rest to peak exercise) in basic or advanced cardiac pumping capability and percent change (rest to peak exercise) in pulmonary oxygen uptake ($\dot{V}O_2$). Solid line is the model goodness-of-fit line for the regression, whereas the upper and lower dotted lines are 95% confidence limits of the model goodness-of-fit line. (A and B) Across heart failure patients with reduced ejection fraction or heart failure patients with preserved ejection fraction (see caption in Fig. 2 for basic demographics). (C and D) Heart failure patients with reduced ejection fraction (HFrEF) (see caption in Fig. 1 for basic demographics). (E and F) Heart failure patients with preserved ejection fraction (HFpEF) (see caption in Fig. 1 for basic demographics).

Thus, these data demonstrate, with immediate clinical implications, that exercise hemodynamic pumping capability of the heart plays a key role in the classification of HF severity [44–48,51,60,89,215–218,190].

IMPAIRED CARDIAC PUMPING CAPABILITY IS EXACERBATED BY INTERACTIONS WITH NON CARDIAC-CENTRIC FACTORS IN HEART FAILURE

In complementing the primary focus of this chapter, which underscores the prognostic impact of quantifying exercise cardiac pumping capability for the clinical classification of HF, it is pertinent to acknowledge there are other well-supported lines of evidence suggesting noncardio-centric factors are influential and additive in contributing to exercise intolerance while also being linked to key prognostic outcomes [33,186,219–224,85]. For example, an encouraging area of study aimed at elucidating the peripheral pathophysiological origins of the signs and symptoms of HF comes from work focusing on the "muscle hypothesis."

The "muscle hypothesis," initially termed by the combined groups of Piepoli et al. [95,186,225–227] and Clark et al. [228–230], is supported by an evolving body of human- and animal-based evidence. This hypothesis suggests that pathophysiology of the musculoskeletal-neural system, linked to dysregulated somatic and autonomic nervous system activity, provokes impaired cardiovascular control during exercise in HF [220–223,231–235]. An accumulating body of evidence has demonstrated that abnormal exercise cardiovascular control is a sequela of impaired nervous system mechanisms underscored by the "muscle hypothesis," which involves a complex series of skeletal muscle changes unique to HF including muscle structure (e.g., ↑ proportion of glycolytic type II fibers), metabolism (e.g., ↑ contribution of anaerobic metabolism for energy production), and blood biochemistry (e.g., ↑ tissue accumulation of byproducts of metabolism such as hydrogen ion, lactate, and inorganic phosphate). These hypothesized pathological skeletal muscle changes are suggested to be key factors in facilitating the debilitating effects of this proposed mechanism in HF [236–243].

The underlying unifying mechanism linking skeletal muscle and peripheral neural systems in this hypothesis is a set of nerve fibers to which information is translated locally from skeletal muscle to cardiovascular control centers in the brainstem [244–251]. These nerve endings are dispersed within and around vasculature and collagenous tissue of skeletal muscle and are phenotyped as polymodal small-diameter unencapsulated group III and IV nerve fibers belonging to the A-δ and C family of neural afferents, respectively [33,186,219–223,248,252]. When stimulated, neural feedback from this integrated musculoneural system results in augmented sympathetic efferent activity, which contributes to ventilatory inefficiency (rate-mediated ↑ slope of the ventilatory equivalent for carbon dioxide ($\dot{V}_E/\dot{V}CO_2$)), increased systemic vascular resistance, and augmented blood pressure [33,186,219–223]. Accordingly, since it is clear HF patients demonstrate reduced cardiac pumping capability, which is ineffective in recruiting the adequate cardiac pumping reserve necessary to perform any level of sustained physical activity [44–48,51,60,89,215–218,190], a metabolically disproportionate increase in breathing frequency and blood pressure via dysregulated group III/IV muscle afferents further contributes to marked declines in exercise tolerance [33,186,189,219–223,231,232,253,254].

More specifically, by appreciating the integrative systems physiological model of exercise function, it is relevant that increased exercise ventilation for given metabolic load translates to an increased flow-resistive and elastic work of breathing in patients with HF [31,255,256] Those observations highlight that increased work of breathing necessitates an increased delivery of O_2 to the respiratory muscles, which in the face of compromised exercise $\dot{Q}$ is ultimately redirected away from locomotor muscle [31]. This maldistribution or "stealing" of blood flow away from locomotor muscles contributes to early recruitment of nonoxidative pathways for energy production within locomotor muscles setting forth rapid cascading changes intramuscularly (e.g., ↑ H^+ and ↓ pH) [31]. These intramuscular changes result in an optimal environment for increased activation of the group III/IV muscle afferent pathway further stimulating efferent sympathetic activity and continued elevations in ventilation [33,186,219,220,222,223,231,232,253]. Thus, pathophysiology involving group III/IV muscle afferents is a strong candidate mechanism functioning in a disparaging manner to exacerbate effects of poor cardiac pumping capability on exercise intolerance in HF and increased hospitalizations and mortality [33,186,219,220,222–224,85,231,232,253].

SUMMARY

While the primary goal of this chapter is to communicate a strong case for assessing cardiac pumping capability during exercise to properly classify HF syndrome severity, it is important to state that established clinical practice guidelines should not be abandoned. By contrast, this chapter underscores the need to supplement the traditional paradigm of HF classification by providing patient driven evidence-based support for the addition of routine clinical measurement of cardiac hemodynamics during exercise in an effort to optimize classification of syndrome severity.

In addition, because of the prognostic importance of exercise cardiac hemodynamic data during the clinical evaluation process, we have also outlined the major strengths and weaknesses of techniques, which are more recognized and validated options for noninvasive assessment of exercise cardiac pumping capability in clinical and research settings. Again, we acknowledged that there are other noninvasive techniques for assessing exercise cardiac hemodynamics aside from those methods mentioned in this chapter (e.g., impedance cardiography and arterial pulse wave modeling) [139–145]. However, those omitted techniques have not been exhaustively studied nor consistently demonstrated reliability and agreement with "gold" standard approaches across studies in patients with HF, particularly during exercise. Therefore, to emphasize that in the absence of availability of resources dedicated to performing invasive "gold" standard cardiac hemodynamic evaluation while also minimizing unnecessary patient risk, the strongest alternative method for assessing exercise cardiac pumping capability is the inert gas uptake technique. However, in the likelihood that there is a presence of experienced sonographers and readers expertly trained in performing exercise ECHO, this imaging-based technique may also be considered an effective methodological approach for assessing exercise cardiac hemodynamics in HF patients.

Finally, because HF is an integrated multi-organ systemic syndrome complicating our pathophysiological understanding of this epidemic, it is critical to recognize that knowledge gaps remain concerning the identification of novel targetable pathways that ultimately may advance our mechanistic understanding of key prognostic outcomes in these patients. Nevertheless, in the absence of a fully elucidated unifying theory of the pathophysiology exercise intolerance, objective measurement of cardiac pumping capability during exercise is a key component necessary for advancing our understanding of the coupling between exercise intolerance and prognosis in patients with HF.

ACKNOWLEDGMENTS

Funding for this work was supported by National Institutes of Health Grant RO1-HL126638 (TPO) and American Heart Association Grant #16POST30260021 (EHV).

REFERENCES

[1] Heidenreich PA, Albert NM, Allen LA, Bluemke DA, Butler J, Fonarow GC, et al. Forecasting the impact of heart failure in the united states a policy statement from the american heart association. Circ Heart Fail 2013;6(3):606–19.

[2] McMurray JJ, Adamopoulos S, Anker SD, Auricchio A, Böhm M, Dickstein K, et al. ESC Guidelines for the diagnosis and treatment of acute and chronic heart failure 2012. Eur J Heart Fail 2012;14(8):803–69.

[3] Paulus WJ, Tschöpe C, Sanderson JE, Rusconi C, Flachskampf FA, Rademakers FE, et al. How to diagnose diastolic heart failure: a consensus statement on the diagnosis of heart failure with normal left ventricular ejection fraction by the Heart Failure and Echocardiography Associations of the European Society of Cardiology. Eur Heart J 2007;28(20):2539–50.

[4] Bhatia RS, Tu JV, Lee DS, Austin PC, Fang J, Haouzi A, et al. Outcome of heart failure with preserved ejection fraction in a population-based study. N Engl J Med 2006;355(3):260–9.

[5] Zile MR, Brutsaert DL. New concepts in diastolic dysfunction and diastolic heart failure: Part II. Causal mechanisms and treatment. Circulation 2002;105(12):1503–8.

[6] Zile MR, Brutsaert DL. New concepts in diastolic dysfunction and diastolic heart failure: Part I. Diagnosis, prognosis, and measurements of diastolic function. Circulation 2002;105(11):1387–93.

[7] Benjamin EJ, Blaha MJ, Chiuve SE, Cushman M, Das SR, Deo R, et al. Heart disease and stroke statistics—2017 update: a report from the American Heart Association. Circulation 2017;135(10):e146–603.

[8] Roger VL, Weston SA, Redfield MM, Hellermann-Homan JP, Killian J, Yawn BP, et al. Trends in heart failure incidence and survival in a community-based population. JAMA 2004;292(3):344–50.

[9] Dunlay SM, Redfield MM, Weston SA, Therneau TM, Long KH, Shah ND, et al. Hospitalizations after heart failure diagnosis: a community perspective. J Am Coll Cardiol 2009;54(18):1695–702.

[10] Tang AS, Wells GA, Talajic M, Arnold MO, Sheldon R, Connolly S, et al. Cardiac-resynchronization therapy for mild-to-moderate heart failure. N Engl J Med 2010;363(25):2385–95.

[11] Sipahi I, Carrigan TP, Rowland DY, Stambler BS, Fang JC. Impact of QRS duration on clinical event reduction with cardiac resynchronization therapy: meta-analysis of randomized controlled trials. Arch Intern Med 2011;171(16):1454–62.

[12] Saxon LA, Ellenbogen KA. Resynchronization therapy for the treatment of heart failure. Circulation 2003;108(9):1044–8.

[13] Tracy CM, Epstein AE, Darbar D, DiMarco JP, Dunbar SB, Estes NM, et al. 2012 ACCF/AHA/HRS focused update of the 2008 guidelines for device-based therapy of cardiac rhythm abnormalities: a report of the American College of Cardiology Foundation/American Heart Association Task Force on Practice Guidelines. J Am Coll Cardiol 2012;60(14):1297–313.

[14] Merlo M, Pivetta A, Pinamonti B, Stolfo D, Zecchin M, Barbati G, et al. Long-term prognostic impact of therapeutic strategies in patients with idiopathic dilated cardiomyopathy: changing mortality over the last 30 years. Eur J Heart Fail 2014;16(3):317–24.

[15] Gerber Y, Weston SA, Redfield MM, Chamberlain AM, Manemann SM, Jiang R, et al. A contemporary appraisal of the heart failure epidemic in Olmsted County, Minnesota, 2000 to 2010. JAMA Intern Med 2015;175(6):996–1004.

[16] Zile MR, Baicu CF, Gaasch WH. Diastolic heart failure—abnormalities in active relaxation and passive stiffness of the left ventricle. N Engl J Med 2004;350(19):1953–9.

[17] Borlaug BA, Paulus WJ. Heart failure with preserved ejection fraction: pathophysiology, diagnosis, and treatment. Eur Heart J 2011;32(6):670–9.

[18] Belardinelli R, Georgiou D, Cianci G, Purcaro A. Randomized, controlled trial of long-term moderate exercise training in chronic heart failure effects on functional capacity, quality of life, and clinical outcome. Circulation 1999;99(9):1173–82.

[19] Chung ES, Leon AR, Tavazzi L, Sun J-P, Nihoyannopoulos P, Merlino J, et al. Results of the predictors of response to CRT (PROSPECT) trial. Circulation 2008;117(20):2608–16.

[20] Pocock SJ, Ariti CA, McMurray JJ, Maggioni A, Køber L, Squire IB, et al. Predicting survival in heart failure: a risk score based on 39 372 patients from 30 studies. Eur Heart J 2012;34(19):1404–13.

[21] Gaggin HK, Truong QA, Rehman SU, Mohammed AA, Bhardwaj A, Parks KA, et al. Characterization and prediction of natriuretic peptide "non-response" during heart failure management: results from the ProBNP outpatient tailored chronic heart failure (PROTECT) and the NT-proBNP–assisted treatment to lessen serial cardiac readmissions and death (BATTLESCARRED) study. Congest Heart Fail 2013;19(3):135–42.

[22] Gasparini M, Auricchio A, Regoli F, Fantoni C, Kawabata M, Galimberti P, et al. Four-year efficacy of cardiac resynchronization therapy on exercise tolerance and disease progression: the importance of performing atrioventricular junction ablation in patients with atrial fibrillation. J Am Coll Cardiol 2006;48(4):734–43.

[23] Holland DJ, Prasad SB, Marwick TH. Contribution of exercise echocardiography to the diagnosis of heart failure with preserved ejection fraction (HFpEF). Heart 2010;96(13):1024–8.

[24] Haykowsky M, Brubaker P, Kitzman D. Role of physical training in heart failure with preserved ejection fraction. Curr Heart Fail Rep 2012;9(2):101–6.

[25] Haykowsky MJ, Brubaker PH, John JM, Stewart KP, Morgan TM, Kitzman DW. Determinants of exercise intolerance in elderly heart failure patients with preserved ejection fraction. J Am Coll Cardiol 2011;58(3):265–74.

[26] Haykowsky MJ, Brubaker PH, Stewart KP, Morgan TM, Eggebeen J, Kitzman DW. Effect of endurance training on the determinants of peak exercise oxygen consumption in elderly patients with stable compensated heart failure and preserved ejection fraction. J Am Coll Cardiol 2012;60(2):120–8.

[27] Kitzman DW, Brubaker PH, Herrington DM, Morgan TM, Stewart KP, Hundley WG, et al. Effect of endurance exercise training on endothelial function and arterial stiffness in older patients with heart failure and preserved ejection fraction: a randomized, controlled, single-blind trial. J Am Coll Cardiol 2013;62(7):584–92.

[28] Kitzman DW, Brubaker PH, Morgan TM, Stewart KP, Little WC. Exercise training in older patients with heart failure and preserved ejection fraction: a randomized, controlled, single-blind trial. Circ Heart Fail 2010;3(6):659–67.

[29] Olson LJ, Snyder EM, Beck KC, Johnson BD. Reduced rate of alveolar-capillary recruitment and fall of pulmonary diffusing capacity during exercise in patients with heart failure. J Card Fail 2006;12(4):299–306.

[30] Olson TP, Johnson BD. Influence of cardiomegaly on disordered breathing during exercise in chronic heart failure. Eur J Heart Fail 2011;13(3).

[31] Olson TP, Joyner MJ, Dietz NM, Eisenach JH, Curry TB, Johnson BD. Effects of respiratory muscle work on blood flow distribution during exercise in heart failure. J Physiol 2010;588(Pt 13):2487–501.

[32] Olson TP, Snyder EM, Johnson BD. Exercise-disordered breathing in chronic heart failure. Exerc Sport Sci Rev 2006;34(4):194–201.

[33] Olson TP, Joyner MJ, Eisenach JH, Curry TB, Johnson BD. Influence of locomotor muscle afferent inhibition on the ventilatory response to exercise in heart failure. Exp Physiol 2014;99(2):414–26.

[34] Franciosa JA, Leddy CL, Wilen M, Schwartz DE. Relation between hemodynamic and ventilatory responses in determining exercise capacity in severe congestive heart failure. Am J Cardiol 1984;53(1):127–34.

[35] Higginbotham MB, Morris KG, Conn EH, Coleman RE, Cobb FR. Determinants of variable exercise performance among patients with severe left ventricular dysfunction. Am J Cardiol 1983;51(1):52–60.

[36] Weber KT, Kinasewitz GT, Janicki JS, Fishman AP. Oxygen utilization and ventilation during exercise in patients with chronic cardiac failure. Circulation 1982;65(6):1213–23.

[37] Weber KT, Janicki JS. Cardiopulmonary exercise testing for evaluation of chronic cardiac failure. Am J Cardiol 1985;55(2):22A–31A.

[38] Guazzi M, Myers J, Arena R. Cardiopulmonary exercise testing in the clinical and prognostic assessment of diastolic heart failure. J Am Coll Cardiol 2005;46(10):1883–90.

[39] Guazzi M, Myers J, Peberdy MA, Bensimhon D, Chase P, Arena R. Exercise oscillatory breathing in diastolic heart failure: prevalence and prognostic insights. Eur Heart J 2008;29(22):2751–9.

[40] Guazzi M, Myers J, Peberdy MA, Bensimhon D, Chase P, Arena R. Cardiopulmonary exercise testing variables reflect the degree of diastolic dysfunction in patients with heart failure-normal ejection fraction. J Cardiopulm Rehabil Prev 2010;30(3):165–72.

[41] Guazzi M, Reina G, Tumminello G, Guazzi MD. Exercise ventilation inefficiency and cardiovascular mortality in heart failure: the critical independent prognostic value of the arterial CO_2 partial pressure. Eur Heart J 2005;26(5):472–80.

[42] Tabet J-Y, Meurin P, Beauvais F, Weber H, Renaud N, Thabut G, et al. Absence of exercise capacity improvement after exercise training program a strong prognostic factor in patients with chronic heart failure. Circ Heart Fail 2008;1(4):220–6.

[43] Myers J, Prakash M, Froelicher V, Do D, Partington S, Atwood JE. Exercise capacity and mortality among men referred for exercise testing. N Engl J Med 2002;346(11):793–801.

[44] Lang CC, Agostoni P, Mancini DM. Prognostic significance and measurement of exercise-derived hemodynamic variables in patients with heart failure. J Card Fail 2007;13(8):672–9.

[45] Lang CC, Karlin P, Haythe J, Lim TK, Mancini DM. Peak cardiac power output, measured noninvasively, is a powerful predictor of outcome in chronic heart failure. Circ Heart Fail 2009;2(1):33–8.

[46] Williams JM, Cooke GA, Barker D, Patwala A, Wright DJ, et al. How do different indicators of cardiac pump function impact upon the long-term prognosis of patients with chronic heart failure? Am Heart J 2005;150(5):983.

[47] Williams SG, Cooke GA, Wright DJ, Parsons WJ, Riley RL, Marshall P, et al. Peak exercise cardiac power output; a direct indicator of cardiac function strongly predictive of prognosis in chronic heart failure. Eur Heart J 2001;22(16):1496–503.

[48] Williams SG, Tzeng BH, Barker D, Tan LB. Comparison and relation of indirect and direct dynamic indexes of cardiac pumping capacity in chronic heart failure. Am J Cardiol 2005;96(8):1149–50.

[49] Cohen-Solal A, Tabet JY, Logeart D, Bourgoin P, Tokmakova M, Dahan M. A non-invasively determined surrogate of cardiac power ("circulatory power") at peak exercise is a powerful prognostic factor in chronic heart failure. Eur Heart J 2002;23(10):806–14.

[50] Kleber FX, Vietzke G, Wernecke KD, Bauer U, Opitz C, Wensel R, et al. Impairment of ventilatory efficiency in heart failure: prognostic impact. Circulation 2000;101(24):2803–9.

[51] Tan LB. Cardiac pumping capability and prognosis in heart failure. Lancet 1986;2(8520):1360–3.

[52] Szlachcic J, Masse BM, Kramer BL, Tubau J. Correlates and prognostic implication of exercise capacity in chronic congestive heart failure. Am J Cardiol 1985;55(8):1037–42.

[53] Arena R, Myers J, Aslam SS, Varughese EB, Peberdy MA. Peak VO2 and VE/VCO2 slope in patients with heart failure: a prognostic comparison. Am Heart J 2004;147(2):354–60.

[54] Arena R, Myers J, Hsu L, Peberdy MA, Pinkstaff S, Bensimhon D, et al. The minute ventilation/carbon dioxide production slope is prognostically superior to the oxygen uptake efficiency slope. J Card Fail 2007;13(6):462–9.

[55] Chua TP, Ponikowski P, Harrington D, Anker SD, Webb-Peploe K, Clark AL, et al. Clinical correlates and prognostic significance of the ventilatory response to exercise in chronic heart failure. J Am Coll Cardiol 1997;29(7):1585–90.

[56] Francis DP, Shamim W, Davies LC, Piepoli MF, Ponikowski P, Anker SD, et al. Cardiopulmonary exercise testing for prognosis in chronic heart failure: continuous and independent prognostic value from VE/VCO2 slope and peak VO2. Eur Heart J 2000;21(2):154–61.

[57] Roul G, Moulichon ME, Bareiss P, Gries P, Koegler A, Sacrez J, et al. Prognostic factors of chronic heart failure in NYHA class II or III: value of invasive exercise haemodynamic data. Eur Heart J 1995;16(10):1387–98.

[58] Mancini DM, Eisen H, Kussmaul W, Mull R, Edmunds Jr LH, Wilson JR. Value of peak exercise oxygen consumption for optimal timing of cardiac transplantation in ambulatory patients with heart failure. Circulation 1991;83(3):778–86.

[59] Mendoza DD, Cooper HA, Panza JA. Cardiac power output predicts mortality across a broad spectrum of patients with acute cardiac disease. Am Heart J 2007;153(3):366–70.

[60] Mancini KS, Donchez L, Aaronson K. Coupling of hemodynamic measurements with oxygen consumption during exercise does not improve risk stratification in patients with heart failure. Circulation 1996;94(10):2492–6.

[61] Borlaug BA, Nishimura RA, Sorajja P, Lam CS, Redfield MM. Exercise hemodynamics enhance diagnosis of early heart failure with preserved ejection fraction. Circ Heart Fail 2010;3(5):588–95.

[62] Franciosa JA, Park M, Levine TB. Lack of correlation between exercise capacity and indexes of resting left-ventricular performance in heart-failure. Am J Cardiol 1981;47(1):33–9.

[63] Cohn JN, Johnson GR, Shabetai R, Loeb H, Tristani F, Rector T, et al. Ejection fraction, peak exercise oxygen consumption, cardiothoracic ratio, ventricular arrhythmias, and plasma norepinephrine as determinants of prognosis in heart failure. The V-HeFT VA Cooperative Studies Group. Circulation 1993;87(6 Suppl.):VI5–VI16.

[64] Maskin CS, Forman R, Sonnenblick EH, Frishman WH, LeJemtel TH. Failure of dobutamine to increase exercise capacity despite hemodynamic improvement in severe chronic heart failure. Am J Cardiol 1983;51(1):177–82.

[65] Maskin CS, Kugler J, Sonnenblick EH, LeJemtel TH. Acute inotropic stimulation with dopamine in severe congestive heart failure: beneficial hemodynamic effect at rest but not during maximal exercise. Am J Cardiol 1983;52(8):1028–32.

[66] Francis GS, Goldsmith SR, Cohn JN. Relationship of exercise capacity to resting left ventricular performance and basal plasma norepinephrine levels in patients with congestive heart failure. Am Heart J 1982;104(4 Pt 1):725–31.

[67] Hall SG, Garcia J, Larson DF, Smith R. Cardiac power index: staging heart failure for mechanical circulatory support. Perfusion 2012;27(6):456–61.

[68] Wilson JR, Mancini DM, Dunkman WB. Exertional fatigue due to skeletal muscle dysfunction in patients with heart failure. Circulation 1993;87(2):470–5.

[69] Wilson JR, Martin JL, Schwartz D, Ferraro N. Exercise intolerance in patients with chronic heart failure: role of impaired nutritive flow to skeletal muscle. Circulation 1984;69(6):1079–87.

[70] Corra U, Mezzani A, Bosimini E, Scapellato F, Imparato A, Giannuzzi P. Ventilatory response to exercise improves risk stratification in patients with chronic heart failure and intermediate functional capacity. Am Heart J 2002;143(3):418–26.

[71] Marvin HM. Diseases of the heart and blood vessels: nomenclature and criteria for diagnosis. Arch Intern Med 1964;113(6):906–7.

[72] Dolgin M, Association CCNYH. Nomenclature and criteria for diagnosis of diseases of the heart and great vessels. Bostan: The Criteria Committee of the New York Heart Association; Little, Brown & Company; 1994.

[73] Ammar KA, Jacobsen SJ, Mahoney DW, Kors JA, Redfield MM, Burnett JC, et al. Prevalence and prognostic significance of heart failure stages application of the American College of Cardiology/American Heart Association heart failure staging criteria in the community. Circulation 2007;115(12):1563–70.

[74] Hunt SA. ACC/AHA 2005 guidelines update for the diagnosis and management of chronic heart failure in the adult: a report of the American College of Cardiology/American Heart Association Task Force on Practice Guidelines. Circulation 2005;112(12):1825–52.

[75] Yancy C, Jessup M, Bozkurt B, Butler J, Casey Jr D, Drazner M, et al. ACCF/AHA guideline for the management of heart failure: a report of the American College of Cardiology Foundation/American Heart Association Task Force on Practice Guidelines. Circulation 2013;128(16):1810–52.

[76] Dickstein K, Cohen-Solal A, Filippatos G, McMurray JJ, Ponikowski P, Poole-Wilson PA, et al. ESC guidelines for the diagnosis and treatment of acute and chronic heart failure 2008: the Task Force for the diagnosis and treatment of acute and chronic heart failure 2008 of the European Society of Cardiology. Developed in collaboration with the Heart Failure Association of the ESC (HFA) and endorsed by the European Society of Intensive Care Medicine (ESICM). Eur J Heart Fail 2008;10(10):933–89.

[77] Lucas C, Johnson W, Hamilton MA, Fonarow GC, Woo MA, Flavell CM, et al. Freedom from congestion predicts good survival despite previous class IV symptoms of heart failure. Am Heart J 2000;140(6):840–7.

[78] Stevenson LW, Perloff JK. The limited reliability of physical signs for estimating hemodynamics in chronic heart failure. JAMA 1989;261(6):884–8.

[79] van den Broek SA, van Veldhuisen DJ, de Graeff PA, Landsman ML, Hillege H, Lie KI. Comparison between New York Heart Association classification and peak oxygen consumption in the assessment of functional status and prognosis in patients with mild to moderate chronic congestive heart failure secondary to either ischemic or idiopathic dilated cardiomyopathy. Am J Cardiol 1992;70(3):359–63.

[80] Balady GJ, Arena R, Sietsema K, Myers J, Coke L, Fletcher GF, et al. Clinician's Guide to cardiopulmonary exercise testing in adults: a scientific statement from the American Heart Association. Circulation 2010;122(2):191–225.

[81] Wilson JR, Rayos G, Yeoh T-K, Gothard P. Dissociation between peak exercise oxygen consumption and hemodynamic dysfunction in potential heart transplant candidates. J Am Coll Cardiol 1995;26(2):429–35.

[82] Becklake MR, Frank H, Dagenais GR, Ostiguy GL, Guzman CA. Influence of age and sex on exercise cardiac output. J Appl Physiol 1965;20(5):938–47.

[83] Chomsky DB, Lang CC, Rayos GH, Shyr Y, Yeoh TK, Pierson 3rd RN, et al. Hemodynamic exercise testing. A valuable tool in the selection of cardiac transplantation candidates. Circulation 1996;94(12):3176–83.

[84] Chua TP, Clark AI, Amadi AA, Coats AJ. Relation between chemosensitivity and the ventilatory response to exercise in chronic heart failure. J Am Coll Cardiol 1996;27(3):650–7.

[85] Ponikowski PP, Chua TP, Francis DP, Capucci A, Coats AJ, Piepoli MF. Muscle ergoreceptor overactivity reflects deterioration in clinical status and cardiorespiratory reflex control in chronic heart failure. Circulation 2001;104(19):2324–30.

[86] Wilson JR, Wilson JL, Ferraro N. Impaired skeletal muscle nutritive flow during exercise in patients with congestive heart failure: role of cardiac pump dysfunction as determined by the effect of dobutamine. Am J Cardiol 1984;53(9):1308–15.

[87] Lipkin DP, Canepa-Anson R, Stephens MR, Poole-Wilson PA. Factors determining symptoms in heart failure: comparison of fast and slow exercise tests. Br Heart J 1986;55(5):439–45.

[88] Epstein SE, Beiser GD, Stampfer M, Robinson BF, Braunwald E. Characterization of the circulatory response to maximal upright exercise in normal subjects and patients with heart disease. Circulation 1967;35(6):1049–62.

[89] Lang CC, Karlin P, Haythe J, Tsao L, Mancini DM. Ease of noninvasive measurement of cardiac output coupled with peak VO2 determination at rest and during exercise in patients with heart failure. Am J Cardiol 2007;99(3):404–5.

[90] Griffin BP, Shah PK, Ferguson J, Rubin SA. Incremental prognostic value of exercise hemodynamic variables in chronic congestive heart failure secondary to coronary artery disease or to dilated cardiomyopathy. Am J Cardiol 1991;67(9):848–53.

[91] Mancini DM, Coyle E, Coggan A, Beltz J, Ferraro N, Montain S, et al. Contribution of intrinsic skeletal muscle changes to 31P NMR skeletal muscle metabolic abnormalities in patients with chronic heart failure. Circulation 1989;80(5):1338–46.

[92] Mancini DM, Walter G, Reichek N, Lenkinski R, McCully KK, Mullen JL, et al. Contribution of skeletal muscle atrophy to exercise intolerance and altered muscle metabolism in heart failure. Circulation 1992;85(4):1364–73.

[93] Massie B, Conway M, Yonge R, Frostick S, Ledingham J, Sleight P, et al. Skeletal muscle metabolism in patients with congestive heart failure: relation to clinical severity and blood flow. Circulation 1987;76(5):1009–19.

[94] Minotti JR, Christoph I, Oka R, Weiner MW, Wells L, Massie BM. Impaired skeletal muscle function in patients with congestive heart failure. Relationship to systemic exercise performance. J Clin Invest 1991;88(6):2077–82.

[95] Piepoli MF, Kaczmarek A, Francis DP, Davies LC, Rauchhaus M, Jankowska EA, et al. Reduced peripheral skeletal muscle mass and abnormal reflex physiology in chronic heart failure. Circulation 2006;114(2):126–34.

[96] Sullivan MJ, Green HJ, Cobb FR. Skeletal muscle biochemistry and histology in ambulatory patients with long-term heart failure. Circulation 1990;81(2):518–27.

[97] Sullivan MJ, Green HJ, Cobb FR. Altered skeletal muscle metabolic response to exercise in chronic heart failure. Relation to skeletal muscle aerobic enzyme activity. Circulation 1991;84(4):1597–607.

[98] Saur J, Trinkmann F, Doesch C, Scherhag A, Brade J, Schoenberg SO, et al. The impact of pulmonary disease on noninvasive measurement of cardiac output by the inert gas rebreathing method. Lung 2010;188(5):433–40.

[99] Triebwasser J, Johnson R, Burpo R, Campbell J, Reardon W, Blomqvist C. Noninvasive determination of cardiac output by a modified acetylene rebreathing procedure utilizing mass spectrometer measurements. Aviat Space Environ Med 1977;48(3):203–9.

[100] Johnson BD, Beck KC, Proctor DN, Miller J, Dietz NM, Joyner MJ. Cardiac output during exercise by the open circuit acetylene washin method: comparison with direct Fick (1985). J Appl Physiol (1985) 2000;88(5):1650–8.

[101] Agostoni P, Cattadori G, Apostolo A, Contini M, Palermo P, Marenzi G, et al. Noninvasive measurement of cardiac output during exercise by inert gas rebreathing technique: a new tool for heart failure evaluation. J Am Coll Cardiol 2005;46(9):1779–81.

[102] Ferguson RJ, Faulkner JA, Julius S, Conway J. Comparison of cardiac output determined by CO_2 rebreathing and dye-dilution methods. J Appl Physiol 1968;25(4):450–4.

[103] Franciosa JA, Ragan DO, Rubenstone SJ. Validation of the CO_2 rebreathing method for measuring cardiac output in patients with hypertension or heart failure. J Lab Clin Med 1976;88(4):672–82.

[104] Hoeper MM, Maier R, Tongers J, Niedermeyer J, Hohlfeld JM, Hamm M, et al. Determination of cardiac output by the Fick method, thermodilution, and acetylene rebreathing in pulmonary hypertension. Am J Respir Crit Care Med 1999;160(2):535–41.

[105] Trinkmann F, Papavassiliu T, Kraus F, Leweling H, Schoenberg SO, Borggrefe M, et al. Inert gas rebreathing: the effect of haemoglobin based pulmonary shunt flow correction on the accuracy of cardiac output measurements in clinical practice. Clin Physiol Funct Imaging 2009;29(4):255–62.

[106] Klausen K. Comparison of CO_2 rebreathing and acetylene methods for cardiac output. J Appl Physiol 1965;20(4):763–6.

[107] Muiesan G, Sorbini CA, Solinas E, Grassi V, Casucci G, Petz E. Comparison of CO_2-rebreathing and direct Fick methods for determining cardiac output. J Appl Physiol 1968;24(3):424–9.

[108] Beekman RH, Katch V, Marks C, Rocchini AP. Validity of CO_2-rebreathing cardiac output during rest and exercise in young adults. Med Sci Sports Exerc 1984;16(3):306–10.

[109] Liu Y, Menold E, Dullenkopf A, Reissnecker S, Lormes W, Lehmann M, et al. Validation of the acetylene rebreathing method for measurement of cardiac output at rest and during high-intensity exercise. Clin Physiol 1997;17(2):171–82.

[110] Magel JR, Andersen KL. Cardiac output in muscular exercise measured by the CO_2 rebreathing procedure. In: Denolin H, Konig K, Messin R, and Degre S, editors. Ergometry in cardiology, Bohringer; 1968.

[111] Zeidifard E, Godfrey S, Davies EE. Estimation of cardiac output by an N_2O rebreathing method in adults and children. J Appl Physiol 1976;41(3):433–8.

[112] Russell AE, Smith SA, West MJ, Aylward PE, McRitchie RJ, Hassam RM, et al. Automated non-invasive measurement of cardiac output by the carbon dioxide rebreathing method: comparisons with dye dilution and thermodilution. Br Heart J 1990;63(3):195–9.

[113] Asmussen E, Nielsen M. Cardiac output during muscular work and its regulation. Physiol Rev 1955;35(4):778–800.

[114] Fick A. Uber die Messung des Blutquantums in den Herzventrike. Phys Med Ges 1870;2(XVI).

[115] Hsia C, Herazo L, Ramanathan M, Johnson Jr R. Cardiac output during exercise measured by acetylene rebreathing, thermodilution, and Fick techniques. J Appl Physiol 1995;78(4):1612–6.

[116] Lipkin DP, Poole-Wilson PA. Measurement of cardiac output during exercise by the thermodilution and direct Fick techniques in patients with chronic congestive heart failure. Am J Cardiol 1985;56(4):321–4.

[117] McMichael J, Sharpey-Schafer EP. Cardiac output in man by a direct Fick method: effects of posture, venous pressure change, atropine, and adrenaline. Br Heart J 1944;6(1):33–40.

[118] Werko L, Berseus S, Lagerlof H. A comparison of the direct Fick and the Grollman methods for determination of the cardiac output in man. J Clin Invest 1949;28(3):516–20.

[119] Selzer A, Sudrann RB. Reliability of the determination of cardiac output in man by means of the Fick principle. Circ Res 1958;6(4):485–90.

[120] Thomasson B. Cardiac output in normal subjects under standard basal conditions the repeatability of measurements by the Fick method. Scand J Clin Lab Investig 1957;9(4):365–76.

[121] Hamilton W, Riley R, Attyah A, Cournand A, Fowell DM, Himmelstein A, et al. Comparison of the Fick and dye injection methods of measuring the cardiac output in man. Am J Physiol 1948;153(2):309–21.

[122] Grenvik A. Errors of the dye dilution method compared to the direct Fick method in determination of cardiac output in man. Scand J Clin Lab Invest 1966;18(5):486–92.

[123] Kopelman H, LEEG J. The intrathoracic blood volume in mitral stenosis and left ventricular failure. Clin Sci (Lond) 1951;10(3):383–403.

[124] Eliasch H, Lagerlof H, Bucht H, Ek J, Eriksson K, Bergstrom J, et al. Comparison of the dye dilution and the direct Fick methods for the measurement of cardiac output in man. Scand J Clin Lab Invest 1955;7(Suppl 20):73–8.

[125] Taylor SH, Shillingford JP. Clinical applications of Coomassie blue. Br Heart J 1959;21(4):497–504.

[126] Hsia CC. Respiratory function of hemoglobin. N Engl J Med 1998;338(4):239–47.

[127] Stetz CW, Miller RG, Kelly GE, Raffin TA. Reliability of the thermodilution method in the determination of cardiac output in clinical practice. Am Rev Respir Dis 1982;126(6):1001–4.

[128] Fegler G. The reliability of the thermodilution method for determination of the cardiac output and the blood flow in central veins. Q J Exp Physiol Cogn Med Sci 1957;42(3):254–66.

[129] Enghoff E, Sjogren S. Thermal dilution for measurement of cardiac output in the pulmonary artery in man in relation to choice of indicator volume and injection time. Ups J Med Sci 1973;78(1):33–7.

[130] Hoel BL. Some aspects of the clinical use of thermodilution in measuring cardiac output. With particular reference to the Swan-Ganz thermodilution catheters. Scand J Clin Lab Invest 1978;38(4):383–8.

[131] Wessel HU, Paul MH, James GW, Grahn AR. Limitations of thermal dilution curves for cardiac output determinations. J Appl Physiol 1971;30(5):643–52.

[132] Weisel RD, Berger RL, Hechtman HB. Measurement of cardiac output by thermodilution. N Engl J Med 1975;292(13):682–4.

[133] Ganz W, Donoso R, Marcus HS, Forrester JS, Swan HJ. A new technique for measurement of cardiac output by thermodilution in man. Am J Cardiol 1971;27(4):392–6.

[134] Venkataraman K, De Guzman MF, Hafeez Khan A, Haywood LJ. Cardiac output measurement: a comparison of direct Fick, dye dilution and thermodilution methods in stable and acutely Ill patients. J Natl Med Assoc 1976;68(4):281–4.

[135] Fischer AP, Benis AM, Jurado RA, Seely E, Teirstein P, Litwak RS. Analysis of errors in measurement of cardiac output by simultaneous dye and thermal dilution in cardiothoracic surgical patients. Cardiovasc Res 1978;12(3):190–9.

[136] Hodges M, Downs JB, Mitchell LA. Thermodilution and Fick cardiac index determinations following cardiac surgery. Crit Care Med 1975;3(5):182–4.

[137] Babu SC, Piccorelli GO, Shah PM, Stein JH, Clauss RH. Incidence and results of arterial complications among 16,350 patients undergoing cardiac catheterization. J Vasc Surg 1989;10(2):113–6.

[138] Kennedy JW. Complications associated with cardiac catheterization and angiography. Catheter Cardiovasc Interv 1982;8(1):5–11.

[139] Bogert LW, Wesseling KH, Schraa O, Van Lieshout EJ, de Mol BA, van Goudoever J, et al. Pulse contour cardiac output derived from non-invasive arterial pressure in cardiovascular disease. Anaesthesia 2010;65(11):1119–25.

[140] Siebenmann C, Rasmussen P, Sorensen H, Zaar M, Hvidtfeldt M, Pichon A, et al. Cardiac output during exercise: a comparison of four methods. Scand J Med Sci Sports 2015;25(1):e20–7.

[141] Critoph CH, Patel V, Mist B, Thomas MD, Elliott PM. Non-invasive assessment of cardiac output at rest and during exercise by finger plethysmography. Clin Physiol Funct Imaging 2013;33(5):338–43.

[142] Monnet X, Picard F, Lidzborski E, Mesnil M, Duranteau J, Richard C, et al. The estimation of cardiac output by the Nexfin device is of poor reliability for tracking the effects of a fluid challenge. Crit Care 2012;16(5):1.

[143] Belardinelli R, Ciampani N, Costantini C, Blandini A, Purcaro A. Comparison of impedance cardiography with thermodilution and direct Fick methods for noninvasive measurement of stroke volume and cardiac output during incremental exercise in patients with ischemic cardiomyopathy. Am J Cardiol 1996;77(15):1293–301.

[144] Jensen L, Yakimets J, Teo KK. A review of impedance cardiography. Heart Lung J Acute Crit Care 1995;24(3):183–93.

[145] Remmen JJ, Aengevaeren WR, Verheugt FW, van der Werf T, Luijten HE, Anja B, et al. Finapres arterial pulse wave analysis with Modelflow® is not a reliable non-invasive method for assessment of cardiac output. Clin Sci 2002;103(2):143–9.

[146] Lang RM, Bierig M, Devereux RB, Flachskampf FA, Foster E, Pellikka PA, et al. Recommendations for chamber quantification: a report from the American Society of Echocardiography's Guidelines and Standards Committee and the Chamber Quantification Writing Group, developed in conjunction with the European Association of Echocardiography, a branch of the European Society of Cardiology. J Am Soc Echocardiogr 2005;18(12):1440–63.

[147] Ponikowski P, Voors A, Anker S, Bueno H, Cleland J, Coats A, et al. Authors/Task Force Members; Document Reviewers. 2016 ESC Guidelines for the diagnosis and treatment of acute and chronic heart failure: the task force for the diagnosis and treatment of acute and chronic heart failure of the European Society of Cardiology (ESC). Developed with the special contribution of the Heart Failure Association (HFA) of the ESC. Eur J Heart Fail 2016;18(8):891–97530.

[148] Rowland TW, Melanson EL, Popowski BE, Ferrone LC. Test-retest reproducibility of maximum cardiac output by Doppler echocardiography. Am J Cardiol 1998;81(10):1228–30.

[149] Loeppky JA, Hoekenga DE, Greene ER, Luft UC. Comparison of noninvasive pulsed Doppler and Fick measurements of stroke volume in cardiac patients. Am Heart J 1984;107(2):339–46.

[150] Nishimura RA, Callahan MJ, Schaff HV, Ilstrup DM, Miller FA, Tajik AJ, editors. Noninvasive measurement of cardiac output by continuous-wave Doppler echocardiography: initial experience and review of the literature. In: Mayo clinic proceedings. Elsevier; 1984.

[151] Temporelli PL, Scapellato F, Eleuteri E, Imparato A, Giannuzzi P. Doppler echocardiography in advanced systolic heart failure a noninvasive alternative to Swan-Ganz catheter. Circ Heart Fail 2010;3(3):387–94.

[152] Stein JH, Neumann A, Preston LM, Costanzo MR, Parrillo JE, Johnson MR, et al. Echocardiography for hemodynamic assessment of patients with advanced heart failure and potential heart transplant recipients. J Am Coll Cardiol 1997;30(7):1765–72.

[153] Dubin J, Wallerson DC, Cody RJ, Devereux RB. Comparative accuracy of Doppler echocardiographic methods for clinical stroke volume determination. Am Heart J 1990;120(1):116–23.

[154] Huntsman LL, Stewart DK, Barnes SR, Franklin SB, Colocousis JS, Hessel EA. Noninvasive Doppler determination of cardiac output in man clinical validation. Circulation 1983;67(3):593–602.

[155] Christie J, Sheldahl LM, Tristani FE, Sagar KB, Ptacin MJ, Wann S. Determination of stroke volume and cardiac output during exercise: comparison of two-dimensional and Doppler echocardiography, Fick oximetry, and thermodilution. Circulation 1987;76(3):539–47.

[156] Lewis JF, Kuo LC, Nelson JG, Limacher MC, Quinones MA. Pulsed Doppler echocardiographic determination of stroke volume and cardiac output: clinical validation of two new methods using the apical window. Circulation 1984;70(3):425–31.

[157] Rowland T, Whatley Blum J. Cardiac dynamics during upright cycle exercise in boys. Am J Hum Biol 2000;12(6):749–57.

[158] Vinet A, Nottin S, Lecoq AM, Guenon P, Obert P. Reproducibility of cardiac output measurements by Doppler echocardiography in prepubertal children and adults. Int J Sports Med 2001;22(06):437–41.

[159] Marwick TH, Nemec JJ, Pashkow FJ, Stewart WJ, Salcedo EE. Accuracy and limitations of exercise echocardiography in a routine clinical setting. J Am Coll Cardiol 1992;19(1):74–81.

[160] Schinkel AF, Bax JJ, Geleijnse ML, Boersma E, Elhendy A, Roelandt JR, et al. Noninvasive evaluation of ischaemic heart disease: myocardial perfusion imaging or stress echocardiography? Eur Heart J 2003;24(9):789–800.

[161] Lu C, Nicolosi GL, Burelli C, Cassin M, Zardo F, Brieda M, et al. Limitations in the assessment of changes of cardiac output by Doppler echocardiography under various hemodynamic conditions. Am J Cardiol 1992;70(15):1370–4.

[162] Grothues F, Smith GC, Moon JC, Bellenger NG, Collins P, Klein HU, et al. Comparison of interstudy reproducibility of cardiovascular magnetic resonance with two-dimensional echocardiography in normal subjects and in patients with heart failure or left ventricular hypertrophy. Am J Cardiol 2002;90(1):29–34.

[163] Katz WE, Gasior TA, Reddy S, Gorcsan III J. Utility and limitations of biplane transesophageal echocardiographic automated border detection for estimation of left ventricular stroke volume and cardiac output. Am Heart J 1994;128(2):389–96.

[164] Labovitz AJ, Buckingham TA, Habermehl K, Nelson J, Kennedy HL, Williams GA. The effects of sampling site on the two-dimensional echo-Doppler determination of cardiac output. Am Heart J 1985;109(2):327–32.

[165] Ihlen H, Amlie J, Dale J, Forfang KR, Nitter-Hauge S, Otterstad J, et al. Determination of cardiac output by Doppler echocardiography. Br Heart J 1984;51(1):54–60.

[166] Dittmann H, Voelker W, Karsch KR, Seipel L. Influence of sampling site and flow area on cardiac output measurements by Doppler echocardiography. J Am Coll Cardiol 1987;10(4):818–23.

[167] Cander L. Solubility of inert gases in human lung tissue. J Appl Physiol 1959;14(4):538–40.

[168] Cander L, Forster RE. Determination of pulmonary parenchymal tissue volume and pulmonary capillary blood flow in man. J Appl Physiol 1959;14(4):541–51.

[169] Forster RE. Diffusion of gases across the alveolar membrane. Compr Physiol. Supplement 13: Handbook of Physiology, The Respiratory System, Gas Exchange; 2011. p. 71–88. First published in print 1987. https://doi.org/10.1002/cphy.cp030405.

[170] Grollman A. The determination of the cardiac output of man by the use of acetylene. Am J Physiol 1929;88:432–45.

[171] McKelvie RS, Heigenhauser G, Jones NL. Measurement of cardiac output by CO_2 rebreathing in unsteady state exercise. Chest J 1987;92(5):777–82.

[172] Clausen JP, Larsen OA, Trap-Jensen J. Cardiac output in middle-aged patients determined with CO_2 rebreathing method. J Appl Physiol 1970;28(3):337–42.

[173] McLure LE, Brown A, Lee WN, Church AC, Peacock AJ, Johnson MK. Non-invasive stroke volume measurement by cardiac magnetic resonance imaging and inert gas rebreathing in pulmonary hypertension. Clin Physiol Funct Imaging 2011;31(3):221–6.

[174] Power GG. Solubility of O_2 and CO in blood and pulmonary and placental tissue. J Appl Physiol 1968;24(4):468–74.

[175] Hoeper MM, Meyer K, Rademacher J, Fuge J, Welte T, Olsson KM. Diffusion capacity and mortality in patients with pulmonary hypertension due to heart failure with preserved ejection fraction. JACC Heart Fail 2016;4(6):441–9.

[176] Olson TP, Johnson BD, Borlaug BA. Impaired pulmonary diffusion in heart failure with preserved ejection fraction. JACC Heart Fail 2016;4(6):490–8.

[177] Friedman M, Wilkins Jr SA, Rothfeld AF, Bromberg PA. Effect of ventilation and perfusion imbalance on inert gas rebreathing variables. J Appl Physiol Respir Environ Exerc Physiol 1984;56(2):364–9.

[178] Meyer M, Scheid P. Solubility of acetylene in human blood determined by mass spectrometry. J Appl Physiol Respir Environ Exerc Physiol 1980;48(6):1035–7.

[179] Sendroy J, Dillon RT, Van Slyke DD. Studies of gas and electrolyte equilibria in blood XIX. The solubility and physical state of uncombined oxygen in blood. J Biol Chem 1934;105(3):597–632.

[180] Taylor BJ, Olson TP, Chul Ho K, Maccarter D, Johnson BD. Use of noninvasive gas exchange to track pulmonary vascular responses to exercise in heart failure. Clin Med Insights Circ Respir Pulm Med 2013;7:53–60.

[181] Whipp BJ, Higgenbotham MB, Cobb FC. Estimating exercise stroke volume from asymptotic oxygen pulse in humans. J Appl Physiol (1985) 1996;81(6).

[182] Dunselman PH, Kuntze CE, van Bruggen A, Beekhuis H, Piers B, Scaf AH, et al. Value of New York Heart Association classification, radionuclide ventriculography, and cardiopulmonary exercise tests for selection of patients for congestive heart failure studies. Am Heart J 1988;116(6 Pt 1):1475–82.

[183] Cohn JN. The management of chronic heart failure. N Engl J Med 1996;335(7):490–8.

[184] Rickenbacher PR, Trindade PT, Haywood GA, Vagelos RH, Schroeder JS, Willson K, et al. Transplant candidates with severe left ventricular dysfunction managed with medical treatment: characteristics and survival. J Am Coll Cardiol 1996;27(5):1192–7.

[185] Stelken AM, Younis LT, Jennison SH, Miller DD, Miller LW, Shaw LJ, et al. Prognostic value of cardiopulmonary exercise testing using percent achieved of predicted peak oxygen uptake for patients with ischemic and dilated cardiomyopathy. J Am Coll Cardiol 1996;27(2):345–52.

[186] Piepoli M, Ponikowski P, Clark AL, Banasiak W, Capucci A, Coats AJS. A neural link to explain the "muscle hypothesis" of exercise intolerance in chronic heart failure. Am Heart J 1999;137(6):1050–6.

[187] Bhella PS, Prasad A, Heinicke K, Hastings JL, Arbab-Zadeh A, Adams-Huet B, et al. Abnormal haemodynamic response to exercise in heart failure with preserved ejection fraction. Eur J Heart Fail 2011;13(12):1296–304.

[188] Mahapatra S, Nishimura RA, Oh JK, McGoon MD. The prognostic value of pulmonary vascular capacitance determined by Doppler echocardiography in patients with pulmonary arterial hypertension. J Am Soc Echocardiogr 2006;19(8):1045–50.

[189] Van Iterson EH, Snyder EM, Joyner MJ, Johnson BD, Olson TP. Intrathecal fentanyl blockade of afferent neural feedback from skeletal muscle during exercise in heart failure patients: influence on circulatory power and pulmonary vascular capacitance. Int J Cardiol 2015;201:384–93.

[190] Metra M, Faggiano P, D'Aloia A, Nodari S, Gualeni A, Raccagni D, et al. Use of cardiopulmonary exercise testing with hemodynamic monitoring in the prognostic assessment of ambulatory patients with chronic heart failure. J Am Coll Cardiol 1999;33(4):943–50.

[191] Borlaug BA, Melenovsky V, Russell SD, Kessler K, Pacak K, Becker LC, et al. Impaired chronotropic and vasodilator reserves limit exercise capacity in patients with heart failure and a preserved ejection fraction. Circulation 2006;114(20):2138–47.

[192] Franciosa JA, Baker BJ, Seth L. Pulmonary versus systemic hemodynamics in determining exercise capacity of patients with chronic left ventricular failure. Am Heart J 1985;110(4):807–13.

[193] Sullivan MJ, Knight J, Higginbotham M, Cobb F. Relation between central and peripheral hemodynamics during exercise in patients with chronic heart failure. Muscle blood flow is reduced with maintenance of arterial perfusion pressure. Circulation 1989;80(4):769–81.

[194] Levy AM, Tabakin BS, Hanson JS. Cardiac output in normal men during steady-state exercise utilizing dye-dilution technique. Br Heart J 1961;23(4):425.

[195] Gleason WL, Braunwald E. Studies on starling's law of the heart VI. Relationships between left ventricular end-diastolic volume and stroke volume in man with observations on the mechanism of pulsus alternans. Circulation 1962;25(5):841–8.

[196] Braunwald E, Ross Jr J, Gault JH, Mason DT, Mills C, Gabe IT, et al. NIH clinical staff conference. Assessment of cardiac function. Ann Intern Med 1969;70(2):369–99.

[197] Ross Jr J, Braunwald E. Studies on starling's law of the heart. IX. The effects of impeding venous return on performance of the normal and failing human left ventricle. Circulation 1964;30(5):719–27.

[198] Ross J, Gault JH, Mason DT, Linhard JW, Braunwald E. Left ventricular performance during muscular exercise in patients with and without cardiac dysfunction. Circulation 1966;34(4):597–608.

[199] Dexter L, Whittenberger J, Haynes F, Goodale W, Gorlin R, Sawyer C. Effect of exercise on circulatory dynamics of normal individuals. J Appl Physiol 1951;3(8):439–53.

[200] Levine HJ, Neill WA, Wagman RJ, Krasnow N, Gorlin R. The effect of exercise on mean left ventricular ejection rate in man. J Clin Investig 1962;41(5):1050.

[201] Donald K, Bishop J, Cumming G, Wade O. The effect of exercise on the cardiac output and circulatory dynamics of normal subjects. Clin Sci (Lond) 1955;14(1):37–73.

[202] Beck KC, Randolph LN, Bailey KR, Wood CM, Snyder EM, Johnson BD. Relationship between cardiac output and oxygen consumption during upright cycle exercise in healthy humans. J Appl Physiol 2006;101(5):1474–80.

[203] Stringer WW, Hansen JE, Wasserman K. Cardiac output estimated noninvasively from oxygen uptake during exercise. J Appl Physiol (1985) 1997;82(3):908–12.

[204] Bhambhani Y, Norris S, Bell G. Prediction of stroke volume from oxygen pulse measurements in untrained and trained men. Can J Appl Physiol 1994;19(1):49–59.

[205] Raphael C, Briscoe C, Davies J, Whinnett ZI, Manisty C, Sutton R, et al. Limitations of the New York Heart Association functional classification system and self-reported walking distances in chronic heart failure. Heart 2007;93(4):476–82.

[206] Goldman L, Hashimoto B, Cook E, Loscalzo A. Comparative reproducibility and validity of systems for assessing cardiovascular functional class: advantages of a new specific activity scale. Circulation 1981;64(6):1227–34.

[207] Rostagno C, Galanti G, Comeglio M, Boddi V, Olivo G, Serneri GGN. Comparison of different methods of functional evaluation in patients with chronic heart failure. Eur J Heart Fail 2000;2(3):273–80.

[208] Goode KM, Nabb S, Cleland JG, Clark AL. A comparison of patient and physician-rated New York Heart Association class in a community-based heart failure clinic. J Card Fail 2008;14(5):379–87.

[209] Miller-Davis C, Marden S, Leidy NK. The New York Heart Association Classes and functional status: what are we really measuring? Heart Lung J Acute Crit Care 2006;35(4):217–24.

[210] Borlaug BA, Olson TP, Lam CS, Flood KS, Lerman A, Johnson BD, et al. Global cardiovascular reserve dysfunction in heart failure with preserved ejection fraction. J Am Coll Cardiol 2010;56(11):845–54.

[211] Abudiab MM, Redfield MM, Melenovsky V, Olson TP, Kass DA, Johnson BD, et al. Cardiac output response to exercise in relation to metabolic demand in heart failure with preserved ejection fraction. Eur J Heart Fail 2013;15(7):776–85.

[212] Zelis R, Longhurst J, Capone RJ, Mason DT. A comparison of regional blood flow and oxygen utilization during dynamic forearm exercise in normal subjects and patients with congestive heart failure. Circulation 1974;50(1):137–43.

[213] Longhurst J, Gifford W, Zelis R. Impaired forearm oxygen consumption during static exercise in patients with congestive heart failure. Circulation 1976;54(3):477–80.

[214] LeJemtel TH, Maskin CS, Lucido D, Chadwick BJ. Failure to augment maximal limb blood flow in response to one-leg versus two-leg exercise in patients with severe heart failure. Circulation 1986;74(2):245–51.

[215] Cooke GA, Marshall P, al-Timman JK, Wright DJ, Riley R, Hainsworth R, et al. Physiological cardiac reserve: development of a non-invasive method and first estimates in man. Heart 1998;79(3):289–94.

[216] Tan L, Bain R, Littler W. Assessing cardiac pumping capability by exercise testing and inotropic stimulation. Br Heart J 1989;62(1):20–5.

[217] Bain RJ, Tan LB, Murray RG, Davies MK, Littler WA. The correlation of cardiac power output to exercise capacity in chronic heart failure. Eur J Appl Physiol Occup Physiol 1990;61(1-2):112–8.

[218] Cotter G, Moshkovitz Y, Kaluski E, Milo O, Nobikov Y, Schneeweiss A, et al. The role of cardiac power and systemic vascular resistance in the pathophysiology and diagnosis of patients with acute congestive heart failure. Eur J Heart Fail 2003;5(4):443–51.

[219] Amann M, Venturelli M, Ives SJ, Morgan DE, Gmelch B, Witman MA, et al. Group III/IV muscle afferents impair limb blood in patients with chronic heart failure. Int J Cardiol 2014;174(2):368–75.

[220] Scott AC, Davies LC, Coats AJ, Piepoli M. Relationship of skeletal muscle metaboreceptors in the upper and lower limbs with the respiratory control in patients with heart failure. Clin Sci (Lond) 2002;102(1):23–30.

[221] Scott AC, Francis DP, Coats AJ, Piepoli MF. Reproducibility of the measurement of the muscle ergoreflex activity in chronic heart failure. Eur J Heart Fail 2003;5(4):453–61.

[222] Scott AC, Francis DP, Davies LC, Ponikowski P, Coats AJ, Piepoli MF. Contribution of skeletal muscle "ergoreceptors" in the human leg to respiratory control in chronic heart failure. J Physiol 2000;529(3):863–70.

[223] Crisafulli A, Salis E, Tocco F, Melis F, Milia R, Pittau G, et al. Impaired central hemodynamic response and exaggerated vasoconstriction during muscle metaboreflex activation in heart failure patients. Am J Physiol Heart Circ Physiol 2007;292(6):H2988–96.

[224] Ponikowski P, Francis DP, Piepoli MF, Davies LC, Chua TP, Davos CH, et al. Enhanced ventilatory response to exercise in patients with chronic heart failure and preserved exercise tolerance: marker of abnormal cardiorespiratory reflex control and predictor of poor prognosis. Circulation 2001;103(7):967–72.

[225] Piepoli M, Clark AL, Coats AJ. Muscle metaboreceptors in hemodynamic, autonomic, and ventilatory responses to exercise in men. Am J Physiol 1995;269(4 Pt 2):H1428–36.

[226] Piepoli M, Clark AL, Volterrani M, Adamopoulos S, Sleight P, Coats AJ. Contribution of muscle afferents to the hemodynamic, autonomic, and ventilatory responses to exercise in patients with chronic heart failure: effects of physical training. Circulation 1996;93(5):940–52.

[227] Piepoli MF, Dimopoulos K, Concu A, Crisafulli A. Cardiovascular and ventilatory control during exercise in chronic heart failure: role of muscle reflexes. Int J Cardiol 2008;130(1):3–10.

[228] Clark AL. Origin of symptoms in chronic heart failure. Heart 2006;92(1):12–6.

[229] Clark AL, Piepoli M, Coats AJ. Skeletal muscle and the control of ventilation on exercise: evidence for metabolic receptors. Eur J Clin Invest 1995;25(5):299–305.

[230] Clark AL, Poole-Wilson PA, Coats AJ. Exercise limitation in chronic heart failure: central role of the periphery. J Am Coll Cardiol 1996;28(5):1092–102.

[231] Scott AC, Wensel R, Davos CH, Georgiadou P, Kemp M, Hooper J, et al. Skeletal muscle reflex in heart failure patients: role of hydrogen. Circulation 2003;107(2):300–6.

[232] Scott AC, Wensel R, Davos CH, Kemp M, Kaczmarek A, Hooper J, et al. Chemical mediators of the muscle ergoreflex in chronic heart failure: a putative role for prostaglandins in reflex ventilatory control. Circulation 2002;106(2):214–20.

[233] Smith SA, Mammen PP, Mitchell JH, Garry MG. Role of the exercise pressor reflex in rats with dilated cardiomyopathy. Circulation 2003;108(9):1126–32.

[234] Smith SA, Mitchell JH, Naseem RH, Garry MG. Mechanoreflex mediates the exaggerated exercise pressor reflex in heart failure. Circulation 2005;112(15):2293–300.

[235] Smith SA, Williams MA, Mitchell JH, Mammen PP, Garry MG. The capsaicin-sensitive afferent neuron in skeletal muscle is abnormal in heart failure. Circulation 2005;111(16):2056–65.

[236] Segal SS, Kurjiaka DT. Coordination of blood flow control in the resistance vasculature of skeletal muscle. Med Sci Sports Exerc 1995;27(8):1158–64.

[237] Rossiter HB, Ward SA, Howe FA, Kowalchuk JM, Griffiths JR, Whipp BJ. Dynamics of intramuscular 31P-MRS Pi peak splitting and the slow components of PCr and O2 uptake during exercise. J Appl Physiol 2002;93(6):2059–69.

[238] Henneman E, Somjen G, Carpenter DO. Excitability and inhibitability of motoneurons of different sizes. J Neurophysiol 1965;28(3):599–620.

[239] Özyener F, Rossiter H, Ward S, Whipp B. Influence of exercise intensity on the on-and off-transient kinetics of pulmonary oxygen uptake in humans. J Physiol 2001;533(3):891–902.

[240] DeLorey DS, Kowalchuk JM, Paterson DH. Relationship between pulmonary O2 uptake kinetics and muscle deoxygenation during moderate-intensity exercise. J Appl Physiol (1985) 2003;95(1):113–20.

[241] DeLorey DS, Kowalchuk JM, Paterson DH. Effect of age on O(2) uptake kinetics and the adaptation of muscle deoxygenation at the onset of moderate-intensity cycling exercise. J Appl Physiol (1985) 2004;(1):165–72.

[242] Mitchell SH, Steele NP, Leclerc KM, Sullivan M, Levy WC. Oxygen cost of exercise is increased in heart failure after accounting for recovery costs. Chest J 2003;124(2):572–9.

[243] Roston WL, Whipp BJ, Davis JA, Cunningham DA, Effros RM, Wasserman K. Oxygen uptake kinetics and lactate concentration during exercise in humans. Am Rev Respir Dis 1987;135(5):1080–4.

[244] Sato A. Spinal and medullary reflex components of the somatosympathetic reflex discharges evoked by stimulation of the group IV somatic afferents. Brain Res 1973;51:307–18.

[245] Iwamoto GA, Kaufmann M, Botterman BR, Mitchell JH. Effects of lateral reticular nucleus lesions on the exercise pressor reflex in cats. Circ Res 1982;51(3):400–3.

[246] Ciriello J, Calaresu FR. Lateral reticular nucleus: a site of somatic and cardiovascular integration in the cat. Am J Physiol 1977;233(3):R100–9.

[247] Mark AL, Victor RG, Nerhed C, Wallin BG. Microneurographic studies of the mechanisms of sympathetic nerve responses to static exercise in humans. Circ Res 1985;57(3):461–9.

[248] McCloskey D, Mitchell J. Reflex cardiovascular and respiratory responses originating in exercising muscle. J Physiol 1972;224(1):173–86.

[249] Coote JH, Hilton SM, Perez-Gonzalez JF. The reflex nature of the pressor response to muscular exercise. J Physiol 1971;215(3):789–804.

[250] Adreani CM, Hill JM, Kaufman MP. Responses of group III and IV muscle afferents to dynamic exercise. J Appl Physiol 1997;82(6):1811–7.

[251] Kaufman MP, Hayes SG, Adreani CM, Pickar JG. Discharge properties of group III and IV muscle afferents. Adv Exp Med Biol 2002;508:25–32.

[252] Coote JH, Perez-Gonzalez JF. The response of some sympathetic neurones to volleys in various afferent nerves. J Physiol 1970;208(2):261–78.

[253] Scott AC, Wensel R, Davos CH, Georgiadou P, Ceri Davies L, Coats AJ, et al. Putative contribution of prostaglandin and bradykinin to muscle reflex hyperactivity in patients on Ace-inhibitor therapy for chronic heart failure. Eur Heart J 2004;25(20):1806–13.

[254] Van Iterson EH, Snyder EM, Johnson BD, Olson TP. Influence of the metaboreflex on pulmonary vascular capacitance in heart failure. Med Sci Sports Exerc 2016;48(3):353–62.

[255] Lalande S, Luoma CE, Miller AD, Johnson BD. Expiratory loading improves cardiac output during exercise in heart failure. Med Sci Sports Exerc 2012;44(12):2309.

[256] Agostoni P, Pellegrino R, Conca C, Rodarte JR, Brusasco V. Exercise hyperpnea in chronic heart failure: relationships to lung stiffness and expiratory flow limitation. J Appl Physiol 2002;92(4):1409–16.

Chapter 7

Exercise-Based Cardiovascular Therapeutics: From Cellular to Molecular Mechanisms

Siyi Fu*, Qiying Dai[†,‡], Yihua Bei*, Yongqin Li*, Junjie Xiao*

*Cardiac Regeneration and Ageing Lab, School of Life Science, Shanghai University, Shanghai, China, [†]Department of Cardiology, The First Affiliated Hospital of Nanjing Medical University, Nanjing, China, [‡]Metrowest Medical Center, Framingham, MA, United States

CARDIOVASCULAR DISEASE

Cardiovascular disease (CVD) is considered as the number one killer around the world. The 2013 overall rate of death attributable to CVD was 222.9 per 100,000 Americans. The death rates were 269.8 for males and 184.8 for females. The rates were 270.6 for non-Hispanic white males, 356.7 for non-Hispanic white females, 246.6 for non-Hispanic black females, and 136.4 for Hispanic females [1]. Moreover, total coronary and stroke deaths will even increase 18% and 50% by 2030 in United States [2]. The estimated increase stems in part from the aging of the population and is fueled by the recent trends for increasing obesity rates and the concomitant rising rates of hypertension (8% increase over the next decade) and diabetes mellitus (100% increase over next three decades) [3]. CVD could be the result of many risk factors including changed lipid level (including low-density lipoprotein cholesterol, LDL-C; HDL; and triglycerides), hypertension, smoking, diabetes, obesity, and physical inactivity [4]. What's more, CVD itself could also be the cause of lots of other diseases including cancer [5]. In some CVDs such as myocardial infarction, more than a billion cardiomyocytes can be killed, resulting in severe cardiac malfunction. Although widely considered as a nonregenerative organ, increasing studies have discovered a potential limited regenerative capacity of the heart [6–9]. All these findings suggested a new strategy in treating CVDs by enhancing cardiac repair. In this chapter, we would like to discuss physical exercise training in treating CVDs.

EXERCISE TREATMENT

The Oxford Dictionary defines exercise as "activity requiring physical effort, carried out to sustain or improve health and fitness." As a planned, structured, and repetitive physical activity, exercise can be divided as aerobic and anaerobic one such as swimming, jogging, high-intensity interval training (HIIT), cycling, yoga, or even tai chi [10,11]. Endurance proper exercise, short time for high-intensity anaerobic exercise, or long time for low-intensity aerobic exercise have increasingly been found to contribute to health and are considerable ways for physical rehabilitation. There is a vast literature on the beneficial effects of physical exercise on all-cause and CVD-related mortality [12]. Here, we focus on the effects of physical exercise on CVD, and the underlying mechanism will also be discussed.

CARDIOVASCULAR ADAPTIONS TO EXERCISE (FIG. 1)

Cardiomyocyte Hypertrophy and Renewal

In general, cardiac hypertrophy can be divided into pathological hypertrophy and physiological hypertrophy. Both cause heart muscle cells to increase in size. However, pathological hypertrophy could be accompanied by proliferation of fibroblasts and accumulation of extracellular matrix [13,14]. On the contrary, with a reversible growth of cardiac mass, exercise-induced physiological hypertrophy preserves normal physiological cardiac structure and even improves cardiac function [15–17]. In an animal model of voluntary exercise, rats were given free access to a running wheel over a 14-week period, and approximately 20% greater volume was observed in endocardium cells [18]. In professional athletes' heart, both

87

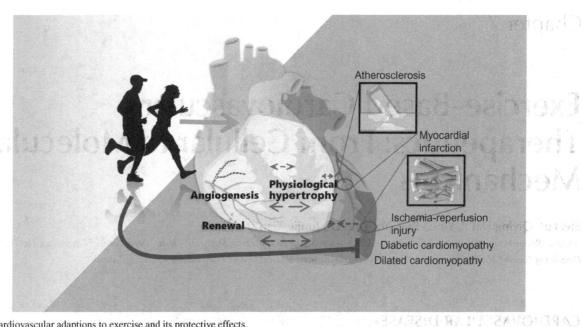

FIG. 1 Cardiovascular adaptions to exercise and its protective effects.

systolic and diastolic function measured by echocardiography show similar condition as the healthy individuals [19], which supports the idea that exercise-induced hypertrophy is physiological hypertrophy.

The idea that adult heart is not a regenerative organ has been widely accepted by the general until successive evidences of cardiomyocyte proliferation and partial heart repair is demonstrated [20–22]. Cardiomyocytes can replace themselves during ischemia, and an increased number of ventricular myocytes can be observed in severe hypertrophic left ventricles [1,23,24]. In 2009, a breakthrough study confirms that cardiomyocytes can renew themselves at a rate of 1% per year in human [25].

Encouraging by the finding, many other studies have been conducted to explore the regenerative activity of the heart. Among them, exercise was found to be one of the stimuli to be able to promote new cardiomyocytes formation [26,27]. However, the source of the new cardiomyocytes remains debatable. Some study considered stem/progenitor cell population to be the source [28–30]. However, others prove that preexisting cardiomyocytes are the dominant source of cardiomyocyte replacement by applying genetic fate-mapping and multiisotope imaging mass spectrometry [31].

Protection of Cardiomyocyte Apoptosis and Necrosis

Aging and reperfusion injury increased mitochondrial-mediated apoptosis [32,33]. Inflammatory and ischemic injury precipitates necrosis, which is commonly seen in some types of CVD patients, especially acute myocardial infarction. For years, studies have shown that physical exercise effectively diminishes apoptosis [32–37].

Angiogenesis

As another way to protect cardiomyocytes, reformation of the coronary vasculature could be affected by physical exercise. A key role in angiogenesis is played by the vascular endothelial growth factor (VEGF) [38]. In an animal study, adult male rats were trained to run on a treadmill for 10 weeks, and the other group of animals were subjected to training followed by 4-week detraining. Markers of angiogenesis like VEGF were found to be increased after exercise training and returned to baseline after detraining [39]. By measuring the coronary collateral flow index (CFI) before and after high-intensity exercise, another group found improvement of blood supply in response to exercise [40].

THE UNDERLYING MECHANISM OF EXERCISE-INDUCED CARDIAC GROWTH

The IGF-1-PI3K(p110α)-AKT Pathway

The IGF (insulin-like growth factor)-1-PI3K (phosphoinositide 3-kinase)-AKT signaling pathway is crucial in physiological cardiac growth induced by exercise. Mainly produced by the liver, IGF-1 is one of the most important growth factors in the genesis of physiological hypertrophy [41]. IGF-1 was found to be responsible for postnatal tissue growth. Furthermore, short-term administration of IGF-1 is beneficial in improving cardiac function in heart failure [42,43]. In a transgenic overexpression IGF-1 receptor (IGF1R) mice model, an increase in cardiomyocyte size was found [43]. Emerging evidence has shown a positive correlation between physical exercise and IGF-1 levels [44]. By studying athlete's heart function, the increased IGF-1 was found to be related to the left ventricular mass index and left ventricular end-diastolic dimension index, suggesting a role of IGF-1 in the induction of physiological heart growth [45]. Several signaling pathways have been investigated to explain this, and one of the well-known is the IGF-1-PI3K-AKT pathway. IGF-1 works by binding its receptor IGF1R. Interestingly, IGF-1 was found to activate both IGF1R and insulin receptor [46]. After IGF-1 binds to its receptor, PI3K pathway, one of major pathways of IGF1R signaling in cardiac myocytes, is activated through the interaction of phosphorylated IRS1/IRS2 with p85 PI3K subunit. Then, p110 catalytic subunit, p70S6K1 and phosphorylation of protein kinase B (also known as AKT), is activated. Among them, PI3K is recognized as the key in the activation process. One study compares cardiac-specific transgenic (Tg) mice with enhanced PI3K (p110α) activity and non-Tg controls, and they found that the former group presented beneficial effect after exercise training [47]. AKT is a serine/threonine kinase involved in the downstream reaction of PI3K pathway, and it has determinative role in this pathway as well. Transgenic mice with cardiomyocyte-specific constitutively activation of AKT (caAKT) demonstrated twofold increase in heart size compared with kinase-deficient AKT (kdAKT) ones regardless of IGF-1 injection [48]. These studies fully demonstrated the essential role of PI3K and AKT in physiological growth and validating the previous hypothesis that IGF-1-PI3K-AKT pathway is required in exercise-induced cardiac hypertrophy [48–54].

Gp130/JAK/STAT Pathway

IL-6 is found to be elevated in human heart failure with multiple components of its glycoprotein (gp) 130 receptor system impaired, indicating an important role of this system in cardiac repair [55]. The two major signaling cascades activated by the gp130 receptor, SHP2/ERK and STAT pathways, have been demonstrated to participate in cardiac development, hypertrophy, protection, and remodeling in response to physiological (exercise and pregnancy) and pathophysiological stimuli (hypertension and fluid overload) [56,57]. The STAT proteins enter the nucleus and bind to the promoter region of target genes and regulate the transcription of the genes once activated [58]. Of note, as a salvage pathway, STAT3-induced cardiac hypertrophy is different from that induced by PI3K pathway. The former protects the heart by helping the heart adapt to the pathological stress. In an animal model of pressure overload-induced left ventricular hypertrophy, STAT3 was found to contribute to cardiac hypertrophy in the presence of IL-6 [59]. Disruption of gp130 was shown to present heart failure in response to mechanical stress accompanied by an increase in apoptosis [60]. Moreover, in the doxorubicin-induced cardiomyopathy model, it was demonstrated that STAT3 transduced not only a hypertrophic but also a protective signal by inhibiting cardiac contractile genes and inducing cardiac protective factors [61].

NRG1-ErbB4-C/EBPβ Pathway

Neuregulin-1 (NRG-1)/ErbB pathway is recognized as another cardioprotective system that becomes activated during stress, most convincingly in response to cardiotoxic chemotherapy [62]. As discussed above, it has been increasingly accepted that the human heart has limited regeneration potential. NRG1 and its receptor, ErbB4 participate in cardiac regeneration. NRG1 induced mononucleated, but not binucleated, cardiomyocyte to divide and ErbB4 was reported to have promotive effect on cardiomyocyte proliferation [63]. It was reported that by using an ErbB signaling inhibitor, exercise-induced cardiac proliferation is blunted [64]. Both animal and human studies have found that NRG-1 administration improved cardiac function in heart failure, but the underlying mechanism needs to be further evaluated [65,66]. As a target of NRG-1-related signaling pathway, CCAAT/enhancer binding protein β (C/EBPβ) is downregulated during endurance exercise training [67]. Both in vitro and in vivo, reduction of C/EBPβ mimics a phenotype of physiological cardiac hypertrophy induced by exercise with both enlargements of size and proliferation of cardiomyocytes. The reduction of C/EBPβ upregulates some important hypertrophy genes like Nkx2.5, Gata4, and Tbx5. Furthermore, C/EBPβ downregulates CBP/p300-interacting protein (CITED4), which promotes the proliferation of cardiomyocytes at baseline and improves cardiac function after ischemic injury [68,69]. The NRG1-ErbB4-C/EBPβ pathway was proved to be regulated by reactive oxygen species as well, indicating its role in cardiac adaptation to oxidative stress [70].

Nitric Oxide (NO) Signaling

The fundamental role of NO in homeostasis of endothelial cells has been well studied. It exerts cardiac protective effects by working together with exercise training. To prove this, a nonselective NO synthase inhibitor was used. Unsurprisingly, treatment group blunted the beneficial effects of exercise training against the increase in heart weigh/body weight and echocardiographic variables [71]. Furthermore, difference in cardiac function after exercise in eNOS$^{-/-}$, eNOS$^{+/-}$, and eNOS$^{+/+}$ mice displayed that the protective effects of exercise disappears in eNOS$^{+/-}$ and eNOS$^{-/-}$ mice, suggesting that the lack of one eNOS allele is sufficient to negate all beneficial effects from exercise [35]. Evidence revealed that the alterations in eNOS phosphorylation status and NO generation were mediated by β3-AR stimulation and that in response to exercise. A deficiency of β3-ARs leads to an exacerbation of myocardial infarction following ischemia-reperfusion injury. By increasing the β3-ARs, exercise protects the heart against this type of injury [72]. During exercise training, NO could be produced by neuronal NO synthase (NOS1) from skeletal muscle, and overexpression of NOS1 may mimic the beneficial effects of exercise [73]. Exercise training involves the regulation of sympathetic nervous system. In hypertensive patients, dysfunction of sympathetic system leads to abnormally elevated blood pressure. Blocking NOS significantly diminished the improvements in functional sympatholysis in exercise-trained hypertensive rats, indicating that functional sympatholysis is impaired in hypertension via a NO-dependent mechanism [74]. Interestingly, compared with traditional land-based training, water-based training is more effective in inducing plasma NO concentration without affecting catecholamine levels, which provides new directions in rehabilitation program [75].

MicroRNAs

MicroRNAs, a novel class of small noncoding RNAs about 22 nucleotides in length, usually interfere the transcription of mRNA. To date, more than one thousand human miRNAs have been identified and reported to participate in a wide range of biological processes, including cellular proliferation, differentiation, aging, and apoptosis [76]. Considering miRNAs' powerful and multiple functions, it's not a surprise that they play a key role in exercise-induced physiological hypertrophy. Numerous complicated connections between the heart growth and miRNAs have been established, giving us new expectation in exercise-induced therapy for CVDs [77,78].

In cardiac hypertrophy induced by transverse aortic constriction (TAC), AKT overexpression, and endurance exercise training, both miR-133 and miR-1 have been found to be decreased. In vitro, cardiac hypertrophy induced by PE or ET-1 can be inhibited by elevating the expression of miR-133 or miR-1. Inhibition of miR-133 in vivo reveals a remarkable cardiac hypertrophy with a reinduction of several fetal genes. RhoA, Cdc42, and Nelf-A/WHSC2, which all have certain effects in cardiac hypertrophy and cardiogenesis, are identified to be the targets of miR-133 [79–81]. The downregulation of miR-214 and improved expression of SERCA2a were proved to be responsible for the enlargement of the left ventricle and enhanced cardiomyocyte contraction during resistance training [82]. Interestingly, miR-222 has been found to be necessary in exercise-induced cardiomyocyte growth and proliferation. By targeting p27 (cyclin-dependent kinase inhibitor 1B), homeodomain-interacting protein kinase ½ (HIPK1/2), and homeobox containing 1 (HMBOX1), miR-222 promotes cardiomyocytes growth, proliferation, and survival in vitro. Meanwhile, inhibition of miR-222 totally prevents cardiac growth in response to exercise with decreasing the key genes of cardiomyocyte proliferation [83].

During tissue injury, miRNAs could also be released into the bloodstream as "circulating" miRNA (c-miRNA). Exercise is one of the most positive and effective means of achieving enhanced physique [84]. Tissue-specific c-miRNAs like miR-1, miR-133a, and miR-208a have been found to increase transiently after exercise. Compared with relatively prolonged elevation in conventional injury marker, these c-miRNAs may better describe acute responsiveness to stress [85]. One study found that different endurance exercise protocols lead to damage of the endothelial cell layers while resistance exercise has no impact by measuring the serum level of miR-126 and miR-133 [86]. These data indicate the potential role of miRNA in directing individualized training regimen.

EXERCISE-BASED PROTECTION FOR CARDIOVASCULAR DISEASE (FIG. 1)

Myocardial Infarction

Among patients who survive the first episode of myocardial infarction, 20% suffer a second cardiovascular event in the first year, and approximately 50% of major coronary events occur in those with a previous hospital discharge diagnosis of ischemic heart disease. Exercise has been revealed to be cardiac protective through various cellular and molecular mechanisms [87,88]. As early as 1978, a 30% decrease in infarct size in chronically exercised rat was found as compared with its sedentary control [89]. In addition, an increase in artery density and changes in metabolism during myocardial infarction after exercise was found [90]. The effect on the arteries involving the infarcted area is correlated to the overexpression of transforming growth factor beta (TGF-β) and downregulation of angiostatin after exercise [91]. Exercise training abolishes the derangement of myocardial glucose and lipid metabolism and further enhances the adaptive increase of mitochondrial biogenesis [92]. On the other hand, stress in cardiac endoplasmic reticulum (ER) through accumulation of misfolded proteins in CVD cannot be neglected. These unfolded protein response (UPR), once activated, could lead to cell apoptosis. Aerobic exercise training is the only resolution available to relieve the ER stress and decrease UPR-induced apoptosis by improving chymotrypsin-like proteasome activity [93]. Moreover, as discussed above, exercise training increase activity of eNOS system, which is related to decreased fibrosis, apoptosis, and improvement in left ventricular function after myocardial infarction [35].

Arrhythmia is one of the most fatal complications in the first 24 h after myocardial infarction. Impaired cardiac control of intracellular diastolic Ca^{2+} gives rise to arrhythmias. In general, exercise training protects against arrhythmia by improving the cardiomyocyte control of diastolic Ca^{2+} [94]. In diabetes, which aggravates ventricular arrhythmias, exercise training normalized the expression profile of key proteins involved in cardiomyocyte calcium handling and reduced inducible ventricular arrhythmias in infarcted diabetic mice [95].

Exercise-based cardiac rehabilitation is proved to improve quality of life and increase health-related benefits [96,97]. In clinic practice, myocardial infarction patients should have exercise regimen based on their cardiac conditions. Peak oxygen uptake, which is a commonly used marker to predict activity capacity, has been reported to be elevated in both high-intensity interval training and moderate-intensity continuous exercise training [97]. In another study, patients after myocardial infarction were assigned to aerobic exercise group and control group. Using cardiac magnetic resonance imaging, the left ventricular mass/end-diastolic volume ratio demonstrated a significant positive remodeling in the exercise group [98].

Myocardial Ischemia-Reperfusion Injury (I/R)

Timely reperfusion is required for myocardial infarction patients though it can also lead to irreversible injury [99]. Moderate aerobic exercise is found to maintain high percentage of preischemia cardiac output during reperfusion, improving recovery and keep coronary flow after ischemia-reperfusion [100,101]. Morphological studies on the contractile function demonstrated that trained myocardium appears to have a greater maximum force-generating ability that may compensate for reduced contractile function induced by a brief period of ischemia [102]. It has been proved that heat-shock protein72 (HSP72) and antioxidant effect induced by exercise are part of the reason for cardiac protection during I/R injury [103–105]. In addition, some studies supported that the antioxidant effect could be achieved by attenuating I/R-induced calpain activation [106]. Other studies also found protein kinase C [107], opioid receptor [108], nitrate reductase [109], adenosine triphosphate (ATP)-sensitive K^+ (KATP) channels [110,111], IL-6 [112], PI3K [113], and HO/NOS [114] to be involved in the process. The cardiac protective effect from exercise works on elderly heart as well. Taking aging into consideration, a study on effects of moderate running training on postischemic recovery of contractile function and coronary perfusion in senescent myocardium has been done, and the beneficial effect of exercise on the elderly was clearly shown [115]. Both in vitro and in vivo studies validate the cardioprotective effect of exercise training [116,117]. Regarding to the timing and the type of exercise, exercise training has most favorable effects if started at the acute phase following MI [118]. Short-term high-intensity interval training (HIIT) was also effective, and this protective effect can be sustained for at least 1 week following the cessation of the training [119].

Atherosclerosis

Both carotid atherosclerosis and coronary atherosclerosis could be prevented and improved by exercise training [120,121].

Apolipoprotein E knockout mouse has been widely used as a model for inducing atherosclerosis. By using this model, exercise training was demonstrated to be an effective approach to stabilize plaque lesions [122], protect from atherosclerosis combined with diet [123], and slow CVD in the presence of kidney disease [124]. Induced NO production could explain the effects in part. Evidence has shown that inflammatory process, which was considered to play a key role in atherosclerosis, could be counteracted by exercise training via the increased NO and eNOS [125]. Moreover, NO pathway could efficiently counterbalance coronary atherosclerosis-induced impairment in body maximal oxygen consumption and exercise capacity [126]. Pretreated with N(G)-nitro-L-arginine methylester (L-NAME), an inhibitor of NO synthase, could block the effects of exercise in atherosclerotic lesions [127]. In addition, exercise also exerts its anti-inflammation activity through modulating inflammatory cytokines like IL-6, IL-8, TNF-α, IFN-γ [128,129], and novel CVD risk factors such as adipokines, visfatin, apelin, caspin, and ghrelin [130]. Moreover, even in established atherosclerotic lesions, exercise training promotes plaque stabilization both in diabetic atherosclerosis [131] and angiotensin II-induced atherosclerosis [132]. Apart from the anti-inflammatory effect of exercise training, modulation endothelial function and increasing the amount of circulating endothelial progenitor cells (EPCs) are also the potential benefits that exercise provides [133].

Exercise training is an effective primary prevention for atherosclerosis. Studies based a cohort of 101 men free of CVD symptoms showed that energy expenditure below 2050 kcal/week was not sufficient in preventing subclinical disorders. The most favorable results of atherosclerosis indexes were associated with energy expenditure 2050–3840 kcal/week [134]. Studying on application of exercise in different patients' subgroups found that in patients with activity impairment like multiple sclerosis, stretching exercise, and gradual increased exercise load improved subclinical atherosclerosis [135]. In women with abdominal obesity, a combination of aerobic and resistance exercises may be preferable to a single exercise mode for the same purpose [136]. Given the complexity and specificity in various patients' subgroups, more studies are highly needed to be conducted to reveal the mechanism underlying the cardiac protective effects of exercise training.

Cardiomyopathy (Diabetic Cardiomyopathy, Dilated Cardiomyopathy, Cardiac Fibrosis)

Diabetes is considered as one of the major causes of cardiomyopathy. A cluster of common features has been recognized in diabetic cardiomyopathy, including early decreased diastolic compliance, left ventricular hypertrophy and increased interstitial fibrosis [137]. Both directly and indirectly, exercise training is effective in preventing and treating diabetic-caused cardiac dysfunction. Most studies focus on type II diabetes-induced cardiomyopathy since it is more prevalent and has more profound metabolic effects. It has been established that enhancement of nonoxidative glucose pathway, lipotoxicity, mitochondrial disturbances, increased activity of reactive oxygen species, impaired endogenous antioxidant capacity, and altered calcium handling play significant roles in diabetic hearts. However, all of them could be influenced by exercise [138]. An 11% increase of peak oxygen uptake was found in diabetic heart after exercise intervention [139]. Exercise training could increase the sarcoplasmic reticulum protein content to mediate its cardioprotective effects in type II diabetes animal model [140]. In addition, improvement in mitochondrial biogenesis [141] and NADH/NADPH oxidase system in diabetic rats after exercise training has also been found [142].

Dilated cardiomyopathy involves enlargement of cardiac chambers, meaning its ability to stretch becomes weaker. As the weakness progresses, heart failure can occur. Short-term exercise training improves resting left ventricular ejection fraction (LVEF) and LVEF with acute physical exercise without altering energy metabolism [143]. In addition, LVEF has a notable positive relation with glutathione peroxidase and negative relation with noradrenaline and adrenaline [144]. Exercise regimen in this might be tailored since exercise capacity is closely related to autonomic modulation and right ventricular dysfunction [145]. More evidence will be necessary to determine the exercise regimen on this group of patients.

Cardiac fibrotic remodeling can result from normal aging or pathological stimuli such as ischemia, necrosis, and long-term high blood pressure. Numerous factors have been identified to play roles in this process. For example, bone morphogenetic protein-4 (BMP4), a mechanosensitive and pro-inflammatory gene, is proved to mediate cardiac hypertrophy, apoptosis, and fibrosis both in vitro and in vivo. The underlying mechanism was through increasing NADPH oxidase 4 expression and ROS-dependent pathways [146]. The other critical factor is miRNA family, which is essential for cardiac growth as we mentioned above. miR-29b and miR-30c, counteracting each other, affect the cardiac fibrotic system. miR-29b was found to attenuate fibroblast response to inflammatory factors, but miR-30c had opposite effects in treatment of factors like leukemia inhibitory factors and insulin-like growth factor 1 [147]. Targeting Smad3 signaling, miR-29b could hinder cardiac fibrotic progress either induced by hypertension [148] or post-infarct cardiac remodeling [149]. miR-29b was also found to be able to inhibit ECM-related genes including Col1a1, Col1a2, and Col3a1 [150], and swimming training has been proved to have a similar effect [151]. Besides that, in patients with obstructive sleep apnea (OSA), exercise training prohibits cardiac fibrosis by inhibiting sodium-hydrogen exchanger-1 (NHE-1) activity [152]. In heart tissue of a model of early ovarian hormone deprivation, physical training was found to prevent increase in fibrosis and promote an increase in the cardiac contractile response [153]. Similar results were reported based on other aging models [154,155].

SUMMARY

Great efforts have been invested in finding effective prevention and treatment for CVD. Emerging evidence supports that exercise training has enormous benefits in improving cardiac function and decreasing premature mortality [156]. Exercise training is the most cost-effective way to prevent cardiovascular related disease. Better understanding the molecular mechanisms underlying exercise-induced cardiac growth and cardiac protection effects would help advance novel therapies for CVDs. In addition, other tools—drugs, patient education, behavioral, and psychological management—must be employed as well [157,158]. In addition, more precise criteria are needed to adjust the exercise regimen based on patients' medical status. By learning all these mechanisms, we are looking forward to the possibility of "exercise in pills" and individualized cardiac rehab based on more detailed biophysiological markers.

ACKNOWLEDGMENTS

This work was supported by the grants from National Natural Science Foundation of China (91639101, 81570362, and 81200169 to JJ Xiao and 81400647 to Y Bei) and the development fund for Shanghai talents (to JJ Xiao).

Competing Financial Interests

The authors declare no competing financial interests.

REFERENCES

[1] Writing Group Members, et al. Heart disease and stroke statistics-2016 update: a report from the American Heart Association. Circulation 2016;133(4):e38–e360.

[2] Pearson-Stuttard J, et al. Modeling future cardiovascular disease mortality in the United States: national trends and racial and ethnic disparities. Circulation 2016;133(10):967–78.

[3] Vasan RS, Benjamin EJ. The future of cardiovascular epidemiology. Circulation 2016;133(25):2626–33.

[4] O'Donnell CJ, Elosua R. Cardiovascular risk factors. Insights from Framingham Heart Study. Rev Esp Cardiol 2008;61(3):299–310.

[5] Koene RJ, et al. Shared risk factors in cardiovascular disease and cancer. Circulation 2016;133(11):1104–14.

[6] Sommese L, et al. Possible muscle repair in the human cardiovascular system. Stem Cell Rev 2017;13(2):170–91.

[7] Graham E, Bergmann O. Dating the heart: exploring cardiomyocyte renewal in humans. Physiology (Bethesda) 2017;32(1):33–41.

[8] Cai WF, et al. Repair injured heart by regulating cardiac regenerative signals. Stem Cells Int 2016;2016:6193419.

[9] Rubin N, et al. Recent advancements in understanding endogenous heart regeneration-insights from adult zebrafish and neonatal mice. Semin Cell Dev Biol 2016;58:34–40.

[10] Caspersen CJ, Powell KE, Christenson GM. Physical activity, exercise, and physical fitness: definitions and distinctions for health-related research. Public Health Rep 1985;100(2):126–31.

[11] Global Recommendations on Physical Activity for Health. WHO: Geneva; 2010.

[12] Joyner MJ, Green DJ. Exercise protects the cardiovascular system: effects beyond traditional risk factors. J Physiol 2009;587(Pt 23):5551–8.

[13] Wakatsuki T, Schlessinger J, Elson EL. The biochemical response of the heart to hypertension and exercise. Trends Biochem Sci 2004;29(11):609–17.

[14] Yeves AM, et al. Physiological cardiac hypertrophy: critical role of AKT in the prevention of NHE-1 hyperactivity. J Mol Cell Cardiol 2014;76:186–95.

[15] McMullen JR, et al. Phosphoinositide 3-kinase (p110alpha) plays a critical role for the induction of physiological, but not pathological, cardiac hypertrophy. Proc Natl Acad Sci U S A 2003;100(21):12355–60.

[16] Yamamoto S, et al. New molecular mechanisms for cardiovascular disease: cardiac hypertrophy and cell-volume regulation. J Pharmacol Sci 2011;116(4):343–9.

[17] Noh J, et al. Phosphoinositide dependent protein kinase 1 is required for exercise-induced cardiac hypertrophy but not the associated mitochondrial adaptations. J Mol Cell Cardiol 2015;89(Pt B):297–305.

[18] Natali AJ, et al. Regional effects of voluntary exercise on cell size and contraction-frequency responses in rat cardiac myocytes. J Exp Biol 2001;204(Pt 6):1191–9.

[19] Maillet M, van Berlo JH, Molkentin JD. Molecular basis of physiological heart growth: fundamental concepts and new players. Nat Rev Mol Cell Biol 2013;14(1):38–48.

[20] Anversa P, Leri A, Kajstura J. Cardiac regeneration. J Am Coll Cardiol 2006;47(9):1769–76.

[21] Nadal-Ginard B, et al. Myocyte death, growth, and regeneration in cardiac hypertrophy and failure. Circ Res 2003;92(2):139–50.

[22] van Amerongen MJ, Engel FB. Features of cardiomyocyte proliferation and its potential for cardiac regeneration. J Cell Mol Med 2008;12(6A):2233–44.

[23] Xiong JW, Chang NN. Recent advances in heart regeneration. Birth Defects Res C Embryo Today 2013;99(3):160–9.

[24] Kuhn B, et al. Periostin induces proliferation of differentiated cardiomyocytes and promotes cardiac repair. Nat Med 2007;13(8):962–9.

[25] Bergmann O, et al. Evidence for cardiomyocyte renewal in humans. Science 2009;324(5923):98–102.

[26] Bostrom P, et al. C/EBPbeta controls exercise-induced cardiac growth and protects against pathological cardiac remodeling. Cell 2010;143(7):1072–83.

[27] Vukusic K, et al. Physical exercise affects slow cycling cells in the rat heart and reveals a new potential niche area in the atrioventricular junction. J Mol Histol 2015;46(4-5):387–98.

[28] Garry DJ, Martin CM. Cardiac regeneration: self-service at the pump. Circ Res 2004;95(9):852–4.

[29] Frati C, et al. Resident cardiac stem cells. Curr Pharm Des 2011;17(30):3252–7.

[30] Zhang Y, et al. Cardiac repair with a novel population of mesenchymal stem cells resident in the human heart. Stem Cells 2015;33(10):3100–13.

[31] Senyo SE, et al. Mammalian heart renewal by pre-existing cardiomyocytes. Nature 2013;493(7432):433–6.

[32] Kwak HB. Effects of aging and exercise training on apoptosis in the heart. J Exerc Rehabil 2013;9(2):212–9.

[33] Quindry J, et al. Exercise training provides cardioprotection against ischemia-reperfusion induced apoptosis in young and old animals. Exp Gerontol 2005;40(5):416–25.

[34] French JP, et al. Exercise-induced protection against myocardial apoptosis and necrosis: MnSOD, calcium-handling proteins, and calpain. FASEB J 2008;22(8):2862–71.

[35] de Waard MC, et al. Beneficial effects of exercise training after myocardial infarction require full eNOS expression. J Mol Cell Cardiol 2010;48(6):1041–9.

[36] Kwak HB, Song W, Lawler JM. Exercise training attenuates age-induced elevation in Bax/Bcl-2 ratio, apoptosis, and remodeling in the rat heart. FASEB J 2006;20(6):791–3.

[37] Lai CH, et al. Exercise training enhanced SIRT1 longevity signaling replaces the IGF1 survival pathway to attenuate aging-induced rat heart apoptosis. Age (Dordr) 2014;36(5):9706.

[38] Bloor CM. Angiogenesis during exercise and training. Angiogenesis 2005;8(3):263–71.

[39] Marini M, et al. Partial persistence of exercise-induced myocardial angiogenesis following 4-week detraining in the rat. Histochem Cell Biol 2008;129(4):479–87.

[40] Mobius-Winkler S, et al. Coronary collateral growth induced by physical exercise: results of the impact of intensive exercise training on coronary collateral circulation in patients with stable coronary artery disease (EXCITE) trial. Circulation 2016;133(15):1438–48. discussion 1448.

[41] Karlowatz RJ, et al. Polymorphisms in the IGF1 signalling pathway including the myostatin gene are associated with left ventricular mass in male athletes. Br J Sports Med 2011;45(1):36–41.

[42] Liu JL, LeRoith D. Insulin-like growth factor I is essential for postnatal growth in response to growth hormone. Endocrinology 1999;140(11):5178–84.

[43] McMullen JR, et al. The insulin-like growth factor 1 receptor induces physiological heart growth via the phosphoinositide 3-kinase (p110alpha) pathway. J Biol Chem 2004;279(6):4782–93.

[44] McMullen JR, Izumo S. Role of the insulin-like growth factor 1 (IGF1)/phosphoinositide-3-kinase (PI3K) pathway mediating physiological cardiac hypertrophy. Novartis Found Symp 2006;274:90–111. discussion 111–7, 152–5, 272–6..

[45] Weeks KL, McMullen JR. The athlete's heart vs. the failing heart: can signaling explain the two distinct outcomes? Physiology (Bethesda) 2011;26(2):97–105.

[46] Ikeda H, et al. Interaction of myocardial insulin receptor and IGF receptor signaling in exercise-induced cardiac hypertrophy. J Mol Cell Cardiol 2009;47(5):664–75.

[47] Weeks KL, et al. Phosphoinositide 3-kinase p110alpha is a master regulator of exercise-induced cardioprotection and PI3K gene therapy rescues cardiac dysfunction. Circ Heart Fail 2012;5(4):523–34.

[48] Shioi T, et al. Akt/protein kinase B promotes organ growth in transgenic mice. Mol Cell Biol 2002;22(8):2799–809.

[49] DeBosch B, et al. Akt1 is required for physiological cardiac growth. Circulation 2006;113(17):2097–104.

[50] Shioi T, et al. The conserved phosphoinositide 3-kinase pathway determines heart size in mice. EMBO J 2000;19(11):2537–48.

[51] Kim J, et al. Insulin-like growth factor I receptor signaling is required for exercise-induced cardiac hypertrophy. Mol Endocrinol 2008;22(11):2531–43.

[52] McMullen JR, et al. Protective effects of exercise and phosphoinositide 3-kinase (p110alpha) signaling in dilated and hypertrophic cardiomyopathy. Proc Natl Acad Sci U S A 2007;104(2):612–7.

[53] Owen KL, Pretorius L, McMullen JR. The protective effects of exercise and phosphoinositide 3-kinase (p110alpha) in the failing heart. Clin Sci (Lond) 2009;116(5):365–75.

[54] Ma Z, et al. Swimming exercise training-induced left ventricular hypertrophy involves microRNAs and synergistic regulation of the PI3K/AKT/mTOR signaling pathway. Eur J Appl Physiol 2013;113(10):2473–86.

[55] Fischer P, Hilfiker-Kleiner D. Survival pathways in hypertrophy and heart failure: the gp130-STAT3 axis. Basic Res Cardiol 2007;102(4):279–97.

[56] Fischer P, Hilfiker-Kleiner D. Role of gp130-mediated signalling pathways in the heart and its impact on potential therapeutic aspects. Br J Pharmacol 2008;153(Suppl. 1):S414–27.

[57] Haghikia A, et al. STAT3 and cardiac remodeling. Heart Fail Rev 2011;16(1):35–47.

[58] Boengler K, et al. The myocardial JAK/STAT pathway: from protection to failure. Pharmacol Ther 2008;120(2):172–85.

[59] Zhao L, et al. Deletion of Interleukin-6 attenuates pressure overload-induced left ventricular hypertrophy and dysfunction. Circ Res 2016;118(12):1918–29.

[60] Yamauchi-Takihara K, Kishimoto T. A novel role for STAT3 in cardiac remodeling. Trends Cardiovasc Med 2000;10(7):298–303.

[61] Kunisada K, et al. Signal transducer and activator of transcription 3 in the heart transduces not only a hypertrophic signal but a protective signal against doxorubicin-induced cardiomyopathy. Proc Natl Acad Sci U S A 2000;97(1):315–9.

[62] Doggen K, et al. Ventricular ErbB2/ErbB4 activation and downstream signaling in pacing-induced heart failure. J Mol Cell Cardiol 2009;46(1):33–8.

[63] Bersell K, et al. Neuregulin1/ErbB4 signaling induces cardiomyocyte proliferation and repair of heart injury. Cell 2009;138(2):257–70.

[64] Cai MX, et al. Exercise training activates neuregulin 1/ErbB signaling and promotes cardiac repair in a rat myocardial infarction model. Life Sci 2016;149:1–9.

[65] Wadugu B, Kuhn B. The role of neuregulin/ErbB2/ErbB4 signaling in the heart with special focus on effects on cardiomyocyte proliferation. Am J Physiol Heart Circ Physiol 2012;302(11):H2139–47.

[66] Garbayo E, et al. Catheter-based intramyocardial injection of FGF1 or NRG1-loaded MPs improves cardiac function in a preclinical model of ischemia-reperfusion. Sci Rep 2016;6:25932.

[67] Nizielski SE, et al. Involvement of transcription factor C/EBP-beta in stimulation of PEPCK gene expression during exercise. Am J Physiol 1996;270(5 Pt 2):R1005–12.

[68] Ryall KA, et al. Phenotypic screen quantifying differential regulation of cardiac myocyte hypertrophy identifies CITED4 regulation of myocyte elongation. J Mol Cell Cardiol 2014;72:74–84.

[69] Bezzerides VJ, et al. CITED4 induces physiologic hypertrophy and promotes functional recovery after ischemic injury. JCI Insight 2016;1(9).

[70] Kuramochi Y, et al. Cardiac endothelial cells regulate reactive oxygen species-induced cardiomyocyte apoptosis through neuregulin-1beta/erbB4 signaling. J Biol Chem 2004;279(49):51141–7.

[71] Ren J, et al. Nitric oxide synthase inhibition abolishes exercise-mediated protection against isoproterenol-induced cardiac hypertrophy in female mice. Cardiology 2015;130(3):175–84.

[72] Calvert JW, et al. Exercise protects against myocardial ischemia-reperfusion injury via stimulation of beta(3)-adrenergic receptors and increased nitric oxide signaling: role of nitrite and nitrosothiols. Circ Res 2011;108(12):1448–58.

[73] Roof SR, et al. Neuronal nitric oxide synthase is indispensable for the cardiac adaptive effects of exercise. Basic Res Cardiol 2013;108(2):332.

[74] Mizuno M, et al. Exercise training improves functional sympatholysis in spontaneously hypertensive rats through a nitric oxide-dependent mechanism. Am J Physiol Heart Circ Physiol 2014;307(2):H242–51.

[75] Laurent M, et al. Training-induced increase in nitric oxide metabolites in chronic heart failure and coronary artery disease: an extra benefit of water-based exercises? Eur J Cardiovasc Prev Rehabil 2009;16(2):215–21.

[76] Suarez Y, Sessa WC. MicroRNAs as novel regulators of angiogenesis. Circ Res 2009;104(4):442–54.

[77] van Rooij E, et al. A signature pattern of stress-responsive microRNAs that can evoke cardiac hypertrophy and heart failure. Proc Natl Acad Sci U S A 2006;103(48):18255–60.

[78] Eulalio A, et al. Functional screening identifies miRNAs inducing cardiac regeneration. Nature 2012;492(7429):376–81.

[79] Care A, et al. MicroRNA-133 controls cardiac hypertrophy. Nat Med 2007;13(5):613–8.

[80] Latronico MV, Catalucci D, Condorelli G. Emerging role of microRNAs in cardiovascular biology. Circ Res 2007;101(12):1225–36.

[81] Tao L, et al. Exercise for the heart: signaling pathways. Oncotarget 2015;6(25):20773–84.

[82] Melo SF, et al. Resistance training regulates cardiac function through modulation of miRNA-214. Int J Mol Sci 2015;16(4):6855–67.

[83] Liu X, et al. miR-222 is necessary for exercise-induced cardiac growth and protects against pathological cardiac remodeling. Cell Metab 2015;21(4):584–95.

[84] Xu T, et al. Circulating microRNAs in response to exercise. Scand J Med Sci Sports 2015;25(2):e149–54.

[85] Baggish AL, et al. Rapid upregulation and clearance of distinct circulating microRNAs after prolonged aerobic exercise. J Appl Physiol (1985) 2014;116(5):522–31.

[86] Uhlemann M, et al. Circulating microRNA-126 increases after different forms of endurance exercise in healthy adults. Eur J Prev Cardiol 2014;21(4):484–91.

[87] Goldhammer E, et al. Exercise training modulates cytokines activity in coronary heart disease patients. Int J Cardiol 2005;100(1):93–9.

[88] Piepoli MF, et al. Challenges in secondary prevention after acute myocardial infarction: a call for action. Eur J Prev Cardiol 2016;23(18):1914–39.

[89] McElroy CL, Gissen SA, Fishbein MC. Exercise-induced reduction in myocardial infarct size after coronary artery occlusion in the rat. Circulation 1978;57(5):958–62.

[90] Freimann S, et al. Prior exercise training improves the outcome of acute myocardial infarction in the rat. Heart structure, function, and gene expression. J Am Coll Cardiol 2005;45(6):931–8.

[91] Ranjbar K, Rahmani-Nia F, Shahabpour E. Aerobic training and l-arginine supplementation promotes rat heart and hindleg muscles arteriogenesis after myocardial infarction. J Physiol Biochem 2016;72(3):393–404.

[92] Tao L, et al. Exercise training protects against acute myocardial infarction via improving myocardial energy metabolism and mitochondrial biogenesis. Cell Physiol Biochem 2015;37(1):162–75.

[93] Bozi LH, et al. Aerobic exercise training rescues cardiac protein quality control and blunts endoplasmic reticulum stress in heart failure rats. J Cell Mol Med 2016;20(11):2208–12.

[94] Kemi OJ, et al. Exercise training corrects control of spontaneous calcium waves in hearts from myocardial infarction heart failure rats. J Cell Physiol 2012;227(1):20–6.

[95] Rolim N, et al. Aerobic interval training reduces inducible ventricular arrhythmias in diabetic mice after myocardial infarction. Basic Res Cardiol 2015;110(4):44.

[96] Ribeiro F, et al. Exercise-based cardiac rehabilitation increases daily physical activity of patients following myocardial infarction: subanalysis of two randomised controlled trials. Physiotherapy 2017;103(1):59–65.

[97] Kim C, Choi HE, Lim MH. Effect of high interval training in acute myocardial infarction patients with drug-eluting stent. Am J Phys Med Rehabil 2015;94(10 Suppl. 1):S79–86.

[98] Izeli NL, et al. Aerobic training after myocardial infarction: remodeling evaluated by cardiac magnetic resonance. Arq Bras Cardiol 2016;106(4):311–8.

[99] Buja LM. Myocardial ischemia and reperfusion injury. Cardiovasc Pathol 2005;14(4):170–5.

[100] Lennon SL, et al. Exercise and myocardial tolerance to ischaemia-reperfusion. Acta Physiol Scand 2004;182(2):161–9.

[101] Brown DA, et al. Exercise training preserves coronary flow and reduces infarct size after ischemia-reperfusion in rat heart. J Appl Physiol (1985) 2003;95(6):2510–8.

[102] Hwang H, Reiser PJ, Billman GE. Effects of exercise training on contractile function in myocardial trabeculae after ischemia-reperfusion. J Appl Physiol (1985) 2005;99(1):230–6.

[103] Demirel HA, et al. Short-term exercise improves myocardial tolerance to in vivo ischemia-reperfusion in the rat. J Appl Physiol (1985) 2001;91(5):2205–12.

[104] Powers SK, Locke M, Demirel HA. Exercise, heat shock proteins, and myocardial protection from I-R injury. Med Sci Sports Exerc 2001;33(3):386–92.

[105] Hamilton KL, et al. Exercise, antioxidants, and HSP72: protection against myocardial ischemia/reperfusion. Free Radic Biol Med 2003;34(7):800–9.

[106] French JP, et al. Ischemia-reperfusion-induced calpain activation and SERCA2a degradation are attenuated by exercise training and calpain inhibition. Am J Physiol Heart Circ Physiol 2006;290(1):H128–36.

[107] Yamashita N, Baxter GF, Yellon DM. Exercise directly enhances myocardial tolerance to ischaemia-reperfusion injury in the rat through a protein kinase C mediated mechanism. Heart 2001;85(3):331–6.

[108] Dickson EW, et al. Exercise enhances myocardial ischemic tolerance via an opioid receptor-dependent mechanism. Am J Physiol Heart Circ Physiol 2008;294(1):H402–8.

[109] Nicholson CK, et al. Chronic exercise downregulates myocardial myoglobin and attenuates nitrite reductase capacity during ischemia-reperfusion. J Mol Cell Cardiol 2013;64:1–10.

[110] Kraljevic J, et al. Role of KATP channels in beneficial effects of exercise in ischemic heart failure. Med Sci Sports Exerc 2015;47(12):2504–12.

[111] Li Y, et al. Endurance exercise accelerates myocardial tissue oxygenation recovery and reduces ischemia reperfusion injury in mice. PLoS One 2014;9(12):e114205.

[112] McGinnis GR, et al. Interleukin-6 mediates exercise preconditioning against myocardial ischemia reperfusion injury. Am J Physiol Heart Circ Physiol 2015;308(11):H1423–33.

[113] Zhang KR, et al. Long-term aerobic exercise protects the heart against ischemia/reperfusion injury via PI3 kinase-dependent and Akt-mediated mechanism. Apoptosis 2007;12(9):1579–88.

[114] Kupai K, et al. Consequences of exercising on ischemia-reperfusion injury in type 2 diabetic Goto-Kakizaki rat hearts: role of the HO/NOS system. Diabetol Metab Syndr 2015;7:85.

[115] Le Page C, et al. Exercise training improves functional post-ischemic recovery in senescent heart. Exp Gerontol 2009;44(3):177–82.

[116] Portal L, et al. A model of hypoxia-reoxygenation on isolated adult mouse cardiomyocytes: characterization, comparison with ischemia-reperfusion, and application to the cardioprotective effect of regular treadmill exercise. J Cardiovasc Pharmacol Ther 2013;18(4):367–75.

[117] Wang Y, et al. Physical exercise-induced protection on ischemic cardiovascular and cerebrovascular diseases. Int J Clin Exp Med 2015;8(11):19859–66.

[118] Zhang YM, et al. The effects of different initiation time of exercise training on left ventricular remodeling and cardiopulmonary rehabilitation in patients with left ventricular dysfunction after myocardial infarction. Disabil Rehabil 2016;38(3):268–76.

[119] Rahimi M, et al. The effect of high intensity interval training on cardioprotection against ischemia-reperfusion injury in wistar rats. EXCLI J 2015;14:237–46.

[120] Kadoglou NP, Iliadis F, Liapis CD. Exercise and carotid atherosclerosis. Eur J Vasc Endovasc Surg 2008;35(3):264–72.

[121] Sung J, et al. Prevalence of coronary atherosclerosis in asymptomatic middle-age men with high aerobic fitness. Am J Cardiol 2012;109(6):839–43.

[122] Shimada K, et al. Atherosclerotic plaques induced by marble-burying behavior are stabilized by exercise training in experimental atherosclerosis. Int J Cardiol 2011;151(3):284–9.

[123] Cesar L, et al. An essential role for diet in exercise-mediated protection against dyslipidemia, inflammation and atherosclerosis in ApoE(-)/(-) mice. PLoS One 2011;6(2):e17263.

[124] Shing CM, et al. Voluntary exercise decreases atherosclerosis in nephrectomised ApoE knockout mice. PLoS One 2015;10(3):e0120287.

[125] Shimada K, et al. Exercise training reduces severity of atherosclerosis in apolipoprotein E knockout mice via nitric oxide. Circ J 2007;71(7):1147–51.

[126] Wojewoda M, et al. Exercise capacity and cardiac hemodynamic response in female ApoE/LDLR(-/-) mice: a paradox of preserved V'O2max and exercise capacity despite coronary atherosclerosis. Sci Rep 2016;6:24714.

[127] Lee J, Cho JY, Kim WK. Anti-inflammation effect of exercise and Korean red ginseng in aging model rats with diet-induced atherosclerosis. Nutr Res Pract 2014;8(3):284–91.

[128] Pinto A, et al. Effects of physical exercise on inflammatory markers of atherosclerosis. Curr Pharm Des 2012;18(28):4326–49.

[129] Palmefors H, et al. The effect of physical activity or exercise on key biomarkers in atherosclerosis—a systematic review. Atherosclerosis 2014;235(1):150–61.

[130] Kadoglou NP, et al. The differential anti-inflammatory effects of exercise modalities and their association with early carotid atherosclerosis progression in patients with type 2 diabetes. Diabet Med 2013;30(2):e41–50.

[131] Kadoglou NP, et al. The anti-inflammatory effects of exercise training promote atherosclerotic plaque stabilization in apolipoprotein E knockout mice with diabetic atherosclerosis. Eur J Histochem 2013;57(1):e3.

[132] Pellegrin M, et al. Voluntary exercise stabilizes established angiotensin II-dependent atherosclerosis in mice through systemic anti-inflammatory effects. PLoS One 2015;10(11):e0143536.

[133] Lenk K, et al. Role of endothelial progenitor cells in the beneficial effects of physical exercise on atherosclerosis and coronary artery disease. J Appl Physiol (1985) 2011;111(1):321–8.

[134] Kwasniewska M, et al. Long-term effect of different physical activity levels on subclinical atherosclerosis in middle-aged men: a 25-year prospective study. PLoS One 2014;9(1):e85209.

[135] Griffith G, et al. Experimental protocol of a randomized controlled clinical trial investigating exercise, subclinical atherosclerosis, and walking mobility in persons with multiple sclerosis. Contemp Clin Trials 2015;41:280–6.

[136] Choo J, et al. Effects of weight management by exercise modes on markers of subclinical atherosclerosis and cardiometabolic profile among women with abdominal obesity: a randomized controlled trial. BMC Cardiovasc Disord 2014;14:82.

[137] Wang J, et al. Causes and characteristics of diabetic cardiomyopathy. Rev Diabet Stud 2006;3(3):108–17.

[138] Hafstad AD, Boardman N, Aasum E. How exercise may amend metabolic disturbances in diabetic cardiomyopathy. Antioxid Redox Signal 2015;22(17):1587–605.

[139] Sacre JW, et al. A six-month exercise intervention in subclinical diabetic heart disease: effects on exercise capacity, autonomic and myocardial function. Metabolism 2014;63(9):1104–14.

[140] Epp RA, et al. Exercise training prevents the development of cardiac dysfunction in the low-dose streptozotocin diabetic rats fed a high-fat diet. Can J Physiol Pharmacol 2013;91(1):80–9.

[141] Wang H, et al. Exercise prevents cardiac injury and improves mitochondrial biogenesis in advanced diabetic cardiomyopathy with PGC-1alpha and Akt activation. Cell Physiol Biochem 2015;35(6):2159–68.

[142] Sharma NM, et al. Exercise training attenuates upregulation of p47(phox) and p67(phox) in hearts of diabetic rats. Oxid Med Cell Longev 2016;2016:5868913.

[143] Holloway CJ, et al. Exercise training in dilated cardiomyopathy improves rest and stress cardiac function without changes in cardiac high energy phosphate metabolism. Heart 2012;98(14):1083–90.

[144] Simeunovic D, et al. Evaluation of oxidative stress markers and catecholamine changes in patients with dilated cardiomyopathy before and after cardiopulmonary exercise testing. Hellenic J Cardiol 2015;56(5):394–401.

[145] Grander W, et al. Determinants of exercise capacity in dilated cardiomyopathy: a prospective, explorative cohort study. Wien Klin Wochenschr 2012;124(19–20):685–91.

[146] Sun B, et al. Bone morphogenetic protein-4 mediates cardiac hypertrophy, apoptosis, and fibrosis in experimentally pathological cardiac hypertrophy. Hypertension 2013;61(2):352–60.

[147] Abonnenc M, et al. Extracellular matrix secretion by cardiac fibroblasts: role of microRNA-29b and microRNA-30c. Circ Res 2013;113(10):1138–47.

[148] Zhang Y, et al. miR-29b as a therapeutic agent for angiotensin II-induced cardiac fibrosis by targeting TGF-beta/Smad3 signaling. Mol Ther 2014;22(5):974–85.

[149] Yang F, et al. microRNA-29b mediates the antifibrotic effect of tanshinone IIA in postinfarct cardiac remodeling. J Cardiovasc Pharmacol 2015;65(5):456–64.

[150] Zhu JN, et al. Smad3 inactivation and MiR-29b upregulation mediate the effect of carvedilol on attenuating the acute myocardium infarction-induced myocardial fibrosis in rat. PLoS One 2013;8(9):e75557.

[151] Melo SF, et al. Expression of MicroRNA-29 and collagen in cardiac muscle after swimming training in myocardial-infarcted rats. Cell Physiol Biochem 2014;33(3):657–69.

[152] Chen TI, Tu WC. Exercise attenuates intermittent hypoxia-induced cardiac fibrosis associated with sodium-hydrogen exchanger-1 in rats. Front Physiol 2016;7:462.

[153] Felix AC, et al. Aerobic physical training increases contractile response and reduces cardiac fibrosis in rats subjected to early ovarian hormone deprivation. J Appl Physiol (1985) 2015;118(10):1276–85.

[154] Kwak HB, et al. Exercise training reduces fibrosis and matrix metalloproteinase dysregulation in the aging rat heart. FASEB J 2011;25(3):1106–17.

[155] Wright KJ, et al. Exercise training initiated in late middle age attenuates cardiac fibrosis and advanced glycation end-product accumulation in senescent rats. Exp Gerontol 2014;50:9–18.

[156] Golbidi S, Laher I. Molecular mechanisms in exercise-induced cardioprotection. Cardiol Res Pract 2011;2011:972807.

[157] Ellison GM, et al. Physiological cardiac remodelling in response to endurance exercise training: cellular and molecular mechanisms. Heart 2012;98(1):5–10.

[158] Villella M, Villella A. Exercise and cardiovascular diseases. Kidney Blood Press Res 2014;39(2–3):147–53.

Chapter 8

Exercise, Fitness, and Cancer Outcomes

Tolulope A. Adesiyun, Stuart D. Russell

Division of Cardiology, Department of Medicine, Johns Hopkins School of Medicine, Baltimore, MD, United States

---

Lack of activity destroys the good condition of every human being, while movement and methodical physical exercise save it and preserve it

Plato (427–347 BC)

INTRODUCTION

Physical inactivity and subsequent poor cardiorespiratory fitness are major health problems worldwide, particularly in developed countries. Physical activity has clear beneficial effects on health outcomes, including cardiovascular disease and all-cause mortality [1]. Globally, large prospective cohort studies have indicated that sedentary behavior is associated with poor health outcomes, including increased mortality [2–6]. Lee et al. [7] calculated the attributable risk for premature mortality and estimated that physical inactivity worldwide causes 9% of premature mortality, accounting for 5.3 million deaths worldwide in 2008.

Exercise favorably impacts multiple systems and health outcomes, and a graded relationship between exercise and the development of common chronic conditions including cardiovascular disease, diabetes, chronic lung disease, chronic kidney disease, and some cancer has been observed in the existing literature. Physical exercise has also consistently been identified as a central element of rehabilitation for many chronic diseases and has been successful in improving quality of life and reducing all-cause mortality [8].

Cancer is another chronic disease where significant research into preventive and therapeutic interventions has been conducted. In 2009, the American Cancer Society (ACS) estimated that there were nearly 1.5 million new cases of cancer diagnosed in the United States and more than 500,000 people who died from the disease [9]. In the last two decades, it has become clear that exercise plays a vital role in cancer prevention and control. There is growing evidence suggesting that exercise decreases the risk of many cancers, and data support the idea that exercise may extend survival for cancer survivors [9]. Our focus of the chapter is on the impact of exercise and obesity on the risk of developing cancer and the beneficial effects that exercise has after a diagnosis of cancer both during and after therapy.

EXERCISE AND FUTURE CANCER RISK

There is a growing understanding that exercise decreases cancer risk. The past two decades of epidemiological studies have produced a significant amount of evidence demonstrating the benefits of moderate to vigorous physical activity as it relates to the risk of cancer at several organs and other chronic diseases [10]. While multiple studies have assessed the relationship between physical activity and different types of cancers, the most consistent associations between increased physical activity and reduced cancer risk have been observed for colon and breast cancers with case-control and cohort studies supporting a moderate, inverse relation between physical activity and the development of these cancers.

Recently, Kruk et al. published a review article that sought to update the available data on the primary prevention of cancer by physical activity. Table 1 summarizes the supporting evidence of the protective effect of physical exercise in the primary prevention of cancer at specific cancer sites.

Breast Cancer

More than two dozen prospective cohort studies and an even greater number of population-based case-control studies have examined the relation between physical activity and breast cancer risk. Overall, the majority of these studies suggest that physically active women have a lower risk of developing breast cancer than sedentary women

Lifestyle in Heart Health and Disease. https://doi.org/10.1016/B978-0-12-811279-3.00008-2

TABLE 1 Summary of Evidence for the Prevention of Cancer by Physical Activity

Cancer Site	Average Risk Reduction	Level of Scientific Evidence	No. of Studies
Colon	20%–25%	Convincing	>60
Breast (postmenopausal)	20%–30%	Probably	>76
Endometrium	20%–30%	Probably	>20
Breast (postmenopausal)	27%	Limited suggestive	>33
Prostate	10%–20%	Limited suggestive	>20
Lung	20%–40%	Limited suggestive	>20
Ovary	10%–20%	Limited suggestive	>20
Pancreatic	40%–50%	Limited suggestive	>20
Gastric	30%	Limited suggestive	>15

With permission from Kruk J, Czerniak U. Physical activity and its relation to cancer risk: updating the evidence. Asian Pac J Cancer Prev 2013;14(7):3993–4003.

[11]. Physical activity has been associated with a 20%–80% reduction in the risk of future breast cancer [12–27]. Additionally, a number of population-based case-control studies have reported a reduction in risk ranging from 20% to 70% among those who were more physically active [28–39]. A meta-analysis by Lagerros et al. [40] that included 23 studies that were focused on physical activity in adolescence and young adulthood demonstrated a summary relative risk estimate of breast cancer of 0.81 (95% CI, 0.73–0.89) for the highest versus lowest category of physical activity. In this study, every 1 h increase in recreational physical activity per week was associated with a 3% (95% CI 0%–6%) risk reduction for future breast cancer. A more extensive systematic review of recreational activity and breast cancer risk included 19 cohort and 29 case-control studies concluded that the evidence that physical activity reduced future risk of postmenopausal breast cancer by 20%–80% was strong and that each additional 1 h of physical activity per week reduced the risk by 6% (95% CI, 3%–8%) [41]. For premenopausal breast cancer, physical activity had a more modest risk reduction of 15%–20%.

Another published review assessed the association between physical activity and breast cancer risk in 73 studies that had been conducted around the world. Across studies, there was an average 25% risk reduction when comparing physically active women with the least active women. The associations they found were strongest for recreational activity, for activity sustained over the lifetime or done after menopause, and for activity that is of moderate to vigorous intensity and performed regularly [42,43]. A study using data from the Iowa Women's Health study including 41,836 women followed for a total of 554,819 person-years of follow-up reported that there were 2548 incident cases of breast cancer and when compared with low physical activity, high physical activity levels were inversely associated with the risk of breast cancer (RR, 0.86; 95% CI, 0.78–0.96) [23].

An association between physical activity and the risk of developing breast cancer has also been observed in multiethnic populations within the United States [21,38,39]. The data show that this relationship exists among black [38,44], Hispanic [32,39], and Asian-American women as well [36].

Prostate Cancer

Several epidemiological studies have examined physical activity and prostate cancer risk with mixed outcomes. Of the studies, at least 19 have found some suggestion of an inverse relationship between prostate cancer and physical activity [45–63], while 14 studies found no overall association [64–76] and some studies further reported an increased risk of prostate cancer among those who are more physically active.

Oliveria et al. [49] conducted a prospective study at the Cooper Clinic to assess the association between cardiorespiratory fitness and prostate cancer among 12,975 men aged 20–80 years old. The men had previously undergone cardiorespiratory fitness testing with a maximal exercise treadmill test and for the study. Higher cardiorespiratory fitness levels were associated with a decreased probability of the development of prostate cancer (Table 2) even after controlling for age, BMI, and smoking habits. The reported adjusted estimate of the incidence rate ratio declines from 1.1 (95% CI 0.63–1.77) to

TABLE 2 The Association Between Cardiorespiratory Fitness and Prostate Cancer

Fitness Level	<13.7 (N = 3026)	13.7–17.0 (N = 2828)	17–21 (N = 3546)	≥21 (N = 3575)	
No. of cases	40	27	21	6	
Adjusted incidence rate ratio[b] (95% CI)	1.00[a]	1.1 (0.63–1.77)	0.73 (0.41–1.29)	0.26 (0.10–0.63)	P trend = 0.0036

Fitness level expressed in treadmill time (minutes).
[a] Reference category.
[b] Adjusted for age, body mass index, and smoking status.
Adapted from Oliveria SA, et al. The association between cardiorespiratory fitness and prostate cancer. Med Sci Sports Exerc 1996;28(1):97–104.

0.73 (95% CI 0.41–1.29) to 0.26 (95% CI 0.10–0.63) across increasing quartiles of fitness with this trend being statistically significant (P < .004). The protective effect, however, was limited to those less than 60 years old.

Subgroup effects have been studied; however, no consistent effects have been defined. For men with a family history of prostate cancer, however, a 52% reduction in risk for those in the highest quartile of physical activity has been reported [74].

Colon Cancer

A review of 23 publications that included data from 12 prospective cohort studies and 8 case-control studies found an overall consistent inverse association between physical activity and future risk of developing colon cancer [64,69,77–97]. The database of 23 studies included a total of 9747 cases of colon cancer that was almost equally distributed between men and women (4933 men and 4814 women). Across all studies, the median relative risk, when the most active subjects were compared with the least active subjects, was 0.7, and results were similar for men and women with a relative risk of 0.7 in both. Studies have also examined whether physical activity has different associations based on the location of the cancer in the colon (i.e., proximal, mid, and distal segments), but these data have been equivocal [11].

Rectal Cancer

Unlike the data available for the association of physical activity with colon cancer, the data on physical activity and future risk of developing rectal cancer are much more mixed. Most studies reported no significant association [69,79,83,86,93,94,98], while a few have reported a significantly lower risk or borderline significant risk reduction with higher degrees of activity. A recent review of the literature in this area reported a median relative risk reduction of 1.0 when comparing rectal cancer risk among those with the least amount of physical activity with those with the most, suggesting little association [83].

Lung Cancer

A recent paper by Kruk et al. [10] reported on over 20 studies that have been performed to study the relationship between physical activity and lung cancer. Controlling for smoking, their results demonstrated a 20%–40% risk reduction with activity [10]. The physical activity guidelines summary published in 2008 reported a median reduction of 24% for the most active versus least active individuals [11,54,99–108]. The reduction in lung cancer risk was more obvious with the case-control studies (mean relative risk over two studies of 0.61) [106,108] than with cohort studies (median relative risk over eight studies of 0.77) [54,99–105]. Of note, the inverse relationship between physical activity and lung cancer risk was similar for men and women.

Endometrial Cancer

Lee et al. [109] performed a review of cohort and case-control studies that looked at the relationship between physical activity and endometrial cancer risk. They calculated a 30% risk reduction in endometrial cancer risk for those with the most activity when compared with those in the lowest level of physical activity. Additional review performed by Voskuil et al. [110] and Cust and colleagues [111] estimated a similar 20%–30% significant reduction in risk among physically active women. The 30% risk reduction occurred even after adjusting for BMI that is important given that the effect of physical activity on body weight has been hypothesized to mediate the observed association of physical activity and risk of endometrial cancer [11].

Other Cancers

Fewer studies have examined associations between physical activity and future risk of developing other types of cancers. A modest inverse association has been reported in ovarian cancer with a reported risk reduction of 19% [112], and there are limited suggestive data in the pancreatic cancer population [10]. The data for other malignancies such as thyroid, kidney, bladder, and hematopoietic are sparse, making it difficult to draw adequate conclusions.

HYPOTHESIZED MECHANISMS OF PREVENTIVE EFFECT OF EXERCISE ON CANCER

To date, the mechanisms that underlie cancer inhibition with increased physical activity remain unknown (Fig. 1) [113]. There are multiple proposed mechanisms outlined below.

Steroid Hormones—Women

It is known that the sex steroid hormones have powerful mitogenic and proliferative influences, and they have been strongly associated with the development of reproductive cancers in both men and women. The effect of exercise on steroid hormones has frequently been cited as one of the potential mechanisms by which exercise imparts a beneficial effect perhaps by altering the beginning of menstrual function [113–115]. Girls who participate in athletics tend to experience menarche later [116,117] and therefore have a delay in establishing normal ovarian cyclicity [118]. While it has not been conclusively established that exercise leads to later onset of menarche, it is known that early age of menarche is associated with increased breast cancer risk [119,120]. Exercise is also associated with the decreased levels of circulating estrogen and progesterone in adult premenopausal women [121]. Both hormones are powerful mitogens in the breast, and reductions in the circulating levels are associated with decreased proliferation [122].

Immune System

The immune system is also thought to play a role in reducing cancer risk through its ability to recognize and eliminate abnormal cells or through acquired and/or innate immune system components [123]. One hypothesis is that the number or function of natural killer cells, which play a role in tumor suppression, is increased by exercise [124]. In experimental models that have assessed the interaction of exercise, immune parameters, and cancer end points, trained mice have higher NK cell activity, resulting in greater clearance of tumor cells and incidence of tumors [125]. The data from randomized clinical trials are inconclusive, and while some cross-sectional studies have shown differences in components of the innate immune system when comparing exercisers with nonexercisers, others have not [124].

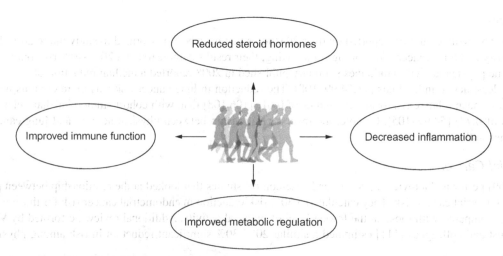

FIG. 1   Mechanisms of preventive effect of exercise on cancer.

Inflammation

Increased levels of pro-inflammatory factors such as C-reactive protein (CRP), interleukin (IL)-6, and tumor necrosis factor-α (TNF-α) and decreased levels of anti-inflammatory factors, such as adiponectin, have been linked with increased cancer risk [126]. Physical activity may reduce systemic inflammation alone or in combination with body weight or composition through reducing macrophage or adipose cell production of inflammatory cytokines in adipose tissue, although exact mechanisms are unknown [124,127]. Cross-sectional studies support an association between chronic physical activity and decreased levels of inflammatory markers such as CRP, IL-6, and TNF-α in men and women; however, intervention studies looking at exercise alone or exercise and diet together have had inconsistent results [124].

Metabolic

An increased risk of breast, colon, pancreatic, endometrial, and stomach cancers has been linked to insulin resistance [128]. Type 2 diabetes or impaired glucose tolerance has also been associated with higher incidences of cancer and mortality [128,129]. Insulin enhances tumor development through its stimulation of cell proliferation or inhibition of apoptosis. Furthermore, it may regulate the production and availability of sex steroid hormones and inhibit hepatic synthesis of sex-hormone-binding globulin [128]. Physical activity leads to improved insulin sensitivity, glucose uptake by skeletal muscle can be increased for up to 12 h [130], and chronic exercise training has been shown to stimulate prolonged improvements in insulin sensitivity [131–133]. Although body composition has been strongly associated with insulin sensitivity, exercise-induced changes in insulin sensitivity can occur from physical activity, independent of the changes in weight or body composition [130,131,134].

OBESITY AND THE RISK OF DEVELOPING CANCER

In 2013, an estimated 4.5 million deaths worldwide were caused by overweight (BMI 23.0–29.9) and obesity (BMI > 30); on the basis of recent estimates, the obesity-related cancer burden may represent up to 9% of the cancer burden among women in North America, Europe, and the Middle East [135,136].

Epidemiological studies have shown that obesity is associated with an increased risk of several cancer types including colon, endometrial, breast (postmenopausal), kidney, esophagus, pancreas, gallbladder, liver, and hematologic cancers [137,138]. In addition, obesity can lead to poorer treatment outcome, worsened prognosis, and increased cancer-related mortality [139]. The International Agency for Research on Cancer (IARC) estimates that 25% of cancer cases worldwide are due to overweight and obesity and a sedentary lifestyle [140]. The IARC concluded in their 2012 report that there was sufficient evidence for a cancer-preventive effect of avoidance of weight gain for cancers of the colon, esophagus (adenocarcinoma), kidney (renal cell), breast (postmenopausal), and corpus uteri [141].

For cancers of the colon, rectum, gastric cardia, liver, gallbladder, pancreas, and kidney and for esophageal adenocarcinoma, statistically significant associations between BMI and cancer risk were reported, with positive dose-response relationships. Relative risks from meta-analyses or pooled analyses were 1.2–1.5 for overweight and 1.5–1.8 for obesity with respect to cancers of the colon [142,143], gastric cardia [144], liver [145], gallbladder [146], pancreas [147], and kidney [148]; the relative risk for esophageal adenocarcinoma was up to 4.8 for a BMI of 40 or more [149]. Positive associations have been observed between adult BMI and postmenopausal breast cancer in numerous studies (relative risk approximately 1.1 per 5 BMI units) [142]. For premenopausal breast cancer, consistent inverse associations have been observed between BMI and risk [142]. Lastly, convincing data also exist that obesity is associated with an increased risk of breast cancer recurrence [150,151].

EXERCISE AND MORTALITY

Physical inactivity is a major health problem worldwide, particularly in developed countries. The medical literature demonstrates the beneficial effects of physical activity on several health outcomes, including cardiovascular disease and all-cause mortality. One study calculated the attributable risk for premature mortality and estimated that physical inactivity worldwide causes 9% of premature mortality, accounting for 5.3 million deaths worldwide in 2008 [7]. They estimated that a 10% reduction in inactivity could avert 533,000 deaths every year.

Large observational studies suggest that regular exercise reduces risk of all-cause mortality for most individuals, including men and women and younger and older populations [1,152–155]. A prospective study was performed where physical fitness and risk of all-cause and cause-specific mortality in 10,224 men and 3120 women were studied. All participants were given a preventive medical examination, and physical fitness was measured by a maximal treadmill exercise test. Average follow-up was slightly more than 8 years, for a total of 110,482 person-years of observation. They observed that age-adjusted all-cause mortality rates declined across physical fitness quintiles from 64.0 per 10,000 person-years in the least-fit men to 18.6 per 10,000 person-years in the most-fit men. Corresponding values for women were 39.5 per 10,000 person-years to 8.5 per 10,000 person-years. These trends remained after statistical adjustment for age, smoking habit, cholesterol level, systolic blood pressure, fasting blood glucose level, parental history of coronary heart disease, and follow-up interval [156]. Further demonstrating the benefit of exercise on decreasing mortality are studies in elite athletes that have found that the physical fitness they develop leads to a 67% lower mortality when compared with the general population [157].

Most of the aforementioned studies looked at physical activity, but cardiorespiratory fitness is another significant measure that has also been shown to have an inverse relationship with all-cause mortality in healthy men and women. A meta-analysis published in 2009 showed that for every 1 MET increase in maximal aerobic capacity, there were 13% and 15% risk reduction of all-cause mortality and cardiovascular disease, respectively. The minimal maximal aerobic capacity value for substantial risk reduction in persons aged 50 years in that study was estimated to be 8 MET for men and 6 MET for women. For men and women older than 60, they found that a minimum of 7 and 5 MET, respectively, was needed to show benefit [1].

WEIGHT GAIN DURING OR AFTER CHEMOTHERAPY AND ITS EFFECTS

Weight gain during chemotherapy and decreased physical activity after diagnosis are additional challenges that those facing a new diagnosis of cancer struggle with, and there are extensive data, particularly in breast cancer, regarding the negative impact of weight gain during chemotherapy. Physical inactivity and overweight/obesity contribute additionally to the primary risk of cancer and CVD [142]. A cancer diagnosis and therapy have further significant unfavorable effects on lifestyle behavior [158]. Weight gain occurs in 50%–96% of all patients with early-stage breast cancer who are receiving adjuvant chemotherapy and usually range from 2.5 to 6.2 kg; however, greater gains are not uncommon [159–163]. This is important because both being overweight at the time of breast cancer diagnosis and weight gain after diagnosis are linked to poorer survival [163–165]. For example, among never-smokers in one study, the all-cause mortality of those who gained weight was about 1.4–1.6 times higher, and among past and current smokers, it was roughly 1.2 times higher when compared with those who did not gain weight [164].

Irwin et al. [166] reported that, on average, patients with early breast cancer decreased their physical activity by 2.0 h per week from before diagnosis, with greater decreases among women receiving combined treatment compared with single-modality adjuvant therapy. Additionally, physical activity levels have been found to be significantly reduced for many women after a diagnosis of breast cancer and remain low after treatment is completed. One prospective observational study in survivors of breast cancer indicated a 50% risk reduction in mortality among women who are regularly active compared with those who remained inactive post diagnosis [163,167]. This is particularly important because weight status at diagnosis and weight gain after diagnosis are associated with worse prognosis including higher recurrence rates and mortality in cancer patients [150,168]. One meta-analysis by Protani et al. published in 2010 that included 43 studies that enrolled women diagnosed with breast cancer between 1963 and 2005 found poorer survival rates among obese women compared with nonobese women with breast cancer for overall (HR, 1.33; 95% CI, 1.21–1.47) and breast cancer specific survival (HR, 1.33; 95% CI, 1.19–1.50) [150]. Another systemic review published by Chlebowski and colleagues that looked at the evidence regarding obesity and breast cancer clinical outcomes found that women with breast cancer who were overweight or gained weight after diagnosis were at greater risk for breast cancer recurrence and death compared with lighter women (Table 3) [168].

The influence of weight gain on prognosis, however, is mixed across studies. Camoriano and colleagues [165] studied 646 lymph-node-positive breast cancer patients treated prospectively in two clinical trials with extended nonanthracycline chemotherapy, chemohormonal therapy, or observation only post mastectomy and found that premenopausal women who gained more than the median weight gain (of 5.9 kg) experienced significantly higher risk of death that persisted after controlling for other factors (multivariate hazard ratio of 1.62); a similar trend in postmenopausal women was not statistically significant.

Kroenke and associates [164] reported results from the Nurses' Health Study (NHS) and noted that patients with greater BMI before breast cancer diagnosis had a higher risk of breast cancer recurrence and death if they were never-smokers or if premenopausal. In addition, patients who were never-smokers who gained weight after treatment had a higher risk of breast cancer recurrence and mortality and all-cause mortality than did never-smokers who had maintained their pretreatment weight and baseline BMI. The effect of weight gain was more pronounced in patients with small tumors and negative lymph nodes in this study, and the effect on survival was seen in normal weight but not overweight (BMI ≥ 25 kg/m^2) subjects [164].

TABLE 3 Breast Cancer Studies of Weight Gain After Diagnosis and Clinical Outcome

First Author	Year, Study Design	Participants	Weight Gain	Results	Comments
Camoriano	1990 Two concurrent, prospective clinical trials of adjuvant breast cancer therapy	N=545 Pre- and postmenopausal women treated with adjuvant therapy for node-positive breast cancer	Median weight gain: premenopausal, 5.9 kg Treated postmenopausal, 3.6 kg Observed postmenopausal, 1.8 kg	(a) 1.5 times greater risk of relapse in premenopausal women gaining more than median weight at 60 weeks (b) 1.8-fold increased risk of death for heavier premenopausal patients who gained more than median weight at 60 weeks	Adjusted for age, tumor size, node status, estrogen receptor status, initial Quetelet index, and nuclear grade
Levine	Prospective cohort study	N=32 Women treated with adjuvant therapy for breast cancer	Mean=1.8 kg (from −3.6 to 10.35 kg) at the end of treatment. At 2 years, mean weight gain of 6 kg	1.36 times higher risk of recurrence among women who gained weight at 2 years versus those with no weight gain (nonsignificant)	Adjusted for physical activity change, demographics, node status, surgery type, and chemotherapy regimen
Nichols	Prospective cohort study	N=3993 Women with invasive nonmetastatic breast cancer 6.3 years follow-up	1.7 kg	Each 5 kg gain associated with a 12% increase in all-cause mortality (P=.004) and 19% increase in CV mortality	Age, state, time between diagnosis and follow-up, breast cancer family history, smoking, recreational physical activity, menopausal status, and cancer stage
Chlebowski	Prospective, multicenter	N=62	2.0 kg for 5-FU group 3.7 kg for CMF group	> 10 kg weight gain in follow-up associated with worse prognosis as none of the women gaining more than 10 kg survived	Overall survival rates: 31% in CMF group and 51% in 5-FU group.
Jeon	Retrospective cohort study	N=108	Mean weight change upon completing chemo was 3.65 kg	Weight gain not associated with relapse-free survival	
Bradshaw	Prospective cohort study	N=1436		>10% weight gain after diagnosis associated with worse survival (hazard ratio, 2.67)	

Adapted from Chlebowski RT, Aiello E, McTiernan A. Weight loss in breast cancer patient management. J Clin Oncol, 2002;20(4):1128–43 and Playdon MC, et al. Weight gain after breast cancer diagnosis and all-cause mortality: systematic review and meta-analysis. J Natl Cancer Inst 2015;107(12):djv275.

In contrast, Caan et al. [169] evaluated a prospective cohort of women in the Life after Cancer Epidemiology (LACE) study drawn from the Kaiser Permanente Northern California and Utah Cancer Registries. Although similar findings regarding patient characteristics associated with weight gain and pretreatment weight associations with mortality were found, no impact of treatment-associated weight gain with mortality was discovered.

Overall, while the effect of weight gain on prognosis is variable among studies, it appears to be significant in a subset of women [169]. In addition to weight gain and decreased physical activity observed with chemotherapy, physical deconditioning may also lead to treatment delay or drug dose reduction [170] that can impact treatment regimen and success. Lastly, weight gain after diagnosis of breast cancer is associated with higher all-cause mortality rates compared with maintaining body weight, and adverse effects are found to be greater for weight gains of 10.0% or higher [171].

A systematic review and meta-analysis by Playdon et al. of 23,832 patients was published in 2015, and they sought to investigate the relationship between weight gain after breast cancer diagnosis and breast cancer-specific, all-cause mortality, and recurrence outcomes. Weight gain was measured over a median time of 1.5 years post diagnosis with a 12% increased risk of all-cause mortality for those with weight gain of more than 5%.

When all-cause mortality was stratified by the amount of weight gain, moderate weight gain (5%–10% increase) was not associated with hazard of mortality (HR, 0.97; 95% CI, 0.86–1.11; $P=.70$); however, the association of mortality with high amounts (>10% change) of weight gain was apparent when compared with maintaining body weight (HR, 1.23; 95% CI, 1.09–1.39; $P<.001$).

EXERCISE DURING CHEMOTHERAPY

Exercise is a nonpharmacological intervention that has been shown to have beneficial effects among patients with cancer. There have been multiple studies that have reported on the safety of exercise and the beneficial impact that it has on cancer patients.

Safety

One of the concerns about any intervention, particularly an exercise regimen in a patient population is safety. One review and meta-analysis that included 82 trials with patients with 11 different types of cancer (mostly breast) surmised that physical exercise in this patient population is well tolerated, feasible, and safe during and following cancer treatment [172].

Breast Cancer

For women with breast cancer, the American College of Sports Medicine round table on exercise guidelines for cancer survivors reported that of the 22 randomized controlled trials they assessed, all concluded that exercise was safe during breast cancer treatment [9,173–182]. Thirteen of these studies reported adverse events, and in all cases, the adverse events were rare and mild [9].

Similarly, the safety of exercise among women who have completed their chemotherapy and radiotherapy treatment has been explored, and of the 15 trials that were found to have commented on the safety of the intervention in one report [9], all 15 found that exercise was indeed safe post treatment [9]. Reported adverse events in those trials were rare, mild, and expected based on the type of activity that was prescribed such as plantar fasciitis from walking or other musculoskeletal injuries.

Prostate Cancer

The impact of exercise in prostate cancer survivors has been assessed in at least 12 intervention studies and 3 observational studies. Of those that reported specifically on safety of exercise interventions, all concluded that exercise was safe in prostate cancer survivors [9].

Other Cancers

Exercise has also been shown to be safe among those with hematologic malignancies who have received a stem cell transplant [24,37,40,46,62,136]. However, limited data exist regarding the safety of exercise among those with gynecologic malignancies and colon cancer.

Exercise and Quality of life

The data regarding the quality of life improvements with exercise have been mixed. There are five randomized controlled trials that showed that supervised exercise (aerobic, resistance, and stretching interventions) improves the quality of life among breast cancer patients who are receiving chemotherapy and radiotherapy [167,173,175,179,183]. A recent systematic review of 56 trials, encompassing 4800 subjects [184] including patients with breast, prostate, gynecologic, hematologic, and other cancers, found that exercise during and after therapy, when compared with control interventions, had a positive impact on overall health-related quality of life. Furthermore, when compared with light exercise, moderate intensity or vigorous exertion during cancer treatment provided greater improvements in health-related quality of life, physical functioning, anxiety, fatigue, and sleep disturbances [43,184]. Further support for the positive impact of exercise on the quality of life was shown by Noble et al. [185] where they enrolled 575 patients with breast cancer in a 24-session program that involved aerobic exercise, resistance training, and stretching exercises. Three hundred eighty-six participants completed the program, and their results showed that a comprehensive physical activity program can significantly improve various quality of life indexes for individuals undergoing treatment for cancer. Lastly, exercise has also been associated with increased chemotherapy completion rates [176]. These findings are summarized in Table 4.

Impact on Aerobic Capacity

It is already well established in the literature that cardiorespiratory fitness and change in fitness strongly predict mortality in healthy adults and those with cardiovascular disease, even after controlling for traditional cardiovascular disease risk factors [186]. There is also evidence that peak oxygen consumption (VO_2) may be important following a diagnosis of cancer such as among patients with lung cancer who will undergo resection [186]. In this population, exercise testing is performed to identify whether physiological reserve is sufficiently high to tolerate the anticipated surgery rather than sufficiently low to justify it. In general, peak $\dot{V}O_2$ of >20 mL/kg/min on incremental cycle ergometry is predictive of the ability to tolerate resection as large as pneumonectomy, whereas values <10 mL/kg/min predict high risk for resection of any extent [187].

Cancer patients, unfortunately, seem to be at a disadvantage at the onset of their therapy. Studies have shown that cardiopulmonary fitness measured by peak VO_2 among cancer patients is consistently approximately 30% below that of age- and sex-matched sedentary individuals without a history of cancer [188,189]. Fortunately, exercise seems to be effective in improving fitness. A meta-analysis of three randomized trials, with a total of 95 patients, among women with early breast cancer found that aerobic training in that patient population was associated with a 3.4 ml/kg/min improvement in oxygen consumption over the course of the intervention [8].

Exercise interventions among those with cancer have ranged from home-based walking programs to structured, supervised fitness sessions including aerobic, resistance, and flexibility activities during chemotherapy and radiation therapy. These interventions have been shown to improve aerobic capacity in this patient population. Among patients with breast cancer who exercised while receiving treatment, 10 randomized controlled trials that were examined showed a significant improvement in aerobic capacity. Similarly, the majority (10/12) of studies that assessed exercise interventions for breast cancer survivors found that the interventions led to an increase in aerobic capacity when compared with controls. A Cochrane review of breast cancer patients performed in 2016 evaluated 15 studies with 1310 women found a standardized mean difference for the pooled data of 0.42 (95% confidence interval of 0.25–0.159). They rated the result of statistically significant improvement in physical fitness as moderate-quality evidence [190]. Similar findings were observed for aerobic and/or resistance training in prostate cancer survivors who were undergoing treatment. The data for colon cancer and hematologic malignancies are also similar [9].

Exercise and Fatigue

Cancer-related fatigue is a persistent, subjective sense of tiredness related to cancer or cancer treatment that interferes with usual functioning [191]. Approximately 50%–90% of cancer patients undergoing cancer treatment experience cancer-related fatigue [192]. Exercise has been proposed as an effective, nonpharmacological intervention to promote psychological well-being during and following cancer treatment. Puetz et al. [193] performed a meta-analysis assessing the differential effects of exercise on cancer-related fatigue during and following treatment. A total of 70 studies were included and found

TABLE 4 Effect of Exercise on Cancer Outcomes

Effects of Exercise During Cancer Therapy
Improved overall health-related quality of life
Improved physical functioning
Increase exercise capacity
Decreased anxiety
Decreased fatigue
Improved sleep habits
Improved chemotherapy completion rate

that exercise training reduced cancer-related fatigue among patients both during and following cancer treatment. During treatment, patients who had lower baseline fatigue scores and higher adherence to exercise regimens experienced the largest improvements. In the posttreatment phase, the greatest improvements were realized when there was a longer duration between treatment completion and exercise initiation and in trials with shorter exercise program lengths. Additionally, the magnitude of the overall mean effect for patients during treatment ($\Delta = 0.32$) and post treatment ($\Delta = 0.38$) is comparable with the effect of (1) exercise interventions on related outcomes in cancer patients, including depression, anxiety, and quality of life; (2) individual or group therapy on cancer-related fatigue; and (3) pharmacotherapy on cancer-related fatigue [193]. Among subgroups by cancer type, exercise was shown to be more effective than control in all except the subgroup of patients with hematologic malignancies [194].

Exercise, Prognosis, Recurrence and Mortality

Few studies have been conducted on the role of physical activity in preventing cancer recurrence or reducing mortality; however, the existing data do show a consistent preventive effect [11].

Studies have been performed among those with breast, colon, and colorectal cancer and suggest that physically active individuals with these cancers may have improved prognosis compared with their sedentary counterparts. Data from the Nurses' Health Study examined the dose-response association of physical activity with overall and breast cancer-specific mortality and recurrence, among 2987 breast cancer survivors over a median of 96 months of follow-up [11,163]. There was a 29% decrease in overall mortality among women who did at least 3 MET-h/week of aerobic activity after diagnosis, with minimal additional protection from greater levels of physical activity. The decreases in breast cancer-specific mortality and recurrence were 50% and 43%, respectively, in women who engaged in at least 9 MET-h/week of physical activity compared with women who did less than 3 MET-h. Activity levels greater than 3 MET-h yielded small additional benefits.

Observational studies have investigated the role of physical activity in colon and colorectal cancer survivorship [11]. The Nurses' Health Study observed an inverse dose-response association of physical activity and overall and colorectal cancer-specific mortality in 554 women with prior diagnoses of colorectal cancer. Women who engaged in at least 18 MET-h/week of physical activity after diagnosis had a 61 and 57% reduced risk of colorectal cancer-specific and overall mortality, respectively, compared with women who did less than 3 MET-h/week [195].

Additionally, there is evidence that being overweight, obesity and weight gain are associated with breast cancer recurrence [11,168,196]. At least 20 prospective observational studies have demonstrated a lower risk of cancer recurrences and improved survival compared with those who are inactive, although the existing data are limited to breast, colorectal, prostate, and ovarian cancer [197]. A dose-response association of physical activity and colon cancer disease-free survival was reported by Meyerhardt et al. [198] in their cohort of 832 male and female patients with stage III colon cancer. They found that 18 MET-h/week or 6h of walking per week at 2.5 mi/h was associated with a 49% reduction in the risk of recurrence. Lastly, some studies have further suggested that physical activity may even increase the rate of completion of chemotherapy [176]. Similarly, a meta-analysis by Ibrahim et al. [199] investigated the relationship between physical activity after breast cancer diagnosis and survival and they found that postdiagnosis exercise was associated with a 34% lower risk of breast cancer deaths, a 41% lower risk of all-cause mortality, and a 24% lower risk of breast cancer recurrence.

One major limitation of the observational trials is that they are comparing patients who have an exercise lifestyle versus a nonexercise lifestyle. It is unclear if beginning exercise during or after therapy will put someone on a different prognostic curve or if the years of exercise that the exercise groups have most likely done can't be replicated by starting exercise at this late time. However, it is clear that the other beneficial effects of exercise during chemotherapy such as reducing weight gain and fatigue and improved quality of life do occur if one starts to exercise.

ONGOING TRIALS

There are several ongoing trials further evaluating the role of exercise and physical activity among those with cancer. A randomized clinical trial in New Hampshire seeks to determine the effect of an exercise intervention on physical activity during chemotherapy among 120 patients with early-stage breast cancer [20]. Another ongoing trial in Australia entitled "Intense Exercise for Survival Among Men with Metastatic Castrate-Resistant Prostate Cancer" is investigating the relationship between high-intensity aerobic and resistance exercise in this population and survival [21]. There are multiple additional trials that are currently enrolling patients geared toward studying the impact of different types of exercise interventions on the various aspects of cancer outcomes.

SUGGESTED AREAS FOR FUTURE RESEARCH

Considerable progress has been made in this area studying the impact of exercise and cancer. It currently is difficult to generalize results due to heterogeneity among trials with regard to the types of exercises prescribed in the various trials, the duration of the intervention, and the outcomes. Additionally, it is unclear if an acute intervention exercise trial or a change in lifestyle to a sustained increase in exercise over years is necessary for improvement. More research on the impact of exercise is also needed in other types of cancer that are not currently well studied including colon cancer, hematologic malignancies, and gynecologic cancers. Lastly, another area of developing interest is the concept of "prehab" where patients go through rehabilitation/exercise prior to undergoing treatment for their malignancy, and further studies in this area are warranted.

REFERENCES

[1] Kodama S, et al. Cardiorespiratory fitness as a quantitative predictor of all-cause mortality and cardiovascular events in healthy men and women: a meta-analysis. JAMA 2009;301(19):2024–35.

[2] Owen N, et al. Sedentary behavior: emerging evidence for a new health risk. Mayo Clin Proc 2010;85(12):1138–41.

[3] Proper KI, et al. Sedentary behaviors and health outcomes among adults: a systematic review of prospective studies. Am J Prev Med 2011;40(2):174–82.

[4] Pavey TG, Peeters GG, Brown WJ. Sitting-time and 9-year all-cause mortality in older women. Br J Sports Med 2015;49(2):95–9.

[5] Leon-Munoz LM, et al. Continued sedentariness, change in sitting time, and mortality in older adults. Med Sci Sports Exerc 2013;45(8):1501–7.

[6] Matthews CE, et al. Physical activity, sedentary behavior, and cause-specific mortality in black and white adults in the Southern Community Cohort Study. Am J Epidemiol 2014;180(4):394–405.

[7] Lee IM, et al. Effect of physical inactivity on major non-communicable diseases worldwide: an analysis of burden of disease and life expectancy. Lancet 2012;380(9838):219–29.

[8] McNeely ML, et al. Effects of exercise on breast cancer patients and survivors: a systematic review and meta-analysis. CMAJ 2006;175(1):34–41.

[9] Schmitz KH, et al. American College of Sports Medicine roundtable on exercise guidelines for cancer survivors. Med Sci Sports Exerc 2010;42(7):1409–26.

[10] Kruk J, Czerniak U. Physical activity and its relation to cancer risk: updating the evidence. Asian Pac J Cancer Prev 2013;14(7):3993–4003.

[11] Physical Activity Guidelines Advisory Committee. Physical Activity Guidelines Advisory Committee report, 2008. Washington, DC: USDHHS; 2008.

[12] Fraser GE, Shavlik D. Risk factors, lifetime risk, and age at onset of breast cancer. Ann Epidemiol 1997;7(6):375–82.

[13] Thune I, et al. Physical activity and the risk of breast cancer. N Engl J Med 1997;336(18):1269–75.

[14] Cerhan JR, et al. Physical activity, physical function, and the risk of breast cancer in a prospective study among elderly women. J Gerontol A Biol Sci Med Sci 1998;53(4):M251–6.

[15] Sesso HD, Paffenbarger Jr. RS, Lee IM. Physical activity and breast cancer risk in the College Alumni Health Study (United States). Cancer Causes Control 1998;9(4):433–9.

[16] Wyshak G, Frisch RE. Breast cancer among former college athletes compared to non-athletes: a 15-year follow-up. Br J Cancer 2000;82(3):726–30.

[17] Dirx MJ, et al. Baseline recreational physical activity, history of sports participation, and postmenopausal breast carcinoma risk in the Netherlands Cohort Study. Cancer 2001;92(6):1638–49.

[18] Lee IM, et al. Physical activity and breast cancer risk: the Women's Health Study (United States). Cancer Causes Control 2001;12(2):137–45.

[19] Breslow RA, et al. Long-term recreational physical activity and breast cancer in the National Health and Nutrition Examination Survey I epidemiologic follow-up study. Cancer Epidemiol Biomarkers Prev 2001;10(7):805–8.

[20] Patel AV, et al. Recreational physical activity and risk of postmenopausal breast cancer in a large cohort of US women. Cancer Causes Control 2003;14(6):519–29.

[21] McTiernan A, et al. Recreational physical activity and the risk of breast cancer in postmenopausal women: the Women's Health Initiative Cohort Study. JAMA 2003;290(10):1331–6.

[22] Rintala P, et al. Physical activity and breast cancer risk among female physical education and language teachers: a 34-year follow-up. Int J Cancer 2003;107(2):268–70.

[23] Bardia A, et al. Recreational physical activity and risk of postmenopausal breast cancer based on hormone receptor status. Arch Intern Med 2006;166(22):2478–83.

[24] Chang SC, et al. Association of energy intake and energy balance with postmenopausal breast cancer in the prostate, lung, colorectal, and ovarian cancer screening trial. Cancer Epidemiol Biomarkers Prev 2006;15(2):334–41.

[25] Tehard B, et al. Effect of physical activity on women at increased risk of breast cancer: results from the E3N cohort study. Cancer Epidemiol Biomarkers Prev 2006;15(1):57–64.

[26] Lahmann PH, et al. Physical activity and breast cancer risk: the European Prospective Investigation into Cancer and Nutrition. Cancer Epidemiol Biomarkers Prev 2007;16(1):36–42.

[27] Dallal CM, et al. Long-term recreational physical activity and risk of invasive and in situ breast cancer: the California teachers study. Arch Intern Med 2007;167(4):408–15.

[28] Bernstein L, et al. Physical exercise and reduced risk of breast cancer in young women. J Natl Cancer Inst 1994;86(18):1403–8.

[29] Friedenreich CM, Rohan TE. Physical activity and risk of breast cancer. Eur J Cancer Prev 1995;4(2):145–51.

[30] McTiernan A, et al. Occurrence of breast cancer in relation to recreational exercise in women age 50–64 years. Epidemiology 1996;7(6):598–604.

[31] Ueji M, et al. Physical activity and the risk of breast cancer: a case-control study of Japanese women. J Epidemiol 1998;8(2):116–22.

[32] Gilliland FD, et al. Physical activity and breast cancer risk in Hispanic and non-Hispanic white women. Am J Epidemiol 2001;154(5):442–50.

[33] Carpenter CL, et al. Effect of family history, obesity and exercise on breast cancer risk among postmenopausal women. Int J Cancer 2003;106(1):96–102.

[34] John EM, Horn-Ross PL, Koo J. Lifetime physical activity and breast cancer risk in a multiethnic population: the San Francisco Bay area breast cancer study. Cancer Epidemiol Biomarkers Prev 2003;12(11 Pt 1):1143–52.

[35] Patel AV, et al. Lifetime recreational exercise activity and risk of breast carcinoma in situ. Cancer 2003;98(10):2161–9.

[36] Yang D, Bernstein L, Wu AH. Physical activity and breast cancer risk among Asian-American women in Los Angeles: a case-control study. Cancer 2003;97(10):2565–75.

[37] Sprague BL, et al. Lifetime recreational and occupational physical activity and risk of in situ and invasive breast cancer. Cancer Epidemiol Biomarkers Prev 2007;16(2):236–43.

[38] Bernstein L, et al. Lifetime recreational exercise activity and breast cancer risk among black women and white women. J Natl Cancer Inst 2005;97(22):1671–9.

[39] Slattery ML, et al. Physical activity and breast cancer risk among women in the southwestern United States. Ann Epidemiol 2007;17(5):342–53.

[40] Lagerros YT, Hsieh SF, Hsieh CC. Physical activity in adolescence and young adulthood and breast cancer risk: a quantitative review. Eur J Cancer Prev 2004;13(1):5–12.

[41] Monninkhof EM, et al. Physical activity and breast cancer: a systematic review. Epidemiology 2007;18(1):137–57.

[42] Lynch BM, Neilson HK, Friedenreich CM. Physical activity and breast cancer prevention. Recent Results Cancer Res 2011;186:13–42.

[43] Lemanne D, Cassileth B, Gubili J. The role of physical activity in cancer prevention, treatment, recovery, and survivorship. Oncology (Williston Park) 2013;27(6):580–5.

[44] Adams-Campbell LL, et al. Strenuous physical activity and breast cancer risk in African-American women. J Natl Med Assoc 2001;93(7–8):267–75.

[45] Paffenbarger Jr. RS, Hyde RT, Wing AL. Physical activity and incidence of cancer in diverse populations: a preliminary report. Am J Clin Nutr 1987;45(1 Suppl.):312–7.

[46] Vena JE, et al. Occupational exercise and risk of cancer. Am J Clin Nutr 1987;45(1 Suppl.):318–27.

[47] Thune I, Lund E. Physical activity and the risk of prostate and testicular cancer: a cohort study of 53,000 Norwegian men. Cancer Causes Control 1994;5(6):549–56.

[48] Steenland K, Nowlin S, Palu S. Cancer incidence in the National Health and Nutrition Survey I. Follow-up data: diabetes, cholesterol, pulse and physical activity. Cancer Epidemiol Biomarkers Prev 1995;4(8):807–11.

[49] Oliveria SA, et al. The association between cardiorespiratory fitness and prostate cancer. Med Sci Sports Exerc 1996;28(1):97–104.

[50] Hartman TJ, et al. Physical activity and prostate cancer in the Alpha-Tocopherol, Beta-Carotene (ATBC) Cancer Prevention Study (Finland). Cancer Causes Control 1998;9(1):11–8.

[51] Giovannucci EL, et al. A prospective study of physical activity and incident and fatal prostate cancer. Arch Intern Med 2005;165(9):1005–10.

[52] Lund Nilsen TI, Johnsen R, Vatten LJ. Socio-economic and lifestyle factors associated with the risk of prostate cancer. Br J Cancer 2000;82(7):1358–63.

[53] Clarke G, Whittemore AS. Prostate cancer risk in relation to anthropometry and physical activity: the National Health and Nutrition Examination Survey I Epidemiological Follow-Up Study. Cancer Epidemiol Biomarkers Prev 2000;9(9):875–81.

[54] Wannamethee SG, Shaper AG, Walker M. Physical activity and risk of cancer in middle-aged men. Br J Cancer 2001;85(9):1311–6.

[55] Norman A, et al. Occupational physical activity and risk for prostate cancer in a nationwide cohort study in Sweden. Br J Cancer 2002;86(1):70–5.

[56] Yu H, Harris RE, Wynder EL. Case-control study of prostate cancer and socioeconomic factors. Prostate 1988;13(4):317–25.

[57] Brownson RC, et al. Physical activity on the job and cancer in Missouri. Am J Public Health 1991;81(5):639–42.

[58] Andersson SO, et al. Early life risk factors for prostate cancer: a population-based case-control study in Sweden. Cancer Epidemiol Biomarkers Prev 1995;4(3):187–92.

[59] Villeneuve PJ, et al. Risk factors for prostate cancer: results from the Canadian National Enhanced Cancer Surveillance System. The Canadian Cancer Registries Epidemiology Research Group. Cancer Causes Control 1999;10(5):355–67.

[60] Bairati I, et al. Lifetime occupational physical activity and incidental prostate cancer (Canada). Cancer Causes Control 2000;11(8):759–64.

[61] Patel AV, et al. Recreational physical activity and risk of prostate cancer in a large cohort of U.S. men. Cancer Epidemiol Biomarkers Prev 2005;14(1):275–9.

[62] Nilsen TI, Romundstad PR, Vatten LJ. Recreational physical activity and risk of prostate cancer: a prospective population-based study in Norway (the HUNT study). Int J Cancer 2006;119(12):2943–7.

[63] Darlington GA, et al. Prostate cancer risk and diet, recreational physical activity and cigarette smoking. Chronic Dis Can 2007;27(4):145–53.

[64] Schnohr P, et al. Physical activity in leisure-time and risk of cancer: 14-year follow-up of 28,000 Danish men and women. Scand J Public Health 2005;33(4):244–9.

[65] Severson RK, et al. A prospective analysis of physical activity and cancer. Am J Epidemiol 1989;130(3):522–9.

[66] Hsing AW, et al. Occupation, physical activity, and risk of prostate cancer in Shanghai, People's Republic of China. Cancer Causes Control 1994;5(2):136–40.

[67] Liu S, et al. A prospective study of physical activity and risk of prostate cancer in US physicians. Int J Epidemiol 2000;29(1):29–35.

[68] Putnam SD, et al. Lifestyle and anthropometric risk factors for prostate cancer in a cohort of Iowa men. Ann Epidemiol 2000;10(6):361–9.

[69] Pukkala E, et al. Cancer incidence among Finnish world class male athletes. Int J Sports Med 2000;21(3):216–20.

[70] Le Marchand L, Kolonel LN, Yoshizawa CN. Lifetime occupational physical activity and prostate cancer risk. Am J Epidemiol 1991;133(2):103–11.

[71] Lee IM, Sesso HD, Paffenbarger Jr. RS. A prospective cohort study of physical activity and body size in relation to prostate cancer risk (United States). Cancer Causes Control 2001;12(2):187–93.

[72] Dosemeci M, et al. Occupational physical activity, socioeconomic status, and risks of 15 cancer sites in Turkey. Cancer Causes Control 1993;4(4):313–21.

[73] Whittemore AS, et al. Prostate cancer in relation to diet, physical activity, and body size in blacks, whites, and Asians in the United States and Canada. J Natl Cancer Inst 1995;87(9):652–61.

[74] Friedenreich CM, et al. Case-control study of lifetime total physical activity and prostate cancer risk. Am J Epidemiol 2004;159(8):740–9.

[75] Zeegers MP, Dirx MJ, van den Brandt PA. Physical activity and the risk of prostate cancer in the Netherlands cohort study, results after 9.3 years of follow-up. Cancer Epidemiol Biomarkers Prev 2005;14(6):1490–5.

[76] Littman AJ, Kristal AR, White E. Recreational physical activity and prostate cancer risk (United States). Cancer Causes Control 2006;17(6):831–41.

[77] Giovannucci E, et al. Physical activity, obesity, and risk for colon cancer and adenoma in men. Ann Intern Med 1995;122(5):327–34.

[78] Calton BA, et al. Physical activity and the risk of colon cancer among women: a prospective cohort study (United States). Int J Cancer 2006;119(2):385–91.

[79] Chao A, et al. Amount, type, and timing of recreational physical activity in relation to colon and rectal cancer in older adults: the Cancer Prevention Study II Nutrition Cohort. Cancer Epidemiol Biomarkers Prev 2004;13(12):2187–95.

[80] Colbert LH, et al. Physical activity in relation to cancer of the colon and rectum in a cohort of male smokers. Cancer Epidemiol Biomarkers Prev 2001;10(3):265–8.

[81] Larsson SC, et al. Physical activity, obesity, and risk of colon and rectal cancer in a cohort of Swedish men. Eur J Cancer 2006;42(15):2590–7.

[82] Lee IM, et al. Physical activity and risk of colon cancer: the Physicians' Health Study (United States). Cancer Causes Control 1997;8(4):568–74.

[83] Lee KJ, et al. Physical activity and risk of colorectal cancer in Japanese men and women: the Japan Public Health Center-based prospective study. Cancer Causes Control 2007;18(2):199–209.

[84] Mai PL, et al. Physical activity and colon cancer risk among women in the California Teachers Study. Cancer Epidemiol Biomarkers Prev 2007;16(3):517–25.

[85] Martinez ME, et al. Leisure-time physical activity, body size, and colon cancer in women. Nurses' Health Study Research Group. J Natl Cancer Inst 1997;89(13):948–55.

[86] Thune I, Lund E. Physical activity and risk of colorectal cancer in men and women. Br J Cancer 1996;73(9):1134–40.

[87] Hou L, et al. Commuting physical activity and risk of colon cancer in Shanghai, China. Am J Epidemiol 2004;160(9):860–7.

[88] Isomura K, et al. Physical activity and colorectal cancer: the Fukuoka Colorectal Cancer Study. Cancer Sci 2006;97(10):1099–104.

[89] Levi F, et al. Occupational and leisure-time physical activity and the risk of colorectal cancer. Eur J Cancer Prev 1999;8(6):487–93.

[90] Slattery ML, et al. Energy balance and colon cancer—beyond physical activity. Cancer Res 1997;57(1):75–80.

[91] Slattery ML, et al. Physical activity and colon cancer: a public health perspective. Ann Epidemiol 1997;7(2):137–45.

[92] Slattery ML, Potter JD. Physical activity and colon cancer: confounding or interaction? Med Sci Sports Exerc 2002;34(6):913–9.

[93] Tavani A, et al. Physical activity and risk of cancers of the colon and rectum: an Italian case-control study. Br J Cancer 1999;79(11–12):1912–6.

[94] Tang R, et al. Physical activity, water intake and risk of colorectal cancer in Taiwan: a hospital-based case-control study. Int J Cancer 1999;82(4):484–9.

[95] White E, Jacobs EJ, Daling JR. Physical activity in relation to colon cancer in middle-aged men and women. Am J Epidemiol 1996;144(1):42–50.

[96] Slattery ML, et al. Physical activity and colorectal cancer. Am J Epidemiol 2003;158(3):214–24.

[97] Zhang Y, et al. Occupational and leisure-time physical activity and risk of colon cancer by subsite. J Occup Environ Med 2006;48(3):236–43.

[98] Mao Y, et al. Physical inactivity, energy intake, obesity and the risk of rectal cancer in Canada. Int J Cancer 2003;105(6):831–7.

[99] Steindorf K, et al. Physical activity and lung cancer risk in the European Prospective Investigation into Cancer and Nutrition Cohort. Int J Cancer 2006;119(10):2389–97.

[100] Sinner P, et al. The association of physical activity with lung cancer incidence in a cohort of older women: the Iowa Women's Health Study. Cancer Epidemiol Biomarkers Prev 2006;15(12):2359–63.

[101] Thune I, Lund E. The influence of physical activity on lung-cancer risk: a prospective study of 81,516 men and women. Int J Cancer 1997;70(1):57–62.

[102] Lee IM, Sesso HD, Paffenbarger Jr. RS. Physical activity and risk of lung cancer. Int J Epidemiol 1999;28(4):620–5.

[103] Colbert LH, et al. Physical activity and lung cancer risk in male smokers. Int J Cancer 2002;98(5):770–3.

[104] Bak H, et al. Physical activity and risk for lung cancer in a Danish cohort. Int J Cancer 2005;116(3):439–44.

[105] Alfano CM, et al. Physical activity in relation to all-site and lung cancer incidence and mortality in current and former smokers. Cancer Epidemiol Biomarkers Prev 2004;13(12):2233–41.

[106] Kubik A, et al. Lung cancer risk among nonsmoking women in relation to diet and physical activity. Neoplasma 2004;51(2):136–43.

[107] Kubik AK, et al. Dietary habits and lung cancer risk among non-smoking women. Eur J Cancer Prev 2004;13(6):471–80.

[108] Mao Y, et al. Physical activity and the risk of lung cancer in Canada. Am J Epidemiol 2003;158(6):564–75.

[109] Lee IM OY. Physical activity. In: Schottenfeld D, Fraumeni JF, editors. Cancer epidemiology and prevention. 3rd ed. New York: Oxford University Press; 2006. p. 449–67.

[110] Voskuil DW, et al. Physical activity and endometrial cancer risk, a systematic review of current evidence. Cancer Epidemiol Biomarkers Prev 2007;16(4):639–48.

[111] Cust AE, et al. Physical activity and endometrial cancer risk: a review of the current evidence, biologic mechanisms and the quality of physical activity assessment methods. Cancer Causes Control 2007;18(3):243–58.

[112] Olsen CM, et al. Obesity and the risk of epithelial ovarian cancer: a systematic review and meta-analysis. Eur J Cancer 2007;43(4):690–709.

[113] Westerlind KC. Physical activity and cancer prevention-mechanisms. Med Sci Sports Exerc 2003;35(11):1834–40.

[114] Hulka BS, Brinton LA. Hormones and breast and endometrial cancers: preventive strategies and future research. Environ Health Perspect 1995;103(Suppl. 8):185–9.

[115] Whelan EA, et al. Menstrual cycle patterns and risk of breast cancer. Am J Epidemiol 1994;140(12):1081–90.

[116] Bernstein L, et al. The effects of moderate physical activity on menstrual cycle patterns in adolescence: implications for breast cancer prevention. Br J Cancer 1987;55(6):681–5.

[117] Warren MP. The effects of exercise on pubertal progression and reproductive function in girls. J Clin Endocrinol Metab 1980;51(5):1150–7.

[118] Apter D. Hormonal events during female puberty in relation to breast cancer risk. Eur J Cancer Prev 1996;5(6):476–82.

[119] Loucks AB. Effects of exercise training on the menstrual cycle: existence and mechanisms. Med Sci Sports Exerc 1990;22(3):275–80.

[120] Vihko RK, Apter DL. The epidemiology and endocrinology of the menarche in relation to breast cancer. Cancer Surv 1986;5(3):561–71.

[121] Bullen BA, et al. Induction of menstrual disorders by strenuous exercise in untrained women. N Engl J Med 1985;312(21):1349–53.

[122] Pike MC, et al. Estrogens, progestogens, normal breast cell proliferation, and breast cancer risk. Epidemiol Rev 1993;15(1):17–35.

[123] Jakobisiak M, Lasek W, Golab J. Natural mechanisms protecting against cancer. Immunol Lett 2003;90(2-3):103–22.

[124] Wetmore CM UC. Mechanisms associating physical activity with cancer incidence: exercise and immune function. In: McTiernan A, editor. Cancer prevention and management through exercise and weight control. Boca Raton: CRC Press; 2006. p. 157–76.

[125] MacNeil B, Hoffman-Goetz L. Chronic exercise enhances in vivo and in vitro cytotoxic mechanisms of natural immunity in mice. J Appl Physiol (1985) 1993;74(1):388–95.

[126] Il'yasova D, et al. Circulating levels of inflammatory markers and cancer risk in the health aging and body composition cohort. Cancer Epidemiol Biomarkers Prev 2005;14(10):2413–8.

[127] Campbell KL, McTiernan A. Exercise and biomarkers for cancer prevention studies. J Nutr 2007;137(1 Suppl.):161s–9s.

[128] Kaaks R, Lukanova A. Energy balance and cancer: the role of insulin and insulin-like growth factor-I. Proc Nutr Soc 2001;60(1):91–106.

[129] Wolf I, et al. Diabetes mellitus and breast cancer. Lancet Oncol 2005;6(2):103–11.

[130] Boule NG, et al. Effects of exercise on glycemic control and body mass in type 2 diabetes mellitus: a meta-analysis of controlled clinical trials. JAMA 2001;286(10):1218–27.

[131] Frank LL, et al. Effects of exercise on metabolic risk variables in overweight postmenopausal women: a randomized clinical trial. Obes Res 2005;13(3):615–25.

[132] Ross R, et al. Reduction in obesity and related comorbid conditions after diet-induced weight loss or exercise-induced weight loss in men. A randomized, controlled trial. Ann Intern Med 2000;133(2):92–103.

[133] Ross R, et al. Exercise-induced reduction in obesity and insulin resistance in women: a randomized controlled trial. Obes Res 2004;12(5):789–98.

[134] Duncan GE, et al. Exercise training, without weight loss, increases insulin sensitivity and postheparin plasma lipase activity in previously sedentary adults. Diabetes Care 2003;26(3):557–62.

[135] Arnold M, et al. Obesity and cancer: an update of the global impact. Cancer Epidemiol 2016;41:8–15.

[136] Peltonen M, Carlsson LM. Body fatness and cancer. N Engl J Med 2016;375(20):2007–8.

[137] World Cancer Research Fund, American Institute for Cancer Research. In: AICR, editor. Food, nutrition, physical activity, and the prevention of cancer: a global perspective. Washington, DC: AICR; 2007.

[138] Lichtman MA. Obesity and the risk for a hematological malignancy: leukemia, lymphoma, or myeloma. Oncologist 2010;15(10):1083–101.

[139] Vucenik I, Stains JP. Obesity and cancer risk: evidence, mechanisms, and recommendations. Ann N Y Acad Sci 2012;1271:37–43.

[140] Vainio H, Kaaks R, Bianchini F. Weight control and physical activity in cancer prevention: international evaluation of the evidence. Eur J Cancer Prev 2002;11(Suppl. 2):S94–100.

[141] International Agency for Research on Cancer. IARC handbooks of cancer prevention: weight control and physical activity. Lyon, France: IARC.

[142] Renehan AG, et al. Body-mass index and incidence of cancer: a systematic review and meta-analysis of prospective observational studies. Lancet 2008;371(9612):569–78.

[143] Ma Y, et al. Obesity and risk of colorectal cancer: a systematic review of prospective studies. PLoS One 2013;8(1):e53916.

[144] Chen Y, et al. Body mass index and risk of gastric cancer: a meta-analysis of a population with more than ten million from 24 prospective studies. Cancer Epidemiol Biomarkers Prev 2013;22(8):1395–408.

[145] Chen Y, et al. Excess body weight and the risk of primary liver cancer: an updated meta-analysis of prospective studies. Eur J Cancer 2012;48(14):2137–45.

[146] World Cancer Research Fund International. Continuous Update Project: diet, nutrition, physical activity and gallbladder cancer; 2015. Available from: http://www.wcrf.org/sites/default/files/Gallbladder-Cancer-2015-Report.pdf.

[147] Genkinger JM, et al. A pooled analysis of 14 cohort studies of anthropometric factors and pancreatic cancer risk. Int J Cancer 2011;129(7):1708–17.

[148] Wang F, Xu Y. Body mass index and risk of renal cell cancer: a dose-response meta-analysis of published cohort studies. Int J Cancer 2014;135(7):1673–86.

[149] Hoyo C, et al. Body mass index in relation to oesophageal and oesophagogastric junction adenocarcinomas: a pooled analysis from the International BEACON Consortium. Int J Epidemiol 2012;41(6):1706–18.

[150] Protani M, Coory M, Martin JH. Effect of obesity on survival of women with breast cancer: systematic review and meta-analysis. Breast Cancer Res Treat 2010;123(3):627–35.

[151] Patterson RE, et al. Physical activity, diet, adiposity and female breast cancer prognosis: a review of the epidemiologic literature. Maturitas 2010;66(1):5–15.

[152] Andersen LB, et al. All-cause mortality associated with physical activity during leisure time, work, sports, and cycling to work. Arch Intern Med 2000;160(11):1621–8.

[153] Paffenbarger Jr. RS, et al. The association of changes in physical-activity level and other lifestyle characteristics with mortality among men. N Engl J Med 1993;328(8):538–45.

[154] Gregg EW, et al. Relationship of changes in physical activity and mortality among older women. JAMA 2003;289(18):2379–86.

[155] Wen CP, et al. Minimum amount of physical activity for reduced mortality and extended life expectancy: a prospective cohort study. Lancet 2011;378(9798):1244–53.

[156] Blair SN, et al. Physical fitness and all-cause mortality. A prospective study of healthy men and women. JAMA 1989;262(17):2395–401.

[157] Garatachea N, et al. Elite athletes live longer than the general population: a meta-analysis. Mayo Clin Proc 2014;89(9):1195–200.

[158] Scott JM, et al. Exercise therapy as treatment for cardiovascular and oncologic disease after a diagnosis of early-stage cancer. Semin Oncol 2013;40(2):218–28.

[159] Rock CL, Demark-Wahnefried W. Nutrition and survival after the diagnosis of breast cancer: a review of the evidence. J Clin Oncol 2002;20(15):3302–16.

[160] Wahnefried W, Rimer BK, Winer EP. Weight gain in women diagnosed with breast cancer. J Am Diet Assoc 1997;97(5):519–29.

[161] Rock CL, et al. Factors associated with weight gain in women after diagnosis of breast cancer. Women's Healthy Eating and Living Study Group. J Am Diet Assoc 1999;99(10):1212–21.

[162] Atalay C, Kucuk AI. The impact of weight gain during adjuvant chemotherapy on survival in breast cancer. Ulus Cerrahi Derg 2015;31(3):124–7.

[163] Holmes MD, et al. Physical activity and survival after breast cancer diagnosis. JAMA 2005;293(20):2479–86.

[164] Kroenke CH, et al. Weight, weight gain, and survival after breast cancer diagnosis. J Clin Oncol 2005;23(7):1370–8.

[165] Camoriano JK, et al. Weight change in women treated with adjuvant therapy or observed following mastectomy for node-positive breast cancer. J Clin Oncol 1990;8(8):1327–34.

[166] Irwin ML, et al. Physical activity levels before and after a diagnosis of breast carcinoma: the Health, Eating, Activity, and Lifestyle (HEAL) study. Cancer 2003;97(7):1746–57.

[167] Mutrie N, et al. Benefits of supervised group exercise programme for women being treated for early stage breast cancer: pragmatic randomised controlled trial. BMJ 2007;334(7592):517.

[168] Chlebowski RT, Aiello E, McTiernan A. Weight loss in breast cancer patient management. J Clin Oncol 2002;20(4):1128–43.

[169] Caan BJ, et al. Pre-diagnosis body mass index, post-diagnosis weight change, and prognosis among women with early stage breast cancer. Cancer Causes Control 2008;19(10):1319–28.

[170] Griffith K, et al. Impact of a walking intervention on cardiorespiratory fitness, self-reported physical function, and pain in patients undergoing treatment for solid tumors. Cancer 2009;115(20):4874–84.

[171] Playdon MC, et al. Weight gain after breast cancer diagnosis and all-cause mortality: systematic review and meta-analysis. J Natl Cancer Inst 2015;107(12):275.

[172] Speck RM, et al. An update of controlled physical activity trials in cancer survivors: a systematic review and meta-analysis. J Cancer Surviv 2010;4(2):87–100.

[173] Adamsen L, et al. Effect of a multimodal high intensity exercise intervention in cancer patients undergoing chemotherapy: randomised controlled trial. BMJ 2009;339:b3410.

[174] Cadmus LA, et al. Exercise and quality of life during and after treatment for breast cancer: results of two randomized controlled trials. Psychooncology 2009;18(4):343–52.

[175] Campbell A, et al. A pilot study of a supervised group exercise programme as a rehabilitation treatment for women with breast cancer receiving adjuvant treatment. Eur J Oncol Nurs 2005;9(1):56–63.

[176] Courneya KS, et al. Effects of aerobic and resistance exercise in breast cancer patients receiving adjuvant chemotherapy: a multicenter randomized controlled trial. J Clin Oncol 2007;25(28):4396–404.

[177] Demark-Wahnefried W, et al. Results of a diet/exercise feasibility trial to prevent adverse body composition change in breast cancer patients on adjuvant chemotherapy. Clin Breast Cancer 2008;8(1):70–9.

[178] Drouin JS, et al. Random control clinical trial on the effects of aerobic exercise training on erythrocyte levels during radiation treatment for breast cancer. Cancer 2006;107(10):2490–5.

[179] Headley JA, Ownby KK, John LD. The effect of seated exercise on fatigue and quality of life in women with advanced breast cancer. Oncol Nurs Forum 2004;31(5):977–83.

[180] Schwartz AL, Winters-Stone K, Gallucci B. Exercise effects on bone mineral density in women with breast cancer receiving adjuvant chemotherapy. Oncol Nurs Forum 2007;34(3):627–33.

[181] Schwartz AL, Winters-Stone K. Effects of a 12-month randomized controlled trial of aerobic or resistance exercise during and following cancer treatment in women. Phys Sportsmed 2009;37(3):62–7.

[182] Segal R, et al. Structured exercise improves physical functioning in women with stages I and II breast cancer: results of a randomized controlled trial. J Clin Oncol 2001;19(3):657–65.

[183] Lahart IM, et al. Randomised controlled trial of a home-based physical activity intervention in breast cancer survivors. BMC Cancer 2016;16:234.

[184] Mishra SI, et al. Exercise interventions on health-related quality of life for people with cancer during active treatment. Clin Otolaryngol 2012;37(5):390–2.

[185] Noble M, et al. UW WELL-FIT: the impact of supervised exercise programs on physical capacity and quality of life in individuals receiving treatment for cancer. Support Care Cancer 2012;20(4):865–73.

[186] Jones LW, et al. Effect of exercise training on peak oxygen consumption in patients with cancer: a meta-analysis. Oncologist 2011;16(1):112–20.

[187] Arena R, Sietsema KE. Cardiopulmonary exercise testing in the clinical evaluation of patients with heart and lung disease. Circulation 2011;123(6):668–80.

[188] Haykowsky MJ, et al. Adjuvant trastuzumab induces ventricular remodeling despite aerobic exercise training. Clin Cancer Res 2009;15(15):4963–7.

[189] Jones LW, et al. Safety and feasibility of cardiopulmonary exercise testing in patients with advanced cancer. Lung Cancer 2007;55(2):225–32.

[190] Furmaniak AC, Menig M, Markes MH. Exercise for women receiving adjuvant therapy for breast cancer. Cochrane Database Syst Rev 2016;9:Cd005001.

[191] Mock V, et al. NCCN practice guidelines for cancer-related fatigue. Oncology (Williston Park) 2000;14(11a):151–61.

[192] Campos MP, et al. Cancer-related fatigue: a practical review. Ann Oncol 2011;22(6):1273–9.

[193] Puetz TW, Herring MP. Differential effects of exercise on cancer-related fatigue during and following treatment: a meta-analysis. Am J Prev Med 2012;43(2):e1–24.

[194] Cramp F, Byron-Daniel J. Exercise for the management of cancer-related fatigue in adults. Cochrane Database Syst Rev 2012;11:Cd006145.

[195] Meyerhardt JA, et al. Physical activity and survival after colorectal cancer diagnosis. J Clin Oncol 2006;24(22):3527–34.

[196] Trentham-Dietz A, et al. Breast cancer risk factors and second primary malignancies among women with breast cancer. Breast Cancer Res Treat 2007;105(2):195–207.

[197] Rock CL, et al. Nutrition and physical activity guidelines for cancer survivors. CA Cancer J Clin 2012;62(4):242–74.

[198] Meyerhardt JA, et al. Impact of physical activity on cancer recurrence and survival in patients with stage III colon cancer: findings from CALGB 89803. J Clin Oncol 2006;24(22):3535–41.

[199] Ibrahim EM, Al-Homaidh A. Physical activity and survival after breast cancer diagnosis: meta-analysis of published studies. Med Oncol 2011;28(3):753–65.

Chapter 9

Exercise Prescription for Hypertension: New Advances for Optimizing Blood Pressure Benefits

Hayley V. MacDonald*, Linda S. Pescatello[†]
*Department of Kinesiology, The University of Alabama, Tuscaloosa, AL, United States, [†]Department of Kinesiology, University of Connecticut, Storrs, CT, United States

INTRODUCTION

Hypertension is a Major Public Health Problem

Hypertension is the most common, costly, and modifiable cardiovascular disease (CVD) risk factor in the United States and world [1,2]. Among 188 countries, high systolic blood pressure (BP) is the *leading* global risk factor and has accounted for 10.4 million deaths and 208.1 million years of life lost over the past two decades [3,4]. In the United States, the number of hypertension-related deaths increased 62% from the year 2000 to 2013 (i.e., from 245,220 to 396,675 deaths) [5]. The estimated direct and indirect cost of hypertension is $51.2 billion, and this figure is projected to increase to $274 billion by the year 2030 [1]. For these reasons, the Department of Health and Human Services established hypertension as a high-priority, leading health indicator within the Healthy People 2020 national objectives by aiming to (a) increase the proportion of adults with hypertension whose BP is under control by 18% and (b) reduce the proportion of adults diagnosed with hypertension by 10% by the year 2020 [6].

Early diagnosis, including accurate and repetitive BP measurements using standard professional methodology and procedures, and effective antihypertensive therapeutic interventions are essential for meeting the Healthy People 2020 national objectives. Untreated and poorly controlled hypertension contribute to and accelerate pathological processes that lead to increased risk of CVD, shorter life expectancy free of CVD, more years lived with CVD, and increased of risk of mortality [2,7–9]. Adoption of a healthy lifestyle is fundamental for the primary prevention, treatment, and control of hypertension. Given that the relationship between BP and CVD risk is linear, continuous, and consistent starting at 115/75 mmHg, early and aggressive lifestyle intervention is critical in order to prevent or delay the rapid, progressive rise in BP [2,4,10].

Antihypertensive Effects of Exercise

Participation in regular exercise is a key modifiable determinant of hypertension and is recognized as a cornerstone therapy for the primary prevention, treatment, and control of high BP [4,11–14]. Large-scale, prospective studies consistently demonstrate that physical activity and cardiorespiratory fitness are inversely associated with the development of hypertension [15–17]. Similar trends have also been reported for muscular fitness (i.e., muscular strength and endurance) and the development of CVD risk factors, including hypertension [18–21]. Among those with hypertension, higher levels of muscular strength had a protective effect against all-cause mortality, independent of cardiorespiratory fitness [19,22]. Recent meta-analyses of randomized controlled intervention trials conclude that regular, aerobic exercise lowers resting BP 5–7 mmHg, while dynamic resistance exercise lowers resting BP 2–3 mmHg among individuals with hypertension [23,24]. The magnitude of these BP reductions following aerobic and resistance exercise can lower CVD risk by 20%–30% [25–27] and 6%–14% [9,25,27], respectively. Furthermore, aerobic exercise can lower BP by a magnitude that rivals those obtained with first-line antihypertensive medications [25–28]. Exercising as little as one day per week is as effective (or even more so) than pharmacotherapy for reducing all-cause mortality among those with hypertension [29]. In addition, when lifestyle modifications are executed appropriately, additional coexisting CVD risk factors may experience benefit, translating to an even greater reduction in overall CVD risk. Indeed, a recent meta-analysis of major exercise and drug trials showed no statistically

Lifestyle in Heart Health and Disease. https://doi.org/10.1016/B978-0-12-811279-3.00009-4

detectable difference between exercise and drug interventions for the secondary prevention of coronary heart disease and prediabetes, and physical activity interventions were more effective than drug interventions for the secondary prevention of stroke mortality [30].

For these reasons, the Seventh Report of the Joint National Committee on Prevention, Detection, Evaluation, and Treatment of High Blood Pressure (JNC 7) [26], 2014 Evidence-Based Guidelines for the Management of High Blood Pressure in Adults (JNC 8) [10], AHA/ACC 2013 Lifestyle Work Group [11], European Society of Hypertension and European Society of Cardiology [31], Canadian Hypertension Education Program [32], and the ACSM [13] universally endorse aerobic exercise for the primary prevention and treatment of hypertension. Although the recommended exercise prescription (Ex R_x) for individuals with hypertension in terms of the *Frequency* (how often?), *Intensity* (how hard?), *Time* (how long?), *Type* (what kind?), *Volume* (how much?), and *Progression* (or FITT-VP) principle varies slightly across the various aforementioned organizational guidelines (Table 1), the general consensus is that adults with prehypertension to established hypertension should participate in 30–60 min/day of moderate-intensity aerobic exercise on most, if not all, days of the week to total 150 min/week of exercise (or more) supplemented by moderate-intensity dynamic resistance training on 2–3 days per week [23].

KEY TERMINOLOGY AND BASIC CONCEPTS

Hypertension

Hypertension is a health condition defined by a transitory or sustained elevation of systemic arterial BP to a level likely to induce cardiovascular damage or result in other adverse health consequences [38]. The JNC 7 [26] defines hypertension as a resting systolic BP of 140 mmHg or greater, a resting diastolic BP of 90 mmHg or greater, taking antihypertensive medication, being told by a physician or health professional on at least two occasions that one has high BP, or any combination of these criteria. Based on this definition, hypertension affects $\approx$ 86 million Americans (34%) [1,39] and $\approx$ 1.5 billion adults (31%) worldwide [2,40].

Prehypertension

Another $\approx$ 90 million of Americans (36%) [1,41] have prehypertension, defined as resting systolic BP ranging from 120 to 139 mmHg and/or diastolic BP ranging from 80 to 89 mmHg. The progression from prehypertension to hypertension occurs rapidly such that about one in four adults with prehypertension will develop hypertension within 5 years [1,9,42,43]. Individuals with prehypertension are 30% more likely than individuals with normal BP to have two or more concomitant risk-enhancing factors (other than high BP) outside of the normal range that includes dyslipidemia, abdominal obesity, hyperinsulinemia and/or insulin resistance, impaired fasting glucose, endothelial dysfunction, among others [41,44]. Prehypertension is also associated with a 10.5%, 4.8%, and 14.6% increased risk of CVD, heart disease, and stroke mortality, respectively [1,45,46]. Therefore, the JNC 7 included this additional BP classification to identify individuals at heightened risk of developing hypertension in the future and to strengthen the recommendations for therapeutic lifestyle intervention among adults with prehypertension [26,44].

Of note, the most current update of this report, the JNC 8 [10], does not specifically address the classification of hypertension or prehypertension in adults *per se*; rather, it provides guidance to clinicians on medication treatment thresholds and goals in the management of hypertension. Therefore, the JNC 7 BP classification scheme is clinically meaningful, remains widely accepted [47], and is the BP classification scheme incorporated in this chapter.

Law of Initial Values

The direction of the response of a body function to an agent depends to a large degree on the initial level of that function. Therefore, BP reductions should be the greatest for those with highest resting BP [48,49].

The Blood Pressure Response to Acute and Chronic Exercise

The BP reductions following acute exercise are immediate, but short-term, persisting for up to 24 h after the exercise bout [50]. This response is termed postexercise hypotension or PEH [13,51,52]. The BP reductions following chronic exercise or exercise training are the long-term BP adaptations that accrue over time.

TABLE 1 The Existing Professional Exercise Recommendations Among Adults with Hypertension

FITT of the Ex R$_x$	Professional Committee/Organization					
	Eighth Report of the Joint National Committee [10] and AHA/ACC Lifestyle Work Group [11]	Seventh Report of the Joint National Committee [26]	American Heart Association [33]	American College of Sports Medicine [13]	European Society of Hypertension/European Society of Cardiology [31]	Canadian Hypertension Education Program [32]
Frequency (how often?)	3–4 sessions per week ≥12 weeks	Most days of the week	Most days of the week	Most, preferably all, days of the week	5–7 days per week	4–7 days per week in addition to habitual, daily activity
Intensity (how hard?)	Moderate to vigorous[a]	None specified	Moderate to high >40%–60% of maximum	Moderate 40%–59% of $\dot{V}O_{2R}$	Moderate[a]	Moderate[a]
Time (how long?)	40 min per session	≥30 min per day	150 min per week	30–60 min continuous or accumulated in bouts ≥10 min each	≥30 min per day	Accumulation of 30–60 min per day
Type (what kind?) Primary	Aerobic	Aerobic	Aerobic	Aerobic	Aerobic	Dynamic exercise (Aerobic)
Evidence rating	"High"[b] grade B[b], class IIa level of evidence A[c]	NA	Class I level of evidence A[c]	Evidence category A[d,e] Evidence category B[d,e]	Class I level of evidence A–B[f]	Grade D[g]
Adjuvant	NA	NA	Dynamic RT	Dynamic RT 2–3 days week^{-1}, moderate 60%–80% 1RM, 8–12 repetitions	Dynamic RT 2–3 days week^{-1}	Dynamic, isometric, or handgrip RT
Evidence rating	NA	NA	Class IIa level of evidence B[c]	Evidence category B[d,h]	NA	Grade D[g]

AHA/ACC, American Heart Association/American College of Cardiology; Ex R$_x$, exercise prescription; NA, not applicable; RT, resistance training; 1RM, one-repetition maximum; $\dot{V}O_{2R}$, oxygen consumption reserve.

[a]*Moderate intensity, 40%–59% of $\dot{V}O_{2R}$ or an intensity that causes noticeable increases in heart rate and breathing; vigorous or high intensity, ≥ 60% of $\dot{V}O_{2R}$ or an intensity that causes substantial increases in heart rate and breathing.*

[b]*Evidence statement: "aerobic exercise lowers blood pressure (BP)" was rated high[d]; evidence recommendation for the frequency, intensity, and time (FIT) of the Ex R$_x$ to lower BP was rated grade B (adapted from [34]) or moderate, corresponding to class IIa level of evidence A[c].*

[c]*Criteria from the AHA [33].*

[d]*Criteria from the National Heart, Lung, and Blood Institute [35].*

[e]*The strength of evidence for aerobic exercise was rated category B[d] for its immediate effects (i.e., postexercise hypotension) and category A[d] for its long-term effects (i.e., chronic effects); the FIT of the Ex R$_x$ to lower BP was rated category B[d].*

[f]*Criteria from the European Society of Cardiology [36].*

[g]*Evidence grading was assigned based on the underlying level of evidence [37]; grade A is the strongest evidence (i.e., based on high-quality studies), and grade D is the weakest evidence (i.e., based on low-power imprecise studies or expert opinion alone); "higher-intensity exercise is not more effective" was assigned grade D.*

[h]*The strength of evidence for the immediate effects of dynamic RT was rated category C[c].*

Modified from Pescatello LS, MacDonald HV, Ash GI, Lambert LM, Farquhar WB, Arena R, et al. Assessing the existing professional exercise recommendations for hypertension: A review and recommendations for future research priorities. Mayo Clin Proc 2015;90(6):801–12.

The relationship between the BP response to acute and chronic exercise is unclear; however, they appear to be related [13,48,53–60]. Several recently published studies support the notion that the reductions in BP experienced immediately following a single bout of aerobic exercise are similar to those experienced after aerobic exercise training; an observation that suggests the BP benefits attributed to chronic exercise are largely the result of PEH [23,55,56,61]. Of note, initial studies examining whether PEH could be used to predict the BP response to exercise training focused exclusively on aerobic (or cardiovascular endurance) exercise [55,56]. Since then, researchers have published similar observations for dynamic resistance [59,60] and isometric resistance [58] exercise. Collectively, these studies support the notion that PEH may account for a significant portion of the magnitude of the BP reduction attributed to exercise training. Further research in a larger, more diverse sample of adults with hypertension is needed to substantiate this premise.

Exercise Prescription (Ex Rx)

An Ex R_x is the process whereby the recommended physical activity program is designed in a systematic and individualized manner in terms of the *Frequency* (how often?), *Intensity* (how hard?), *Time* (how long?), *Type* (what kind?), *Volume* (how much?), and *Progression* or *FITT-VP* principle. The optimal Ex R_x for the primary prevention and treatment of hypertension will be discussed in detail later in this chapter. Presently, the ACSM recommends the following *FITT-VP* Ex R_x for individuals with hypertension (Table 2): [65]

- *Frequency*: For aerobic exercise, on most, preferably all days of the week supplemented by resistance exercise 2–3 days per week and flexibility exercise 2–3 days per week.

TABLE 2 The Current Exercise Prescription for Adults With Hypertension [12,13]

FITT-VP Principle of the Ex R_x	ACSM Recommendations
Frequency (how often?)	5–7 days per week
Intensity (how hard?)[a]	Moderate (40%–59% of $\dot{V}O_{2R}$ or 12–13 on a scale of 6 (no exertion) to 20 (maximal exertion) level of physical exertion or an intensity that causes noticeable increases in heart rate and breathing)
Time (how long?)	≥ 30–60 min/day, one continuous bout or multiple bouts of at least 10 min each
Type (what kind?) *Primary*	Aerobic exercise; prolonged, rhythmic activities using large muscle groups (e.g., walking, cycling, and swimming)
Adjuvant 1	Muscle strengthening *F*, 2–3 days per week (nonconsecutive) *I*, moderate to vigorous intensity (60%–80% of 1RM); major muscle groups *T*, 8–10 exercises; 2–4 set of 8–12 repetitions
Adjuvant 2	Flexibility *F*, ≥ 2–3 days per week *I*, stretch to the point of feeling tightness or slight discomfort *T*, ≥ 10 min/day; ≥ 4 repetitions per muscle group; hold each static stretch for 10–30 s
Adjuvant 3[b]	Neuromotor *F*, ≥ 2–3 days per week *I*, undetermined *T*, ≥ 20–30 min/day
Volume (how much?) [c]	≥ 150 min/week or 700–2000 kcal/week
Progression	Progress gradually, avoiding large increases in any of the components of the Ex R_x; increase exercise duration over first 4–6 weeks and then increase frequency, intensity, and time (or some combination of these) to achieve recommended quantity and quality of exercise over next 4–8 months

Note: ACSM, American College of Sports Medicine; FITT-VP, Frequency, Intensity, Time, Type, Volume, and Progression of exercise; Ex R_x, exercise prescription; 1RM, one-repetition maximum; $\dot{V}O_{2R}$, oxygen consumption reserve

[a]*Vigorous-intensity aerobic exercise (i.e., ≥ 60% $\dot{V}O_{2R}$ or ≥ 14 on a scale of 6–20) [62–64] may elicit greater and more extensive benefits and may be introduced after exercise preparticipation health screening and gradual progression.*
[b]*Balance (neuromotor) training is recommended for older adults, individuals who are at substantial risk of falling, and is likely to benefit younger adults as well.*
[c]*For greater and more extensive benefits, progress exercise volume to total 60 min/day and 300 min/week of moderate intensity.*

- *Intensity:* Moderate (i.e., 40%–59% of oxygen consumption reserve ($\dot{V}O_{2R}$)) or 12–13 on a scale of 6 (no exertion) to 20 (maximal exertion) level of physical exertion [62–64] or an intensity that causes noticeable increases in heart rate (HR) and breathing for aerobic exercise, moderate to vigorous (60–80% of one-repetition maximum (1RM)) for resistance, and stretch to the point of feeling tightness or slight discomfort for flexibility.
- *Time:* For aerobic exercise, a minimum of 30 min or up to 60 min/day for continuous or accumulated aerobic exercise. If intermittent, begin with a minimum of 10 min bouts.
- *Type:* For aerobic exercise, emphasis should be placed on prolonged, rhythmic activities using large muscle groups such as walking, cycling, or swimming. Dynamic resistance exercise may supplement aerobic exercise and should consist of at least one set of 8–12 repetitions of 8–10 different exercises targeting the major muscle groups. For flexibility, hold each muscle 10–30 s for 2–4 repetitions per muscle group. Balance (neuromotor) exercise training 2–3 days per week is also recommended as adjuvant exercise in individuals at high risk for fall (i.e., older adults) and is likely to benefit younger adults as well.
- *Volume:* To total at least 150 min/week or 700–2,000 kcal/week of moderate-intensity aerobic exercise.
- *Progression:* Progress gradually, avoiding large increases in any of the components of the FITT. Increase exercise duration over the first 4–6 weeks and then increase frequency, intensity, and time (or some combination of these) to achieve the recommended volume of 700–2000 kcal/week over the next 4–8 months. Progression may be individualized based on tolerance and preference in a conservative manner.

PURPOSES OF THIS CHAPTER

The purposes of this chapter are to (1) overview the current consensus on the effects of acute (immediate, short-term, or PEH) and chronic (long-term or training) aerobic, resistance, and combined aerobic and resistance exercise or *concurrent* exercise on BP among individuals with prehypertension or established hypertension; (2) discuss new and emerging research on the effects of acute and chronic aerobic, resistance, and concurrent exercise on BP that has the potential to alter the way in which exercise is prescribed to prevent, treat, and control hypertension; and (3) present Ex R_x recommendations and special considerations for individuals with hypertension that consider this new and emerging research.

SYSTEMATIC REVIEW METHODS

For this review chapter, we have combined and updated the comprehensive search strategies used in our recently published systematic reviews on this topic [14,23,24,66]. This chapter summarizes the most relevant literature on the BP response to the acute and chronic aerobic, dynamic resistance, and concurrent exercise published to date. Briefly, a Boolean search was conducted in PubMed (Medline) from 1 January 2015 to 1 March 2017 using a combination of the following inclusive search terms: ("exercise" OR "aerobic*" OR "resistance*" OR "concurrent*" OR "training") AND ("blood pressure" OR "hypertension" OR "hypotension" OR "prehypertension"). We restricted our review to controlled exercise interventions involving humans aged ≥ 18 years with prehypertension or established hypertension. In addition, we excluded studies that included adults with disease(s) unrelated to CVD (e.g., cancer and HIV/AIDS) or involved adjuvant diet or weight loss therapies (e.g., dietary modifications and weight loss drugs) in addition to exercise. Our search yielded 485 potentially relevant reports (of these, 21 were systematic reviews or meta-analyses). After applying our selection criteria, ≈93 studies remained. From these, the authors self-selected the studies most relevant to the purposes of this chapter. These studies, in addition to those from our earlier reviews [14,23,24,66], are discussed in greater detail in the sections that follow.

AEROBIC EXERCISE AND BLOOD PRESSURE EFFECTS

Acute, Immediate, or Short-Term Effects or Postexercise Hypotension

PEH is now considered an expected physiological response to exercise [13,14,48,50–53,57,67–81]. For this reason, individuals with hypertension are encouraged to exercise on most days of the week in order to benefit from the acute effects of aerobic exercise on BP [13,65]. Several reviews [82–85], and more recently, a meta-analysis [86], have summarized the BP response to acute aerobic exercise. Overall, these studies concluded that a single bout of aerobic exercise lowers resting systolic BP 6–11 mmHg and diastolic BP 4–5 mmHg among adults with high BP and that the magnitude of PEH depends on several factors, including the characteristics of the sample and the intensity and duration of the aerobic exercise performed. Despite the heterogeneity in the magnitude of PEH, the extant literature appears to support that reductions in BP immediately following aerobic exercise are most pronounced in individuals who stand to benefit the most (i.e., those with higher

BP compared with normal BP) [13,48,85–87]. New and emerging evidence regarding the acute effects of aerobic exercise and its application to the current Ex R_x for hypertension will be discussed in the sections that follow.

Despite the volume of literature, a limited number of studies have directly compared aspects of the FIT of an acute aerobic exercise intervention on PEH among adults with high BP [69,88–94]. As a result, it remains unclear which factors, that is, aerobic exercise intensity, duration, total work performed [95], or how the exercise is conducted (i.e., continuous, interval, intermittent or fractionized), are most influential in determining magnitude and duration of PEH.

MacDonald et al. [92] investigated the influence of exercise duration on PEH among physically active adults ($n=8$) with high normal to stage 1 hypertension (133/79 mmHg). They found that both short (10 min) and long (30 min) acute aerobic exercise performed at moderate to vigorous intensity (70% of $\dot{V}O_{2peak}$) reduced systolic and diastolic BP an average of $\approx$ 10/8 mmHg compared with baseline for 60 min following exercise [92]. Similarly, Guidry et al. [91] compared the effects of a short (15 min) and long (30 min) acute aerobic exercise bout performed at light (40% of $\dot{V}O_{2max}$) or moderate (60% of $\dot{V}O_{2max}$) intensity on PEH among 45 white, middle-aged overweight men with high normal to stage 1 hypertension. Both short- and long-duration acute aerobic exercise reduced systolic BP an average of 4–6 mmHg compared with control for the remainder of the day, independent of exercise intensity [91].

Ciolac and colleagues [90] were one of the first research groups to investigate whether the magnitude of PEH differed between acute aerobic exercise performed in a continuous bout at a constant intensity or in repeated bouts of high versus low intensity (i.e., interval exercise). Ciolac et al. randomized men and women ($n=52$) to either 40 min of continuous aerobic exercise at 60% of HR reserve or 40 min of interval aerobic exercise consisting of alternating bouts of 2 min at 50% of HR reserve with 1 min at 80% of HR reserve. Once again, the greatest BP reductions were observed among individuals with a resting BP greater than the study median of 131/99 mmHg compared with those below the study median. Interestingly, BP was reduced to 4–8 mmHg in the continuous exercise group and 5–6 mmHg for systolic BP only in the interval exercise group over 24 h, suggesting that moderate- to vigorous-intensity acute aerobic exercise performed continuously at a constant intensity or in intervals elicits PEH.

Similarly, Carvalho et al. [89] compared the effects of continuous (42 min at $\approx$ 67% of $\dot{V}O_{2max}$) and interval (42 min of alternating bouts consisting of 4 min at $\approx$ 83% of $\dot{V}O_{2max}$ with 2 min of recovery at 40% of $\dot{V}O_{2max}$) aerobic exercise on PEH among 20 sedentary, older women with stage 1 hypertension ($\approx$143/89 mmHg). Carvalho et al. found that BP was lower following both the continuous ($\approx$10–11 mmHg) and interval ($\approx$15–16 mmHg) aerobic exercise sessions compared with control ($P<.05$), persisting throughout the day up to 20 h postexercise. Furthermore, these authors reported that BP was 4–5 mmHg lower over the awake hours following interval than continuous aerobic exercise ($P<.05$). Collectively, these results confirm that interval or continuous aerobic exercise can elicit PEH [89,90], but that BP-lowering benefits may be maximized when aerobic exercise is performed in intervals that include more vigorous levels of physical exertion (i.e., the magnitude of the BP reductions is intensity-dependent) [89].

More recently, Bhammar et al. [88] compared the effects of fractionized aerobic exercise (3×10 min bouts) spread throughout the day (morning, midday, and afternoon) and a single bout of continuous exercise (1×30 min bout) performed at 60%–65% of $\dot{V}O_{2peak}$ on ambulatory BP among individuals with prehypertension ($n=11$). Bhammer et al. concluded that fractionized exercise was as at least as effective as continuous exercise in eliciting PEH, reducing systolic BP 3–4 mmHg compared with control throughout the day until the following morning. Taken together, these results suggest that short, intermittent bouts of aerobic exercise interspersed throughout the day elicit PEH to magnitudes that equal a single bout of continuous aerobic exercise [88].

Less is known about the antihypertensive effects of very short (<10 min) bouts of aerobic exercise. Miyashita et al. [93] compared the BP response to 30 min of running at 70% of $\dot{V}O_{2max}$ performed in either a single continuous bout (1×30 min) or multiple, very short bouts (10×3 min) among young men with prehypertension ($n=7$). Miyashita et al. reported reductions in systolic BP of 6 and 8 mmHg, respectively, that persisted for 24 h ($Ps<.01$). These findings suggest that short bouts (i.e., 3 min) of vigorous-intensity running interspersed throughout the day elicit PEH, and the antihypertensive effects of very short bouts of vigorous-intensity aerobic exercise are similar in magnitude and duration to those observed following a single bout of continuous vigorous-intensity aerobic exercise [93]. This new and emerging research supports the notion that PEH can be a low-threshold phenomenon in terms of the duration of exercise, but that the minimum duration needed to produce the effect appears to be *inversely* related to the intensity of the exercise. That is, very short bouts of aerobic exercise (i.e., < 10 min in duration) can elicit PEH but only when performed at more vigorous levels of physical exertion. More importantly, if adults with hypertension are willing and able to tolerate higher levels of physical exertion, these short bouts of vigorous-intensity aerobic exercise interspersed throughout the day can offer a viable therapeutic lifestyle option for BP control.

There is growing evidence to suggest that the magnitude of the BP reductions that result from acute aerobic exercise occur as a direct function of intensity such that the greater the intensity, the greater the BP reduction [23,96]. Eicher et al. [69] examined the antihypertensive effects of three bouts of acute aerobic exercise performed at light (40% of $\dot{V}O_{2max}$), moderate

(60% of $\dot{V}O_{2max}$), and vigorous (100% of $\dot{V}_{2max}$) intensity in men with prehypertension to stage 1 hypertension ($n=45$). The authors found that BP is decreased by 1.5/0.6 mmHg for every 10% increase in relative $\dot{V}O_{2max}$, suggesting that more vigorous levels of physical exertion lowered BP to a greater extent than lower levels of physical exertion for individuals willing and able to tolerate more intense levels of exercise [69].

A recent meta-analysis of 65 acute aerobic, dynamic resistance, and concurrent exercise studies [86] seems to provide confirmation for some of the new and emerging evidence discussed to this point. Briefly, Carpio-Rivera et al. [86] examined the acute effects of aerobic exercise on BP among adults with normal BP to stage 1 hypertension and found that resting BP on average was reduced ≈4–6 mmHg compared with control with no statistically significant differences among the three modalities (i.e., ≈3–4 and ≈3–7 mmHg following resistance and concurrent exercise, respectively). Carpio-Rivera et al. [86] also identified moderator patterns with respect to population characteristics, exercise intensity and duration, and how the aerobic exercise was performed. Interestingly, they observed greater BP reductions (expressed as mean effect size (ES)±SD) among physically active versus sedentary samples (systolic BP only, -0.71 ± 0.98 vs -0.46 ± 0.79, $P=.002$) and among samples consisting of all men versus all women (systolic BP only, -0.68 ± 0.94 vs -0.27 ± 0.60, $P=.004$). In addition to these findings, Carpio-Rivera et al. [86] also documented that the magnitude of PEH was greater following aerobic exercise performed at higher (e.g., running and jogging) versus lower (e.g., walking) intensities (from -0.77 to -1.66 vs from -0.19 to -0.53, $P=.01$) and when performed intermittently or incrementally versus continuously (systolic BP only, from -0.67 to -1.12 vs -0.50, $P=.004$). This meta-analysis appears to confirm that short, intermittent bouts of aerobic exercise interspersed throughout the day elicits PEH to magnitudes that are equal to or exceed those following a single bout of continuous aerobic exercise [88,93]. Nonetheless, this meta-analysis is not without limitations. It does not appear that the authors accounted for nonindependent ES estimates arising from multiple treatment arms (i.e., multiple ESs nested within a single study), which can over- or underinflate the mean effect of acute exercise on PEH [97]. Furthermore, Carpio-Rivera et al. relied on subgroup (univariate) analyses to explore potential moderators of PEH, thereby limiting their ability to explore multiple variables simultaneously, in a single model, to determine which moderators explain unique variance in the BP response to exercise [98–100]. As a result, it is unclear which sample or acute aerobic exercise intervention characteristics (i.e., aerobic exercise intensity, duration, or how the exercise is performed) are more influential in determining magnitude and duration of PEH.

In summary, a single, isolated session of aerobic exercise results in an immediate reduction in BP on the order of 5–7 mmHg among individuals with hypertension (i.e., PEH), with the greatest reductions occurring among those with the highest BP (i.e., upward of 8–11 mmHg) [13,23,48,85–87]. PEH is a low-threshold phenomenon in terms of exercise duration, as short durations of exercise produce PEH [88,90–92]. However, the minimum duration needed to produce the effect is dependent on the intensity of the exercise [93] and, at this time, remains to be determined. Nonetheless, these short bouts of exercise accumulated over the course of a day can have the same beneficial impact on BP as one continuous bout of exercise [86,88,93]. New and emerging research indicates that exercise intensity is an important determinant of PEH such that increasing levels of exertion lower BP in a dose-response pattern [23,69,86,94].

Chronic, Training, or Long-Term Effects

Meta-analyses of studies investigating the antihypertensive effect of chronic aerobic exercise training among individuals with hypertension have concluded that dynamic aerobic or endurance exercise training reduces resting office and 24 h ambulatory BP 5–7 [101–104] and 3–4 mmHg [105,106], respectively, among individuals with high BP. Of note, the participants in these meta-analyses were generally white and/or middle-aged, limiting the generalizability of the results to more diverse populations [24]. Nonetheless, one clear pattern that has emerged from these meta-analyses is that resting BP is lower due to aerobic exercise training and that the magnitude of the reduction is greatest for those with the highest BP.

A meta-analysis by Kelley et al. [102] examined the effect of aerobic exercise training among 2543 adults with prehypertension. On average, the training programs were performed 3 days per week for 40 min at vigorous-intensity (67% of $\dot{V}O_{2max}$) for 23 weeks. The authors reported BP reductions of 2/1 mmHg among those with normal BP and BP reductions of 5/6 mmHg among individuals with hypertension [102]. Similarly, Cornelissen and Fagard [101] meta-analyzed 72 trials that included 3936 middle-aged adults with prehypertension. On average, the aerobic training programs were performed 3 days per week for 40 min at moderate- to vigorous-intensity (65% of $\dot{V}O_{2max}$) for 16 weeks. The authors reported average BP reductions of 2–4 mmHg, with greater reductions among samples with hypertension (5–7 mmHg) compared with those with normal BP (2 mmHg) [101]. More recently, Cornelissen et al. [105] meta-analyzed 15 randomized controlled trials that included 633 middle-aged adults with normal ambulatory BP (128/81 mmHg) [31]. On average, aerobic exercise training performed 3 days per week for 40 min at moderate- to vigorous-intensity (50%–75% of HR reserve) for 15 weeks lowered daytime ambulatory BP ≈3 mmHg compared with control. The authors reported no differences in daytime ambulatory BP among samples with hypertension (3–4 mmHg) and normal BP (2–3 mmHg). To summarize, the magnitude of BP reductions following chronic aerobic exercise appears to be similar in magnitude to those reductions seen after an acute session and, for the most part, follows the same law of initial values such that those with the highest BP experience the greatest reductions after training.

Although many other meta-analyses have examined the influence of aerobic exercise training on BP, most have failed to provide insight into how population characteristics and/or the FIT of the exercise intervention moderates the antihypertensive effects of aerobic exercise [24,106]. One of the few who identified sample characteristics as a moderator of the antihypertensive effects of exercise was Whelton et al. [28]. These authors examined a large group ($n = 2419$) of racially/ethnically diverse patients ($n = 1935$ Whites, $n = 391$ Asians, and $n = 93$ Blacks) and reported BP reductions of 3/3 mmHg for Whites, 6/7 mmHg for Asians, and 11/3 mmHg for Blacks [28]. More recently, Cornelissen and Smart [107] examined the effect of aerobic exercise training lasting at least 4 weeks in duration and observed that samples consisting of all men experienced BP reductions that were greater in magnitude than those of all women (3–5 vs 1 mmHg), concluding that sex may influence the BP response to exercise training. In addition to this finding, Cornelissen and Smart also identified several moderators related to the FIT of the aerobic exercise intervention, such that exercise training < 24 weeks reduced BP to a greater extent than training programs ≥ 24 weeks (3–6 vs 1–2 mmHg, respectively) [107]. They also documented that the magnitude of the BP reductions was greatest in those who trained 30–45 min per session and who accumulated a weekly exercise volume of < 210 min compared with a weekly volume ≥ 210 min [107]. Last, exercise intensity appeared to alter the BP response to aerobic exercise training such that BP reductions were greatest with moderate- to vigorous-intensity aerobic exercise training compared with lower-intensity aerobic exercise (4–5/2–3 vs ~1 mmHg, respectively).

High-intensity interval training (HIIT) is characterized by brief periods of very vigorous-intensity aerobic exercise (>90% of $\dot{V}O_{2max}$) separated by recovery periods of lower-intensity exercise or rest [108]. A recent review suggests that HIIT may be superior to continuous, moderate-intensity aerobic exercise in lowering BP, and the magnitude of reductions following HIIT appears to be higher among individuals with hypertension versus normal BP (8 vs 3 mmHg, respectively) [23]. HIIT holds promise for some people with hypertension because it allows individuals to perform brief periods of vigorous-intensity exercise that would not be tolerable for longer periods of time. In addition, HIIT can also yield an equal amount of work (i.e., energy expenditure) compared with continuous, moderate-intensity exercise in a shorter amount of time [23,108,109].

In summary, the BP reductions experienced after regular, aerobic exercise training appear to be equivalent in magnitude to those observed with PEH (i.e., 5–7 mmHg) [101,102,104]. Similar to PEH, BP reductions with training appear to be greatest in those with the highest BP [101–104,106], and exercise intensity appears to be an important moderator of the BP response to aerobic exercise training, with BP reductions occurring in a dose-response fashion [107]. HIIT shows promise as a viable alternative to the current ACSM Ex R$_x$ recommendations for hypertension; however, further investigation is warranted among individuals with hypertension to more definitively determine the benefit-to-risk ratio of exercising at vigorous-intensity among a population that is predisposed to heightened CVD risk [23,108]. Last, the FITT components of the aerobic exercise training intervention (i.e., the duration, intensity, and weekly volume of exercise) and population characteristics (i.e., sex and race/ethnicity) appear to moderate the BP response to aerobic exercise training and warrant confirmation in future randomized controlled trials [28,107].

DYNAMIC RESISTANCE EXERCISE AND BLOOD PRESSURE EFFECTS

Acute, Immediate, or Short-Term Effects or Postexercise Hypotension

It was previously thought that individuals with hypertension should avoid resistance exercise due to reports of marked elevations in BP while exercising and following the Valsalva maneuver [110,111]. The Valsalva maneuver is characterized by a strenuous and prolonged expiratory effort when the glottis is closed, resulting in decreased venous return and increase in peripheral venous pressures, quickly followed by increased venous return to the heart and subsequent elevations in arterial pressure [110]. Increases in BP as high as 480/350 mmHg have been recorded among bodybuilders during a single bout of heavy resistance exercise performed at or above 80% of 1RM until concentric failure [111]. However, such BP surges are known to return to initial values within 10 s of the last repetition of each set [111].

Nevertheless, to the best of our knowledge, it remains to be answered whether BP surges of this magnitude are harmful. In fact, there is an established, but growing, body of literature that demonstrates that muscular strength is inversely associated with the development of hypertension and other CVD risk factors [18–21] and the risk for CVD as well as all-cause mortality [19,22,112,113]. Furthermore, there is an increasing number of studies [114–120], in addition to reviews [83,84] and meta-analyses [86,121], that have reported immediate reductions in BP following acute resistance exercise.

For example, Melo et al. [122] found that light-intensity resistance exercise (40% of 1RM) reduced BP 12/6 mmHg compared with control, an effect that persisted for $\approx$ 10h following the session. They also reported that the magnitude of PEH was greater among those with higher ambulatory BP (i.e., >135/85 mmHg); BP was reduced an additional $\approx$1.5/2.7 mmHg for every 5 mmHg increase in resting BP beginning at $\approx$110/70 mmHg [122]. Similarly, Moraes and colleagues [116] observed reductions in ambulatory BP of 8 mmHg that lasted up to 8h following a single bout of resistance exercise performed at moderate-intensity (70% of 1RM). Consistent with previous studies, Brito et al. [115] found that dynamic resistance exercise elicited PEH 90 min after the bout and that the reductions in BP were greater following resistance exercise performed at 80% (33/15 mmHg) than 50% (23/7 mmHg) of 1RM, respectively. Together, these studies suggest that while BP during resistance exercise may increase, BP levels appear to immediately decrease back to levels below that of baseline, BP reductions persist for several hours after the bout (i.e., PEH), and the magnitude of these reductions appears to be the greatest among those with the highest BP.

A limited number of studies have examined aspects of the FIT of an acute resistance exercise intervention on PEH, and when they have, the results have been mixed. For example, some studies have shown that vigorous-intensity (80% of 1RM) acute resistance exercise results in greater BP reductions than light- to moderate-intensity (50% of 1RM) resistance exercise ($\approx$34/16 vs $\approx$24/7 mmHg) [115,119], while other studies have reported reductions of similar magnitude following light- (40% of 1RM) and vigorous- (80% of 1RM) intensity acute resistance exercise (14/1–2 mmHg) [114].

Furthermore, the volume of resistance exercise (i.e., the number of exercises, repetitions, and sets of a given exercise) appears to influence the magnitude of PEH. Scher et al. [117] examined the effect of low versus high volume on PEH among older adults ($n=16$) with treated hypertension (130/76 mmHg). All subjects performed 1 set of 20 repetitions at light-intensity (40% of 1RM) for each station in the 10-exercise circuit; however, the number of total circuits (laps) differed such that subjects completed two sessions that consisted of low (1 lap, 20 min) and high (2 laps, 40 min) volume. The authors reported that both low and high volumes of resistance exercise elicited PEH for 60 min in the laboratory compared with control ($Ps<.05$), but the magnitude of these reductions was greater after high- rather than low-volume resistance exercise (8/6 vs 10/7 mmHg, respectively) ($P<.05$). Furthermore, only high-volume resistance exercise reduced awake and 24h ambulatory systolic BP compared with control, and again, these reductions were greater than those observed after lower volumes of resistance exercise ($Ps<.05$).

Despite the limited number of PEH studies involving dynamic resistance exercise, two meta-analyses have been conducted to date and concluded that resistance exercise can acutely lower BP 3–5 mmHg among adults with normal BP to stage 1 hypertension [86,121]. Briefly, Cassonatto et al. [121] meta-analyzed 30 studies (81 interventions) that included 646 adults with normal BP ($n=505$) and stage 1 hypertension ($n=141$). Overall, the authors reported that resting office BP was reduced to 5.3/4.7 mmHg compared with control 90 min following exercise. They also found that dynamic resistance exercise elicited small, but significant reductions in 24h ambulatory (1–2 mmHg). Consistent with previous reports in the aerobic exercise and PEH literature [13,48,85–87], Cassonatto et al. observed that BP reductions followed the law of initial values such that samples with hypertension experienced greater reductions in BP compared with samples with normal BP (9/5–6 vs 3/3 mmHg; $P<.01$). Furthermore, systolic BP reductions were greater following a bout of resistance exercise that involved larger muscle groups (targeted with either single- or multijoint movements) than smaller muscle groups (3–5 vs 1–2 mmHg, $P<.05$). In a separate meta-analysis, Carpio-Rivera et al. [86] reported that PEH magnitude was associated with the number of resistance exercises performed in a given session ($r\approx-0.20$, $Ps\leq.006$) and the number of sets performed for a given exercise ($r=-0.47$, $P<.001$) such that a greater number of resistance exercises and sets per exercise elicited greater reductions in BP.

To summarize, acute resistance exercise can lead to remarkable surges in BP while exercising; however, whether this phenomenon is harmful to overall cardiovascular health is unknown [111]. In fact, most acute resistance exercise studies have reported immediate reductions in BP following acute resistance exercise that appear to be clinically meaningful [114–119,123] and are most pronounced in individuals who stand to benefit the most (i.e., those with higher BP compared with normal BP) [86,121,122]. At this time, it is unclear whether other patient characteristics or aspects of the FIT of the acute resistance exercise intervention influence PEH.

Chronic, Training, or Long-Term Effects

Randomized controlled trials and meta-analyses [107,124–129] have concluded that dynamic resistance exercise training reduces resting BP (i.e., $\approx$2–3 mmHg) but to a smaller degree than aerobic exercise training ($\approx$5–7 mmHg). Of note, the participants in these meta-analyses were generally white and/or middle-aged with normal BP and prehypertension, limiting the generalizability of the results to more diverse populations and to those with hypertension [24]. An early meta-analysis by Cornelissen et al. [126] pooled data from 9 trials (12 interventions) to examine the influence of resistance exercise training

(3 days per week at 61% of 1RM for 16 weeks) on BP and reported reductions of 3.2/3.5 mmHg among 341 individuals with prehypertension. More recently, both intervention studies and meta-analyses have shown BP reductions following dynamic resistance training to be similar to those that result from aerobic exercise training. For example, Mota et al. [123] examined the influence of a 16-week, moderate-intensity (i.e., 70% of 1RM) dynamic resistance training program on women with treated hypertension and reported BP reductions of 14/4 mmHg. Likewise, Moraes et al. [116] found similar BP reductions of 16/12 mmHg after 12 weeks of moderate-intensity (60% of 1RM) dynamic resistance training among men with hypertension. Most recently, we meta-analyzed 64 controlled studies (71 interventions) involving middle-aged adults ($n=2344$), the majority of who were white (57%) with prehypertension (126/76 mmHg) [129]. Overall, the dynamic resistance training interventions were of moderate-intensity and were performed 2–3 days per week for 14 weeks. On average, dynamic resistance training elicited BP reductions of $\approx$3/2 mmHg. However, subsequent moderator analyses revealed that greater BP reductions were found among individuals with hypertension (6/5 mmHg) and prehypertension (3/3 mmHg) compared with normal BP (0/1 mmHg).

Of note, we found that the antihypertensive effects of dynamic resistance training appear to be moderated by race/ethnicity such that, among nonwhite samples with hypertension, BP was reduced by 14/10 mmHg, a magnitude that is approximately twice that reported following aerobic exercise training (i.e., 5–7 mmHg) [129]. These promising findings suggest that, for some populations, dynamic resistance training elicits BP reductions comparable with or greater than those achieved with aerobic exercise training, and for those patients (i.e., nonwhite samples with hypertension), dynamic resistance exercise may serve as a viable stand-alone antihypertensive lifestyle therapeutic option [129].

Presently, the recommendations for resistance exercise training among individuals with hypertension are to engage in moderate-intensity, dynamic resistance exercise 2–3 days per week as a supplement to aerobic exercise training. Upon more careful scrutiny of the literature [23] and the new findings by MacDonald et al. [129], there is suggestive evidence that dynamic resistance exercise can be an alternative stand-alone exercise modality option for patients with hypertension. Additional randomized controlled intervention trials are needed to determine if these novel findings prove to be true and to better understand which FIT features of the dynamic resistance exercise training program would yield the greatest BP benefit.

In summary, it has long been thought that dynamic resistance exercise training reduces BP $\approx$ 2–3 mmHg. However, new and emerging research has demonstrated that dynamic resistance exercise has an even more beneficial influence on BP among those diagnosed with hypertension [116,129,130]. Additionally, of the few meta-analyses that have explored the influence of population characteristics, they have shown that initial BP levels and race/ethnicity can influence the magnitude of reductions experienced after dynamic resistance exercise training and warrant additional investigation [129]. Based on the available evidence, it is unclear whether other aspects of the FIT of the Ex R_x recommendations influence the magnitude of BP reductions following dynamic resistance exercise training.

CONCURRENT EXERCISE AND BLOOD PRESSURE EFFECTS

Concurrent exercise is defined as aerobic and resistance exercise being performed in close proximity to each other (i.e., in a single exercise session or within a couple hours of one another) [61,66]. Few studies have investigated the acute and chronic effects of concurrent exercise on BP among adults with hypertension. In fact, we are only aware of 5 PEH [131–135] and 11 chronic [136] studies that have included adults with prehypertension to stage 1 hypertension.

Acute, Immediate, or Short-Term Effects or Postexercise Hypotension

As noted above, five studies have investigated the acute effects of concurrent or combined aerobic and resistance exercise among adults with high BP and reported BP reductions ranging 6–12 mmHg for systolic BP and 3–17 mmHg for diastolic BP, respectively [66,131–135]. Based on the limited number of small studies conducted to date, it is likely premature to comment on the population or FIT characteristics that may modulate the duration and magnitude of PEH following acute concurrent exercise. At this time, however, PEH does not appear to be influenced by intervention characteristics such as the concurrent exercise intensity, volume, or order of the aerobic and resistance exercise components.

For example, acute concurrent exercise consisting of moderate- to vigorous-intensity aerobic exercise and moderate-intensity resistance exercise (70%–75% of 1RM) lowered BP 6–11/3–5 mmHg, an effect that persisted for 60–90 min following the bout [131,133,134]. In addition, two other research teams found that moderate- to vigorous-intensity aerobic exercise performed with light-intensity resistance exercise (40%–50% of 1RM) also lowered BP by 7–8/3 [135] and 12/17 mmHg [132], respectively, for 30–180 min after the bout. Based on the results of these studies, it appears that concurrent exercise consisting of moderate- to vigorous-intensity aerobic exercise and light- or moderate-intensity resistance exercise elicits PEH, and the magnitude of these reductions rivals those observed after aerobic exercise among adults with hypertension.

The volume of the resistance exercise performed during a concurrent exercise session does not appear to influence the magnitude of PEH that occurs after a single bout of concurrent exercise. Tibana et al. [131] examined the effect of 30 min of moderate-intensity aerobic exercise (70% of HR reserve) followed by either low- (one set) or high- (three sets) volume resistance exercise (8–12 repetitions for 6 exercises at 80% of 10RM workload) among 16 women with normal to high BP. Systolic BP was reduced to 7–9 mmHg compared with control for 90 min following both concurrent exercise sessions. These results suggest that low-volume resistance combined with aerobic exercise elicits PEH to a similar magnitude as concurrent exercise that involves higher volumes of resistance exercise.

It also appears that the order of the aerobic and resistance exercise components of the concurrent exercise session and the type of resistance exercise (i.e., eccentric vs traditional) do not influence PEH among adults with hypertension. Menêses et al. [135] found that aerobic exercise (30 min at 50%–60% of HR reserve) performed before or after dynamic resistance exercise (3 sets of 10 repetitions for 7 exercises at 50% of 1RM) elicited PEH among middle-aged women ($n = 19$) with controlled hypertension (130/68 mmHg), and the magnitude of these reductions was not different between conditions (7–8/3 mmHg). Dos Santos et al. [134] found that concurrent exercise consisting of either eccentric or traditional (concentric) resistance exercise (3 sets of 10 repetitions for 7 exercises at 100% or 70% of 1RM, respectively) followed by aerobic exercise (20 min at 65–75% of target HR) elicited PEH to a similar magnitude (4/5 vs 3/4 mmHg, respectively) among older women with uncontrolled hypertension (166/91 mmHg).

On the other hand, how the concurrent exercise session is performed (i.e., single bout vs multiple bouts interspersed throughout the day) does appear to modulate PEH. In a recent study, Azevêdo et al. [133] examined the effects of fractionized concurrent exercise spread throughout the day (morning and evening) compared with a single bout of concurrent performed in the morning or evening among middle-aged women with hypertension ($n = 11$). The authors found that a single session of concurrent exercise (3 sets of 10 repetitions for 4 exercises at 75% of 8RM workload followed by 20 min of moderate- to vigorous-intensity aerobic exercise) performed in the morning or evening, but not fractionized throughout the day, reduced systolic BP 10.7 and 6.3 mmHg compared with control. Furthermore, reductions in systolic BP were greater following a single bout of concurrent exercise performed in the morning compared with fractionized concurrent exercise (−10.7 vs +3.3 mmHg, respectively).

Finally, a recent meta-analysis by Carpio-Rivera et al. [86] concluded that acute concurrent exercise elicited PEH by a magnitude of ≈3–7 mmHg, a magnitude that rivals those reported to occur as a result of acute and chronic aerobic exercise among adults with hypertension (i.e., 5–7 mmHg). In fact, the authors reported that BP was reduced by a similar magnitude following a single bout of acute aerobic (6.2/3.8 mmHg), resistance (3.5/2.7 mmHg), and concurrent (7.3/2.9 mmHg) exercise ($Ps > .05$). Unfortunately, this meta-analysis did not provide any additional insight to whether patient characteristics or aspects of the acute concurrent exercise intervention influences PEH.

To summarize, a single bout of concurrent exercise elicits PEH among middle-aged to older adults with high BP, and the magnitude of these reductions appears to be similar to those observed after aerobic exercise. Moreover, based on the limited number of small studies conducted to date, the currently available data do not suggest that PEH is modulated by aspects of the acute concurrent exercise intervention (i.e., exercise intensity, volume, or the order of the aerobic and resistance exercise components). However, further investigation is warranted among larger samples of adults with hypertension to determine whether aspects of the FIT of the acute concurrent exercise intervention influence PEH. At this time, it is unclear whether patient characteristics influence PEH.

Chronic, Training, or Long-Term Effects

The ACSM does not currently provide any guidelines on concurrent exercise training for adults with hypertension; however, new and emerging evidence suggests that concurrent exercise training may be as effective as aerobic exercise training as antihypertensive therapy among individuals with hypertension [23]. Hayashino et al. [137] performed a meta-analysis of 42 trials, 14 of which were concurrent exercise training trials involving middle-aged adults with type 2 diabetes mellitus and hypertension (36%). The authors reported BP reductions of 1.7/2.3 mmHg following aerobic training, 2.8/2.3 mmHg following dynamic resistance training, and 3.2/1.9 mmHg following concurrent exercise training, with no statistically significant differences among the three modalities [137]. Cornelissen and Smart [107] reported similar findings in their meta-analysis of 93 trials, such that BP was reduced to 3.5/2.5 mmHg following aerobic exercise training (105 interventions), 1.8/3.2 mmHg following dynamic resistance exercise training (29 interventions), and 2.2 mmHg (systolic BP only) following concurrent exercise training (14 interventions). Once again, no differences were noted among the three modality groups.

Most recently, Corso et al. [136] pooled 68 trials (76 interventions) and examined the influence of concurrent exercise training on BP. Concurrent exercise training was performed at moderate-intensity (aerobic = 55% of $\dot{V}O_{2max}$; resistance = 60% of 1RM), 3 days per week for ≈60 min for 20 weeks. The authors found that, on average, BP was significantly lowered by 3.2/2.5 mmHg. However, among individuals with hypertension and in trials of higher study quality that examined BP as the primary outcome, BP reductions were as great as 9.2/7.7 mmHg [136]. BP reductions of this magnitude are clinically meaningful and, if confirmed, could result in an expansion of the existing professional exercise recommendations for hypertension to include concurrent exercise training. Interestingly, as stated within the current exercise recommendations for hypertension, individuals with hypertension would almost *always* be engaging in concurrent exercise training (i.e., aerobic exercise supplemented by dynamic resistance training). Therefore, it is imperative that future trials better explore the combined influence of aerobic and dynamic resistance exercise training as antihypertensive lifestyle therapy among individuals with hypertension to determine whether patient characteristics or aspects of the FIT of the Ex R_x influence the BP response to chronic concurrent exercise.

In summary, new and emerging research examining the influence of aerobic exercise training in combination with dynamic resistance exercise (i.e., concurrent exercise training) is promising as it appears that concurrent exercise may elicit chronic BP reductions similar in magnitude to those seen after aerobic exercise training [23,66,136]. However, future randomized controlled trials are needed before concurrent exercise training can be integrated in the prescription of exercise for individuals with hypertension.

THE INTERACTIVE EFFECTS OF EXERCISE AND ANTIHYPERTENSIVE MEDICATIONS ON RESTING BLOOD PRESSURE

Despite increases in the awareness, treatment, and control of hypertension, the rate of improvement has slowed in recent years. Indeed, data from the 2011–14 National Health and Nutrition Examination Survey revealed that 84% of adults in the United States aged 20 years with hypertension were aware of their condition, yet only 76% of these individuals were receiving pharmacological treatment [1]. Moreover, nearly half (46%) of individuals with treated hypertension were not properly controlled [1]. Participation in regular aerobic and, now, based on the new and emerging evidence discussed in this chapter, dynamic resistance and concurrent exercise can reduce BP by 5–7 mmHg among individuals with hypertension; BP reductions that rival the magnitude of those obtained with first-line antihypertensive medications and lower CVD risk by 20%–30% [25–27]. As such, participation in regular exercise is recognized as a cornerstone therapy for the primary prevention, treatment, and control of high BP [4,11–14]. Nonetheless, for those patients who cannot achieve goal BP with lifestyle management alone, BP-lowering medication should be initiated [10]. Indeed, most individuals who require treatment for hypertension need two or more antihypertensive medications to achieve goal BP [1,10,26].

It has been shown that antihypertensive medications in conjunction with exercise facilitate greater improvements in health outcomes and risk factors than with exercise alone [138]. For example, acute [60,87,89,90,94,114,115,117,122,130, 134,135,139,140] and chronic [130,134,141–145] exercise studies have reported a potentially additive or synergistic effect when combined with long-term antihypertensive drug therapy among adults with treated hypertension. In contrast, there is some evidence [146,147], including meta-analyses [28,86,105,106,129], that have found no such effect. Taken together, the currently available literature seems to support there is an interactive effect between exercise and antihypertensive medications, but it is less clear whether they interact synergistically to potentiate or enhance BP reductions when compared to the magnitude of reduction reported for each separately.

Few comprehensive studies have been designed specifically to investigate this question [148,149]. In fact, we are aware of only seven human studies [138,150–155] that have investigated the potential additive effects of drug therapy to exercise and with mixed results. For example, three studies [150,153,155] used a double-blind, randomized crossover, placebo-controlled design to examine the separate and combined effects of antihypertensive medication and acute exercise on PEH among sedentary men with hypertension. Interestingly, they found that antihypertensive medication had an additive effect on PEH following acute aerobic [150,155], but not dynamic resistance [153], exercise. Four other studies investigated the potentially synergistic interaction between antihypertensive medications and exercise training on resting BP among middle-aged adults with hypertension [138,151,152,154]. Both Ammar et al. [138] and Maruf et al. [152] reported an additive BP-lowering effect following 12 weeks of aerobic exercise training with concomitant antihypertensive medication compared with antihypertensive medication only [138,152]. In contrast, Radaelli et al. [154] found that 3 weeks of aerobic exercise training (5 days per week, 60% of $\dot{V}O_{2max}$, 20 min/day) did not reduce BP compared with antihypertensive medication, and their combined effect offered no additional BP benefit. Similarly, Keleman et al. [151] found that 10 weeks of concurrent aerobic and resistance exercise yielded modest reductions in BP and that concomitant antihypertensive medication provided no additional BP benefit.

To summarize, many antihypertensive medications affect the physiological response to acute and chronic exercise and should be taken into consideration during exercise testing and Ex R_x recommendations (see Table A.1. in the ACSM's Guidelines for Exercise Testing and Prescription [65] for a comprehensive summary of the effect of common medications at rest and in response to exercise). There is a paucity of studies designed specifically to investigate the potentially additive or synergistic effect of antihypertensive medication and exercise. Based on this limited literature, exercise appears to provide additional BP-lowering benefit for adults who are already receiving pharmacological treatment for their high BP; however, whether their interaction produces a *truly* additive or synergistic effect remains to be determined. Nonetheless, these findings have clinical relevance given that most adults require two or more antihypertensive medications to achieve a goal BP [1,10,26]. The addition of regular exercise to antihypertensive drug therapy may help more adults not only reach their goal BP, but do so without increasing the number of medications in their treatment plan. Additional studies designed specifically to investigate the separate and combined effects of exercise and antihypertensive medications are warranted.

CLINICAL IMPLICATIONS AND NEW ADVANCES IN EXERCISE PRESCRIPTION FOR OPTIMIZING BLOOD PRESSURE BENEFITS

The FITT-VP Ex R_x recommendations that follow are based upon the new and emerging findings discussed in this chapter as they relate to the current consensus of knowledge regarding the effects of acute and chronic aerobic, dynamic resistance, and concurrent exercise on hypertension as presented in this chapter. The modified Ex R_x for adults with hypertension we propose is summarized in Table 3:

Frequency:

- Aerobic exercise should be performed on most, preferably all, days of the week (i.e., 5–7 days per week) in combination with dynamic resistance exercise 2–3 days per week on nonconsecutive days.
- A combination of aerobic and dynamic resistance exercise can be performed on separate days (i.e., *combined* aerobic and resistance exercise) or during the same exercise session (i.e., *concurrent* exercise).
- Aerobic and resistance exercise should be supplemented by flexibility exercise 2–3 days per week.

This recommendation is made due to the well-established immediate and sustained BP-lowering effects of acute aerobic exercise or PEH [13,23,48,69,85–88,90,91,93], and the new and emerging evidence that supports acute dynamic resistance and concurrent exercise can elicit PEH to a similar magnitude as aerobic exercise among adults with hypertension [66,86,114–119,121–123,131–135]. Put simply, BP is lower on days when individuals with hypertension exercise than when they do not exercise. Also, individuals with hypertension are often overweight to obese and have additional CVD risk factors (i.e., insulin resistance, dyslipidemia, and metabolic syndrome) [156–159]. Therefore, large amounts of caloric expenditure should be emphasized [160] while maintaining lean mass, muscular strength, and function [21,112,161,162].

TABLE 3 The Modified Exercise Prescription for Adults With Hypertension Based on New and Emerging Evidence

FITT-VP Principle of the Ex R_x	Modified ACSM Recommendations
Frequency (how often?)	• Aerobic exercise, 5–7 days per week • Dynamic resistance exercise, ≥2–3 days per week (nonconsecutive) • A combination of aerobic and dynamic resistance exercise can be performed on separate days (i.e., *combined* aerobic and resistance exercise) or during the same exercise session (i.e., *concurrent* exercise) What's new: • The addition of dynamic resistance exercise as another viable stand-alone antihypertensive therapeutic option
Intensity (how hard?)[a]	• Aerobic exercise, moderate to vigorous (40%–≥60% of $\dot{V}O_{2R}$ or 12–≥14 on a scale of 6 (no exertion) to 20 (maximal exertion) level of physical exertion or an intensity that causes noticeable increases in heart rate and breathing) • Dynamic resistance exercise, moderate (60%–80% of 1RM or 12–14 on a scale of 6 (no exertion) to 20 (maximal exertion) level of physical exertion) • Concurrent exercise, a combination of moderate- to vigorous-intensity aerobic exercise and moderate-intensity resistance exercise as described above What's new: • The expansion of the recommendation for aerobic exercise intensity to include higher levels of physical exertion (i.e., vigorous-intensity aerobic exercise)

Continued

TABLE 3 The Modified Exercise Prescription for Adults With Hypertension Based on New and Emerging Evidence—cont'd

FITT-VP Principle of the Ex Rₓ	Modified ACSM Recommendations
Time (how long?)	• Aerobic exercise, ≥30–60 min/day; one continuous bout or multiple bouts of ≥10 min (moderate-intensity, i.e., 40%–59% of $\dot{V}O_{2R}$) or ≥3–10 min (vigorous-intensity, i.e., ≥60% of $\dot{V}O_{2R}$) each • Dynamic resistance exercise, 8–10 exercises, 1–4 sets of 8–12 repetitions • Concurrent exercise, ≥20–30 min/day of aerobic exercise and dynamic resistance exercise consisting of 4–8 exercises, 1–3 sets of 8–12 repetitions What's new: • The inclusion of multiple "very short bouts" (i.e., 3–10 min) of aerobic exercise, performed at vigorous-intensity, that can be interspersed throughout the day • Specific recommendations regarding the aerobic and resistance exercises when performed concurrently in a single exercise session
Type (what kind?) Primary	• Aerobic exercise, prolonged, rhythmic activities using large muscle groups (e.g., walking, cycling, and swimming) performed continuously at a constant intensity or with repeated bouts alternating between high versus low (≥70% vs <40% of $\dot{V}O_{2R}$) intensity (i.e., interval exercise) • Dynamic resistance exercise, a combination of multi- and single-joint exercises targeting the major muscle groups of the upper and lower body using traditional or circuit resistance training • Concurrent exercise, a combination of the aerobic and resistance exercises described above performed in any order (i.e., aerobic exercise can be performed before or after resistance exercise) What's new: • The recommendation for the type of aerobic exercise has been expanded to include interval exercise • Specific recommendations regarding the order of aerobic and resistance exercises when performed concurrently in a single exercise session
Adjuvant 1	Flexibility F, ≥2–3 days per week I, stretch to the point of feeling tightness or slight discomfort T, ≥10 min/day; ≥4 repetitions per muscle group; hold each static stretch for 10–30 s
Adjuvant 2[b]	Neuromotor F, ≥2–3 days per week I, undetermined T, ≥20–30 min/day
Volume (how much?) [c]	≥150 min/week or 700–2000 kcal/week of total aerobic and resistance exercise
Progression	Progress gradually, avoiding large increases in any of the components of the Ex Rₓ; increase exercise duration over first 4–6 weeks and then increase frequency, intensity, and time (or some combination of these) to achieve recommended quantity and quality of exercise over next 4–8 months

Note: ACSM, American College of Sports Medicine; FITT-VP, Frequency, Intensity, Time, Type, Volume, and Progression of exercise; Ex Rₓ, exercise prescription; 1RM, one-repetition maximum; $\dot{V}O_{2R}$, oxygen consumption reserve

[a]Vigorous-intensity aerobic exercise (i.e., ≥60% of $\dot{V}O_{2R}$ or ≥14 on a scale of 6–20) [62–64] appears to elicit greater and more extensive benefits than lower levels of physical exertion for individuals who are willing and able to tolerate more intense levels of exercise and may be introduced after exercise preparticipation health screening and gradual progression.
[b]Balance (neuromotor) training is recommended for older adults, individuals who are at substantial risk of falling, and is likely to benefit younger adults as well.
[c]For greater and more extensive benefits, progress exercise volume to total 60 min/day and 300 min/week of moderate to vigorous intensity.

Intensity:

• Aerobic exercise: moderate- to vigorous-intensity (i.e., 40%–≥60% of $\dot{V}O_{2R}$, 12–≥14 rating of perceived physical exertion on the Borg 6–20 scale [62–64]) or an intensity that causes noticeable increases in HR and breathing.
• Dynamic resistance exercise: moderate-intensity (60%–80% of 1RM, 12–14 rating of perceived physical exertion on the Borg 6–20 scale [62–64]).

Due to the growing evidence that greater BP reductions can be achieved with greater levels of physical exertion [23,69,86,94], the aerobic exercise intensity recommendation has been expanded to include vigorous intensity if the patient or client is willing and able to tolerate higher levels of physical exertion. For dynamic resistance and concurrent exercise, it appears that moderate-intensity resistance exercise is efficacious for reducing BP among adults with hypertension [114–117,119,122].

*T*ime:

- Aerobic exercise: 30–60 min/day of continuous or intermittent (i.e., fractionized) exercise. If intermittent, bouts should be ≥10 min (moderate-intensity exercise i.e., 40%–59% of $\dot{V}O_{2R}$) or ≥3–10 min (vigorous-intensity exercise, i.e., ≥60%–80% of $\dot{V}O_{2R}$) in duration depending on the level of physical exertion and accumulate to total 30–60 min/day.
- Dynamic resistance exercise: should consist of at least one set of 8–12 repetitions for 8–10 exercises targeting the major muscle groups.
- Concurrent exercise: ≥ 20–30 min/day of continuous aerobic exercise and dynamic resistance exercise consisting of at least one set of 8–12 repetitions for 4–8 exercises targeting the major muscle groups.

This recommendation is consistent with existing evidence that PEH is a low-threshold phenomenon regarding the time (duration) of the acute aerobic exercise bout but has been expanded to consider the interaction between time and intensity. When several short bouts of aerobic exercise are interspersed throughout the day, PEH offers a viable therapeutic lifestyle option for BP control among individuals with high BP [86,88,93]. Bouts of at least 10 min are recommended for moderate-intensity aerobic exercise, while bouts of <10 min (i.e., ≥3–10 min) may be recommended for more vigorous-intensity aerobic exercise. At this time, there is no compelling evidence to support that intermittent or fractionized resistance or concurrent exercise [133] can offer the same PEH benefit reported with a single, continuous bout of resistance or concurrent exercise, respectively.

*T*ype:

- For aerobic exercise, emphasis should be placed on prolonged, rhythmic activities using large muscle groups such as walking, cycling, or swimming. Aerobic activities can be performed continuously at a constant intensity or in repeated bouts alternating between high versus low (≥70% vs <40% of $\dot{V}O_{2R}$) intensity (i.e., *interval exercise*).
- For dynamic resistance exercise, emphasis should be placed on multi- and single-joint exercises that target the major muscle groups of the upper and lower body. Resistance exercises can be performed using a conventional or circuit (i.e., lighter weights, higher repetitions, with minimal rest between exercises) resistance training protocol.
- For concurrent exercise, a combination of the aerobic and resistance exercises described above can be performed in the same exercise session, in any order (i.e., aerobic exercise can be performed before *or* after resistance exercise).
- For older adults or individuals who are at substantial risk of falling, neuromotor (balance) training 2–3 days per week is also recommended as adjuvant exercise.

Aerobic exercise training has consistently been shown to lower BP among adults with hypertension, and now, new and emerging research supports that dynamic resistance and concurrent exercise training can elicit comparable BP reductions. Therefore, we have expanded this recommendation to include dynamic resistance exercise and concurrent exercise as viable stand-alone antihypertensive therapy that should be performed in addition to aerobic exercise. Consistent with our expanded recommendation for aerobic exercise intensity, we have also modified this recommendation to include interval aerobic exercise. BP reductions following aerobic interval training (e.g., HIIT) are similar to or exceed those observed with continuous, constant intensity aerobic exercise [23,90,94,108]. Furthermore, interval exercise allows adults with hypertension to experience the health and BP benefits associated with higher levels of physical exertion that would not be tolerable with longer duration exercise [23,108]. The recommendation for dynamic resistance exercise has been expanded with specific information regarding the resistance training protocol. Existing and new and emerging evidence has shown that adults with hypertension experience similar BP benefit with conventional and circuit resistance exercise [13,86,126,139,140,163,164]. Finally, there is no strong evidence to support that the order of the aerobic and resistance exercise preformed in a single concurrent exercise session influences PEH or the BP reductions that occur with training.

GAPS IN THE LITERATURE AND FUTURE RESEARCH NEEDS IN THE EXERCISE PRESCRIPTION FOR HYPERTENSION

It should be noted that these Ex R_x recommendations are limited by the methodological quality of the studies upon which the evidence is based [23,24]. Major limitations in the current state of the literature include small sample sizes, assessing study populations with normal BP rather than hypertension, not accounting for major confounders to the BP response to exercise that include timing of the last bout of exercise and detraining effects, and the lack of standard protocols for the assessment of BP and the exercise intervention. As a result of these limitations, the effectiveness of exercise as antihypertensive lifestyle therapy among individuals with hypertension has been underestimated [23,24,165]. Furthermore, large randomized clinical trials that examine both the acute and chronic BP-lowering effects of exercise among diverse populations are needed before professional organizations can definitively determine the optimal Ex R_x for individuals with hypertension.

CONCLUSION

Hypertension is one of the most important CVD risk factors due to its high prevalence and significant medical costs [1,2]. Indeed, over $\approx 34\%$ of adults in the United States [1,39] and $\approx 31\%$ worldwide [2,40] have hypertension. Both the JNC 8 [10] and ACSM [13] recommend aerobic exercise supplemented by dynamic resistance exercise as initial lifestyle therapy for individuals with hypertension because it lowers BP 5–7 and 2–3 mmHg, respectively, among those with hypertension. BP reductions of this magnitude can decrease the risk of developing CVD by 20%–30% [25–27] and rival those obtained with first-line antihypertensive medications [25–28] and with other types of lifestyle therapy [26]. New and emerging research has shown dynamic resistance and concurrent exercise to elicit BP reductions comparable with or greater than those achieved with aerobic exercise training and may serve as viable stand-alone antihypertensive therapeutic options for some patients. We have expanded the current exercise recommendations for hypertension in Table 3 to include this, as well as the other new and emerging research, discussed in this chapter. Future research that addresses the existing research gaps is needed to confirm these findings.

KEY POINTS

- Hypertension is the most common, modifiable, and costly CVD risk factor.
- Adults with hypertension are encouraged to engage in 30 min/day (or more) of moderate-intensity aerobic exercise on most, if not all, days of the week in addition to moderate-intensity dynamic resistance exercise 2–3 days per week to total 150 min/week of total exercise (or more) to prevent and control high BP (Table 3).
- The antihypertensive effects of acute and chronic aerobic exercise are a low-duration phenomenon with intermittent durations appearing as effective as continuous durations in lowering BP.
- Exercise intensity is an important moderator of the antihypertensive effects of acute and chronic aerobic exercise with individuals with the highest BP experiencing the greatest BP benefit. Furthermore, vigorous-intensity exercise appears to elicit greater and more extensive benefits than lower levels of physical exertion for individuals who are willing and able to tolerate more intense levels of exercise and may be introduced after exercise preparticipation health screening and gradual progression (Table 3).
- Acute resistance exercise may result in marked elevations in BP; however, these BP surges appear to immediately decrease back to levels below that of baseline; the magnitudes of the BP reductions are clinically meaningful and are most pronounced in individuals who stand to benefit the most (i.e., those with higher BP compared with normal BP).
- It has long been thought that dynamic resistance exercise training reduces BP ≈ 2–3 mmHg. However, new and emerging research has demonstrated that dynamic resistance exercise has an even more beneficial influence on BP among those diagnosed with hypertension, reductions that are comparable to or greater than those achieved with aerobic exercise training. Based on this new and emerging research, we recommend that adults with hypertension perform dynamic resistance training *in addition to* aerobic exercise as stand-alone antihypertensive lifestyle therapy (Table 3).
- New and emerging research also indicates that aerobic and resistance exercise performed *concurrently* reduces BP to levels similar to that of aerobic exercise training. Future research is needed to confirm these findings.
- Despite the volume of literature on exercise and hypertension, there remains a critical need to identify patient and FITT exercise intervention characteristics that influence the BP response to acute and chronic exercise so that exercise can be more precisely prescribed as antihypertensive therapy.

ABBREVIATIONS

ACSM	American College of Sports Medicine
AHA/ACC	American Heart Association/American College of Cardiology
BP	blood pressure
CVD	cardiovascular disease
Ex R$_x$	exercise prescription
ES	effect size
FITT-VP	frequency, intensity, time, type, volume, and progression
JNC 8	The Eighth Report of the Joint National Committee on Prevention, Detection, Evaluation, and Treatment of High Blood Pressure
JNC 7	The Seventh Report of the Joint National Committee on Prevention, Detection, Evaluation, and Treatment of High Blood Pressure
HR	heart rate
HIIT	high-intensity interval training
1RM	one-repetition maximum
PEH	postexercise hypotension
RPE	rating of perceived exertion
$\dot{V}O_{2max}$	maximal oxygen consumption
$\dot{V}O_{2peak}$	peak oxygen consumption
$\dot{V}O_{2R}$	oxygen consumption reserve

REFERENCES

[1] Benjamin EJ, Blaha MJ, Chiuve SE, Cushman M, Das SR, Deo R, et al. Heart disease and stroke statistics-2017 update: a report from the American Heart Association. Circulation 2017;135(10):e146–603.

[2] Olsen MH, Angell SY, Asma S, Boutouyrie P, Burger D, Chirinos JA, et al. A call to action and a lifecourse strategy to address the global burden of raised blood pressure on current and future generations: the Lancet Commission on hypertension. Lancet 2016;388(10060):2665–712.

[3] GBD 2013 Risk Factors Collaborators, Forouzanfar MH, Alexander L, Anderson HR, Bachman VF, Biryukov S, et al. Global, regional, and national comparative risk assessment of 79 behavioural, environmental and occupational, and metabolic risks or clusters of risks in 188 countries, 1990–2013: a systematic analysis for the Global Burden of Disease Study 2013. Lancet 2015;386(10010):2287–323.

[4] Forouzanfar MH, Liu P, Roth GA, Ng M, Biryukov S, Marczak L, et al. Global Burden of Hypertension and Systolic Blood Pressure of at Least 110 to 115 mm Hg, 1990–2015. JAMA 2017;317(2):165–82.

[5] Kung HC, Xu J. Hypertension-related Mortality in the United States, 2000–2013. NCHS Data Brief 2015;193:1–8.

[6] Office of Disease Prevention and Health Promotion. Heart disease and stroke. In: Healthy People 2020. Retrieved from https://www.healthypeople.gov/2020/topics-objectives/topic/heart-disease-and-stroke; 2016.

[7] WHO. A Global Brief on Hypertension: Silent Killer, Global Public Health Crisis. Switzerland: WHO/DCO/WHD/2013.2; 2013. 1 p. Available From: WHO Press, World Health Organization.

[8] Franco OH, Peeters A, Bonneux L, de Laet C. Blood pressure in adulthood and life expectancy with cardiovascular disease in men and women: life course analysis. Hypertension 2005;46(2):280–6.

[9] Lewington S, Clarke R, Qizilbash N, Peto R, Collins R, Prospective Studies Collaboration. Age-specific relevance of usual blood pressure to vascular mortality: a meta-analysis of individual data for one million adults in 61 prospective studies. Lancet 2002; 360(9349):1903–13.

[10] James PA, Oparil S, Carter BL, Cushman WC, Dennison-Himmelfarb C, Handler J, et al. 2014 Evidence-based guideline for the management of high blood pressure in adults: report from the panel members appointed to the Eighth Joint National Committee (JNC 8). JAMA 2013;311(5):507–20.

[11] Eckel RH, Jakicic JM, Ard JD, De Jesus JM, Houston Miller N, Hubbard VS, et al. 2013 AHA/ACC guideline on lifestyle management to reduce cardiovascular risk: a report of the American college of cardiology/American heart association task force on practice guidelines. J Am Coll Cardiol 2014;63(25 Pt B):2960–84.

[12] Pescatello LS, American College of Sports Medicine. ACSM's guidelines for exercise testing and prescription. 9th Philadelphia, PA: Wolters Kluwer/Lippincott Williams & Wilkins Health; 2014. 456 pp.

[13] Pescatello LS, Franklin BA, Fagard R, Farquhar WB, Kelley GA, Ray CA, American College of Sports Medicine. American College of Sports Medicine position stand: exercise and hypertension. Med Sci Sports Exerc 2004;36(3):533–53.

[14] Pescatello LS, MacDonald HV, Johnson BT. The effects of aerobic exercise on hypertension: current consensus and emerging research. In: Pescatello LS, editor. Effects of exercise on hypertension: from cells to physiological systems. Switzerland: Springer International Publishing; 2015. p. 3–24 [chapter 1].

[15] Barlow CE, LaMonte MJ, Fitzgerald SJ, Kampert JB, Perrin JL, Blair SN. Cardiorespiratory fitness is an independent predictor of hypertension incidence among initially normotensive healthy women. Am J Epidemiol 2006;163(2):142–50.

[16] Banda JA, Clouston K, Sui X, Hooker SP, Lee CD, Blair SN. Protective health factors and incident hypertension in men. Am J Hypertens 2010;23(6):599–605.

[17] Swift DL, Lavie CJ, Johannsen NM, Arena R, Earnest CP, O'Keefe JH, et al. Physical activity, cardiorespiratory fitness, and exercise training in primary and secondary coronary prevention. Circ J 2013;77(2):281–92.

[18] Maslow AL, Sui X, Colabianchi N, Hussey J, Blair SN. Muscular strength and incident hypertension in normotensive and prehypertensive men. Med Sci Sports Exerc 2010;42(2):288–95.

[19] Artero EG, Lee DC, Lavie CJ, Espana-Romero V, Sui X, Church TS, et al. Effects of muscular strength on cardiovascular risk factors and prognosis. J Cardiopulm Rehabil Prev 2012;32(6):351–8.

[20] Crump C, Sundquist J, Winkleby MA, Sundquist K. Interactive effects of physical fitness and body mass index on the risk of hypertension. JAMA Intern Med 2016;176(2):210–6.

[21] Vaara JP, Fogelholm M, Vasankari T, Santtila M, Hakkinen K, Kyrolainen H. Associations of maximal strength and muscular endurance with cardiovascular risk factors. Int J Sports Med 2014;35(4):356–60.

[22] Artero EG, Lee DC, Ruiz JR, Sui X, Ortega FB, Church TS, et al. A prospective study of muscular strength and all-cause mortality in men with hypertension. J Am Coll Cardiol 2011;57(18):1831–7.

[23] Pescatello LS, MacDonald HV, Lamberti L, Johnson BT. Exercise for hypertension: a prescription update integrating existing recommendations with emerging research. Curr Hypertens Rep 2015;17(11):87. https://doi.org/10.1007/s11906-015-0600-y.

[24] Johnson BT, MacDonald HV, Bruneau Jr ML, Goldsby TU, Brown JC, Huedo-Medina TB, et al. Methodological quality of meta-analyses on the blood pressure response to exercise: a review. J Hypertens 2014;32(4):706–23.

[25] ALLHAT Officers and Coordinators for the ALLHAT Collaborative Research Group. Major outcomes in high-risk hypertensive patients randomized to angiotensin-converting enzyme inhibitor or calcium channel blocker vs diuretic: the Antihypertensive and Lipid-Lowering Treatment to Prevent Heart Attack Trial (ALLHAT). JAMA 2002;288(23):2981–97.

[26] Chobanian AV, Bakris GL, Black HR, Cushman WC, Green LA, Izzo Jr JL, et al. Seventh report of the Joint National Committee on Prevention, Detection, Evaluation, and Treatment of High Blood Pressure. Hypertension 2003;42(6):1206–52.

[27] Whelton PK, He J, Appel LJ, Cutler JA, Havas S, Kotchen TA, et al. Primary prevention of hypertension: clinical and public health advisory from The National High Blood Pressure Education Program. JAMA 2002;288(15):1882–8.

[28] Whelton SP, Chin A, Xin X, He J. Effect of aerobic exercise on blood pressure: a meta-analysis of randomized, controlled trials. Ann Intern Med 2002;136(7):493–503.

[29] Brown RE, Riddell MC, Macpherson AK, Canning KL, Kuk JL. The joint association of physical activity, blood-pressure control, and pharmacologic treatment of hypertension for all-cause mortality risk. Am J Hypertens 2013;26(8):1005–10.

[30] Naci H, Ioannidis JP. Comparative effectiveness of exercise and drug interventions on mortality outcomes: metaepidemiological study. BMJ 2013;347:f5577.

[31] Mancia G, Fagard R, Narkiewicz K, Redon J, Zanchetti A, Bohm M, et al. 2013 ESH/ESC practice guidelines for the management of arterial hypertension. Blood Press 2014;23(1):3–16.

[32] Dasgupta K, Quinn RR, Zarnke KB, Rabi DM, Ravani P, Daskalopoulou SS, et al. The 2014 Canadian hypertension education program recommendations for blood pressure measurement, diagnosis, assessment of risk, prevention, and treatment of hypertension. Can J Cardiol 2014;30(5):485–501.

[33] Brook RD, Appel LJ, Rubenfire M, Ogedegbe G, Bisognano JD, Elliott WJ, et al. Beyond medications and diet: alternative approaches to lowering blood pressure: a scientific statement from the American Heart Association. Hypertension 2013;61(6):1360–83.

[34] U.S. Preventive Services Task Force Procedure Manual [Internet] [cited 2014 September 11]. Available from http://www.uspreventiveservicestaskforce.org/uspstf08/methods/procmanual.pdf.

[35] Clinical guidelines on the identification, evaluation, and treatment of overweight and obesity in adults—the evidence report. National Institutes of Health. Obes Res 1998;6(2):51S–209S.

[36] Writing ESC guidelines: recommendations for guidelines production [Internet] [cited 2014 September 11]. Available from http://www.escardio.org/guidelines-surveys/esc-guidelines/about/Pages/rules-writing.aspx.

[37] McAlister FA. The Canadian hypertension education program—a unique Canadian initiative. Can J Cardiol 2006;22(7):559–64.

[38] Stedman TL. Stedman's medical dictionary. 28th ed. Philadelphia, PA: Lippincott Williams & Wilkins; 2006.

[39] Nwankwo T, Yoon SS, Burt V, Gu Q. Hypertension among adults in the United States: National Health and Nutrition Examination Survey 2011–2012. NCHS Data Brief 2013;133:1–8.

[40] Mills KT, Bundy JD, Kelly TN, Reed JE, Kearney PM, Reynolds K, et al. Global disparities of hypertension prevalence and control: a systematic analysis of population-based studies from 90 countries. Circulation 2016;134(6):441–50.

[41] Gupta AK, McGlone M, Greenway FL, Johnson WD. Prehypertension in disease-free adults: a marker for an adverse cardiometabolic risk profile. Hypertens Res 2010;33(9):905–10.

[42] Fields LE, Burt VL, Cutler JA, Hughes J, Roccella EJ, Sorlie P. The burden of adult hypertension in the United States 1999 to 2000: a rising tide. Hypertension 2004;44(4):398–404.

[43] Muntner P, Woodward M, Mann DM, Shimbo D, Michos ED, Blumenthal RS, et al. Comparison of the Framingham Heart Study hypertension model with blood pressure alone in the prediction of risk of hypertension: the multi-ethnic study of atherosclerosis. Hypertension 2010;55(6):1339–45.

[44] Egan BM, Lackland DT, Jones DW. Prehypertension: an opportunity for a new public health paradigm. Cardiol Clin 2010;28(4):561–9.

[45] Huang Y, Su L, Cai X, Mai W, Wang S, Hu Y, et al. Association of all-cause and cardiovascular mortality with prehypertension: a meta-analysis. Am Heart J 2014;167(2). 160–8.e1.

[46] Vasan RS, Larson MG, Leip EP, Evans JC, O'Donnell CJ, Kannel WB, et al. Impact of high-normal blood pressure on the risk of cardiovascular disease. N Engl J Med 2001;345(18):1291–7.

[47] Fernandez C, Sander GE, Giles TD. Prehypertension: defining the transitional phenotype. Curr Hypertens Rep 2016;18(1):2. https://doi.org/10.1007/s11906-015-0611-8.

[48] Pescatello LS, Kulikowich JM. The aftereffects of dynamic exercise on ambulatory blood pressure. Med Sci Sports Exerc 2001;33(11):1855–61.

[49] Wilder J. The law of initial value in neurology and psychiatry; facts and problems. J Nerv Ment Dis 1957;125(1):73–86.

[50] Brandao Rondon MU, Alves MJ, Braga AM, Teixeira OT, Barretto AC, Krieger EM, et al. Postexercise blood pressure reduction in elderly hypertensive patients. J Am Coll Cardiol 2002;39(4):676–82.

[51] Kenney MJ, Seals DR. Postexercise hypotension. Key features, mechanisms, and clinical significance. Hypertension 1993;22(5):653–64.

[52] Pescatello LS, Guidry MA, Blanchard BE, Kerr A, Taylor AL, Johnson AN, et al. Exercise intensity alters postexercise hypotension. J Hypertens 2004;22(10):1881–8.

[53] Fitzgerald W. Labile hypertension and jogging: new diagnostic tool or spurious discovery? Br Med J (Clin Res Ed) 1981;282(6263):542–4.

[54] Haskell WL. J.B. Wolffe Memorial Lecture. Health consequences of physical activity: understanding and challenges regarding dose-response. Med Sci Sports Exerc 1994;26(6):649–60.

[55] Hecksteden A, Grutters T, Meyer T. Association between postexercise hypotension and long-term training-induced blood pressure reduction: a pilot study. Clin J Sport Med 2013;23(1):58–63.

[56] Liu S, Goodman J, Nolan R, Lacombe S, Thomas SG. Blood pressure responses to acute and chronic exercise are related in prehypertension. Med Sci Sports Exerc 2012;44(9):1644–52.

[57] Wilcox RG, Bennett T, Brown AM, Macdonald IA. Is exercise good for high blood pressure? Br Med J (Clin Res Ed) 1982;285(6344):767–9.

[58] Devereux GR, Wiles JD, Howden R. Immediate post-isometric exercise cardiovascular responses are associated with training-induced resting systolic blood pressure reductions. Eur J Appl Physiol 2015;115(2):327–33.

[59] Tibana RA, de Sousa NM, da Cunha Nascimento D, Pereira GB, Thomas SG, Balsamo S, et al. Correlation between acute and chronic 24-hour blood pressure response to resistance training in adult women. Int J Sports Med 2015;36(1):82–9.

[60] Moreira SR, Cucato GG, Terra DF, Ritti-Dias RM. Acute blood pressure changes are related to chronic effects of resistance exercise in medicated hypertensives elderly women. Clin Physiol Funct Imaging 2016;36(3):242–8.

[61] Pescatello LS, MacDonald HV, Ash GI, Lambert LM, Farquhar WB, Arena R, et al. Assessing the existing professional exercise recommendations for hypertension: a review and recommendations for future research priorities. Mayo Clin Proc 2015;90(6):801–12.

[62] Borg GA. Perceived exertion. Exerc Sport Sci Rev 1974;2:131–53.

[63] Borg G, Ljunggren G, Ceci R. The increase of perceived exertion, aches and pain in the legs, heart rate and blood lactate during exercise on a bicycle ergometer. Eur J Appl Physiol Occup Physiol 1985;54(4):343–9.

[64] Borg G, Hassmen P, Lagerstrom M. Perceived exertion related to heart rate and blood lactate during arm and leg exercise. Eur J Appl Physiol Occup Physiol 1987;56(6):679–85.

[65] Riebe D, American College of Sports Medicine. ACSM's guidelines for exercise testing and prescription. 10th ed. Philadelphia, PA: Wolters Kluwer/Lippincott Williams & Wilkins Health; 2017.

[66] MacDonald HV, Farinatti PV, Lamberti LM, Pescatello LS. The effects of concurrent exercise on hypertension: current consensus and emerging research. In: Pescatello LS, editor. Effects of exercise on hypertension: from cells to physiological systems. Switzerland: Springer International Publishing; 2015. p. 47–86 [chapter 3].

[67] Bennett T, Wilcox RG, Macdonald IA. Post-exercise reduction of blood pressure in hypertensive men is not due to acute impairment of baroreflex function. Clin Sci (Lond) 1984;67(1):97–103.

[68] Cleroux J, Kouame N, Nadeau A, Coulombe D, Lacourciere Y. Aftereffects of exercise on regional and systemic hemodynamics in hypertension. Hypertension 1992;19(2):183–91.

[69] Eicher JD, Maresh CM, Tsongalis GJ, Thompson PD, Pescatello LS. The additive blood pressure lowering effects of exercise intensity on postexercise hypotension. Am Heart J 2010;160(3):513–20.

[70] Floras JS, Sinkey CA, Aylward PE, Seals DR, Thoren PN, Mark AL. Postexercise hypotension and sympathoinhibition in borderline hypertensive men. Hypertension 1989;14(1):28–35.

[71] Floras JS, Hara K. Sympathoneural and haemodynamic characteristics of young subjects with mild essential hypertension. J Hypertens 1993;11(6):647–55.

[72] Halliwill JR, Buck TM, Lacewell AN, Romero SA. Postexercise hypotension and sustained postexercise vasodilatation: what happens after we exercise? Exp Physiol 2013;98(1):7–18.

[73] Luttrell MJ, Halliwill JR. Recovery from exercise: vulnerable state, window of opportunity, or crystal ball? Front Physiol 2015;6:204.

[74] Wallace JP, Bogle PG, King BA, Krasnoff JB, Jastremski CA. The magnitude and duration of ambulatory blood pressure reduction following acute exercise. J Hum Hypertens 1999;13(6):361–6.

[75] Taylor-Tolbert NS, Dengel DR, Brown MD, McCole SD, Pratley RE, Ferrell RE, et al. Ambulatory blood pressure after acute exercise in older men with essential hypertension. Am J Hypertens 2000;13(1 Pt 1):44–51.

[76] Paulev PE, Jordal R, Kristensen O, Ladefoged J. Therapeutic effect of exercise on hypertension. Eur J Appl Physiol Occup Physiol 1984;53(2):180–5.

[77] Quinn TJ. Twenty-four hour, ambulatory blood pressure responses following acute exercise: impact of exercise intensity. J Hum Hypertens 2000;14(9):547–53.

[78] Hagberg JM, Montain SJ, Martin 3rd WH. Blood pressure and hemodynamic responses after exercise in older hypertensives. J Appl Physiol (1985) 1987;63(1):270–6.

[79] Kraul J, Chrastek J, Adamirova J. The hypotensive effect of physical activity. In: Rabb W, editor. Prevention of ischemic heart disease: principles and practice. Springfield, IL: Charles C. Thomas; 1966.

[80] MacDonald JR, Hogben CD, Tarnopolsky MA, MacDougall JD. Post exercise hypotension is sustained during subsequent bouts of mild exercise and simulated activities of daily living. J Hum Hypertens 2001;15(8):567–71.

[81] Pescatello LS, Fargo AE, Leach Jr CN, Scherzer HH. Short-term effect of dynamic exercise on arterial blood pressure. Circulation 1991;83(5):1557–61.

[82] MacDonald JR. Potential causes, mechanisms, and implications of post exercise hypotension. J Hum Hypertens 2002;16(4):225–36.

[83] Cardoso Jr CG, Gomides RS, Queiroz AC, Pinto LG, da Silveira Lobo F, Tinucci T, et al. Acute and chronic effects of aerobic and resistance exercise on ambulatory blood pressure. Clinics (Sao Paulo) 2010;65(3):317–25.

[84] Gomes Anunciacao P, Doederlein Polito M. A review on post-exercise hypotension in hypertensive individuals. Arq Bras Cardiol 2011;96(5):e100–9.

[85] Oliveira Marques-Silvestre AC, do Socorro Brasileiro-Santos M, Sarmento de Oliveira A, Thiago Maciel da Silva F, da Cruz Santos A, da Cruz Santos A. Magnitude of acute aerobic post-exercise hypotension: a systematic review of randomized trials. Motricidade 2014;10(3):99–111.

[86] Carpio-Rivera E, Moncada-Jimenez J, Salazar-Rojas W, Solera-Herrera A. Acute effects of exercise on blood pressure: a meta-analytic investigation. Arq Bras Cardiol 2016;106(5):422–33.

[87] Ciolac EG, Guimaraes GV, D'Avila VM, Bortolotto LA, Doria EL, Bocchi EA. Acute aerobic exercise reduces 24-h ambulatory blood pressure levels in long-term-treated hypertensive patients. Clinics (Sao Paulo) 2008;63(6):753–8.

[88] Bhammar DM, Angadi SS, Gaesser GA. Effects of fractionized and continuous exercise on 24-h ambulatory blood pressure. Med Sci Sports Exerc 2012;44(12):2270–6.

[89] Carvalho RS, Pires CM, Junqueira GC, Freitas D, Marchi-Alves LM. Hypotensive response magnitude and duration in hypertensives: continuous and interval exercise. Arq Bras Cardiol 2015;104(3):234–41.

[90] Ciolac EG, Guimaraes GV, D Avila VM, Bortolotto LA, Doria EL, Bocchi EA. Acute effects of continuous and interval aerobic exercise on 24-h ambulatory blood pressure in long-term treated hypertensive patients. Int J Cardiol 2009;133(3):381–7.

[91] Guidry MA, Blanchard BE, Thompson PD, Maresh CM, Seip RL, Taylor AL, et al. The influence of short and long duration on the blood pressure response to an acute bout of dynamic exercise. Am Heart J 2006;151(6):1322.e5–1322.e12.

[92] MacDonald JR, MacDougall JD, Hogben CD. The effects of exercise duration on post-exercise hypotension. J Hum Hypertens 2000;14(2):125–9.

[93] Miyashita M, Burns SF, Stensel DJ. Accumulating short bouts of running reduces resting blood pressure in young normotensive/pre-hypertensive men. J Sports Sci 2011;29(14):1473–82.

[94] Santana HA, Moreira SR, Asano RY, Sales MM, Cordova C, Campbell CS, et al. Exercise intensity modulates nitric oxide and blood pressure responses in hypertensive older women. Aging Clin Exp Res 2013;25(1):43–8.

[95] Jones H, George K, Edwards B, Atkinson G. Is the magnitude of acute post-exercise hypotension mediated by exercise intensity or total work done? Eur J Appl Physiol 2007;102(1):33–40.

[96] Ciolac EG. High-intensity interval training and hypertension: maximizing the benefits of exercise? Am J Cardiovasc Dis 2012;2(2):102–10.

[97] Gleser LJ, Olkin I. Stochastically dependent effect sizes. In: Cooper H, Hedges LV, Valentine JC, editors. The handbook of research synthesis and meta-analysis. 2nd ed. New York, NY: Russell Sage Foundation; 2009. p. 357–76.

[98] Steel PD, Kammeyer-Mueller JD. Comparing meta-analytic moderator estimation techniques under realistic conditions. J Appl Psychol 2002;87(1):96–111.

[99] Viechtbauer W. Accounting for heterogeneity via random-effects models and moderator analyses in meta-analysis. Z Psychol/J Psychol 2007;215(2):104–21.

[100] Visweswaran C, Sanchez JI. Moderator search in meta-analysis: a review and cautionary note on existing approaches. Educ Psychol Meas 1998;58(1):77–87.

[101] Cornelissen VA, Fagard RH. Effects of endurance training on blood pressure, blood pressure-regulating mechanisms, and cardiovascular risk factors. Hypertension 2005;46(4):667–75.

[102] Kelley GA, Kelley KA, Tran ZV. Aerobic exercise and resting blood pressure: a meta-analytic review of randomized, controlled trials. Prev Cardiol 2001;4(2):73–80.

[103] Kelley GA, Kelley KS, Tran ZV. Walking and resting blood pressure in adults: a meta-analysis. Prev Med 2001;33(2 Pt 1):120–7.

[104] Fagard RH. Physical fitness and blood pressure. J Hypertens Suppl 1993;11(5):S47–52.

[105] Cornelissen VA, Buys R, Smart NA. Endurance exercise beneficially affects ambulatory blood pressure: a systematic review and meta-analysis. J Hypertens 2013;31(4):639–48.

[106] Sosner P, Guiraud T, Gremeaux V, Arvisais D, Herpin D, Bosquet L. The ambulatory hypotensive effect of aerobic training: a reappraisal through a meta-analysis of selected moderators. Scand J Med Sci Sports 2017;27(3):327–41.

[107] Cornelissen VA, Smart NA. Exercise training for blood pressure: a systematic review and meta-analysis. J Am Heart Assoc 2013;2(1):e004473.

[108] Kessler HS, Sisson SB, Short KR. The potential for high-intensity interval training to reduce cardiometabolic disease risk. Sports Med 2012;42(6):489–509.

[109] Gibala MJ, Little JP, Macdonald MJ, Hawley JA. Physiological adaptations to low-volume, high-intensity interval training in health and disease. J Physiol 2012;590(5):1077–84.

[110] Porth CJ, Bamrah VS, Tristani FE, Smith JJ. The Valsalva maneuver: mechanisms and clinical implications. Heart Lung 1984;13(5):507–18.

[111] MacDougall JD, Tuxen D, Sale DG, Moroz JR, Sutton JR. Arterial blood pressure response to heavy resistance exercise. J Appl Physiol (1985) 1985;58(3):785–90.

[112] Crump C, Sundquist J, Winkleby MA, Sundquist K. Interactive effects of aerobic fitness, strength, and obesity on mortality in men. Am J Prev Med 2017;52(3):353–61.

[113] Volaklis KA, Halle M, Meisinger C. Muscular strength as a strong predictor of mortality: a narrative review. Eur J Intern Med 2015;26(5):303–10.

[114] Cavalcante PA, Rica RL, Evangelista AL, Serra AJ, Figueira Jr A, Pontes Jr FL, et al. Effects of exercise intensity on postexercise hypotension after resistance training session in overweight hypertensive patients. Clin Interv Aging 2015;10:1487–95.

[115] Brito Ade F, de Oliveira CV, Santos Mdo S, Santos Ada C. High-intensity exercise promotes postexercise hypotension greater than moderate intensity in elderly hypertensive individuals. Clin Physiol Funct Imaging 2014;34(2):126–32.

[116] Moraes MR, Bacurau RF, Simoes HG, Campbell CS, Pudo MA, Wasinski F, et al. Effect of 12 weeks of resistance exercise on post-exercise hypotension in stage 1 hypertensive individuals. J Hum Hypertens 2012;26(9):533–9.

[117] Scher LM, Ferriolli E, Moriguti JC, Scher R, Lima NK. The effect of different volumes of acute resistance exercise on elderly individuals with treated hypertension. J Strength Cond Res 2011;25(4):1016–23.

[118] Simao R, Fleck SJ, Polito M, Monteiro W, Farinatti P. Effects of resistance training intensity, volume, and session format on the postexercise hypotensive response. J Strength Cond Res 2005;19(4):853–8.

[119] de Freitas BA, Brasileiro-Santos Mdo S, Coutinho de Oliveira CV, Sarmento da Nobrega TK, Lucia de Moraes Forjaz C, da Cruz Santos A. High-intensity resistance exercise promotes postexercise hypotension greater than moderate intensity and affects cardiac autonomic responses in women who are hypertensive. J Strength Cond Res 2015;29(12):3486–93.

[120] Prista A, Macucule CF, Queiroz AC, Silva Jr ND, Cardoso Jr. CG, Tinucci T, et al. A bout of resistance exercise following the 2007 AHA guidelines decreases asleep blood pressure in Mozambican men. J Strength Cond Res 2013;27(3):786–92.

[121] Casonatto J, Goessler KF, Cornelissen VA, Cardoso JR, Polito MD. The blood pressure-lowering effect of a single bout of resistance exercise: a systematic review and meta-analysis of randomised controlled trials. Eur J Prev Cardiol 2016;23(16):1700–14.

[122] Melo CM, Alencar Filho AC, Tinucci T, Mion Jr D, Forjaz CL. Postexercise hypotension induced by low-intensity resistance exercise in hypertensive women receiving captopril. Blood Press Monit 2006;11(4):183–9.

[123] Mota MR, Oliveira RJ, Terra DF, Pardono E, Dutra MT, de Almeida JA, et al. Acute and chronic effects of resistance exercise on blood pressure in elderly women and the possible influence of ACE I/D polymorphism. Int J Gen Med 2013;6:581–7.

[124] Kelley GA. Dynamic resistance exercise and resting blood pressure in adults: a meta-analysis. J Appl Physiol 1997;82(5):1559–65.

[125] Kelley GA, Kelley KS. Progressive resistance exercise and resting blood pressure: a meta-analysis of randomized controlled trials. Hypertension 2000;35(3):838–43.

[126] Cornelissen VA, Fagard RH. Effect of resistance training on resting blood pressure: a meta-analysis of randomized controlled trials. J Hypertens 2005;23(2):251–9.

[127] Cornelissen VA, Fagard RH, Coeckelberghs E, Vanhees L. Impact of resistance training on blood pressure and other cardiovascular risk factors: a meta-analysis of randomized, controlled trials. Hypertension 2011;58(5):950–8.

[128] Rossi AM, Moullec G, Lavoie KL, Gour-Provencal G, Bacon SL. The evolution of a Canadian Hypertension Education Program recommendation: the impact of resistance training on resting blood pressure in adults as an example. Can J Cardiol 2013;29(5):622–7.

[129] MacDonald HV, Johnson BT, Huedo-Medina TB, Livingston J, Forsyth KC, Kraemer WJ, et al. Dynamic resistance training as stand-alone antihypertensive lifestyle therapy: a meta-analysis. J Am Heart Assoc 2016;5(10). https://doi.org/10.1161/JAHA.116.003231.

[130] Mota MR, de Oliveira RJ, Dutra MT, Pardono E, Terra DF, Lima RM, et al. Acute and chronic effects of resistive exercise on blood pressure in hypertensive elderly women. J Strength Cond Res 2013;27(12):3475–80.

[131] Tibana RA, Nascimento Dda C, de Sousa NM, de Almeida JA, Moraes MR, Durigan JL, et al. Similar hypotensive effects of combined aerobic and resistance exercise with 1 set versus 3 sets in women with metabolic syndrome. Clin Physiol Funct Imaging 2015;35(6):443–50.

[132] Anunciacao PG, Farinatti PT, Goessler KF, Casonatto J, Polito MD. Blood pressure and autonomic responses following isolated and combined aerobic and resistance exercise in hypertensive older women. Clin Exp Hypertens 2016;38(8):710–4.

[133] Azevedo LM, de Souza AC, Santos LE, Miguel Dos Santos R, de Fernandes MO, Almeida JA, et al. Fractionated concurrent exercise throughout the day does not promote acute blood pressure benefits in hypertensive middle-aged women. Front Cardiovasc Med 2017;4:6.

[134] Dos Santos ES, Asano RY, Gomes Filho I, Nilson Lima L, Panelli P, Nascimento DD, et al. Acute and chronic cardiovascular response to 16 weeks of combined eccentric or traditional resistance and aerobic training in elderly hypertensive women: a randomized controlled trial. J Strength Cond Res 2017;28(11):3073–84. https://doi.org/10.1519/JSC.0000000000000537.

[135] Meneses AL, Forjaz CL, de Lima PF, Batista RM, Monteiro MD, Ritti-Dias RM. Influence of endurance and resistance exercise order on the postexercise hemodynamic responses in hypertensive women. J Strength Cond Res 2015;29(3):612–8. https://doi.org/10.1519/JSC.0000000000000676.

[136] Corso LM, Macdonald HV, Johnson BT, Farinatti P, Livingston J, Zaleski AL, et al. Is concurrent training efficacious antihypertensive therapy? A meta-analysis. Med Sci Sports Exerc 2016;48(12):2398–406.

[137] Hayashino Y, Jackson JL, Fukumori N, Nakamura F, Fukuhara S. Effects of supervised exercise on lipid profiles and blood pressure control in people with type 2 diabetes mellitus: a meta-analysis of randomized controlled trials. Diabetes Res Clin Pract 2012;98(3):349–60.

[138] Ammar T. Effects of aerobic exercise on blood pressure and lipids in overweight hypertensive postmenopausal women. J Exerc Rehabil 2015;11(3):145–50.

[139] Mota MR, Pardono E, Lima LC, Arsa G, Bottaro M, Campbell CS, et al. Effects of treadmill running and resistance exercises on lowering blood pressure during the daily work of hypertensive subjects. J Strength Cond Res 2009;23(8):2331–8.

[140] Brito AF, Alves NF, Araujo AS, Goncalves MC, Silva AS. Active intervals between sets of resistance exercises potentiate the magnitude of postexercise hypotension in elderly hypertensive women. J Strength Cond Res 2011;25(11):3129–36.

[141] Cano-Montoya J, Ramirez-Campillo R, Martinez C, Sade-Calles F, Salas-Parada A, Alvarez C. Interaction between antihypertensive therapy and exercise training therapy requires drug regulation in hypertensive patients. Rev Med Chil 2016;144(2):152–61.

[142] Cozza IC, Di Sacco TH, Mazon JH, Salgado MC, Dutra SG, Cesarino EJ, et al. Physical exercise improves cardiac autonomic modulation in hypertensive patients independently of angiotensin-converting enzyme inhibitor treatment. Hypertens Res 2012;35(1):82–7.

[143] Goldie CL, Brown CA, Hains SM, Parlow JL, Birtwhistle R. Synergistic effects of low-intensity exercise conditioning and beta-blockade on cardiovascular and autonomic adaptation in pre- and postmenopausal women with hypertension. Biol Res Nurs 2013;15(4):433–42.

[144] Dimeo F, Pagonas N, Seibert F, Arndt R, Zidek W, Westhoff TH. Aerobic exercise reduces blood pressure in resistant hypertension. Hypertension 2012;60(3):653–8.

[145] Motoyama M, Sunami Y, Kinoshita F, Kiyonaga A, Tanaka H, Shindo M, et al. Blood pressure lowering effect of low intensity aerobic training in elderly hypertensive patients. Med Sci Sports Exerc 1998;30(6):818–23.

[146] Cade R, Mars D, Wagemaker H, Zauner C, Packer D, Privette M, et al. Effect of aerobic exercise training on patients with systemic arterial hypertension. Am J Med 1984;77(5):785–90.

[147] Cononie CC, Graves JE, Pollock ML, Phillips MI, Sumners C, Hagberg JM. Effect of exercise training on blood pressure in 70- to 79-yr-old men and women. Med Sci Sports Exerc 1991;23(4):505–11.

[148] Maruf FA, Salako BL, Akinpelu AO. Can aerobic exercise complement antihypertensive drugs to achieve blood pressure control in individuals with essential hypertension? J Cardiovasc Med (Hagerstown) 2014;15(6):456–62.

[149] Wallace JP. Exercise in hypertension. A clinical review. Sports Med 2003;33(8):585–98.

[150] Beaulieu M, Nadeau A, Lacourciere Y, Cleroux J. Post-exercise reduction in blood pressure in hypertensive subjects: effects of angiotensin converting enzyme inhibition. Br J Clin Pharmacol 1993;36(4):331–8.

[151] Kelemen MH, Effron MB, Valenti SA, Stewart KJ. Exercise training combined with antihypertensive drug therapy. Effects on lipids, blood pressure, and left ventricular mass. JAMA 1990;263(20):2766–71.

[152] Maruf FA, Akinpelu AO, Salako BL, Akinyemi JO. Effects of aerobic dance training on blood pressure in individuals with uncontrolled hypertension on two antihypertensive drugs: a randomized clinical trial. J Am Soc Hypertens 2016;10(4):336–45.

[153] Queiroz AC, Sousa Jr. JC, Silva Jr ND, Tobaldini E, Ortega KC, De Oliveira EM, et al. Captopril does not potentiate post-exercise hypotension: a randomized crossover study. Int J Sports Med 2017;38(4):270–7. https://doi.org/10.1055/s-0042-123044.

[154] Radaelli A, Piepoli M, Adamopoulos S, Pipilis A, Clark SJ, Casadei B, et al. Effects of mild physical activity, atenolol and the combination on ambulatory blood pressure in hypertensive subjects. J Hypertens 1992;10(10):1279–82.

[155] Wilcox RG, Bennett T, Macdonald IA, Broughton-Pipkin F, Baylis PH. Post-exercise hypotension: the effects of epanolol or atenolol on some hormonal and cardiovascular variables in hypertensive men. Br J Clin Pharmacol 1987;24(2):151–62.

[156] Arena R, Daugherty J, Bond S, Lavie CJ, Phillips S, Borghi-Silva A. The combination of obesity and hypertension: a highly unfavorable phenotype requiring attention. Curr Opin Cardiol 2016;31(4):394–401.

[157] Chapman MJ, Sposito AC. Hypertension and dyslipidaemia in obesity and insulin resistance: pathophysiology, impact on atherosclerotic disease and pharmacotherapy. Pharmacol Ther 2008;117(3):354–73.

[158] Kannel WB. Risk stratification in hypertension: new insights from the Framingham Study. Am J Hypertens 2000;13(1 Pt 2):3S–10S.

[159] Ward BW, Schiller JS, Goodman RA. Multiple chronic conditions among US adults: a 2012 update. Prev Chronic Dis [Internet] 2014;11(130389). https://doi.org/10.5888/pcd11.130389.

[160] Donnelly JE, Blair SN, Jakicic JM, Manore MM, Rankin JW, Smith BK, et al. American College of Sports Medicine Position Stand. Appropriate physical activity intervention strategies for weight loss and prevention of weight regain for adults. Med Sci Sports Exerc 2009;41(2):459–71.

[161] Hurley BF, Hanson ED, Sheaff AK. Strength training as a countermeasure to aging muscle and chronic disease. Sports Med 2011;41(4):289–306.

[162] Hurley BF, Gillin AR. Can resistance training play a role in the prevention or treatment of hypertension? In: Pescatello LS, Pescatello LS, editors. Effects of exercise on hypertension: from cells to physiological systems. Switzerland: Springer International Publishing; 2015. p. 25–46 [chapter 2].

[163] Moraes MR, Bacurau RF, Ramalho JD, Reis FC, Casarini DE, Chagas JR, et al. Increase in kinins on post-exercise hypotension in normotensive and hypertensive volunteers. Biol Chem 2007;388(5):533–40.

[164] Braith RW, Stewart KJ. Resistance exercise training: its role in the prevention of cardiovascular disease. Circulation 2006;113(22):2642–50.

[165] Pescatello LS, Corso LML, MacDonald HV, Thompson PDT, Taylor BA, Panza GA, et al. Small sample sizes confound understanding of cardiometabolic responses to exercise. Exerc Sport Sci Rev 2017;45(3):173–80. https://doi.org/10.1249/JES.0000000000000115.

Chapter 10

Exercise and Cardiovascular Disease: Emphasis on Efficacy, Dosing, and Adverse Effects and Toxicity

Michael D. Morledge*, Sergey Kachur[†], Carl J. Lavie*, Parham Parto*, James H. O'Keefe[‡], Richard V. Milani*

*Department of Cardiovascular Diseases, John Ochsner Heart and Vascular Institute, Ochsner Clinical School-University of Queensland School of Medicine, New Orleans, LA, United States, [†]Department of Graduate Medical Education, Ocala Regional Medical Center, Ocala, FL, United States, [‡]St. Luke's Mid America Heart Institute, University of Missouri-Kansas City, Kansas City, MO, United States

INTRODUCTION

Daily physical activity (PA), regular exercise training (ET), and higher levels of cardiorespiratory fitness (CRF) have profound benefits on lowering cardiovascular (CV) disease (CVD) and all-cause morbidity and mortality [1–7]. PA levels do not meet national guidelines in 50%–80% of the population in the United States [8] and in a substantial percentage of the population throughout most of the world. Physical inactivity is currently one of the greatest threats to health in the United States and most of the Westernized world [3,5–7,9].

ET is increasingly prescribed by physicians as first-line therapy for hyperlipidemia (HLD) [10], hypertension (HTN) [11], type 2 diabetes mellitus (T2DM) [12], and CVD [13]. Similar to a drug, ET has indications and contraindications, as well as a dose-response range [14]. Evolving evidence suggests that high doses of ET beyond a threshold may provide similar or even less overall benefit compared with that provided by low doses of ET [15–20]. There is also evidence that high doses of ET may be associated with increased risk of coronary heart disease (CHD), atrial fibrillation (AF), and sudden cardiac death (SCD) [21–23].

In this chapter, we discuss the evidence for the tremendous benefits of ET, specifically running, for protection of various chronic diseases, including CVD and all-cause mortality. We make recommendations for ET dose range associated with maximal health benefits while diminishing the risk of toxicity. We also discuss the potential cardiotoxicity of high doses of ET.

EFFICACY

Data analysis from several large studies estimates that 80% of chronic disease [24] and specifically 80% of myocardial infarction (MI) [25] can be prevented with modification of lifestyle behaviors, specifically engaging in regular PA, adhering to healthy dietary principles, maintaining a healthy weight, and not smoking.

The positive effects of exercise on CV health have been established clearly over the past century [2,5,6,26–28]. Regular moderate doses of ET have been demonstrated to be effective for primary and secondary prevention of CVD in both men and women [2,5,6]. Considerable evidence suggests that physical inactivity may be the greatest current threat to health [3,5–7,9]. A minority of adults in the United States and much of Western civilization are meeting minimal PA recommendations per national guidelines [8]. There has been a progressive decline in occupational and household PA over the past half century. This trend has contributed significantly to the increase in obesity, chronic disease, and adverse CVD outcomes [7,29–32].

Both individual studies and metaanalyses have documented a strong inverse relationship between PA levels and all-cause as well as CVD mortality [3,5,28,33–35]. Low levels of PA are associated with higher prevalence of most CVD risk factors, including HTN, obesity, DLD, metabolic syndrome (MetS), T2DM, and depression [3,5,36–39]. High levels of PA and ET are associated with reductions in morbidity and mortality beyond the effects of traditional risk factors. This suggests that other factors, including autonomic function, preconditioning, and endothelial function may be significant contributors [1,3,5,6].

Lifestyle in Heart Health and Disease. https://doi.org/10.1016/B978-0-12-811279-3.00010-0

Amounts and intensities of PA are the major determinants of CRF, which may be one of the strongest modulators of CVD risk [2–5]. Preserved levels of CRF are associated with favorable prognosis in patients with chronic diseases, including HLD, HTN, and T2DM [3–5]. Patients with these disorders but with higher levels of CRF have a considerably better prognosis than patients without these disorders but with lower levels of CRF [3–5].

DOSING

The most widely accepted, well-established chronic disease practice guidelines uniformly call for lifestyle changes, including ET, as first-line therapy [10–13]. US guidelines for ET recommend at least 150 min/week of moderate PA, 75 min/week of vigorous PA, or an equivalent of a combination of both [3,40]. Evolving evidence suggests that the considerable benefits of ET may be attained at levels well below those suggested by national guidelines. Running, for example, is a popular and practical form of ET from which much of this evidence is based [16,17].

Recent running studies have demonstrated a U-shaped relationship between running doses and reduction in CVD and all-cause mortality. These prospective observational studies include the Copenhagen City Heart Study (CCHS) [18,19], the National Runners' and Walkers' Health Studies (NRWHS) [41], and the Aerobics Center Longitudinal Study (ACLS) [15–17].

The CCHS followed 1,878 runners and 16,827 nonrunners for up to 35 years [18,19]. The runners had a 44% reduction in all-cause mortality compared with nonrunners, with an average increase in life expectancy of 6.2 years in men and 5.6 years in women [18]. A recent analysis of this cohort showed maximal reduction in all-cause mortality at low doses of running with slow-to-moderate running speeds, frequency of 2–3 times per week, and 1–2.4 h/week of running [19]. Very high doses of running were associated with trends of worse survival compared with nonrunners and low- and moderate-dose runners [19] (Fig. 1).

Data from the NRWHS of 2377 patients with CHD with a history of MI also showed progressive reduction in CVD and all-cause mortality with increasing ET, up to a point [41]. High ET doses of >30 miles/week running or >46 miles/week walking were no longer associated with a mortality benefit [41]. This was equivalent to >7.2 metabolic equivalent of task-h/day (MET-h/day) [41] (Figs. 2 and 3).

Data from the ACLS of 13,016 runners and 42,121 nonrunners, with 3413 deaths, followed for an average of 15 years, have been reported recently [15–17]. The runners had a 30% reduction in all-cause mortality and 45% reduction in CVD mortality compared with nonrunners, with an average increase in life expectancy of 3 years [17]. Persistent runners had the greatest reduction in mortality, whereas those who started running but stopped or those who were not running at baseline but subsequently started running had about half of the reduction in mortality [17] (Fig. 4). Maximal reduction in CVD and all-cause mortality occurred at low doses of running at less than 6 miles/week, less than 52 min of running per week, and running 1–2 times per week, equivalent to <506 MET-min/week [17]. Similar mortality reduction occurred at increasing doses of running. High doses of running at >20 miles/week, >176 min/week, and >6 times per week, equivalent to >1840 MET-min/week, were associated with declining mortality benefits [17] (Fig. 5).

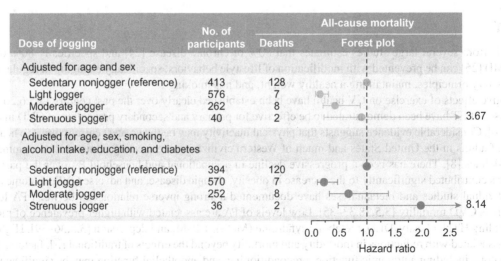

FIG. 1 Forest plot indicating all-cause mortality in light, moderate, and strenuous runners compared with sedentary nonrunners. *Reproduced with permission from Schnohr P, O'Keefe JH, Marott JL, Lange P, Jensen GB. Dose of jogging and long-term mortality: the Copenhagen City Heart Study. J Am Coll Cardiol. 2015;65(5):411–9.*

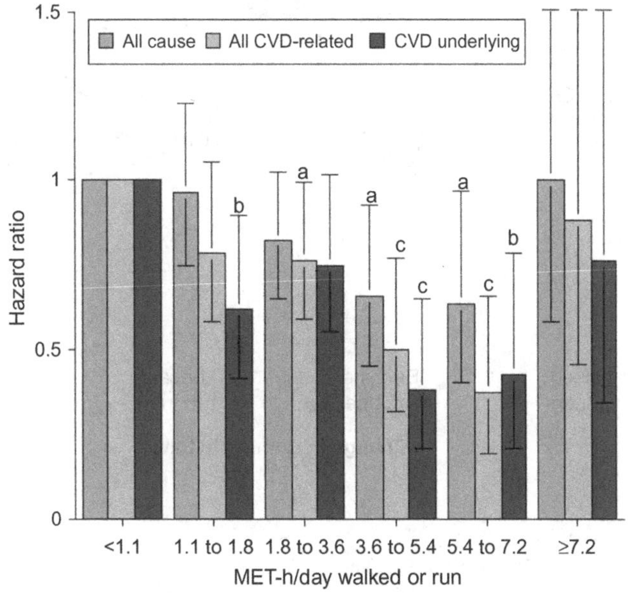

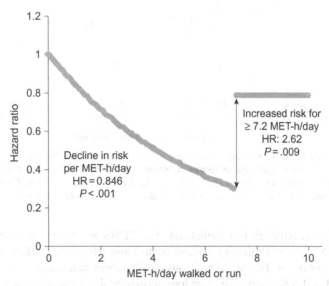

FIG. 2 Categorical model. Cox proportional survival analyses of the risk of CVD-related mortality versus MET-hours per day run or walked. *From Williams PT, Thompson PD. Increased cardiovascular disease mortality associated with excessive exercise in heart attack survivors. Mayo Clin Proc 2014;89(9):1187–94.*

FIG. 3 Continuous model. Cox proportional survival analyses of the risk of CVD-related mortality versus MET-hours per day run or walked. *From Williams PT, Thompson PD. Increased cardiovascular disease mortality associated with excessive exercise in heart attack survivors. Mayo Clin Proc 2014;89(9):1187–94.*

ACLS data showed that runners have higher levels of CRF than nonrunners and that CRF levels in runners progressively increased with increased doses of running [17] (Fig. 6). CRF is a strong morbidity and mortality predictor, and most evidence indicates better survival with metabolic equivalent (MET) levels greater than 10 [42–44], with some evidence of progressively better survival with even higher CRF levels [45]. However, the ACLS data indicate that among runners who already have high levels of CRF even when running at lower volumes, low doses of running provide maximal reduction in CVD and all-cause mortality.

More recently the ACLS examined those within the highest quintile of runners [15]. This group had lower CVD and all-cause mortality benefits than lower-dose runners. Furthermore, the highest tertile of these (the top 7% of all runners) had a mortality and CVD risk that was comparable to that of the non-runner subgroup and was significantly higher than the lower-dose runners [15] (Fig. 7). This loss of benefit occurred at doses of running at >31 miles/week, >270 min/week, and >7 times per week, equivalent to >2944 MET-min/week [15]. Although compelling, these data do not demonstrate causation, and are derived from a small sample size (*N* < 1000) with wide confidence intervals.

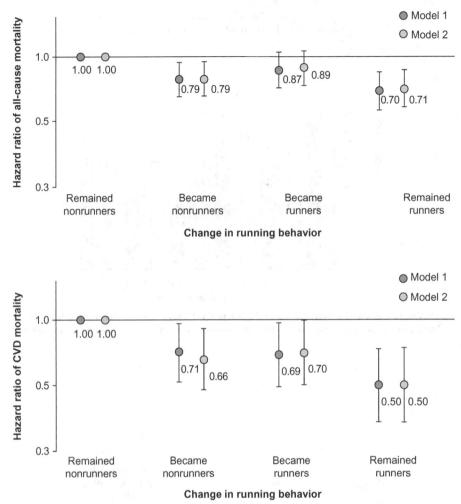

FIG. 4 Model 1 was adjusted for baseline age (years), sex, examination year, and interval between the baseline and last examinations (years). Model 2 was adjusted for model 1 plus baseline smoking status (never, former, or current), alcohol consumption (heavy drinker or not), other physical activities except running (0, 1–499, or ≥500 MET-min/week), and parental cardiovascular disease (CVD; yes or no). *From Lee DC, Pate RR, Lavie CJ, Sui X, Church TS, Blair SN. Leisure-time running reduces all-cause and cardiovascular mortality risk. J Am Coll Cardiol 2014;64(5):472–81.*

Similarly, both the CCHS and NRWHS had limitations. The CCHS had relatively small numbers of deaths and low statistical power. The NRWHS had no nonrunning inactive control group but used inadequate exercisers of <1.07 MET-h/day as the reference group for analyses. The ACLS had considerably larger number of runners and deaths.

The CCHS, NRWHS, and ACLS all show consistent long-term mortality benefits of running. Low-dose running generally provides maximal mortality benefits in all three of the studies, and high-dose running provides decreased mortality benefits. Further studies are needed to determine whether there is an optimum upper limit of running beyond which additional running produces adverse health effects.

Current guidelines recommend a minimum of 150 min/week of moderate PA or 75 min/week of vigorous PA, such as running, for health benefits [3,40]. The CCHS, NRWHS, and ACLS demonstrate that running, even at relatively low doses (5–10 min/day), below the current minimum guidelines for vigorous PA, is sufficient for substantial mortality benefits. These studies and other analyses [20,46] also support maximal benefits of running at 40 min/day or less. This suggests that the therapeutic dose range for running is likely 5–40 min/day and most likely at the lower ends of this range.

In general, running for 5 min may equal to approximately 15 min of walking, and running for 25 min may equal to approximately 105 min of walking, suggesting that 3–4 times the duration of walking is needed to achieve the same benefits as of running [46] (Fig. 8).

These findings have clinical and public health importance, because time is one of the greatest barriers to exercise. Lower time requirements have the potential to motivate a larger number of people to adopt running as a means to gain health benefits. Initiating daily vigorous PA, such as running, may be difficult for the majority of the population who are

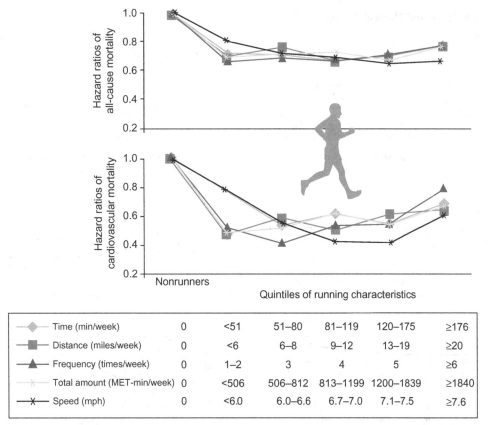

FIG. 5 Hazard ratios (HRs) of all-cause and cardiovascular disease mortality stratified by running characteristics (weekly running time, distance, frequency, total amount, and speed). Participants were classified into six groups: nonrunners (reference group) and five quintiles of each running characteristic. All HRs were adjusted for baseline age (years), sex, examination year, smoking status (never, former, or current), alcohol consumption (heavy drinker or not), other physical activities except running (0, 1–499, or ≥500 MET-min/week), and parental history of cardiovascular disease (yes or no). All *P* values for HRs across running characteristics were less than .05 for all-cause and cardiovascular mortality except for running frequency of 6 times/week or more (*P* = .11) and speed of less than 6.0 mph (P1 = .10) for cardiovascular mortality. MET, metabolic equivalent. *Reproduced with permission from Lee DC, Pate RR, Lavie CJ, Sui X, Church TS, Blair SN. Leisure-time running reduces all-cause and cardiovascular mortality risk. J Am Coll Cardiol. 2014;64(5):472–81.*

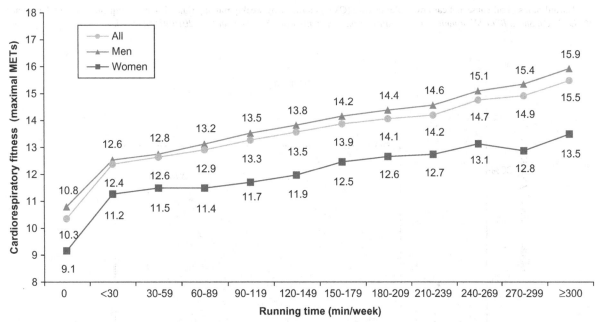

FIG. 6 Cardiorespiratory fitness level was estimated from the final treadmill speed and grade during the maximal exercise test in a subsample of 50,995 participants. All *P* values for a linear trend across weekly running time were less than .001 after adjustment for age and sex (not in sex-stratified analyses). MET, metabolic equivalent. *Reproduced with permission from Lee DC, Pate RR, Lavie CJ, Sui X, Church TS, Blair SN. Leisure-time running reduces all-cause and cardiovascular mortality risk. J Am Coll Cardiol. 2014;64(5):472–81.*

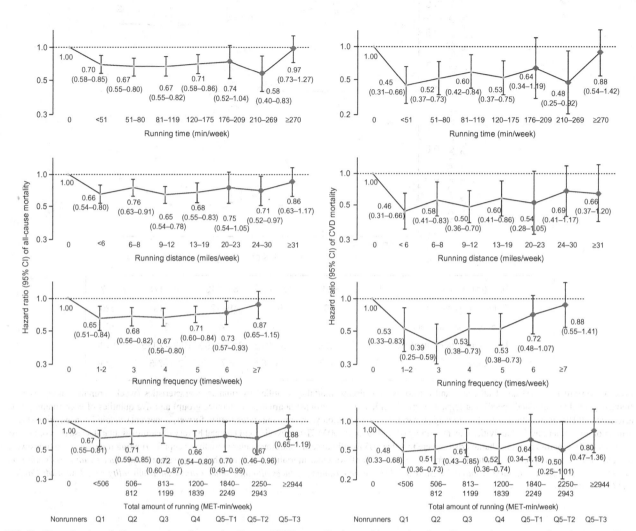

FIG. 7   Hazard ratios of all-cause and cardiovascular disease (CVD) mortality by weekly running time, distance, frequency, and total amount. *From Lee DC, Lavie CJ, Sui X, Blair SN. Running and mortality: is more actually worse? Mayo Clin Proc 2016;91(4):534–6.*

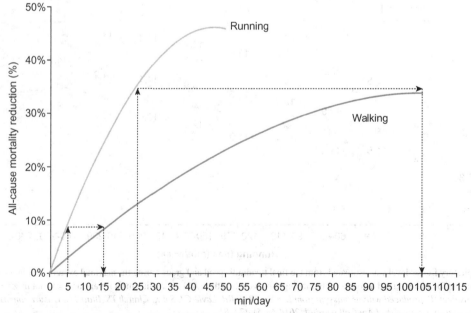

FIG. 8   A 5 min run generates the same benefits as a 15 min walk, and a 25 min run is equivalent to a 105 min walk. *Reproduced with permission from Wen CP, Wai JP, Tsai MK, Chen CH. Minimal amount of exercise to prolong life: to walk, to run, or just mix it up? J Am Coll Cardiol. 2014;64(5):482–4.*

inactive, and they may find it useful to begin with low-intensity PA, such as walking, and then transition to more vigorous PA. Implementing a gradual transition to vigorous PA has the dual advantage of removing barriers to initiation and reducing injury risk. To these patients, we should emphasize "even a little is great" [15].

ADVERSE EFFECTS/TOXICITY

Higher levels of PA/ET are associated with numerous health benefits in comparison to a sedentary lifestyle including protection against CVD and all-cause mortality. However, high doses of running and other forms of excessive endurance exercise (EEE) carry their own cardiovascular risks. Marathons, ultramarathons, ironman-distance triathlons, and very-long-distance bicycle races are all associated with levels well above 40 min/day of vigorous PA. This kind of ET has been associated with potential cardiotoxicity, including adverse effects on cardiac structure and function and increased risks of CHD, AF, and SCD [21–23].

Effects on CV Structure and Function

There are many potential adverse effects of EEE on cardiac structure and function [21–23] (Fig. 9). In animal studies, high levels of ET have resulted in atrial and ventricular enlargement, fibrosis, and propensity for high-grade ventricular arrhythmias (VA) [47] (Fig. 10). These adverse structural changes and VA largely reverse following detraining [47].

Similar adverse structural remodeling following EEE has been noted in humans and is most evident right side of the heart [14,23,48–51]. Serological markers of cardiac damage, including troponin, creatine kinase-myocardial band, and B-type natriuretic peptide, have been documented to increase in up to 50% of participants during and after marathon running [52,53] (Fig. 11). Approximately a third of marathon runners experience marked increases in atrial size and dilation of the ventricular chambers, most commonly characterized by dilation of the right ventricle (RV) and dysfunction of the interventricular septum [52] (Figs. 12 and 13). These abnormalities appear to largely resolve within the first week after the marathon and without obvious permanent adverse effects on most marathon participants. However, in a significant minority of athletes, long-term continuous exposure to EEE may lead to the development of chronic structural changes, including chronic dilation of the RV and right atrium with patchy myocardial scarring [52] (Fig. 14). These abnormalities are often asymptomatic and probably accrue over many years but may predispose to potentially serious arrhythmias such as AF and/or VA [14,23,48–51].

Impact of EEE on CHD

Recent studies suggest that long-distance runners may have increased levels of atherosclerosis and CHD as measured by higher levels of coronary artery calcium and more plaque on intravascular ultrasound [48,54] (Fig. 15). The mechanism

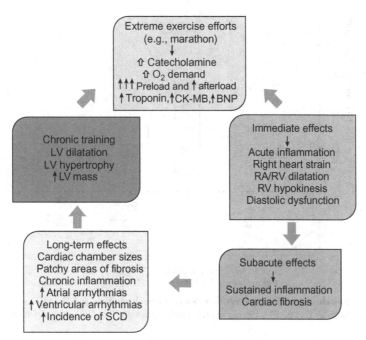

FIG. 9 Proposed pathogenesis of cardiomyopathy in endurance athletes. *From O'Keefe JH, Patil HR, Lavie CJ, Magalski A, Vogel RA, McCullough PA. Potential adverse cardiovascular effects from excessive endurance exercise. Mayo Clin Proc 2012;87(6):587–95.*

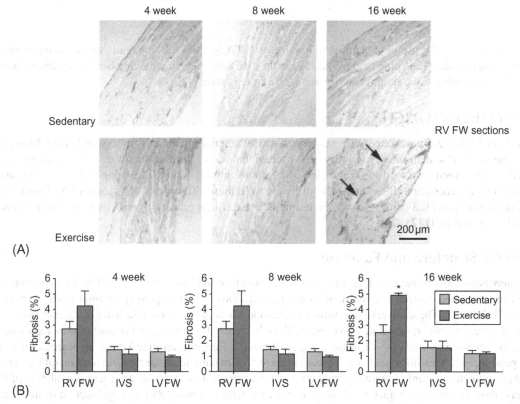

FIG. 10 (A) Picrosirius-stained photomicrographs of RV sections. By 16 weeks, the RVs of exercising rats show widespread interstitial collagen deposition with disarray of myocardial architecture (arrows). *From Benito B, Gay-Jordi G, Serrano-Mollar A, et al. Cardiac arrhythmogenic remodeling in a rat model of long-term intensive exercise training. Circulation 2011;123(1):13–22.*

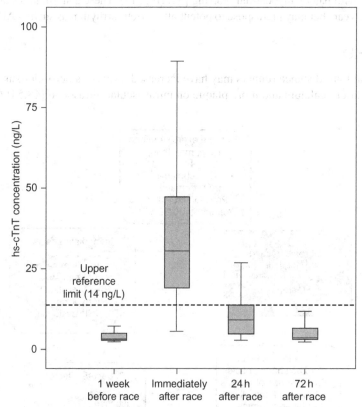

FIG. 11 High-sensitivity cardiac troponin T (hs-cTnT) concentrations before, immediately after, and 24 and 72 h after marathon race. *Reproduced with permission from Scherr J, Braun S, Schuster T, et al. 72-h kinetics of high-sensitive troponin T and inflammatory markers after marathon. Med Sci Sports Exerc 2011;43(10):1819–27.*

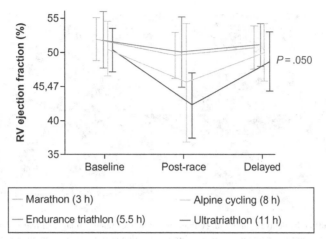

FIG. 12 Duration-dependent effect of endurance events on right ventricular (RV) ejection fraction. *Reproduced with permission from La Gerche A, Burns AT, Mooney DJ, et al. Exercise-induced right ventricular dysfunction and structural remodelling in endurance athletes. Eur Heart J 2012;33(8):998–1006.*

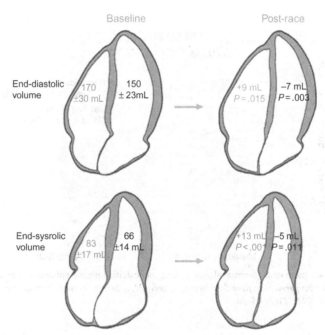

FIG. 13 Differential effect of prolonged intense exercise on RV and LV volumes. Baseline volumes are shown on the left, and the changes in volume after the race are shown on the right. RV volumes increased in the postrace setting, whereas LV volumes decreased, resulting in a decrease in RVEF but not in LVEF. *Reproduced with permission from La Gerche A, Burns AT, Mooney DJ, et al. Exercise-induced right ventricular dysfunction and structural remodelling in endurance athletes. Eur Heart J 2012;33(8):998–1006.*

for increased severity of coronary atherosclerosis with EEE appears to be independent of traditional CHD risk factors, as runners have better overall CHD risk profiles [21]. Increased CHD risk as a result of very-high-dose running is supported by the analysis of data from the NRWS [41] and ACLS [16] running studies, where high-dose running groups appeared to lose most of the protection against CVD mortality noted at more moderate running doses (Fig. 16). The ACLS study population excluded individuals with previous MI or other CVD, and the NRWS study population consisted of individuals with CHD with a history of MI, suggesting that high-dose ET may not be ideal for individuals with or without history of CVD.

Impact of EEE on Risk of AF

Many studies and meta-analyses have reported a U-shaped relationship between PA/ET and risk of AF [14,55–58] (Fig. 17). Risk of AF appears to be reduced in those with low-moderate ET and increased with high ET as compared with sedentary

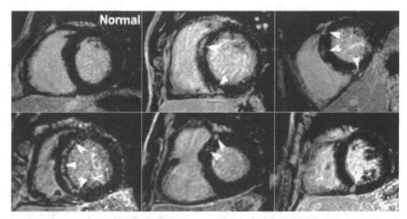

FIG. 14 Myocardial scarring as detected by delayed gadolinium enhancement in five athletes. Images of five athletes in whom focal delayed gadolinium enhancement was identified in the interventricular septum (arrows), compared with a normal study in an athlete (top left). *Reproduced with permission from La Gerche A, Burns AT, Mooney DJ, et al. Exercise-induced right ventricular dysfunction and structural remodelling in endurance athletes. Eur Heart J 2012;33(8):998–1006.*

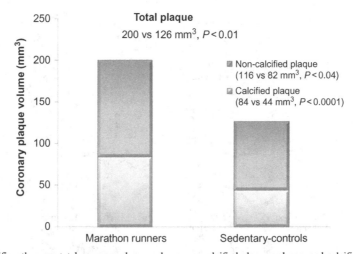

FIG. 15 Marathoners had significantly more total coronary plaque volume, noncalcified plaque volume, and calcified plaque volume compared with those of control subjects. *Reproduced with permission from Schwartz R, Kraus SM, Schwartz J, et al. Increased coronary artery plaque volume among male marathon runners. Missouri Med 2014;111(2):85–90.*

controls. However, not all studies have shown strong statistically significant association [59]. Potential mechanisms underlying AF development in chronic EEE include increased vagal and sympathetic tone, bradycardia, inflammatory changes, atrial wall fibrosis, and increased atrial size [21]. Left atrial enlargement has been found in up to 20% of competitive athletes, and this may be a predictor for AF. [60,61].

Risk of SCD with EEE

Perhaps, the most serious consequence of EEE is SCD. Over the past 35 years, the number of Americans participating in a marathon has risen 25-fold [14,23]; in 2015, over a half-million runners completed a marathon in the United States [62] (Fig. 18). A recent study reviewed all marathons and half marathons in the United States during a 10-year period and reported 59 cardiac arrests in 10.9 million runners, a rate of 0.54/100,000 participants [63]. The true occurrence may be two- to fourfold higher than this as half-marathon runners were included in the study and SCD was accounted for during the marathon but not in the postrace period [48,64,65]. Nevertheless, the overall mortality in marathons appears to be relatively low. The causes of SCD during or after EEE in individuals younger than 30 years old most commonly includes genetic causes: hypertrophic cardiomyopathy, anomalous coronary arteries, dilated cardiomyopathy, and congenital long QT syndrome. In athletes older than 30 years, CHD, acute MI [66] and ischemia [67–73] are the predominant causes of exercise-related SCD [23]. The high-catecholamine state of EEE competitions superimposed on preexisting ET-induced structural myocardial abnormalities is the most logical explanation for cases of SCD after common causes are excluded [21].

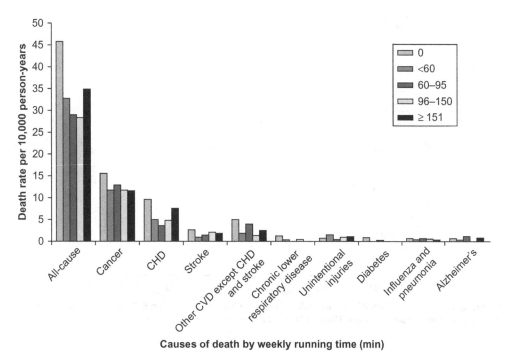

FIG. 16 Participants were classified into five groups: nonrunners and four quartiles of average weekly running time (in min). Death rates were adjusted for baseline age, sex, and examination year, and causes of death were divided into all-cause, cancer, CHD, stroke, other CVDs, chronic respiratory disease, unintentional injuries, diabetes, influenza, and pneumonia, and Alzheimer's disease. CHD, coronary heart disease; CVD, cardiovascular disease. *Reproduced with permission from Lee DC, Lavie CJ, Vedanthan R. Optimal dose of running for longevity: is more better or worse? J Am Coll Cardiol. 2015;65(5):420–2.*

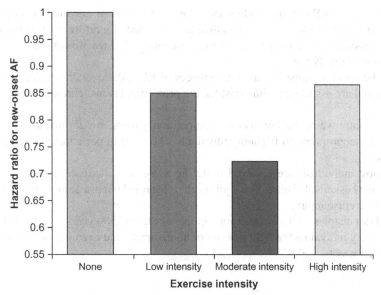

FIG. 17 Risk of new-onset atrial fibrillation among 5446 older adults (>65 years) as a function of exercise intensity. *Reproduced with permission from O'Keefe JH, Franklin B, Lavie CJ. Exercising for health and longevity vs peak performance: different regimens for different goals. Mayo Clin Proc 2014;89(9):1171–5.*

EEE Recommendations

The risks of EEE appear to be relatively low, and athletes should not be discouraged from participation in sporting events [21–23,74,75]. Most participants in these events are quite healthy overall with high levels of CRF. Specific guidelines and recommendations do not exist for athletes older than 50 years of age with known CHD and CVD risk factors or prior MIs. Coronary artery calcium scanning, exercise stress testing, low-dose aspirin therapy, and statin therapy may be recommended in this population [7,76].

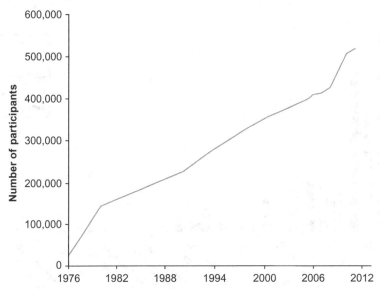

FIG. 18   Marathon running trends in the United States. *Reproduced with permission from Schwartz R, Kraus SM, Schwartz J, et al. Increased coronary artery plaque volume among male marathon runners. Missouri Med 2014;111(2):85–90.*

CONCLUSIONS

From a population perspective, lack of PA is much more prevalent than is EEE. Most US adults do not meet minimum ET recommendations per US guidelines, and we have made a call to action that all clinicians should promote PA throughout the health-care system.

The CCHS, NRWHS, and ACLS running studies and other analyses show maximal benefits of running occur at low doses, certainly maximizing at 40 min/day or well less than this cut point. Therefore, EEE is not needed to maximize protection against CVD and all-cause mortality. EEE at levels of running well above 40 min/day is associated with some risks, although these risks appear relatively low.

Although there may be some negative health consequences of EEE, the overall benefits of running far outweigh the risk for most individuals and are associated with considerable protection against chronic diseases and CVD and all-cause mortality.

People run not only to improve health but also for competition, fitness, weight management, stress relief, socialization, and fun. It would be inappropriate to frighten individuals who want to participate in marathons, triathlons, or even ultraendurance races [74].

Typical sports that most individuals are involved in during adolescence and some professionally through adulthood include football, soccer, and basketball. These do not all involve sustained intense aerobic PA, which is the main pattern of concern regarding cardiac overuse injury.

We would like to end this chapter with a quote from Hippocrates from 2500 years ago that still seems prudent and wise today. "If we could give every individual the right amount of nourishment and exercise, not too little and not too much, we would have found the safest way to health."

REFERENCES

[1] DeFina LF, Haskell WL, Willis BL, et al. Physical activity versus cardiorespiratory fitness: two (partly) distinct components of cardiovascular health? Prog Cardiovasc Dis 2015;57(4):324–9.

[2] Franklin BA, Lavie CJ, Squires RW, Milani RV. Exercise-based cardiac rehabilitation and improvements in cardiorespiratory fitness: implications regarding patient benefit. Mayo Clin Proc 2013;88(5):431–7.

[3] Lavie CJ, Arena R, Swift DL, et al. Exercise and the cardiovascular system: clinical science and cardiovascular outcomes. Circ Res 2015;117(2):207–19.

[4] Myers J, McAuley P, Lavie CJ, Despres JP, Arena R, Kokkinos P. Physical activity and cardiorespiratory fitness as major markers of cardiovascular risk: their independent and interwoven importance to health status. Prog Cardiovasc Dis 2015;57(4):306–14.

[5] Swift DL, Lavie CJ, Johannsen NM, et al. Physical activity, cardiorespiratory fitness, and exercise training in primary and secondary coronary prevention. Circ J 2013;77(2):281–92.

[6] Vuori IM, Lavie CJ, Blair SN. Physical activity promotion in the health care system. Mayo Clin Proc 2013;88(12):1446–61.

[7] Lavie CJ, Lee DC, Sui X, et al. Effects of running on chronic diseases and cardiovascular and all-cause mortality. Mayo Clin Proc 2015;90(11):1541–52.

[8] Ward BW, Clarke TC, Nugent CN, Schiller J. Early release of selected estimates based on data from the 2015 National Health Interview Survey. National Center for Health Statistics; 2016.

[9] Blair SN. Physical inactivity: the biggest public health problem of the 21st century. Br J Sports Med 2009;43(1):1–2.

[10] Stone NJ, Robinson JG, Lichtenstein AH, et al. 2013 ACC/AHA guideline on the treatment of blood cholesterol to reduce atherosclerotic cardiovascular risk in adults: a report of the American College of Cardiology/American Heart Association Task Force on Practice Guidelines. Circulation 2014;129(25 Suppl 2):S1–45.

[11] James PA, Oparil S, Carter BL, et al. 2014 evidence-based guideline for the management of high blood pressure in adults: report from the panel members appointed to the Eighth Joint National Committee (JNC 8). JAMA 2014;311(5):507–20.

[12] Garber AJ, Abrahamson MJ, Barzilay JI, et al. Consensus statement by the American Association of Clinical Endocrinologists and American College of Endocrinology on the comprehensive type 2 diabetes management algorithm—2016 executive summary. Endocr Pract 2016;22(1):84–113.

[13] Lavie CJ, Arena R, Franklin BA. Cardiac rehabilitation and healthy life-style interventions: rectifying program deficiencies to improve patient outcomes. J Am Coll Cardiol 2016;67(1):13–5.

[14] O'Keefe JH, Franklin B, Lavie CJ. Exercising for health and longevity vs peak performance: different regimens for different goals. Mayo Clin Proc 2014;89(9):1171–5.

[15] Lee DC, Lavie CJ, Sui X, Blair SN. Running and mortality: is more actually worse? Mayo Clin Proc 2016;91(4):534–6.

[16] Lee DC, Lavie CJ, Vedanthan R. Optimal dose of running for longevity: is more better or worse? J Am Coll Cardiol 2015;65(5):420–2.

[17] Lee DC, Pate RR, Lavie CJ, Sui X, Church TS, Blair SN. Leisure-time running reduces all-cause and cardiovascular mortality risk. J Am Coll Cardiol 2014;64(5):472–81.

[18] Schnohr P, Marott JL, Lange P, Jensen GB. Longevity in male and female joggers: the Copenhagen City Heart Study. Am J Epidemiol 2013;177(7):683–9.

[19] Schnohr P, O'Keefe JH, Marott JL, Lange P, Jensen GB. Dose of jogging and long-term mortality: the Copenhagen City Heart Study. J Am Coll Cardiol 2015;65(5):411–9.

[20] Wen CP, Wai JPM, Tsai MK, et al. Minimum amount of physical activity for reduced mortality and extended life expectancy: a prospective cohort study. Lancet 2011;378(9798):1244–53.

[21] Lavie CJ, O'Keefe JH, Sallis RE. Exercise and the heart—the harm of too little and too much. Curr Sports Med Rep 2015;14(2):104–9.

[22] O'Keefe JH, Lavie CJ, Guazzi M. Part 1: potential dangers of extreme endurance exercise: how much is too much? Part 2: screening of school-age athletes. Prog Cardiovasc Dis 2015;57(4):396–405.

[23] O'Keefe JH, Patil HR, Lavie CJ, Magalski A, Vogel RA, McCullough PA. Potential adverse cardiovascular effects from excessive endurance exercise. Mayo Clin Proc 2012;87(6):587–95.

[24] Ford ES, Bergmann MM, Kroger J, Schienkiewitz A, Weikert C, Boeing H. Healthy living is the best revenge: findings from the European Prospective Investigation Into Cancer and Nutrition-Potsdam study. Arch Intern Med 2009;169(15):1355–62.

[25] Yusuf S, Hawken S, Ounpuu S, et al. Effect of potentially modifiable risk factors associated with myocardial infarction in 52 countries (the INTERHEART study): case-control study. Lancet (London, England) 2004;364(9438):937–52.

[26] Morris JN, Heady JA, Raffle PA, Roberts CG, Parks JW. Coronary heart-disease and physical activity of work. Lancet (London, England) 1953;265(6795):1053–7.

[27] Morris JN, Heady JA, Raffle PA, Roberts CG, Parks JW. Coronary heart-disease and physical activity of work. Lancet (London, England) 1953;265(6796):1111–20.

[28] Paffenbarger Jr RS, Hyde RT, Wing AL, Hsieh CC. Physical activity, all-cause mortality, and longevity of college alumni. N Engl J Med 1986;314(10):605–13.

[29] Archer E. The childhood obesity epidemic as a result of nongenetic evolution: the maternal resources hypothesis. Mayo Clin Proc 2015;90(1):77–92.

[30] Archer E, Lavie CJ, McDonald SM, et al. Maternal inactivity: 45-year trends in mothers' use of time. Mayo Clin Proc 2013;88(12):1368–77.

[31] Archer E, Shook RP, Thomas DM, et al. 45-Year trends in women's use of time and household management energy expenditure. PLoS ONE 2013;8(2). e56620.

[32] Church TS, Thomas DM, Tudor-Locke C, et al. Trends over 5 decades in U.S. occupation-related physical activity and their associations with obesity. PLoS ONE 2011;6(5):e19657.

[33] Barengo NC, Hu G, Lakka TA, Pekkarinen H, Nissinen A, Tuomilehto J. Low physical activity as a predictor for total and cardiovascular disease mortality in middle-aged men and women in Finland. Eur Heart J 2004;25(24):2204–11.

[34] Hu FB, Willett WC, Li T, Stampfer MJ, Colditz GA, Manson JE. Adiposity as compared with physical activity in predicting mortality among women. N Engl J Med 2004;351(26):2694–703.

[35] Hu G, Eriksson J, Barengo NC, et al. Occupational, commuting, and leisure-time physical activity in relation to total and cardiovascular mortality among Finnish subjects with type 2 diabetes. Circulation 2004;110(6):666–73.

[36] Hu G, Barengo NC, Tuomilehto J, Lakka TA, Nissinen A, Jousilahti P. Relationship of physical activity and body mass index to the risk of hypertension: a prospective study in Finland. Hypertension (Dallas, TX: 1979) 2004;43(1):25–30.

[37] Kriska AM, Saremi A, Hanson RL, et al. Physical activity, obesity, and the incidence of type 2 diabetes in a high-risk population. Am J Epidemiol 2003;158(7):669–75.

[38] Lahti-Koski M, Pietinen P, Heliovaara M, Vartiainen E. Associations of body mass index and obesity with physical activity, food choices, alcohol intake, and smoking in the 1982–1997 FINRISK studies. Am J Clin Nutr 2002;75(5):809–17.

[39] Rennie KL, McCarthy N, Yazdgerdi S, Marmot M, Brunner E. Association of the metabolic syndrome with both vigorous and moderate physical activity. Int J Epidemiol 2003;32(4):600–6.

[40] US Department of Health and Human Services. 2008 physical activity guidelines for Americans. Washington, DC: US Department of Health and Human Services; 2008.

[41] Williams PT, Thompson PD. Increased cardiovascular disease mortality associated with excessive exercise in heart attack survivors. Mayo Clin Proc 2014;89(9):1187–94.

[42] Abudiab M, Aijaz B, Konecny T, et al. Use of functional aerobic capacity based on stress testing to predict outcomes in normal, overweight, and obese patients. Mayo Clin Proc 2013;88(12):1427–34.

[43] Barnes JN, Joyner MJ. Physical activity and cardiovascular risk: 10 metabolic equivalents or bust. Mayo Clin Proc 2013;88(12):1353–5.

[44] Kodama S, Saito K, Tanaka S, et al. Cardiorespiratory fitness as a quantitative predictor of all-cause mortality and cardiovascular events in healthy men and women: a meta-analysis. JAMA 2009;301(19):2024–35.

[45] Feldman DI, Al-Mallah MH, Keteyian SJ, et al. No evidence of an upper threshold for mortality benefit at high levels of cardiorespiratory fitness. J Am Coll Cardiol 2015;65(6):629–30.

[46] Wen CP, Wai JP, Tsai MK, Chen CH. Minimal amount of exercise to prolong life: to walk, to run, or just mix it up? J Am Coll Cardiol 2014;64(5):482–4.

[47] Benito B, Gay-Jordi G, Serrano-Mollar A, et al. Cardiac arrhythmogenic remodeling in a rat model of long-term intensive exercise training. Circulation 2011;123(1):13–22.

[48] McCullough P, Lavie C. Coronary artery plaque and cardiotoxicity as a result of extreme endurance exercise. Mo Med 2014;111:91–4.

[49] O'Keefe JH, Lavie CJ. Run for your life ... at a comfortable speed and not too far. Heart 2013;99(8):516–9.

[50] O'Keefe JH, Schnohr P, Lavie CJ. The dose of running that best confers longevity. Heart 2013;99(8):588–90.

[51] Patil HR, O'Keefe JH, Lavie CJ, Magalski A, Vogel RA, McCullough PA. Cardiovascular damage resulting from chronic excessive endurance exercise. Mo Med 2012;109(4):312–21.

[52] La Gerche A, Burns AT, Mooney DJ, et al. Exercise-induced right ventricular dysfunction and structural remodelling in endurance athletes. Eur Heart J 2012;33(8):998–1006.

[53] Scherr J, Braun S, Schuster T, et al. 72-h kinetics of high-sensitive troponin T and inflammatory markers after marathon. Med Sci Sports Exerc 2011;43(10):1819–27.

[54] Schwartz R, Kraus SM, Schwartz J, et al. Increased coronary artery plaque volume among male marathon runners. Missouri Med 2014;111(2):85–90.

[55] Abdulla J, Nielsen JR. Is the risk of atrial fibrillation higher in athletes than in the general population? A systematic review and meta-analysis. Europace: Eur Pacing Arrhythmias Cardiac Electrophysiol: J Work Groups Cardiac Pacing Arrhythmias Cardiac Cell Electrophysiol Eur Soc Cardiol 2009;11(9):1156–9.

[56] Menezes AR, Lavie CJ, Dinicolantonio JJ, et al. Cardiometabolic risk factors and atrial fibrillation. Rev Cardiovasc Med 2013;14(2-4):e73–81.

[57] Menezes AR, Lavie CJ, DiNicolantonio JJ, et al. Atrial fibrillation in the 21st century: a current understanding of risk factors and primary prevention strategies. In: Paper presented at: Mayo Clinic Proceedings 2013; 2013.

[58] Mozaffarian D, Furberg CD, Psaty BM, Siscovick D. Physical activity and incidence of atrial fibrillation in older adults: the cardiovascular health study. Circulation 2008;118(8):800–7.

[59] Ofman P, Khawaja O, Rahilly-Tierney CR, et al. Regular physical activity and risk of atrial fibrillation: a systematic review and meta-analysis. Circulation Arrhythmia Electrophysiol 2013;6(2):252–6.

[60] Kirchhof P, Lip GY, Van Gelder IC, et al. Comprehensive risk reduction in patients with atrial fibrillation: emerging diagnostic and therapeutic options—a report from the 3rd Atrial Fibrillation Competence NETwork/European Heart Rhythm Association consensus conference. Europace: Eur Pacing Arrhythmias Cardiac Electrophysiol: J Work Groups Cardiac Pacing Arrhythmias Cardiac Cell Electrophysiol Eur Soc Cardiol 2012;14(1):8–27.

[61] Mont L, Sambola A, Brugada J, et al. Long-lasting sport practice and lone atrial fibrillation. Eur Heart J 2002;23(6):477–82.

[62] Running USA. 2015 Running USA Annual Marathon Report. 2016; http://www.runningusa.org/marathon-report-2016 [Accessed 16 January 2017].

[63] Kim JH, Malhotra R, Chiampas G, et al. Cardiac arrest during long-distance running races. N Engl J Med 2012;366(2):130–40.

[64] Redelmeier DA, Greenwald JA. Competing risks of mortality with marathons: retrospective analysis. BMJ (Clin Res Ed) 2007;335(7633):1275–7.

[65] Roberts WO, Roberts DM, Lunos S. Marathon related cardiac arrest risk differences in men and women. Br J Sports Med 2013;47(3):168–71.

[66] Albano AJ, Thompson PD, Kapur NK. Acute coronary thrombosis in Boston marathon runners. N Engl J Med 2012;366(2):184–5.

[67] Maron BJ, Pelliccia A. The heart of trained athletes: cardiac remodeling and the risks of sports, including sudden death. Circulation 2006;114(15):1633–44.

[68] Maron BJ, Pelliccia A, Spirito P. Cardiac disease in young trained athletes. Insights into methods for distinguishing athlete's heart from structural heart disease, with particular emphasis on hypertrophic cardiomyopathy. Circulation 1995;91(5):1596–601.

[69] Pelliccia A, Culasso F, Di Paolo FM, Maron BJ. Physiologic left ventricular cavity dilatation in elite athletes. Ann Intern Med 1999;130(1):23–31.

[70] Pelliccia A, Maron BJ, Di Paolo FM, et al. Prevalence and clinical significance of left atrial remodeling in competitive athletes. J Am Coll Cardiol 2005;46(4):690–6.

[71] Pelliccia A, Maron BJ, Spataro A, Proschan MA, Spirito P. The upper limit of physiologic cardiac hypertrophy in highly trained elite athletes. N Engl J Med 1991;324(5):295–301.

[72] Pluim BM, Zwinderman AH, van der Laarse A, van der Wall EE. The athlete's heart. A meta-analysis of cardiac structure and function. Circulation 2000;101(3):336–44.

[73] Spirito P, Pelliccia A, Proschan MA, et al. Morphology of the "athlete's heart" assessed by echocardiography in 947 elite athletes representing 27 sports. Am J Cardiol 1994;74(8):802–6.

[74] Levine BD. Can intensive exercise harm the heart? The benefits of competitive endurance training for cardiovascular structure and function. Circulation 2014;130(12):987–91.

[75] Ruiz JR, Joyner M, Lucia A. CrossTalk opposing view: Prolonged intense exercise does not lead to cardiac damage. J Physiol 2013;591(20):4943–5.

[76] Chugh SS, Weiss JB. Sudden cardiac death in the older athlete. J Am Coll Cardiol 2015;65(5):493–502.

Chapter 11

The Effect of Exercise Training in Systolic and Diastolic Function

Ricardo Fontes-Carvalho*[,†,‡], Eduardo M. Vilela*, Pedro Gonçalves-Teixeira*[,†]

*Cardiology Department, Gaia Hospital Center, Vila Nova de Gaia, Portugal, †Department of Surgery and Physiology, University of Porto, Porto, Portugal, ‡Unidade de Investigação Cardiovascular (UnIC), Porto, Portugal

INTRODUCTION

Exercise, with all its different expressions, is a physiological human activity. Regular physical activity is one of the foundations of a healthy lifestyle and cardiovascular (CV) disease prevention, being associated with a reduction in all-cause and CV mortality [1].

Several studies have shown that physical activity has a positive impact on the reduction of the prevalence of CV disease [2]. Moreover, exercise training should be used as an important therapeutic tool in several CV diseases, from myocardial infarction (MI) to heart failure (HF) [3,4]. Exercise induces several beneficial adaptations on the heart, vessels, and muscu-lature, which are influenced both by type and duration of exercise and the context in which it is practiced [5]. In accordance with these observations, both the American Heart Association (AHA)/American College of Cardiology (ACC) and the European Association of Cardiology (ESC) issued recommendations for the performance of exercise to reduce the burden of CV disease [1,5].

This chapter will explore the mechanisms by which exercise can influence the CV system, focusing on favorable and potentially deleterious effects. We will review the importance of exercise training in different CV diseases, with a special emphasis on its interaction with left ventricular (LV) function, namely, systolic and diastolic function parameters.

CARDIOVASCULAR RESPONSE TO EXERCISE TRAINING IN HEALTHY INDIVIDUALS

Brief Exercise Physiology

The acute increase in metabolic demands after the onset of physical activity is the key mechanism leading to subsequent CV system adaptations, which occur by a complex integration of neural, chemical, and mechanical factors [6].

Peripheral impulses from receptors located throughout the skeletal muscle and vascular system coordinate with control regions in ventrolateral medulla and somatomotor centers of the brain, to meet body energy requirements, which ultimately modulate autonomic tone and different aspects of cardiac output [7].

The oxygen (O_2) required to fuel oxidative phosphorylation in muscle is first uptaken at a specific rate by the lungs, referred to as $\dot{V}O_2$. This can be used as an indirect measure of energy expenditure during effort. According to the Fick equation (cardiac output $= \dot{V}O_2$/arteriovenous O_2 difference), we can present $\dot{V}O_2$ as the product of cardiac output and arte-riovenous O_2 difference (A-$\dot{V}O_2$). Ultimately, the energy requirements during exercise depend on increasing the O_2 delivery to the muscle by increasing cardiac output (the product of cardiac stroke volume and heart rate) and/or by increasing the A-VO_2 [8]. Important additional mechanisms are also involved in O_2 delivery, O_2 extraction, and blood flow redistribution during exercise, but they surpass the scope of this chapter.

During exercise, as the skeletal muscle pump is engaged, increased venous return enhances diastolic filling, leading to an increase in systolic emptying, driven by the Frank-Starling mechanism. Stroke volume increases, which is also enhanced by a positive inotropic effect due to the activation of several neurohumoral stimuli during exercise. However, it is known that stroke volume plateaus early in the increasing cardiac output curve. Therefore, changes in heart rate seem to account for a greater percentage of the increase in cardiac output [9].

Lifestyle in Heart Health and Disease. https://doi.org/10.1016/B978-0-12-811279-3.00011-2

In the acute setting, the integrity of diastolic function is pivotal for optimal left ventricle relaxation and filling. Several neurohormonal stimuli have a positive lusitropic effect, improving myocardial relaxation. On the long term, exercise training also induces several molecular and structural adaptations leading to enhanced diastolic function, as further detailed below.

Exercise training can improve functional capacity not only by its positive effects on systolic and diastolic functions but also by improving cardiorespiratory fitness, assessed by the estimation of an individual's maximum $\dot{V}O_2$ ($\dot{V}O_{2\,max}$). Individuals with higher exercise capacity (higher $\dot{V}O_{2\,max}$) have a larger stroke volume at any given exercise, requiring a slower heart rate to generate the necessary cardiac output. This is better explained considering the relatively fixed upper limit of the components of the Fick equation, namely, heart rate and the A-VO$_2$, making $\dot{V}O_{2\,max}$ an indirect measure of maximal cardiac pump capacity [8]. In response to maximal exercise, cardiac output can increase up to fourfold in sedentary individuals and as much as eightfold in trained athletes [9]. The long-term benefits of exercise on functional capacity can result from combined actions on the heart, vasculature, and muscle as detailed in Fig. 1.

Cardiovascular Effects of Different Types and Intensity of Exercise

Exercise training is usually artificially categorized by the degree of dynamic (aerobic) or static (strength) exercise. In theory, different types of exercise can lead to different CV adaptations.

Endurance training typically poses more demand on aerobic respiration and requires larger increases in cardiac output. The resultant characteristic structural remodeling consists of eccentric left ventricular (LV) hypertrophy, with increased left ventricle compliance and enhanced early diastolic filling [10–13]. In contrast, static (strength or isometric) exercise leads to more pressure than volume load to the heart, increased systemic vascular resistance, and greater cardiac afterload. This is typically accompanied by concentric LV hypertrophy and possible increase in late left ventricle diastolic function [10]. However, these traditional morphological changes may vary according to age, gender, training intensity, genetic background, and different exercise demands (not always strictly static or dynamic). Therefore, this classic hypertrophic response to specific patterns of exercise training is now controversial [14].

Several studies have shown that in healthy individuals, a minimum frequency and intensity of exercise can counteract the aging-associated effects in myocardial compliance and distensibility [15,16]. This could explain how lifelong exercise training protects against common CV diseases in which LV stiffening has an important pathophysiological role, such as heart failure with preserved ejection fraction (HFpEF). It is not well established, though, what the optimal "dose" or intensity of exercise should be, toward reaching this goal.

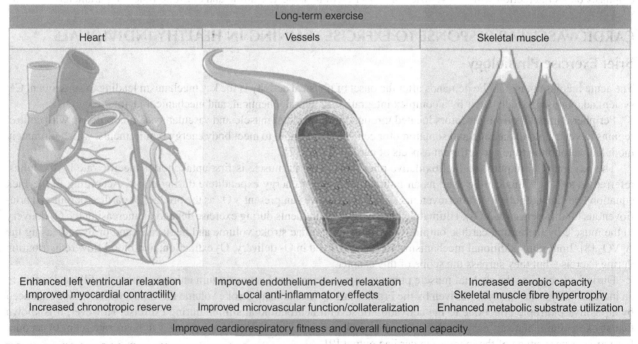

FIG. 1 Possible beneficial effects of long-term exercise training on the heart, peripheral vasculature, and skeletal muscle.

Chronic Effects of Exercise Training in the Cardiovascular System: From Molecular Mechanisms to "Whole-Pump" Effects

Hipertrophy

Highly trained individuals frequently have moderate cardiac chamber enlargement and subtle increases in wall thickness, which is sometimes referred to as the "athlete's heart." At a cellular level, there is an increase in cardiomyocyte length and width, which result from different molecular pathways, comprising complex transcriptional and translational features that are beyond the scope of this chapter [17]. In clinical studies, a period of 8 weeks of regular aerobic training has been found to be necessary to reach a hypertrophic response plateau, whereas a period of 3–6 weeks of detraining leads to regression of such changes [18].

Calcium Homeostasis

The excitation-contraction coupling in the cardiomyocytes is dependent upon a meticulously orchestrated process of calcium (Ca^{2+})-induced Ca^{2+} release, after depolarization of membrane action potential.

Regular exercise training can improve intracellular calcium homeostasis, which can partially explain beneficial effects on systolic and diastolic function. Proposed mechanisms by which exercise can improve myocardial contractility are (i) a faster systolic rise time of Ca^{2+} inward current, (ii) a greater amplitude of Ca^{2+} current and greater peak systolic $[Ca^{2+}]_i$, and (iii) improved myofilament Ca^{2+} sensitivity [17,19]. It is known that the amount of Ca^{2+} released from intracellular storage sites and myofilaments sensibility to calcium determines the extent (fractional shortening) and force of contraction [20]. Moreover, better calcium handling induced by exercise can also improve myocardial relaxation, through (i) upregulation of Ca^{2+} ATPase SERCA2a, which leads to faster reuptake of free cytoplasmatic Ca^{2+} and lower diastolic $(Ca^{2+})_i$; (ii) chronic activation of enzymes that phosphorylate phospholamban, thus removing its inhibition on SERCA2a; and (iii) dilution of cytoplasmatic Ca^{2+} due to cellular volume expansion in hypertrophy [15,17,19].

Relaxation is further enhanced by other non-calcium-dependent adaptations, such as modification of collagen isoform composition toward more compliant variants, and decreases in circulation levels of inflammatory mediators, thus mitigating inflammatory-driven myocardial fibrosis [15,17,19].

Cardioprotective Changes

Exercise training is highly protective against cardiomyocyte apoptosis [19,21]. This effect can slow the progression of some CV diseases and delay the physiological detrimental effects of aging in cardiac function.

Also, there is growing evidence supporting exercise-mediated cardioprotection after exposure to ischemia, by several mechanisms including improved antioxidative protection [19,22].

Endocrine-Metabolic Pathways

Several new studies have shown that obesity can directly influence cardiac structure and function, especially diastolic dysfunction, through systemic, paracrine, and endocrine mechanisms [23,24]. This has been recently recognized as a new clinical entity, called obesity cardiomyopathy [25]. Some studies have suggested that exercise can influence these obesity-related effects on myocardial structure and function by decreasing obesity and insulin resistance [26].

Interestingly, many metabolic risk factors [23,24] and adipose-tissue-associated signaling pathways have been recently linked to diastolic dysfunction, providing exercise training a theoretical benefit in its prevention.

Deleterious Effects of Exercise Training on the Cardiovascular System

It is now demonstrated that strenuous exercise training can also have a deleterious effect on the CV system, as detailed in Fig. 2. This suggests a U-shaped relation between exercise and CV risk. Those effects mainly relate to structural, rhythm, and neurohumoral changes imposed by "excessive" physical activity [26].

For example, data from marathon runners have shown a relevant increase in cardiac biomarkers (such as troponin) in these individuals, possibly indicating myocardial injury caused by extreme hemodynamic challenges and/or cardiac fatigue [27–29]. Also, highly trained athletes showed an increased prevalence of coronary artery disease, which is thought to result from inflammation and increased vascular oxidative stress induced by intensive exercise [27].

There is evidence that light-to-moderate exercise may decrease the incidence of atrial fibrillation. However, recent data have consistently shown a clear association between intense endurance exercise training and an increased risk of atrial fibrillation [30].

Cardioprotective effects

- Antiapoptotic effects
- Protection against ischemia/reperfusion injury
- Antioxidative effects
- CV risk factor modification
- Slower atherosclerotic progression
- ↓ fat mass and visceral fat

Healthy individuals

Effects on systolic function

- ↑ systolic [Ca²⁺]ᵢ release
- Improved myofilament Ca²⁺ sensitivity

Effects on diastolic function

- Improved [Ca²⁺]ᵢ diastolic re-uptake
- Prevention of age-related decline in LV compliance and distensibility
- ↓ inflammatory-driven myocardial fibrosis

Effects on functional capacity

- ↑ VO₂ₘₐₓ
- Improved vascular/endothelial function
- Improved skeletal muscle efficiency
- ↑ Oxygen transport capacity of blood cells

Cardiovascular disease

Effects on systolic function

- Favourable LV remodeling
  - Reduced LV diameter
- Improved LV ejection fraction (*)

Effects on diastolic function

- Improved LV relaxation
- Reduced LV stiffness

Effects on functional capacity

- Improved cardiorespiratory fitness
- Improved quality of life

POSSIBLE DELETERIOUS EFFECTS

- Myocardial damage, troponin release (extreme exercise training)
- Increased incidence of AF (highly trained athletes)
- Increased prevalence of CAD (*)
- Impaired healing of inflammatory cardiomyopathies
- Arrhythmogenic remodeling/triggering for arrhythmic events
- Enhances phenotypic expression of some genetic cardiomyopathies (may include SCD)

FIG. 2 Positive and deleterious effects of exercise training in cardiovascular health and disease. *Note*: * = Controversial data. *(Ca²⁺)ᵢ*, intracellular calcium; *CAD*, coronary artery disease; *CV*, cardiovascular; *LV*, left ventricle; *max*, maximum; *SCD*, sudden cardiac death.

Finally, exercise can have serious deleterious effects in patients with myocarditis or inflammatory cardiomyopathies, by imposing an important hemodynamic burden to inflamed myocardium [31]. Moreover, in some genetic cardiomyopathies, heavy exercise training can also enhance its phenotypic expression [31].

EXERCISE TRAINING AFTER ACUTE MYOCARDIAL INFARCTION

After myocardial infarction (MI), exercise training reduces cardiovascular (CV) mortality, decreases hospitalizations, and improves quality of life, being an important hallmark in the contemporary treatment of MI [1]. Despite the wealth of data showing the benefits of exercise training in this setting, the involved mechanisms are not entirely clear. Moreover, the timing, intensity, type, and duration of the "ideal" exercise-training program are debatable, as detailed below.

The Effect of Exercise Training on Systolic Function

After MI, exercise training has yielded mixed results in systolic function. In a seminal work by Cobb et al., exercise training improved exercise capacity, body mass index, and lipid profile, but no effect was observed on ejection fraction (EF), end-diastolic volume, or wall motion pattern [32]. Several contemporary studies have also shown similar results, reporting improvements in exercise capacity, but not on systolic function parameters [33–35]. On the contrary, data from an early study by Jugdutt et al. described possible detrimental effects of exercise training in a subgroup of individuals with anterior Q-wave MI [36]. These results have been questioned by more recent analyses showing that exercise training can be performed safely after MI, without significant adverse effects. There are also studies showing an improvement in LV systolic function in MI survivors [37,38].

A meta-analysis by Haykowsky et al. attempted to address the effect of exercise on systolic function [39]. Although there was an important degree of heterogeneity among studies, the authors observed a beneficial effect of exercise training on left ventricular remodeling, which was dependent on the interventions timing and duration [39]. A more recent meta-analysis performed by Zhang et al. also yielded important data [40]. These authors found that in clinically stable post-MI survivors, exercise in an earlier phase was associated with an improvement in LV dimensions and EF. In contrast, no significant effect was observed when exercise training was initiated more than 28 days after the event [40].

Several important caveats should be addressed when interpreting these results. Firstly, the high level of heterogeneity which was observed between different studies [40]. Secondly, the different types of exercise-training protocols (type of exercise, timing of initiation, and time duration) could have significantly influence the results [41]. Finally, the parameters which were used to assess systolic function. In most of these studies, systolic function was assessed using EF, which is known to be highly dependent on preload and afterload conditions [4,33]. In a recent randomized trial, exercise training did not improve systolic function assessed by tissue-Doppler-derived systolic velocities [33]. On the contrary, another study reported a significant improvement in LV function, assessed by global longitudinal and circumferential strain [37].

The Effect of Exercise Training on Diastolic Function

In patients after MI, diastolic function is now an important predictor of adverse CV events [42,43]. Moreover, diastolic function parameters have a good correlation with exercise capacity, independently of systolic function [44,45].

However, few studies have assessed the effect of exercise training on diastolic function parameters after an MI [33,34]. One such study by Yu et al. showed an improvement in diastolic function parameters in patients submitted to exercise training [34].In another small study of 40 patients after MI, the exercise-training group showed an improvement in E/A ratio and a decrease in NT-proBNP levels [35]. Interestingly, patients had improved exercise capacity, despite unchanged EF, similarly to the findings reported by Yu et al. [34]. Nevertheless, a recent randomized study in 188 patients after AMI, which evaluated diastolic function with several contemporary parameters, failed to demonstrate a significant improvement in diastolic function after exercise training, combining endurance and resistance training [33]. As previously discussed, the type, initiation, and duration of exercise could be important considerations when assessing its impact on diastolic function.

The Effect of Exercise Training on Functional Capacity

As previously detailed, the benefits of exercise training on systolic and diastolic function after MI are controversial. On the contrary, there is substantial evidence that exercise training significantly improves functional capacity after MI independently of cardiac function [40]. This effect was more significant when exercise training was started shortly after the event [40]. Interestingly, the long-term effect of this intervention may also be influenced by the cardiopulmonary status at baseline and by the medication used after the event [46].

Regarding the type of exercise-training protocol, a large randomized trial including 200 patients (115 after AMI) showed a similar improvement in exercise capacity with both interval and continuous training [47]. However, the effect of different exercise-training protocols needs further investigation in the future [48,49].

Considering that exercise training significantly improves cardiopulmonary status, in the absence of systolic and diastolic function changes, this effect seems to result from extracardiac benefits of exercise [33], such as adaptations in skeletal muscle performance, adaptations in autonomic nervous system, and local improvements in oxygen uptake [48,49] as detailed in Fig. 1.

Exercise Training as Part of an Integrated Cardiac Rehabilitation Program

In the last years, the treatment of MI has improved markedly, but there are still important unmet needs in these patients [1,50]. Despite contemporary management strategies, many patients fail to achieve optimal risk factor management and do not adopt a healthy lifestyle after the event [51,52].

Cardiac rehabilitation (CR) programs (in which exercise training plays a pivotal role) have evolved considerably over the last decades, now constituting a holistic approach to the MI patient [53,54]. This paradigm is highlighted in the current guidelines of both the American College of Cardiology/American Heart Association (ACC/AHA) and the European Society of Cardiology (ESC) [1,55,56].

Coronary heart disease (CHD) is one of the classical indications for cardiac rehabilitation [57]. Despite the beneficial effects associated with CR in older studies, however, questions have arisen concerning its role in contemporary settings [58,59]. A recent meta-analysis conducted by Anderson et al. showed that in CHD, CR was associated with a significant reduction in CV mortality and hospital admissions [58]. Another meta-analysis assessing only patients in a contemporary setting also showed a mortality benefit in patients enrolled in CR programs [60].

In spite of their usefulness, there are still many hindrances to the application of CR programs in different patient populations [51,53,61]. Future studies should focus on the impact of these programs on less represented individuals, such as women and the elderly [58].

EXERCISE TRAINING IN PATIENTS WITH HEART FAILURE

Heart failure (HF) is a complex syndrome, with protean manifestations [62]. Even in optimally managed patients, the morbidity and mortality are high [62]. According to left ventricular ejection fraction (LVEF), this syndrome is usually divided into HF with reduced EF (HFrEF; EF < 40%), HF with preserved EF (HFpEF; EF ≥ 50%), and a recently developed category of HF with midrange EF (HFmrEF; EF between 40% and 49%) [62]. This division is important as it represents different patient profiles and different responses to therapy.

Exercise training can have important benefits in HF patients [1,56,62,63], across the whole spectrum of HF [62,64]. It should be noted, however, that exercise seems to elicit distinct hemodynamic responses in patients with HFrEF when compared with HFpEF, as discussed below [65]. As detailed in Fig. 1, beyond the cardiac benefits per se, exercise training can have several important extracardiac effects—on the peripheral circulation, skeletal muscle, and ventilation response to exercise—that can explain the improvement in functional capacity seen in several trials assessing the effect of exercise in HF [66,67].

More interestingly, several studies have shown that regular exercise can reduce the risk of developing HF [68]. Physically inactive individuals have worse functional capacity and an increased risk for diastolic dysfunction [69]. This is in accordance with a previous study that showed that individuals with low levels of physical activity seem to be at increased risk of developing HF [70]. A recent meta-analysis comprising over 370,000 individuals reported an inverse, dose-response relationship between physical activity and HF risk [71]. The mechanisms by which exercise training can influence the development and the progression of HF are still not fully ascertained.

Effects of Exercise Training in Patients With HFrEF

Over the last decades, there have been substantial improvements in the treatment of HFrEF [62]. Since the first descriptions of the possible benefits of exercise in individuals with HF about 30 years ago [72], this intervention has significantly evolved and should be now an integral part of management in these patients [56,62].

Important data regarding the benefits of exercise training in HFrEF come from the seminal HF-ACTION study [73]. Although in a primary analysis there were no significant reductions in mortality or hospitalizations, these benefits were observed after adjustment for certain prognostic variables. As acknowledged by the authors, several points should be taken into consideration when interpreting these results, such as the patient population studied, the design of the program and background therapy, and the blinding status [67,73]. The HF-ACTION also showed that exercise training was safe in this population and associated with significant improvement in functional capacity [73]. The beneficial effects of exercise training on exercise capacity were also observed in several other studies [64,74,75].

Regarding the effects of exercise on systolic function in patients with HFrEF, the benefits are not so consistent. In a small study including 99 patients, there was no improvement in EF, despite significant benefits in functional capacity and quality of life [75]. On the contrary, other studies reported significant improvements in systolic function [64,74]. In a study that tried to assess the long-term benefits of a 10-year exercise program, there was no significant improvement in EF until the fourth year of follow-up, but then, the exercise-training group showed better EF [74,76]. A recent study showed a beneficial effect of exercise on EF, LV remodeling, and exercise capacity in patients with severe HF, with a mean EF of around 25% [77].

In HFrEF, the assessment of diastolic function is also an important predictor of reduced exercise capacity and increased mortality [78]. In a study by Sandri et al. that included patients with HFrEF, endurance training during 4 weeks improved both EF and diastolic function parameters [78]. Improvements in diastolic function parameters were also reported in a study of HF patients with different degrees of LV dysfunction [77], though this effect was not seen in other studies.

In terms of different exercise-training protocols, interest has arisen concerning the use of interval training in HF patients [79]. However, a large recent randomized trial did not confirm the superiority of high-intensity interval training in HFrEF [80]. The optimal exercise-training program to use in this patient population is an important area of research for the future.

Finally, some concerns were raised regarding the practice of exercise in patients with HF and an implantable cardioverter-defibrillator (ICD). Data from the HF-ACTION study showed that in these patients, exercise training was not associated with an increased risk of ICD shocks [81]. Moreover, a meta-analysis in HF patients with ICDs showed that exercise is associated with improved cardiopulmonary performance and even a reduced risk of ICD shocks [82].

Effects of Exercise Training in Patients With HFpEF

HFpEF is a distinct syndrome, with clinical, etiologic, and phenotypic heterogeneity [62]. Contrary to HFrEF, several pharmacological interventions proved unsuccessful in the management of this syndrome [62,83].

Diastolic dysfunction, which is an important predictor of reduced exercise capacity, has a central role in the pathophysiology of HFpEF [44,45,62,84]. Some data suggested that exercise training can improve diastolic function parameters in patients with HF [64]. However, this effect was not shown in other studies, generating debate [85,86].

This question was assessed in a study by Edelmann et al., which showed that in patients with HFpEF, exercise training improved functional capacity and diastolic function parameters, with a significant improvement in the E/e' ratio and a reduction in left atrial size [87]. In another recent randomized trial including 100 patients, the authors reported no significant benefits in diastolic function parameters in the exercise group [88].

However, in HFpEF, there is robust and consistent evidence supporting the benefits of exercise in functional capacity and quality of life, as demonstrated in two recent meta-analysis [3,80]. It is important to reinforce that in HFpEF, exercise training is one of the few therapeutic interventions that have shown a significant benefit.

These observations are in accordance with the multifactorial nature of HFpEF where both cardiac and extracardiac factors play a significant role in the pathophysiology of the disease [62]. The effect of exercise training on skeletal muscle, oxygen uptake, and body mass index should be further explored in patients with HFpEF [62].

The Effects of Cardiac Rehabilitation in Heart Failure

HF is a classical indication for CR programs, as highlighted by both the American College of Cardiology/American Heart Association (ACC/AHA) and the European Society of Cardiology (ESC) [57]. When assessing the effect of exercise-based CR in HF patients, Sagar et al. found a beneficial effect in morbidity, especially reduced hospitalizations and better quality of life [89]. Though a significant one-year survival benefit was not demonstrated, there was a trend toward a reduction in all-cause mortality in studies with longer follow-up [89], highlighting that in HF, longer duration CR programs can be associated with the best outcomes [90].

As HF prevalence increases with age, with older patients many times presenting with other comorbidities (and frailty), the comprehensive nature of CR makes it a crucial ancillary tool in the treatment of this group of patients [54].

In summary, in HF patients, exercise training is associated with significant improvements in functional capacity, both in HFrEF and HFpEF. Most of these benefits seem to be mediated by extracardiac and systemic mechanisms. However, in HFrEF, most studies suggested that exercise could slightly improve systolic function. In HFpEF, a disease with significant clinical heterogeneity, there is evidence that exercise training can also possibly improve diastolic function, at least in some subgroups of patients.

Given the complex nature of the heart failure syndrome, future research should focus on the individualization of the exercise-training protocol, focusing on the identification of the patients most likely to improve and the definition of the best type, intensity, and duration of training.

CONCLUSION

Physical activity plays a critical role in the prevention of cardiovascular disease. More recently, its importance has extended as an important therapeutic tool in several disease entities, such as heart failure and after myocardial infarction.

Even in patients with cardiovascular disease, exercise training clearly improves functional capacity in several clinical scenarios. On the contrary, the benefits of exercise on systolic and diastolic function can be protean and depend on the clinical context; the baseline characteristics; and the timing, type, intensity, and duration of the exercise-training program.

Despite these issues, exercise training as part of a comprehensive cardiac rehabilitation program should have a fundamental part in the management of several cardiovascular diseases. As medicine and cardiology move toward an ever more personalized approach, so does the notion that exercise (as both a preventive and therapeutic agent) needs to be tailored to the individual patient, to provide the optimal benefit.

REFERENCES

[1] Piepoli MF, Hoes AW, Agewall S, Albus C, Brotons C, Catapano AL, et al. 2016 European Guidelines on cardiovascular disease prevention in clinical practice. Atherosclerosis 2016;252:207–74.

[2] Maessen MFH, Verbeek ALM, Bakker EA, Thompson PD, Hopman MTE, Eijsvogels TMH. Lifelong exercise patterns and cardiovascular health. Mayo Clin Proc 2016;91(6):745–54.

[3] Pandey A, Parashar A, Kumbhani DJ, Agarwal S, Garg J, Kitzman D, et al. Exercise training in patients with heart failure and preserved ejection fraction: meta-analysis of randomized control trials. Circ Heart Fail 2015;8(1):33–40.

[4] Giallauria F, Acampa W, Ricci F, Vitelli A, Torella G, Lucci R, et al. Exercise training early after acute myocardial infarction reduces stress-induced hypoperfusion and improves left ventricular function. Eur J Nucl Med Mol Imaging 2013;40(3):315–24.

[5] Eckel RH, Jakicic JM, Ard JD, De Jesus JM, Houston Miller N, Hubbard VS, et al. 2013 AHA/ACC guideline on lifestyle management to reduce cardiovascular risk: a report of the American College of Cardiology/American Heart Association Task Force on Practice Guidelines. J Am Coll Cardiol 2014;63(25 (Pt B)):2960–84.

[6] Crisafulli A, Marongiu E, Ogoh S. Cardiovascular reflexes activity and their interaction during exercise. BioMed Res Int 2015;2015:394183.

[7] Rowell LB, O'Leary DS. Reflex control of the circulation during exercise: chemoreflexes and mechanoreflexes. J Appl Physiol (Bethesda, MD, 1985) 1990;69(2):407–18.

[8] Opondo MA, Sarma S, Levine BD. The cardiovascular physiology of sports and exercise. Clin Sports Med 2015;34(3):391–404.

[9] Hurst JW, Walsh RA, Fuster V, Fang JC, editors. Hurst's the heart manual of cardiology. 13th New York: McGraw-Hill; 2013. 882 p.

[10] Weiner RB, Baggish AL. Exercise-induced cardiac remodeling. Prog Cardiovasc Dis 2012;54(5):380–6.

[11] Granger CB, Karimeddini MK, Smith VE, Shapiro HR, Katz AM, Riba AL. Rapid ventricular filling in left ventricular hypertrophy: I. Physiologic hypertrophy. J Am Coll Cardiol 1985;5(4):862–8.

[12] Levy WC, Cerqueira MD, Abrass IB, Schwartz RS, Stratton JR. Endurance exercise training augments diastolic filling at rest and during exercise in healthy young and older men. Circulation 1993;88(1):116–26.

[13] Matsuda M, Sugishita Y, Koseki S, Ito I, Akatsuka T, Takamatsu K. Effect of exercise on left ventricular diastolic filling in athletes and nonathletes. J Appl Physiol 1983;55(2):323–8.

[14] Maron BJ, Pelliccia A. The heart of trained athletes: cardiac remodeling and the risks of sports, including sudden death. Circulation 2006;114(15):1633–44.

[15] Bhella PS, Hastings JL, Fujimoto N, Shibata S, Carrick-Ranson G, Palmer MD, et al. Impact of lifelong exercise "dose" on left ventricular compliance and distensibility. J Am Coll Cardiol 2014;64(12):1257–66.

[16] Brinker SK, Pandey A, Ayers CR, Barlow CE, DeFina LF, Willis BL, et al. Association of cardiorespiratory fitness with left ventricular remodeling and diastolic function: the cooper center longitudinal study. JACC Heart Fail 2014;2(3):238–46.

[17] Kemi OJ, Wisløff U. Mechanisms of exercise-induced improvements in the contractile apparatus of the mammalian myocardium. Acta Physiol (Oxf, Engl) 2010;199(4):425–39.

[18] Maron BJ, Pelliccia A, Spataro A, Granata M. Reduction in left ventricular wall thickness after deconditioning in highly trained Olympic athletes. Br Heart J 1993;69(2):125–8.

[19] Gielen S, Schuler G, Adams V. Cardiovascular effects of exercise training: molecular mechanisms. Circulation 2010;122(12):1221–38.

[20] Bers DM. Cardiac excitation-contraction coupling. Nature 2002;415(6868):198–205.

[21] Kwak H-B, Song W, Lawler JM. Exercise training attenuates age-induced elevation in Bax/Bcl-2 ratio, apoptosis, and remodeling in the rat heart. FASEB J Off Publ Fed Am Soc Exp Biol 2006;20(6):791–3.

[22] Ascensão A, Ferreira R, Magalhães J. Exercise-induced cardioprotection–biochemical, morphological and functional evidence in whole tissue and isolated mitochondria. Int J Cardiol 2007;117(1):16–30.

[23] Fontes-Carvalho R, Fontes-Oliveira M, Sampaio F, Mancio J, Bettencourt N, Teixeira M, et al. Influence of epicardial and visceral fat on left ventricular diastolic and systolic functions in patients after myocardial infarction. Am J Cardiol 2014;114(11):1663–9.

[24] Fontes-Carvalho R, Gonçalves A, Severo M, Lourenço P, Rocha Gonçalves F, Bettencourt P, et al. Direct, inflammation-mediated and blood-pressure-mediated effects of total and abdominal adiposity on diastolic function: EPIPorto study. Int J Cardiol 2015;191:64–70.

[25] Bozkurt B, Colvin M, Cook J, Cooper LT, Deswal A, Fonarow GC, et al. Current diagnostic and treatment strategies for specific dilated cardiomyopathies: a scientific statement from the American Heart Association. Circulation 2016;134(23):e579–646.

[26] Sharma S, Merghani A, Mont L. Exercise and the heart: the good, the bad, and the ugly. Eur Heart J 2015;36(23):1445–53.

[27] Predel H-G. Marathon run: cardiovascular adaptation and cardiovascular risk. Eur Heart J 2014;35(44):3091–8.

[28] Vilela EM, Bettencourt-Silva R, Nunes JPL, Ribeiro VG. BNP and NT-proBNP elevation after running—a systematic review. Acta Cardiol 2015;70(5):501–9.

[29] Vilela EM, Bastos JCC, Rodrigues RP, Nunes JPL. High-sensitivity troponin after running—a systematic review. Neth J Med 2014;72(1):5–9.

[30] Morseth B, Graff-Iversen S, Jacobsen BK, Jørgensen L, Nyrnes A, Thelle DS, et al. Physical activity, resting heart rate, and atrial fibrillation: the Tromsø study. Eur Heart J 2016;37(29):2307–13.

[31] Mont L, Pelliccia A, Sharma S, Biffi A, Borjesson M, Brugada Terradellas J, et al. Pre-participation cardiovascular evaluation for athletic participants to prevent sudden death: position paper from the EHRA and the EACPR, branches of the ESC. Endorsed by APHRS, HRS, and SOLAECE. Eur J Prev Cardiol 2017;24(1):41–69.

[32] Cobb FR, Williams RS, McEwan P, Jones RH, Coleman RE, Wallace AG. Effects of exercise training on ventricular function in patients with recent myocardial infarction. Circulation 1982;66(1):100–8.

[33] Fontes-Carvalho R, Azevedo AI, Sampaio F, Teixeira M, Bettencourt N, Campos L, et al. The effect of exercise training on diastolic and systolic function after acute myocardial infarction: a randomized study. Medicine (Baltimore) 2015;94(36):e1450.

[34] Yu C-M, Li LS-W, Lam M-F, Siu DC-W, Miu RK-M, Lau C-P. Effect of a cardiac rehabilitation program on left ventricular diastolic function and its relationship to exercise capacity in patients with coronary heart disease: experience from a randomized, controlled study. Am Heart J 2004;147(5):e24.

[35] Giallauria F, Lucci R, De Lorenzo A, D'Agostino M, Del Forno D, Vigorito C. Favourable effects of exercise training on N-terminal pro-brain natriuretic peptide plasma levels in elderly patients after acute myocardial infarction. Age Ageing 2006;35(6):601–7.

[36] Jugdutt BI, Michorowski BL, Kappagoda CT. Exercise training after anterior Q wave myocardial infarction: importance of regional left ventricular function and topography. J Am Coll Cardiol 1988;12(2):362–72.

[37] Acar RD, Bulut M, Ergün S, Yesin M, Akçakoyun M. Evaluation of the effect of cardiac rehabilitation on left atrial and left ventricular function and its relationship with changes in arterial stiffness in patients with acute myocardial infarction. Echocardiogr (Mt Kisco, NY) 2015;32(3):443–7.

[38] Giallauria F, Galizia G, Lucci R, D'Agostino M, Vitelli A, Maresca L, et al. Favourable effects of exercise-based cardiac rehabilitation after acute myocardial infarction on left atrial remodeling. Int J Cardiol 2009;136(3):300–6.

[39] Haykowsky M, Scott J, Esch B, Schopflocher D, Myers J, Paterson I, et al. A meta-analysis of the effects of exercise training on left ventricular remodeling following myocardial infarction: start early and go longer for greatest exercise benefits on remodeling. Trials 2011;12:92.

[40] Zhang Y-M, Lu Y, Tang Y, Yang D, Wu H-F, Bian Z-P, et al. The effects of different initiation time of exercise training on left ventricular remodeling and cardiopulmonary rehabilitation in patients with left ventricular dysfunction after myocardial infarction. Disabil Rehabil 2016;38(3):268–76.

[41] Prado DML, Rocco EA, Silva AG, Rocco DF, Pacheco MT, Silva PF, et al. Effects of continuous vs interval exercise training on oxygen uptake efficiency slope in patients with coronary artery disease. Braz J Med Biol Res Rev 2016;49(2):e4890.

[42] Nagueh SF, Smiseth OA, Appleton CP, Byrd BF, Dokainish H, Edvardsen T, et al. Recommendations for the evaluation of left ventricular diastolic function by echocardiography: an update from the American Society of Echocardiography and the European Association of Cardiovascular Imaging. J Am Soc Echocardiogr Off Publ Am Soc Echocardiogr 2016;29(4):277–314.

[43] Møller JE, Pellikka PA, Hillis GS, Oh JK. Prognostic importance of diastolic function and filling pressure in patients with acute myocardial infarction. Circulation 2006;114(5):438–44.

[44] Fontes-Carvalho R, Sampaio F, Teixeira M, Rocha-Gonçalves F, Gama V, Azevedo A, et al. Left ventricular diastolic dysfunction and E/E′ ratio as the strongest echocardiographic predictors of reduced exercise capacity after acute myocardial infarction. Clin Cardiol 2015;38(4):222–9.

[45] Grewal J, McCully RB, Kane GC, Lam C, Pellikka PA. Left ventricular function and exercise capacity. JAMA 2009;301(3):286–94.

[46] Heldal M, Sire S, Dale J. Randomised training after myocardial infarction: short and long-term effects of exercise training after myocardial infarction in patients on beta-blocker treatment. A randomized, controlled study. Scand Cardiovasc J 2000;34(1):59–64.

[47] Conraads VM, Pattyn N, De Maeyer C, Beckers PJ, Coeckelberghs E, Cornelissen VA, et al. Aerobic interval training and continuous training equally improve aerobic exercise capacity in patients with coronary artery disease: the SAINTEX-CAD study. Int J Cardiol 2015;179:203–10.

[48] Beller GA, Murray GC, ErkenBrack SK. Influence of exercise training soon after myocardial infarction on regional myocardial perfusion and resting left ventricular function. Clin Cardiol 1992;15(1):17–23.

[49] Kida K, Osada N, Akashi YJ, Sekizuka H, Omiya K, Miyake F. The exercise training effects of skeletal muscle strength and muscle volume to improve functional capacity in patients with myocardial infarction. Int J Cardiol 2008;129(2):180–6.

[50] Piepoli MF, Corrà U, Dendale P, Frederix I, Prescott E, Schmid JP, et al. Challenges in secondary prevention after acute myocardial infarction: a call for action. Eur J Prev Cardiol 2016;23(18):1994–2006.

[51] Kotseva K, Wood D, De Bacquer D, De Backer G, Ryden L, Jennings C, et al. EUROASPIRE IV: a European Society of Cardiology survey on the lifestyle, risk factor and therapeutic management of coronary patients from 24 European countries. Eur J Prev Cardiol 2016;23(6):636–48.

[52] Farkouh ME, Boden WE, Bittner V, Muratov V, Hartigan P, Ogdie M, et al. Risk factor control for coronary artery disease secondary prevention in large randomized trials. J Am Coll Cardiol 2013;61(15):1607–15.

[53] Mampuya WM. Cardiac rehabilitation past, present and future: an overview. Cardiovasc Diagn Ther 2012;2(1):38–49.

[54] Schopfer DW, Forman DE. Growing relevance of cardiac rehabilitation for an older population with heart failure. J Card Fail 2016;22(12):1015–22.

[55] Smith SC, Benjamin EJ, Bonow RO, Braun LT, Creager MA, Franklin BA, et al. AHA/ACCF secondary prevention and risk reduction therapy for patients with coronary and other atherosclerotic vascular disease: 2011 update: a guideline from the American Heart Association and American College of Cardiology Foundation endorsed by the World Heart Federation and the Preventive Cardiovascular Nurses Association. J Am Coll Cardiol 2011;58(23):2432–46.

[56] Yancy CW, Jessup M, Bozkurt B, Butler J, Casey DE, Drazner MH, et al. ACCF/AHA guideline for the management of heart failure: a report of the American College of Cardiology Foundation/American Heart Association Task Force on Practice Guidelines. J Am Coll Cardiol 2013;62(16):e147–239.

[57] Piepoli MF, Corrà U, Benzer W, Bjarnason-Wehrens B, Dendale P, Gaita D, et al. Secondary prevention through cardiac rehabilitation: from knowledge to implementation. A position paper from the Cardiac Rehabilitation Section of the European Association of Cardiovascular Prevention and Rehabilitation. Eur J Cardiovasc Prev Rehabil 2010;17(1):1–17.

[58] Anderson L, Oldridge N, Thompson DR, Zwisler A-D, Rees K, Martin N, et al. Exercise-based cardiac rehabilitation for coronary heart disease: cochrane systematic review and meta-analysis. J Am Coll Cardiol 2016;67(1):1–12.

[59] Oldridge NB, Guyatt GH, Fischer ME, Rimm AA. Cardiac rehabilitation after myocardial infarction. Combined experience of randomized clinical trials. JAMA 1988;260(7):945–50.

[60] Rauch B, Davos CH, Doherty P, Saure D, Metzendorf M-I, Salzwedel A, et al. The prognostic effect of cardiac rehabilitation in the era of acute revascularisation and statin therapy: a systematic review and meta-analysis of randomized and non-randomized studies—the Cardiac Rehabilitation Outcome Study (CROS). Eur J Prev Cardiol 2016;24.

[61] Sumner J, Grace SL, Doherty P. Predictors of cardiac rehabilitation utilization in England: results from the national audit. J Am Heart Assoc 2016;5(10).

[62] Ponikowski P, Voors AA, Anker SD, Bueno H, Cleland JGF, Coats AJS, et al. ESC Guidelines for the diagnosis and treatment of acute and chronic heart failure: the task force for the diagnosis and treatment of acute and chronic heart failure of the European Society of Cardiology (ESC). Developed with the special contribution of the Heart Failure Association (HFA) of the ESC. Eur J Heart Fail 2016;18(8):891–975.

[63] Piepoli MF, Conraads V, Corrà U, Dickstein K, Francis DP, Jaarsma T, et al. Exercise training in heart failure: from theory to practice. A consensus document of the Heart Failure Association and the European Association for Cardiovascular Prevention and Rehabilitation. Eur J Heart Fail 2011;13(4):347–57.

[64] Alves AJ, Ribeiro F, Goldhammer E, Rivlin Y, Rosenschein U, Viana JL, et al. Exercise training improves diastolic function in heart failure patients. Med Sci Sports Exerc 2012;44(5):776–85.

[65] Zile MR, Kjellstrom B, Bennett T, Cho Y, Baicu CF, Aaron MF, et al. Effects of exercise on left ventricular systolic and diastolic properties in patients with heart failure and a preserved ejection fraction versus heart failure and a reduced ejection fraction. Circ Heart Fail 2013;6(3):508–16.

[66] Fleg JL, Cooper LS, Borlaug BA, Haykowsky MJ, Kraus WE, Levine BD, et al. Exercise training as therapy for heart failure: current status and future directions. Circ Heart Fail 2015;8(1):209–20.

[67] Tabet J-Y, Meurin P, Driss AB, Weber H, Renaud N, Grosdemouge A, et al. Benefits of exercise training in chronic heart failure. Arch Cardiovasc Dis 2009;102(10):721–30.

[68] Rahman I, Bellavia A, Wolk A, Orsini N. Physical activity and heart failure risk in a prospective study of men. JACC Heart Fail 2015;3(9):681–7.

[69] Matta S, Chammas E, Alraies C, Abchee A, AlJaroudi W. Association between sedentary lifestyle and diastolic dysfunction among outpatients with normal left ventricular systolic function presenting to a tertiary referral center in the middle east. Clin Cardiol 2016;39(5):269–75.

[70] Young DR, Reynolds K, Sidell M, Brar S, Ghai NR, Sternfeld B, et al. Effects of physical activity and sedentary time on the risk of heart failure. Circ Heart Fail 2014;7(1):21–7.

[71] Pandey A, Garg S, Khunger M, Darden D, Ayers C, Kumbhani DJ, et al. Dose-response relationship between physical activity and risk of heart failure: a meta-analysis. Circulation 2015;132(19):1786–94.

[72] Sullivan MJ, Higginbotham MB, Cobb FR. Exercise training in patients with severe left ventricular dysfunction. Hemodynamic and metabolic effects. Circulation 1988;78(3):506–15.

[73] O'Connor CM, Whellan DJ, Lee KL, Keteyian SJ, Cooper LS, Ellis SJ, et al. Efficacy and safety of exercise training in patients with chronic heart failure: HF-ACTION randomized controlled trial. JAMA 2009;301(14):1439–50.

[74] Belardinelli R, Georgiou D, Cianci G, Purcaro A. 10-year exercise training in chronic heart failure: a randomized controlled trial. J Am Coll Cardiol 2012;60(16):1521–8.

[75] Belardinelli R, Georgiou D, Cianci G, Purcaro A. Randomized, controlled trial of long-term moderate exercise training in chronic heart failure: effects on functional capacity, quality of life, and clinical outcome. Circulation 1999;99(9):1173–82.

[76] Whellan DJ. Long-term exercise training and adherence: it is not just exercise. J Am Coll Cardiol 2012;60(16):1529–30.

[77] Höllriegel R, Winzer EB, Linke A, Adams V, Mangner N, Sandri M, et al. Long-term exercise training in patients with advanced chronic heart failure: sustained benefits on left ventricular performance and exercise capacity. J Cardiopulm Rehabil Prev 2016;36(2):117–24.

[78] Sandri M, Kozarez I, Adams V, Mangner N, Höllriegel R, Erbs S, et al. Age-related effects of exercise training on diastolic function in heart failure with reduced ejection fraction: the Leipzig Exercise Intervention in Chronic Heart Failure and Aging (LEICA) Diastolic Dysfunction Study. Eur Heart J 2012;33(14):1758–68.

[79] Haykowsky MJ, Timmons MP, Kruger C, McNeely M, Taylor DA, Clark AM. Meta-analysis of aerobic interval training on exercise capacity and systolic function in patients with heart failure and reduced ejection fractions. Am J Cardiol 2013;111(10):1466–9.

[80] Ellingsen Ø, Halle M, Conraads VM, Støylen A, Dalen H, Delagardelle C, et al. High intensity interval training in heart failure patients with reduced ejection fraction. Circulation 2017;135(9):839–49.

[81] Piccini JP, Hellkamp AS, Whellan DJ, Ellis SJ, Keteyian SJ, Kraus WE, et al. Exercise training and implantable cardioverter-defibrillator shocks in patients with heart failure: results from HF-ACTION (heart failure and a controlled trial investigating outcomes of exercise training). JACC Heart Fail 2013;1(2):142–8.

[82] Pandey A, Parashar A, Moore C, Ngo C, Salahuddin U, Bhargava M, et al. Safety and efficacy of exercise training in patients with an implantable cardioverter-defibrillator. JACC Clin Electrophysiol 2016. [Internet, Online, cited 2017 Jan. 9] Available from: http://linkinghub.elsevier.com/retrieve/pii/S2405500X16302110.

[83] Pfeffer MA, Braunwald E. Treatment of heart failure with preserved ejection fraction: reflections on its treatment with an aldosterone antagonist. JAMA Cardiol 2016;1(1):7.

[84] LeWinter MM, Meyer M. Mechanisms of diastolic dysfunction in heart failure with a preserved ejection fraction: If it's not one thing it's another. Circ Heart Fail 2013;6(6):1112–5.

[85] Smart N, Haluska B, Jeffriess L, Marwick TH. Exercise training in systolic and diastolic dysfunction: effects on cardiac function, functional capacity, and quality of life. Am Heart J 2007;153(4):530–6.

[86] Kitzman DW, Brubaker PH, Morgan TM, Stewart KP, Little WC. Exercise training in older patients with heart failure and preserved ejection fraction: a randomized, controlled, single-blind trial. Circ Heart Fail 2010;3(6):659–67.

[87] Edelmann F, Gelbrich G, Düngen H-D, Fröhling S, Wachter R, Stahrenberg R, et al. Exercise training improves exercise capacity and diastolic function in patients with heart failure with preserved ejection fraction: results of the Ex-DHF (exercise training in diastolic heart failure) pilot study. J Am Coll Cardiol 2011;58(17):1780–91.

[88] Kitzman DW, Brubaker P, Morgan T, Haykowsky M, Hundley G, Kraus WE, et al. Effect of caloric restriction or aerobic exercise training on peak oxygen consumption and quality of life in obese older patients with heart failure with preserved ejection fraction: a randomized clinical trial. JAMA 2016;315(1):36–46.

[89] Sagar VA, Davies EJ, Briscoe S, Coats AJS, Dalal HM, Lough F, et al. Exercise-based rehabilitation for heart failure: systematic review and meta-analysis. Open Heart 2015;2(1):e000163.

[90] Taylor RS, Sagar VA, Davies EJ, Briscoe S, Coats AJS, Dalal H, et al. Exercise-based rehabilitation for heart failure. Cochrane Database Syst Rev 2014;4. CD003331.

Chapter 12

Lifestyle and Heart Diseases in Choice Experiments

José M. Grisolía

Department of Applied Economics, Universidad de Las Palmas de Gran Canaria, Las Palmas de GC, Spain

INTRODUCTION

A healthy lifestyle consists of eating a healthy diet, doing regular physical activity (PA), and avoiding toxic habits such as smoking. An unhealthy diet is associated with an increase in obesity and cardiovascular diseases [1], and is related to diabetes, certain cancers and morbidity disabilities [2].

Governments, agencies, and industry are engaged in a battle to promote healthy behavior and to reduce heart diseases, and other major health problems. They do this by applying interventions that encourage people to follow a healthy lifestyle. From a traditional perspective, increasing price through taxation of unhealthy items has always been used [3,4]. Information and education campaigns in schools, community centers, and workplace are essential [5]. A more recent alternative consists of guiding people in a subtle way, using indirect suggestions to make them choose a healthy alternative. This is called libertarian paternalism [6] or nudge [7], and is usually based on the default option. Furthermore, direct incentives might work; there are experiments that apply financial incentives to encourage people to do regular PA or to give up smoking [8–10].

Choice experiments (CE) are used to analyze individuals' choices when they are facing a closed set of alternatives (see, for instance, Train [11]). CE alternatives are decomposed into characteristics—the so-called attributes. This is very useful when we are analyzing lifestyles since, by definition, a lifestyle is composed of a set of behaviors, habits, or tasks. The interesting point of CE is that it is usually combined with stated preferences (SP) (see, for instance, Louviere et al. [12]) that consists of designing hypothetical choice scenarios, with a set of alternatives. Individuals are asked to *state* which one they would prefer. SP methods of data collection are a good way to analyze *ex-ante* policy interventions without the limitations of the real world. In addition, models are more efficient and provide more information since the researcher has full control of the explanatory variables.

On the other hand, a CE provides willingness to pay (WTP) for every characteristic of the experiment, offering a set of prices that can be inserted into a cost-benefit analysis [13] and thus, policy makers can decide which strategy is more effective. The use of DCM in health care have has been pioneered by Ryan et al. [14] and Ryan and Gerard [15].

In this chapter I will review CE applied to the analysis of lifestyle interventions. In the next section I will explain the methodology of CE, followed by a review of the relevant experiments in the area of lifestyles, and finally, I draw some conclusions.

METHODOLOGY

Choice models [16] are used in situations when there is a closed set of J alternatives. Each alternative is defined by its utility U_j. It is assumed that individual q would select the alternative with the highest utility. However, there must be inconsistencies or a lack of information in the model; therefore, it is necessary to add an error term in the utility, and this implies that we will have to work in probabilistic rather than deterministic terms (Train, 2011). Therefore, the utility U_{qj} is decomposed into two parts: a deterministic utility V_{qj} that can be measured and the error term ε:

$$U_{qj} = V_{qj} + \varepsilon_{qj} \tag{1}$$

Notice that (1) depicts the utility of alternative j for individual q. The deterministic part is measured by decomposing each alternative into x_{qj} attributes [17]. These attributes are characteristics that describe the good. With regards to lifestyle, for instance, these could be the type of diet and time spent regularly in physical exercise. At the same time, each attribute is divided

Lifestyle in Heart Health and Disease. https://doi.org/10.1016/B978-0-12-811279-3.00012-4

into levels. These levels and attributes are combined following particular rules and creating artificial scenarios that are presented to each individual. This is called experimental design (for instance, [18]). Every attribute is weighted by parameters called *beta* that are the ultimate object of the estimation. We can split the deterministic part into a sum-product of attributes and its weights:

$$U_j = \beta_{0j} + \beta_{1j}\chi_{1j} + \cdots + \beta_{kj}\chi_{kj} + \varepsilon_{qj} \tag{2}$$

The estimated parameters would be those which maximize the likelihood function. The assumptions made about the type of error would lead to different models. The simplest is called multinomial logit (MNL), which assumes that error distributes extreme value type I. In this case, McFadden [19] demonstrated that the probability of choosing alternative j for individual q is calculated as

$$P_{jq} = \frac{\exp\left(\beta' x_{jK}\right)}{\sum_k \exp\left(\beta' x_{jK}\right)} \forall q, k \in J \tag{3}$$

This is the logit probability. In an MNL model, estimated parameter betas are considered fixed, that is, there is no heterogeneity in tastes and these have to be explored by inserting socioeconomic variables. Mixed logit (ML) is a more flexible model that assumes that betas are distributed among the population, and therefore, the role of the researcher is to estimate the parameters (mean and variance) that describe its underlying distribution $f(\beta|\theta)$. Thus, in the case of ML, parameters $\tilde{\beta}_q$ are considered random (unlike fixed in a MNL). For an ML model, the probability of choosing alternative j for individual q has a nonclosed form expression [20]:

$$P_{iq} = \int L_{iq}(\beta) f(\beta \mid \theta) d\theta \tag{4}$$

where $Liq(\beta)$ is the MNL probability of choice for a fixed value of β. A very useful alternative is to consider beta as discrete rather than a continuous distribution, leading to a latent class model that is becoming common in health economics [21].

Willingness to Pay

The output of a DCM is the estimated beta parameters, which show the importance of every attribute in the utility and the willingness to pay (WTP) for each attribute. In its simplest form, WTP is the marginal rate of substitution between any attribute k and the marginal utility of income. Since this is unknown, it is used as a parameter that represents a cost associated with the choice scenario:

$$\text{WTP}_k = -\frac{\beta_k}{\beta_{cost}}$$

However, this is only correct in the case of MNL. In the case of ML, if the cost parameter is random, there are problems in using this estimation [22], and other methods are recommended [23], particularly willingness-to-pay space [24], which consists in estimating WTP as another parameter in a nonlinear utility function.

CHOICE EXPERIMENTS ON LIFESTYLE INTERVENTIONS

Choice experiments have been used to explore health policies and their effectiveness. In what follows, I will present some of the most important issues addressed in relation to diet, exercise, and lifestyle choices using DCM.

Diet

A healthy diet is essential for preventing heart diseases and for well-being in general. Public agencies have responded to this with different set of interventions. For instance, in the United Kingdom, the Food Standard Agency (FSA) established the campaign of the "six per day of salt intake," and the Department for Public Health launched the "five-a-day" campaign to promote the consumption of fruits and vegetables. These campaigns have been accompanied by interventions related to how the information is presented to people so that they can make the right choice.

From an individual perspective, choosing food is a complicated process. Research based on CE has uncovered the main factors that determine choices. In general, in developed countries, as well as price, individuals consider other characteristics such as sensory attributes, convenience, and healthiness (see, for instance, [25]). For Kamphuis et al. [26], individuals seem to be driven by healthiness, taste, price, and travel time to the grocery store. Healthiness is a remarkable explanation, and it is more important for the richer and more educated individuals. Annunziata and Vecchio [27] analyze choice in the context of functional foods and conclude that the most important attributes are the base product and the health claim. These and other examples support the idea that healthiness seems to be very important in people's choices. How can this attribute be made more explicit to help individuals to make the right choice? There are several strategies tested with choice experiments: health claims [28], labels, and, within labels, the traffic light system. Table 1 summarizes these works.

Labels informing consumers about nutritional components can be described as *nutrition content claims*. In addition, *health claims* are "statements linking food components to a desired state of health" [34] such as "reduces cholesterol," "reduces the risk of heart disease," "light," and "fat-free." Both nutritional information and health claims positively influence purchase intention [35]. Some works with CE also confirm this: in testing how to make people eat more fruit, Schnettler et al. [36] discovered that health claims that incorporate messages such "contain antioxidants" and "prevent diseases" seem to be popular ways to increase fruit consumption. Grisolía et al. [37] include the use of a health claim in a choice experiment combining stated and revealed preferences about eating different sea urchin dishes; Øvrum et al. [30] applied a choice experiment with a sample of 408 individuals in Norway. The experiment about choosing semi hard cheese with different nutrition contents was related to saturated fat. Half of the sample received health information about saturated fat, and this influenced their choices. The authors consider that health information is a way to reduce educational differences in terms of nutrition.

A *nutrition content claim* shows content that is different from that of a *health claim*. Which one has more impact on decisions? Using a choice experiment applied to 400 Spanish shoppers, Gracia et al. [31] found that consumers are more

TABLE 1 Summary of Choice Experiments on Diet

Authors	Method	Attributes	Conclusions
Balcombe et al. [29]	A mail survey distributed to 3000 UK households getting 477 responses	• Salt • Sugar • Fat • Saturates • Price	TLS is an efficient way of communicating health information about food
Kamphuis et al. [26]	Sample of 399 adults in Eindhoven (the Netherlands)	• Taste • Healthiness • Price • Preparation time • Travel time to shop	Healthiness, taste, price, and travel time to shop are the most important factors in choosing food. Healthiness is more important for educated people
Øvrum et al. [30]	Sample of 408. DCM with RP: experiment. They apply a treatment with half of the participants receiving information about saturated fats	• Price • Regular saturated fat/low saturated fat • Regular-fat/low-fat cheese • Conventional/organic cheese	Information did affect people's choices and works to reduce the gap of uneducated people
Gracia et al. [31]	A sample of 400 food shoppers in Aragon (Spain)	• Price: 2/3 € per box • Brand: Well-known/unknown • Nutritional fact panel label: Yes/No • Nutritional claim label: "light claim," Yes/No	Nutrition claims are more important in determining people's choices
Montandon and Colli [32]	A sample of 80 individuals. Cape Town, South Africa	• GDA • TLS • Health claim • Price	TLS system is the best orientation to costumers
Loureiro and Rahmani [33])	119 participants recruited among university students. Using best-worse method and applying a treatment of informing about calories	They display real menus from a fast-food restaurant	Information about calories does not have an impact on actual behavior

influenced by nutrition labels than health claim. They offered them two sets of breakfast cookies combining labels and a health claim as attributes. Individuals were more influenced by labels showing a double WTP premium. Nevertheless, this work analyzes only a particular health claim: "light."

One way to enhance nutrition claims is the traffic light system (TLS) [29]. This is a nutritional food labeling system that indicates the levels of four key nutrients (fat, sugar, saturates, and salt) where the colors, from green to red, represent the relative level of content per 100 g. This is a simple and clear way to inform consumers about the healthiness of any processed food. This system has been adopted by the United Kingdom. At present, regulations about nutrition labels in the European Union are based on Regulation (EU) No 1169/2011 which come into force in 2014. This regulation makes labels easier to read, with clear information about allergens and other nutritional facts. However, TLS is a voluntary system.

An alternative to TLS is the Guideline Daily Amount (GDA) symbols (see Fig. 1 below). This is a system developed in the United Kingdom in 1998. It divides the content of the food into five basic elements—calories, sugars, fat, saturated fat, and salt—and compares this content with the recommended portion of these ingredients for the average person. Recently, the GDA system has been replaced by the Reference Intake (RI) system.

Montandon et al. [38] were applying a CE to analyze the three systems: TLS, color-coded Guideline Daily Amount (GDA) symbols, and the health endorsement logo (see Fig. 2 below). The health endorsement logo is a health claim that is certified by an official institution. In this case, it was the National Heart Foundation. Using a sample of 80 preselected shoppers in Cape Town (South Africa), they asked them to choose between pairs of fast-food options combining the three systems using a CE. They conclude that the TLS is the best for the orientation of people's choices.

Scarborough et al. [39] analyze an important question: Which of the five nutrients represented in TLS has more impact upon individuals' choices? Which colors have a greater effect? They recruited supermarket shoppers in the United Kingdom and launched an online experiment with a large sample of 187 individuals. They conducted a paired comparison choice experiment and a logistic regression concluding that a food product with a high red rating is more likely to dissuade people from purchasing it than a product with a high green rating is to attract people to buy it. It seems that salt and saturated fat are the two items that individuals pay the most attention to.

A different policy could be to inform customers about calorie content with a calorie labeling system. A best-worst system [40,41] was applied on a sample of students in Spain. Their aim was to notice how information about calories affected their behavior. They also intended to trace their real choices compared with their stated ones. The context was choosing

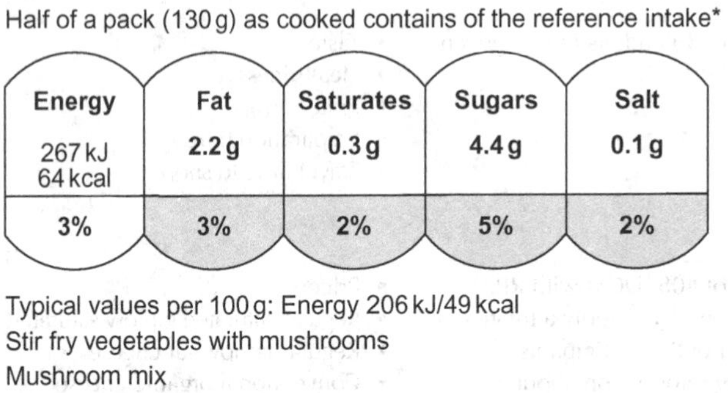

Half of a pack (130 g) as cooked contains of the reference intake*

Energy	Fat	Saturates	Sugars	Salt
267 kJ 64 kcal	2.2 g	0.3 g	4.4 g	0.1 g
3%	3%	2%	5%	2%

Typical values per 100 g: Energy 206 kJ/49 kcal
Stir fry vegetables with mushrooms
Mushroom mix

FIG. 1 GDA label.

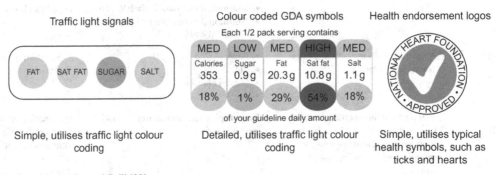

FIG. 2 Examples from Montandon and Colli [32].

within fast-food alternatives. They concluded that providing calorie information did not have impact on actual behavior. Nevertheless, calorie labeling has been demonstrated to have an impact on certain group of individuals such as women [42], interested individuals [43], and those who are on diet [44].

Physical Activity

A second habit that constitutes an essential part of any healthy lifestyle is practicing moderate physical activity (PA). There is extensive evidence on the benefits of a regular physical activity in preventing cardiovascular diseases [45]. Despite all the efforts made by government and public agencies, PA remains low in most developed countries, and inactivity can be considered a pandemic [46].

The US guidelines [47] set a general recommendation of at least 150 minutes per week of moderate-intensity aerobic physical activity. The 2013 report shows that 51.6% of individuals met the aerobic activity guideline. This report demonstrates the importance of urban facilities to promote physical activity. These facilities included places such as parks, playgrounds, sidewalks, walking paths, and biking paths. The British Heart Foundation in their 2015 report [48] states "our data suggest that levels of sedentary behavior remain stubbornly high in the UK." The UK guidelines set recommendations of regular PA for all adults aged 19–64 and specifically by age range. The simplest message is to do 30 minutes of fast walking. Nevertheless, they found that 67% of men in England and Scotland meet the guidelines.

It is difficult to find CE based entirely on physical activity (see Table 2). Most studies that address the issue of physical activity are part of a broader lifestyle intervention where there is also diet and other aspects to consider. Usually,

TABLE 2 Summary of Choice Experiments on Physical Activity

Author(s)	Method	Attributes	Conclusions
Franco et al. [49]	220 older people with previous history of falls or mobility problems	• Exercise type • Time spent on exercise per day frequency • Transport type • Travel time • Out-of-pocket costs • Reduction in the chance of falling and improvement in the ability to undertake tasks inside and outside of home	The key point to make a PA program popular is the accessibility because travel time has a strong negative impact and they showed their preference for home-based exercise
Farooqui et al. [50]	A choice experiment with a sample of 1000 Singaporeans over 50 years old	• Average number of sessions per week: 1, 2, 3 • Cost to travel to sessions: free, $2, $5 • Time to travel to sessions (minutes): 15, 25, 30, 45 • Incentive payment at 6 months: $150, $200, $300, $450 • Type of incentive: cash, medisave credit, supermarket voucher, sporting goods voucher • Enrollment fee: free, $20, $50	Financial incentives worked as a way to increase participation in PA programs. People prefer cash among the incentives. 1 week per month program and 450SGD as incentive
Roux et al. [51]	A sample of 164 overweight adults	• Cost ($): 250, 500, 750, 1000 • Travel time required to attend (minutes): 15, 30, 45, 60 • Extent of physician involvement: none, monthly, counseling every 2 weeks • Components: diet only, diet and exercise; diet, exercise and self-management • Focus: general group, group-specific, personal	Travel time is more important for those who are nonobese. General individuals show a high WTP for a 3-month weight-loss program
Ryan et al. [52]	A sample of 504 adults in the United Kingdom	• Weight change • Short-term and long-term health gains • Time spent on the intervention • Financial costs incurred	People are more interested in doing exercise rather than changing their eating habits. This reinforces the need for financial compensation so they can change their lifestyles

physical activity interventions that use CE are addressed to specific segments of the population, in particular older people. Experiments involving individuals over 65 are important, as the number of people in this age group is expected to triple in the next 30 years [53]. These works usually comprise attributes to describe PA programs such as access (travel time, convenience, and at home or not), type of exercises, cost, and potential health benefits.

Franco et al. [49] carried out a systematic review of PA interventions for older people. Attributes might be summarized as follows: firstly the benefits of attending sessions—improvements in the ability to undertake tasks at home, improvements in the ability to leave home to carry out other activities and socialize, and reducing their chances of falling; secondly the type of exercise; and thirdly convenience, that is, how to access the sessions. The surprising finding of this work is that the older population seems to be less interested in health benefits paying more attention to the type of exercise. On the other hand, transport seems to be a key finding:

"The two attribute levels with the highest utility identified in the present study, *exercise at home* and *no need to use transport*, give support to the idea that easy access to exercise programs is generally preferred (…) reflecting participants' aversion to travel long distances and to pay for high-cost exercise programs. In the context of chronic conditions, provision of healthcare services at home or close to home has recently gained growing attention in different countries."

Farooqui et al. [50] applied a CE in Singapore over a sample of people over 50. The experiment proposed they to take part in a 6-month program that implied 45 minutes of walking and 15 minutes of stretching. The activity comes with an enrollment fee, and the work also included a travel time attribute. Not surprisingly, individuals preferred the less costly and most paid activity. A $500 cash incentive would lead to a 58.5% enrollment rate. The conclusion is that cash works as a powerful incentive.

Physical activity loyalty (PAL) is a scheme designed to provoke a behavioral change through points and rewards as a result of meeting certain targets of PA. Hunter et al. [54] explain a workplace intervention of this style in Belfast (United Kingdom). In this paper, they describe the experiment, though results are not yet shown: a group of volunteers are encouraged to do PA during 6 months. Their exercises will be tracked by sensors placed in their workplace and the vicinity. Their activity is transformed into points, and subjects can redeem these points into vouchers.

Lifestyle in General

A final possibility is to combine both diet and exercise, creating a complete lifestyle change. There is some CE applied in this area. The issues discussed here are about the composition of lifestyle interventions and, therefore, public preferences and how to make them more effective (see Table 3).

Physical Activity Versus Diet

As lifestyle is usually composed of diet and exercise, it is interesting to explore which habit is preferred. Grisolía et al. [56] analyzed lifestyle choices in the context of cardiovascular diseases. They interviewed a representative sample of 493 individuals in Northern Ireland aged 40–65 years. Researchers employed the Nottingham QUICK risk algorithm [59] to calculate the real risk of a heart attack for every participant. This risk was used to set a tailored status quo in percentage, which was explained in simple terms to each respondent. Individuals were then asked about their diet, following an adaptation to the Irish food of the Block screener for fat [60] given the impracticality of obtaining reliable information about diet during the interview. The Block screener is a questionnaire of eating frequency of 17 selected items characterized by a high fat or sugar content; 17 items were then sufficient to undertake the project. With this information about diet and the baseline of their actual risk, individuals were offered, using a stated preference experiment, improvements in terms of risk of CVD as a result of diet and exercise. It is worthy noting that the experiment was fully tailored since the status quo was set specifically to each respondent using his actual risk and eating habits. Apart from WTP for reducing the risk of CVD, results show that overweight and obese individuals prefer to do exercise rather than change their diets.

Ryan et al. [52] explore an intervention to convince people to change their lifestyle with a program of PA and diet. In terms of attributes, this program is defined as type of program, benefits in terms of weight change, health benefits, time spent, and cost. Although there is some evidence that it would be easier to convince people to follow a diet rather than carry out PA, this work suggests exactly the opposite: people prefer physical activity.

Roux et al. [51] use a sample of 164 overweight adults enrolled in a weight-loss program. They use a CE to explore participant's preferences. Their attributes include convenience (travel time to program); cost and service characteristics (focus of program, whether the program includes a doctor who would check out the participant progress); and the composition (diet; diet and exercise; or diet, exercise, and self-management). First, they found a large WTP for a 3-month weight-loss program (up to 600$). Depending on individual's needs, their attribute preferences were different—for

TABLE 3 Summary of Choice Experiments and Lifestyles

Author(s)	Method	Attributes	Conclusions
Giles et al. [55]	356 adults, following a quota sample. From a panel in the United Kingdom Web questionnaire	• Smoking cessation • Regular physical activity • Attending vaccination • Attending screening	Cash rewards were as preferred as no rewards at all
Grisolía et al. [56]	493 individuals in Northern Ireland aged 40–65 years. Random parameter logit	• Diet (grams of fat) • Exercise (minutes) • Cost • Risk of a fatal heart attack	People prefer exercise. WTP for reducing by 1% points the risk of a fatal CVD event in the next 10 years equal to 0.632 GBP per week. This implies 282.85 GBP in 10 years
Grisolía et al. [21]	A sample of 384 individuals aged 40–65 in Northern Ireland. A latent class model	• Eating frequency of selected items • Exercise • Cost • Risk of fatal heart attack	There is a skeptical segment that refuses to follow any healthy recommendation
Wanders et al. [57]	A sample of 206 patients with type 2 diabetes mellitus in the Netherlands	• Menu schedule: flexible, general, elaborate. • Physical activity schedule: flexible, general, elaborate • Consult structure: individual, in a group with five others, in a group with 10 others • Expected outcome: no weights loss, weight loss of 5 kg, weights loss of 10 kg • Money: rewards or costs (75, 150, 225 €)	Out-of-pocket costs worked to reduce the willingness to participate in the intervention. Financial rewards increase participation, but as the rewards increases, participation decreases
Ryan et al. [52]	A sample of 504 adults in the United Kingdom	• Weight change • Short-term and long-term health gains • Time spent on the intervention • Financial costs incurred	People are more interested in doing exercise rather than changing their eating habits. This reinforces the need for a financial compensation so they can change their lifestyles
Owen et al. [58]	55 individuals with metabolic syndrome participate in a 16-week lifestyle intervention. Having of the sample received sessions to instruct them in nutrition	• Diet • Exercise • Cost/rewards	People in the intervention group are more interested in doing exercise rather than diet

instance, nonobese individuals were more interested in reducing travel time to the place of the program. The most important attribute was "program components" (i.e., moving from diet to diet and exercise).

Time and Instruction

Metabolic syndrome is a cluster of factors that increases the risk of cardiovascular diseases, type 2 diabetes mellitus, and other causes of mortality [61]. Therefore, it is in the particular interest of this chapter to analyze the lifestyles related to metabolic syndrome. This is the work done by Owen et al. [58] applying a choice experiment on 55 individuals that suffer from metabolic syndrome. This sample was allocated into an intervention group and a control group. The intervention group received instruction about nutrition, cooking, and PA; one hour of exercise class weekly; full access to a gym; and healthy and free food. This lasted for 16 weeks. After this, a subgroup within the intervention group received a follow-up: a monthly support session for 8 months.

All participants had to fill out a SP questionnaire with several attributes: cost, support, exercise duration and structure, diet, and health outcome (weight loss). The idea was to check if the intervention group offered different responses compared with the control group. Individuals were offered different types of intervention and asked which one they preferred and what would be the likelihood of their continuing the program (doing diet and exercise) for one year. Firstly, individuals placed in the control group paid more attention to diet attributes, whereas those in the intervention group were more interested in exercise. After 16 weeks, preferences moved toward self-directed rather than organized/supervised PA. Success was measured in terms of gaining or losing abdominal fat. In this aspect, gainers of fat during the 16 weeks considered that this was the result of external factors.

Financial Incentives

An increasingly popular approach in health promotion is the use of financial incentives [62]. These are rewards for attaining certain targets in terms of lifestyle. They could be cash, vouchers, discounts, or free gym access. The rational behind this lies in time preferences since individuals value the present pleasure more than future benefits.

> Due to the present-biased time preferences (...) people tend to attach lower value to the delayed and uncertain benefits of healthy behavior or preventive services, while they are attracted by the immediate reward of unhealthy habits (...) Thus, economic theory suggests that financial incentives can be effective as they increase the value of gratification and benefits from healthy behavior or costs of unhealthy habits. ([62], p. 406)

In general, monetary incentives work. They can be used to dissuade people from unhealthy behaviors (out-of-pocket cost—for instance, [1,63,64], on smoking cessation) or to persuade them to engage in good habits. In the first case, taxes or other means can be used to stimulate smoking cessation. In the second case, the incentive could be in the form of a reward to individuals enrolled in lifestyle interventions, provided they achieve certain targets [65–68]. The effect of both types of interventions has been explored by Wanders et al. [57]. They applied a CE aiming to analyze a lifestyle intervention on individuals with type 2 diabetes. The interesting point of their experiment is that they do not analyze volunteers since they might be affected by self-selection bias. In addition, they explore the effect of both costs and rewards. Their results show that costs reduce the willingness to participate in the program but increasing rewards does not lead to increasing participation as expected. The explanation might be connected to the reciprocity of the social exchange theory [69]; in that individuals think that the greater the reward for an activity, the more unpleasant it must be.

Acceptability

Even if we agree that financial incentives work in health promotion, paying for an individual either to quit smoking, to attend screening, or to exercise could be interpreted as rewarding bad behavior or an incentive to cheat [70–72], and thus, raises a case for public acceptability. For this reason, Giles et al. [55] proposed a CE to explore to what extent the general population is against financial reward for these improvements in lifestyle. They use a sampled quota panel representative of the UK population and showed them alternative lifestyle interventions including PA, diet, smoking cessation, and regular health screenings. As reward, along with health benefits, individuals would receive cash, lottery tickets, or vouchers. Respondents were not asked to enroll in these interventions (hypothetically) but to give their approval as a health policy. Participants considered cash incentives to be as bad as no rewards at all; therefore, we might discard any problem of acceptability of financial incentives. Nevertheless, they find that people are reluctant to accept this type of intervention if it only applies to a particular segment of the population (older people, pregnant women, etc.).

Skepticism

Even the best designed intervention could fail if participants are unconvinced about its effect on their health. What if they are skeptical about the connection between behavior and health? This is the issue explored by Grisolía et al. (2015). Using a sample of 384 individuals aged 40–65 in Northern Ireland, the questionnaire was a SP in which they had to elicit changes in their lifestyle comprising diet, exercise, and money and obtaining a reduction in their risk of suffering a fatal CVD disease. With this data, they applied a latent class model (see, for instance, Green and Hensher [73]). This type of model produces segments into the population defined by different socioeconomic characteristics and tastes. They inserted a health locus control (HLC) questionnaire into the model [74,75] using Likert scale. The HLC analyzes to what extent individuals think they are in control of their health. Some individuals might think it depends on luck, genes, or God, and that they could suffer a CVD disease or other health problems no matter what type of lifestyle they follow. Rotter [76] defines HLC as follows:

> When a reinforcement is perceived by the subject as [...] not being entirely contingent upon his action, then, in our culture, it is typically perceived as the result of luck, chance, fate, as under the control of powerful others, or as unpredictable because of the great complexity of the forces surrounding him. When the event is interpreted in this way by an individual, we have labelled this a belief in external control. If the person perceives that the event is contingent upon his own behaviour or his own relatively permanent characteristics, we have termed this a belief in internal control. (p1)

The model generates three classes: *Skeptical pessimistic*, which is the type of individuals that are not interested in any lifestyle changes despite their possible health benefits. They have a determinist sense about their health which, they believe, depends entirely on external factors. This segment has a higher correlation with smokers, men, and uneducated individuals, accounting for 29% of the sample. The second class is called the *wishful thinkers*. They are unhappy with their current situation though they are willing to change their lifestyle with exercise and diet to improve their health. Finally, there is a class of *healthy optimists* that, even though they are satisfied with their situation now, are happy to do more PA and diet. The existence of a skeptical class accounting for almost one-third of the sample shows the difficulties of policy interventions.

CONCLUSIONS

Choice experiments are useful to analyze policy interventions on lifestyles choices. This might be critical in preventing heart diseases. From this review, some ideas about people's preferences and effective policies arise.

Healthiness is a key attribute in explaining people's choices of food. It follows that the right policy is to make the information available in a clear and simple way with the right labeling policy. The most usual labels are nutrition facts and health claims. The first is a declaration of certain basic ingredients: sugar, fat, saturated fat, and salt. The second is a message emphasizing healthy characteristics of the food product. A system of colors such as the traffic light system or using percentages as daily guideline recommendations are effective messages. Another possibility is to inform consumers about calories. Finally, a health claim certified by an official institution—for instance, in the case of heart diseases and for the United Kingdom this could be the British Heart Foundation—is a powerful message to orientate customers. All these labeling strategies have a positive effect in orienting people's choices.

Unlike the case of diet, information is not the key strategy for encouraging people to engage in more physical activity. From the choice experiment literature, it seems that convenience is a key variable for interventions, such as closed facilities and a good neighborhood design for walking, biking, and other moderate exercise. Physical activity loyalty is a promising intervention that uses financial incentives redeemed through points and uses devices to track participant's progress.

In the context of heart diseases, people are willing to do more exercise rather than diet to reduce their risk of suffering a fatal CVD. For patients with metabolic syndrome, which increases the risk of heart disease, the movement toward exercise increases if there is a monitoring intervention. This experiment also shows that individuals usually blame external factors to justify their lack of results.

The most successful interventions are those which use financial incentives. These may be in the form of a cost or reward and cash, rather than vouchers, is the preferred way. Nonetheless, even the best designed policy may fail if policy makers do not consider the existence of a segment of people who refute any connection between their lifestyle and health outcomes and will therefore show a higher resistance to change.

REFERENCES

[1] Borgmeier I, Westenhoefer J. Impact of different food label formats on healthiness evaluation and food choice of consumers: a randomized-controlled study. BMC Public Health 2009;9(1):184.

[2] Etilé F. Social norms, ideal body weight and food attitudes. Health Econ 2007;16(9):945–66. https://doi.org/10.1002/(ISSN)1099-1050.

[3] Chaloupka FJ, Yurekli A, Fong GT. Tobacco taxes as a tobacco control strategy. Tob Control 2012;21(2):172–80.

[4] Wagenaar AC, Salois MJ, Komro KA. Effects of beverage alcohol price and tax levels on drinking: a meta-analysis of 1003 estimates from 112 studies. Addiction 2009;104(2):179–90.

[5] Nutbeam D. Health promotion glossary. Health Promot Int 1998;13(4):349–64.

[6] Thaler RH, Sunstein CR. Libertarian paternalism. Am Econ Rev 2003;93(2):175–9.

[7] Leonard TC, Thaler RH, Sunstein CR. Nudge: improving decisions about health, wealth, and happiness. Constit Polit Econ 2008;19(4):356–60.

[8] Cahill K, Perera R. Competitions and incentives for smoking cessation. Cochrane Database Syst Rev 2011;13(4).

[9] Hall J. Disease prevention, health care, and economics. In: Glied S, Smith PC, editors. The oxford handbook of health economics. Oxford: Oxford University Press; 2011. p. 555–77.

[10] Kane RL, Johnson PE, Town RJ, Butler M. A structured review of the effect of economic incentives on consumers' preventive behavior. Am J Prev Med 2004;27(4):327–52.

[11] Train KE. Discrete choice methods with simulation. Cambridge: Cambridge University Press; 2009.

[12] Louviere JJ, Hensher DA, Swait JD. Stated choice methods: analysis and applications. Cambridge: Cambridge University Press; 2000.

[13] McIntosh E. Using discrete choice experiments within a cost-benefit analysis framework. Pharmacoeconomics 2006;24(9):855–68.

[14] Ryan M, Gerard K, Amaya-Amaya M, editors. Using discrete choice experiments to value health and health care. In: Vol. 11. Springer Science & Business Media, Dordrecht; 2007.

[15] Ryan M, Gerard K. Using discrete choice experiments to value health care programmes: current practice and future research reflections. Appl Health Econ Health Policy 2003;2(1):55–64.

[16] Ortúzar JdeD, Willumsen LG. Modelling transport. 4th ed. Chichester: Wiley; 2011.

[17] Lancaster KJ. A new approach to consumer theory. Journal of Political Economy 1966;74(2):132–57.

[18] Hensher DA, Rose JM, Greene WH. Applied choice analysis: a primer. Cambridge, UK: Cambridge University Press; 2005.

[19] McFadden DL. Conditional logit analysis of qualitative choice behavior. In: Zrembka P, editor. Frontiers in Econometrics. New York: Academic Press; 1974.

[20] Ortúzar J de D, Willumsen LG. Modelling Transport. Chichester: John Wiley & Sons; 2001.

[21] Grisolía JM, Longo A, Hutchinson G, Kee F. Applying health locus of control and latent class modelling to food and physical activity choices affecting CVD risk. Soc Sci Med 2015;132:1–10.

[22] Armstrong P, Garrido R, de Dios Ortúzar J. Confidence intervals to bound the value of time. Transp Res E: Logistics Trans Rev 2001;37(2):143–61.

[23] Sillano M, Ortúzar JD. Willingness-to-pay estimation with mixed logit models: some new evidence. Environ Plann A 2005;37(3):525–50.

[24] Train KE, Weeks M. Discrete choice models in preference space and willingness-to-pay space. In: Scarpa R, Alberini A, editors. Applications of simulation methods in environmental and resource economics. Dordrecht: Springer; 2005. p. 1–16.

[25] Grunert KG. Food quality and safety: consumer perception and demand. Eur Rev Agric Econ 2005;32(3):369–91.

[26] Kamphuis CB, de Bekker-Grob EW, van Lenthe FJ. Factors affecting food choices of older adults from high and low socioeconomic groups: a discrete choice experiment. Am J Clin Nutr 2015;101(4):768–74.

[27] Annunziata A, Vecchio R. Consumer perception of functional foods: a conjoint analysis with probiotics. Food Qual Prefer 2013;28(1):348–55.

[28] Roe B, Levy AS, Derby BM. The impact of health claims on consumer search and product evaluation outcomes: results from FDA experimental data. J Public Policy Market 1999;89–105.

[29] Balcombe K, Fraser I, Di Falco S. Traffic lights and food choice: a choice experiment examining the relationship between nutritional food labels and price. Food Policy 2010;35(3):211–20.

[30] Øvrum A, Alfnes F, Almli VL, Rickertsen K. Health information and diet choices: Results from a cheese experiment. Food Policy 2012;37(5):520–9.

[31] Gracia A, Loureiro M, Nayga Jr. RM. Consumers' valuation of nutritional information: a choice experiment study. Food Qual Prefer 2009;20:463–71.

[32] Montandon AC, Colli C. Effective nutrition labels for fast food consumers. Br Food J 2016;118(10):2534–49.

[33] Loureiro ML, Rahmani D. The incidence of calorie labeling on fast food choices: a comparison between stated preferences and actual choices. Econ Human Biol 2016;22:82–93.

[34] Williams P. Consumer understanding and use of health claims for foods. Nutr Rev 2005;63(7):256–64.

[35] Kozup JC, Creyer EH, Burton S. Making healthful food choices: the influence of health claims and nutrition information on consumer evaluations of packaged food products and restaurant menu items. J Mark 2003;67:19–34.

[36] Schnettler B, Miranda H, Lobos G, Sepulveda J, Orellana L, Mora M, Grunert K. Willingness to purchase functional foods according to their benefits: Consumer profiles in Southern Chile. Br Food J 2015;117(5):1453–73.

[37] Grisolía JM, López F, de Dios Ortuzar J. Sea urchin: from plague to market opportunity. Food Qual Prefer 2012;25(1):46–56.

[38] Montandon AC, Montandon AC, Colli C, Colli C. Effective nutrition labels for fast food consumers. Br Food J 2016;118(10):2534–49.

[39] Scarborough P, Matthews A, Eyles H, Kaur A, Hodgkins C, Raats MM, et al. Reds are more important than greens: how UK supermarket shoppers use the different information on a traffic light nutrition label in a choice experiment. Int J Behav Nutr Phys Act 2015;12(1):151.

[40] Louviere J, Timmermans H. Stated preference and choice models applied to recreation research: a review. Leis Sci 1990;12(1):9–32.

[41] Finn A, Louviere JJ. Determining the appropriate response to evidence of public concern: the case of food safety. J Public Policy Market 1992;12–25.

[42] Krieger JW, Chan NL, Saelens BE, Ta ML, Solet D, Fleming DW. Menu labeling regulations and calories purchased at chain restaurants. Am J Prev Med 2013;44(6):595–604.

[43] Arsenault JE. Can nutrition labeling affect obesity. Choices 2010;25(3):1–4.

[44] Variyam JN. Role of demographics, knowledge, and attitudes in America's eating habits: changes and consequences. In: Frazao E, editor. Agricultural information bulletin 750. Washington, DC: US Department of Agriculture; 1999. p. 281–94.

[45] Franco OH, de Laet C, Peeters A, Jonker J, Mackenbach J, Nusselder W. Effects of physical activity on life expectancy with cardiovascular disease. Arch Intern Med 2005;165(20):2355–60.

[46] Kohl 3rd. HW, Craig CL, Lambert EV, Inoue S, Alkandari JR, Leetongin G, et al. The pandemic of physical inactivity: global action for public health. Lancet 2012;380:294–305.

[47] Physical Activity Guidelines Advisory Committee. Physical Activity Guidelines Advisory Committee Report, 2008. In: U.S. Department of Health and Human Services, editor. Washington, DC: Physical Activity Guidelines Advisory Committee; 2008.

[48] Townsend N, Wickramasinghe K, Williams J, Bhatnagar P, Rayner M. Physical activity statistics 2015. London: British Heart Foundation; 2015.

[49] Franco MR, Howard K, Sherrington C, Ferreira PH, Rose J, Gomes JL, et al. Eliciting older people's preferences for exercise programs: a best-worst scaling choice experiment. J Physiother 2015;61(1):34–41.

[50] Farooqui MA, Tan YT, Bilger M, Finkelstein EA. Effects of financial incentives on motivating physical activity among older adults: results from a discrete choice experiment. BMC Public Health 2014;14(1):141.

[51] Roux L, Ubach C, Donaldson C, Ryan M. Valuing the benefits of weight loss programs: an application of the discrete choice experiment. Obes Res 2004;12(8):1342–51.

[52] Ryan M, Yi D, Avenell A, Douglas F, Aucott L, Van Teijlingen E, Vale L. Gaining pounds by losing pounds: preferences for lifestyle interventions to reduce obesity. Health Econ Policy Law 2015;10(2):161–82.

[53] World Health Organization (WHO). Global health and aging. http://www.who.int/ageing/publications/global_health.pdf [accessed 14.11.13].

[54] Hunter RF, Brennan SF, Tang J, Smith OJ, Murray J, Tully MA, et al. Effectiveness and cost-effectiveness of a physical activity loyalty scheme for behaviour change maintenance: a cluster randomised controlled trial. BMC Public Health 2016;16:618. https://doi.org/10.1186/s12889-016-3244-1.

[55] Giles EL, Becker F, Ternent L, Sniehotta FF, McColl E, Adams J. Acceptability of financial incentives for health behaviours: a discrete choice experiment. PLoS ONE 2016;11(6):e0157403.

[56] Grisolía JM, Longo A, Boeri M, Hutchinson G, Kee F. Trading off dietary choices, physical exercise and cardiovascular disease risks. Soc Sci Med 2013;93:130–8.

[57] Wanders JO, Veldwijk J, de Wit GA, Hart HE, van Gils PF, Lambooij MS. The effect of out-of-pocket costs and financial rewards in a discrete choice experiment: an application to lifestyle programs. BMC Public Health 2014;14(1):870.

[58] Owen K, Pettman T, Haas M, Viney R, Misan G. Individual preferences for diet and exercise programmes: changes over a lifestyle intervention and their link with outcomes. Public Health Nutr 2010;13(02):245–52.

[59] Hippisley-Cox J, Coupland C, Vinogradova Y, Robson J, May M, Brindle P. Derivation and validation of QRISK, a new cardiovascular disease risk score for the United Kingdom: prospective open cohort study. BMJ 2007;335(7611):136.

[60] Block G, Gillespie C, Rosenbaum EH, Jenson C. A rapid food screener to assess fat and fruit and vegetable intake. Am J Prev Med 2000;18(4):284–8.

[61] Kaur J. A comprehensive review on metabolic syndrome. Cardiol Res Pract 2014;2014:1–21.

[62] Tambor M, Pavlova M, Golinowska S, Arsenijevic J, Groot W. Financial incentives for a healthy life style and disease prevention among older people: a systematic literature review. BMC Health Serv Res BMC Ser – Open Inclusive Trust 2016;16(Suppl. 5):426.

[63] Meier KJ, Licari MJ. The effect of cigarette taxes on cigarette consumption, 1955 through 1994. Am J Public Health 1997;87(7):1126–30.

[64] Wilson NA, Thomson GW. Tobacco tax as a health protecting policy: a brief review of the New Zealand evidence. N Z Med J 2005;118(1213):1213.

[65] Schmidt H. Bonuses as incentives and rewards for health responsibility: a good thing? J Med Philos 2008;33(3):198–220.

[66] Sigmon SC, Patrick ME. The use of financial incentives in promoting smoking cessation. Prev Med 2012;55(Suppl):S24–32.

[67] Spence J, Holt N, Dutove J, Carson V. Uptake and effectiveness of the Children's Fitness Tax Credit in Canada: the rich get richer. BMC Public Health 2010;10:356.

[68] Volpp KG, Troxel AB, Pauly MV, Glick HA, Puig A, Asch DA, et al. A randomized, controlled trial of financial incentives for smoking cessation. N Engl J Med 2009;360(7):699–709.

[69] Cook KS, Cheshire C, Rice ER, Nakagawa S. Social exchange theory. In: Handbook of social psychology. Netherlands: Springer; 2013. p. 61–88.

[70] Giles E, McColl E, Sniehotta F, Adams J. Acceptability of financial incentives and penalties for encouraging uptake of healthy behaviours: focus groups. BMC Public Health 2015;2015:15(58).

[71] Giles E, Robalino S, Sniehotta F, Adams J, McColl E. Acceptability of financial incentives for encouraging uptake of healthy behaviours: a critical review using systematic methods. Preventive Medicine 2015;2015:10.

[72] Whelan B, Thomas K, Van Cleemput P, Whitford H, Strong M, Renfrew M, et al. Healthcare providers' views on the acceptability of financial incentives for breastfeeding: a qualitative study. BMC Pregnancy Childbirth 2014;14:355. https://doi.org/10.1186/1471-2393-14-355. PMID: 25296687.

[73] Hensher DA, Greene WH. The mixed logit model: the state of practice. Transportation 2003;30(2):133–76.

[74] Wallston KA, Wallston BS, DeVellis R. Development of the multidimensional health locus of control (MHLC) scales. Health Educ Behav 1978;6(1):160e–70e.

[75] Skinner EA. A guide to constructs of control. J Pers Soc Psychol 1996;71(3):549.

[76] Rotter J. Generalized expectancies for internal versus external control of reinforcement. Physchol Monogr 1966;80(1):609.

FURTHER READING

[1] Centers for Disease Control and Prevention (CDC). Adult participation in aerobic and muscle-strengthening physical activities—United States. MMWR Morb Mortal Wkly Rep 2011;62(17):326.

[2] Coast J, Horrocks S. Developing attributes and levels for discrete choice experiments using qualitative methods. J Health Serv Res Policy 2007;12(1):25–30.

[3] Giles E, Robalino S, McColl E, Sniehotta F, Adams J. Systematic review, meta-analysis and meta-regression of the effectiveness of financial incentives for encouraging healthy behaviours. PLoS ONE 2014;9(3):e90347.11.

[4] Lappalainen R, Kearney J, Gibney M. A pan EU survey of consumer attitudes to food, nutrition and health: an overview. Food Qual Prefer 1998;9(6):467–78.

[5] Marteau T, Ashcroft R, Oliver A. Using financial incentives to achieve healthy behaviour. Br Med J 2009;338. https://doi.org/10.1136/bmj.b141510.

[6] National Center for Chronic Disease Prevention and Health Promotion (2014). State indicator report of physical activity.

[7] Troiano RP, Berrigan D, Dodd KW, Masse LC, Tilert T, McDowell M. Physical activity in the United States measured by accelerometer. Med Sci Sports Exerc 2008;40(1):181–8.

[8] Viney R, Lancsar E, Louviere J. Discrete choice experiments to measure consumer preferences for health and healthcare. Expert Rev Pharmacoecon Outcomes Res 2002;2(4):319–26.

[9] Vos T, Corry J, Haby MM, Carter R, Andrews G. Cost-effectiveness of cognitive–behavioural therapy and drug interventions for major depression. Aust N Z J Psychiatry 2005;39(8):683–92.

[10] Wakefield MA, Loken B, Hornik RC. Use of mass media campaigns to change health behaviour. Lancet 2010;376:1261–71 [how anti-smoking campaigns have been successful].

[11] Williams P. Consumer understanding and use of health claims for foods. Nutr Rev 2005;63(7):256–64.

[12] Bekker-Grob EW, Ryan M, Gerard K. Discrete choice experiments in health economics: a review of the literature. Health Econ 2012;21(2):145–72.

Chapter 13

Lost in Translation: What Does the Physical Activity and Health Evidence Actually Tell Us?

Darren E.R. Warburton, Shannon S.D. Bredin

*Physical Activity Promotion and Chronic Disease Prevention Unit, University of British Columbia (UBC), Vancouver, BC, Canada

INTRODUCTION

Physical inactivity is considered to be the fourth leading risk factor for global mortality accounting for roughly 3.2 million deaths annually [1]. The World Health Organization estimates that one in four adults is not active enough globally [1]. Importantly, the prevalence of physical inactivity is frequently higher than other risk factors. For instance, in Canada, physical inactivity is more prevalent than other risk factors for heart disease (Fig. 1). Similarly, a recent Australian report revealed that from the age of 30, the population risk of heart disease attributable to inactivity was greater than other risk factors (including a high body mass index, smoking, and high blood pressure) [2].

The health benefits of routine physical activity and exercise are well established [3–5]. Habitual physical activity has been associated with a reduced risk for premature mortality and is a highly effective primary and secondary preventative strategy for at least 25 chronic medical conditions [3,6–8]. Most international physical activity guidelines recommend achieving 150 min of MVPA physical activity (or 75 min of vigorous physical activity) [1]. This volume of activity has consistently been shown to be associated with 20%–30% risk reductions for chronic disease and premature mortality. Many agencies have advocated this level of physical activity as the minimal level for health benefits. However, by doing so, a major knowledge translation may have been introduced forming a significant barrier for those that serve to benefit greatly from becoming more physically active (particularly the elderly and those living with chronic medical conditions) [4]. Accordingly, the primary purpose of this chapter is to examine the current evidence relating physical activity to health benefits with a particular emphasis on the minimal and optimal volumes of activity for morbidity and premature mortality. In particular, we look to examine more closely the evidence that may have been "lost in the translation" of international physical activity messaging. We also examine why these knowledge translation errors may have been introduced and the potential impact upon physical activity promotion and uptake at the individual and population levels.

SHAPE OF THE DOSE RESPONSE CURVE: DOES IT MATTER?

It is important to look closely at the shape of the dose-response curve between physical activity/fitness and health outcomes to fully appreciate the potential errors in knowledge translation. There is overwhelming evidence demonstrating a dose-dependent relationship between physical activity/fitness and health indicators (for at least 25 chronic medical conditions) and premature mortality [4] (Fig. 2). The shape of the relationship between physical activity/fitness varies depending upon the medical condition and/or primary outcome [9]; however, generally, a curvilinear relationship exists such that relatively small volumes of physical activity can lead to marked health benefits with an attenuation at the highest volumes of physical activity (Fig. 2). Relative risk reductions of 20%–30% are often observed when examining the relationship between physical activity and morbidity and mortality end points [4]. Even greater risk reductions (i.e., 50% or more) are seen when objective measures of health-related physical fitness (particularly cardiorespiratory fitness) are related to health outcomes [4,7,10].

A volume of activity less than half of what is currently recommended (i.e., less than 75 min of MVPA intensity physical activity) provides marked health benefits in apparently healthy individuals and those living with chronic medical conditions.

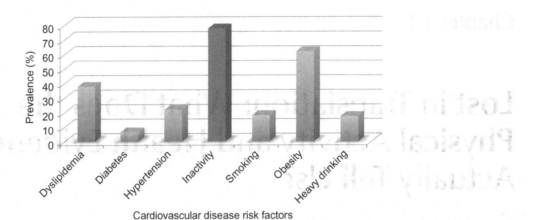

Cardiovascular disease risk factors

FIG. 1 Prevalence of traditional risk factors for cardiovascular disease in Canada according to sex. Dyslipidemia was defined as LDL-C (≥3.5 mmol/L), or a TC:HDL-C ratio ≥5.0, or self-reported use of a lipid-modifying medication. Diabetes (in 12 years and older individuals) was diagnosed via a health professional. Hypertension was defined as a measured systolic blood pressure ≥140 mmHg, or a measured diastolic blood pressure ≥90 mmHg, or a self-reported diagnosis of high blood pressure, or the self-reported use of antihypertensive medication. Physical activity (aged 18 to 79 years) was evaluated via accelerometry. Inactive individuals were considered those who engaged in less than 150 min of MVPA (in 10 min bouts). Smokers (aged 12 and over) included those that reported being a current smoker. Obesity was directly measured by body mass index of ³30 kg/m². Heavy drinking (aged 12 and older) included those who reported having five or more drinks on one occasion, at least once a month in the past 12 months. Data from Canadian Health Measures Survey (2012–13). *From Warburton DE, Bredin SS. Reflections on physical activity and health: what should we recommend? Can J Cardiol 2016;32(4):495–504.*

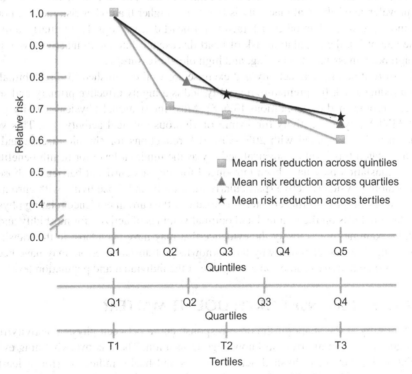

FIG. 2 Relative risk for all-cause mortality across physical activity/fitness categories. Data compiled from studies involving over 1.5 million participants as evaluated in the systematic review of Warburton and colleagues [7]. *From Warburton DE, Bredin SS. Reflections on physical activity and health: what should we recommend? Can J Cardiol 2016;32(4):495–504.*

The evidence is clear; the greatest health benefits are observed when inactive/fit individuals simply become more physically active/fit [4]. An overwhelming body of evidence including epidemiological studies [11–19], randomized controlled trials [20], and systematic reviews/meta-analyses [7,21,22] indicates that health benefits can be accrued at relatively low volumes of physical activity. For instance, Lee and colleagues [15] revealed that as little as 1 h/week of walking was associated with a reduction in coronary heart disease-related morbidity in women. Wisloff et al. [19] also reported that one weekly bout of high-intensity exercise was associated with a lower risk of cardiovascular death relative to those reporting no activity in both men (RR=0.61 and 95% CI=0.49–0.75) and women (RR=0.49 and 95% CI=0.27–0.89). No additional benefit was seen with higher durations or frequency of exercise sessions [19]. Wen and coworkers [23] recently examined the minimum amount of activity required for reduced mortality and extended life expectancy. In this study, 90 min/week (15 min/day) of moderate-intensity exercise was associated with a reduced risk for deaths related to cardiovascular disease, all cancers, diabetes, and all causes. Approximately 15 min/day of exercise was associated with a 14% risk reduction for all-cause mortality, and every additional 15 min of exercise (to a maximum of 100 min/day) provided an additional risk reduction of 4% and 1%, respectively, for all-cause and all-cancer mortality [23]. The greatest health benefits were observed during the first 1–2 h of exercise per week. Lee and colleagues [24] also observed that weekly running of <51 min, <6 miles, 1–2 times per week, <506 MET-min, or <6 mph led to a significant reduction in the risk of premature mortality (in comparison with not running) with a 3-year life expectancy benefit. Runners had 30 and 45% lower risk of all-cause and cardiovascular-related mortality, respectively, in comparison with nonrunners. The authors highlighted that running at slow speeds (< 6 miles/h) for 5–10 min/day can lead to significant health benefits.

Importantly, this research collectively reveals that 150 min/week of MVPA intensity physical activity provides risk reductions for many chronic medical conditions and premature mortality that is closer to the optimal than the minimal levels [4]. For instance, Moore and colleagues [25] revealed that any physical activity level (0.1–3.74 MET-h/week of leisure time MVPA, equivalent to 75 min/week of brisk walking) was associated with 1.8 year of life gained, with high levels of physical activity leading to greater gains (e.g., 4.5 year at 22.5+ MET-h/week or brisk walking for 450+ min/week). Similarly, Arem and colleagues [26] recently reported that any level of physical activity participation (i.e., from 0.1 to <7.5 MET-h/week) lowered the risk for premature mortality by approximately 20%. Engaging in the recommended level of 150 min MVPA had a mortality benefit of 31%, which was closer to the optimal risk reduction (i.e., 39%) observed at 3–5 times the physical activity recommendation. Similar relationships were also observed for mortality related to cardiovascular disease and cancer. Importantly, simply moving from an inactive state to any level of physical activity participation was associated with the greatest relative risk reduction as compared with those reporting no leisure time physical activity.

It is acknowledged that select studies may provide a relatively poor glimpse into the overall relationship between physical activity and health status. Therefore, systematic reviews and meta-analyses of large volumes of data provide important insight into the minimal and optimal dosage of physical activity for health benefits. Our systematic reviews of the literature (involving millions of participants) revealed that there is a clear dose-response relationship between physical activity and premature mortality and several chronic medical conditions [7,21]. Current recommendations of 150 min MVPA were associated with marked health benefits; however, health benefits were clearly observed with remarkably small volumes of physical activity with an attenuation at the highest dosages, with no apparent threshold for benefit. These findings have been supported by other systematic reviews and meta-analyses. For example, Lollgen and colleagues [22] revealed a dose-response relationship between physical activity and all-cause mortality. Inactive/sedentary individuals benefited greatly from engaging in low to moderate exercise intensities with "only a minor additional reduction with further increase in activity level." Sattelmair and colleagues [27] in a meta-analysis not only highlighted the health benefits associated with 150 min/week MVPA but also acknowledged that people that engaged in levels below the minimum recommended amount also had significantly lower risk for coronary heart disease. Those individuals that were physically active at half of the current recommendations (275 kcal/week) demonstrated a 14% lower risk of coronary heart disease (RR=0.86 and 95% CI=0.76–0.97). In a summary for the article, the authors stated "the biggest bang for the buck for coronary heart disease risk reduction occurs at the lower end of the activity spectrum: very modest, achievable levels of physical activity."

In a recent systematic review and Bayesian dose-response meta-analysis, Kyu et al. [28] reported that higher volumes of total physical activity (from multiple domains including leisure time, occupation, domestic, and transportation) are associated with a reduced risk of breast cancer, colon cancer, diabetes, heart disease, and ischemic strokes. Although the findings from this systematic review are difficult to compare with other reviews that focused on leisure time physical activity, it does provide important insight into the importance of being physically active. The authors concluded that total physical activity (not just leisure time) should be several times higher than recommended levels for greater reductions in the risk for these diseases. However, the authors also acknowledged that the major health benefits were seen at the lower levels of activity, with an attenuation of the benefits at physical activity levels higher than 3000–4000 MET-min/week.

Clinical exercise gerontology and medicine research further supports the health benefits that can be accrued at low volumes of physical activity/exercise [4,29–33]. For example, clinical exercise rehabilitation programs commonly provide exercise prescriptions that are well below the 150 min/week MVPA threshold [4,29–32]. A recent observational evaluation of 18,028 adults with type 1 diabetes revealed that self-reported physical activity (at least 1–2 sessions per week) was associated with improved glycemic control, diabetes-related comorbidities, and cardiovascular risk factors without an increase in adverse events [34]. Similarly, another recent epidemiological trial [35] examined the role of physical activity and television watching (and other sedentary behaviors) on the progression from gestational diabetes to type 2 diabetes in 4554 women from the Nurses' Health Study II. The authors revealed that physical activity participation was associated (in a dose-dependent fashion) with a lower risk of type 2 diabetes. Each 5 MET-h/week increment in total physical activity (equivalent to 100 min/week of moderate-intensity physical activity) was associated with a 9% lower risk for type 2 diabetes. An increase in physical activity was also associated with a lower risk of developing type 2 diabetes [35] such that women who increased their physical activity levels by 7.5 MET-h/week or more (equal to 150 min/week moderate-intensity physical activity) had a 47% lower risk for type 2 diabetes. Television watching was also associated (in a dose-dependent fashion) with the risk for type 2 diabetes; however, this association was attenuated after adjustment for body mass index. In a recent systematic review of the literature, Hupin and colleagues [33] revealed that a low dose of physical activity (1–499 MET-min/week (15 min/day)) was associated with a marked reduction (22%) in all-cause mortality in older adults with further benefits at higher dosages (i.e., 28% at current recommendations and 35% at 1000 MET-min/week).

This research collectively reveals that a volume of physical activity of half (or less than) current international physical activity recommendations can lead to a marked reduction risks for multiple chronic medical conditions and premature mortality [4]. It is clear that 150 min/week of MVPA can lead to health benefits; however, a clear threshold for health benefit does not appear to exist and messaging that implies as much does not appear to be evidence-based.

HEALTH-RELATED PHYSICAL FITNESS VS. PHYSICAL ACTIVITY

The importance of health-related physical fitness for the prevention of premature mortality and the primary and secondary prevention of chronic disease is well established [36,37]. In particular, cardiorespiratory fitness has reliably been connected to greater risk reductions for chronic disease and premature mortality [10,16,38–40]. As we discussed recently [4], this may be the result (at least in part) of increased precision of measurement for aerobic fitness versus physical activity [39] and various other environmental, genetic, and constitutional factors [41,42]. Cardiorespiratory fitness has recently been advocated as simple "vital sign" that should be incorporated into cardiovascular risk factor evaluation and treatment [37]. It is not uncommon to observe 50%–60% risk reductions when comparing the most physically fit individuals with the least fit leading many to promote the importance of examining and promoting cardiorespiratory fitness [4,36,37]. For instance, Myers and colleagues in a review of the literature recently demonstrated that every 1 MET increase in cardiorespiratory fitness was associated with a 10%–25% reduction in the risk for premature mortality in men and women [43]. Importantly, these authors reported that current evidence suggests that the premature mortality risk reduction may be even greater in low-fit individuals (i.e., peak MET of <5) who engage in cardiac rehabilitation and improve their peak MET (i.e., approximately 30% per MET improvement) [43]. Research from this team also established that a peak exercise capacity of approximately 5–6 MET is a key point wherein marked changes in mortality risk appear to occur. For instance, Kokkinos and colleagues [44] recently developed age-specific exercise capacity thresholds (i.e., 8–9, 7–8, 6–7, and 5–6 MET for <50, 50–59, 60–69, and ≥70 year, respectively) for mortality risk in males. Thresholds were created based on a specific MET level associated with no increase in mortality risk. Mortality risk was progressively higher in those individuals with a peak MET level below each threshold. The authors revealed that for every 1 MET increase in exercise capacity, mortality risk was 12% lower for the entire cohort, 15% for those <60 years, and 11% for those ≥60 years. These findings are consistent with the findings of Paterson and colleagues that demonstrated a threshold for independent living at the age of 85 years of approximately 15 and 18 mL kg^{-1} min^{-1}, respectively, for women and men [45]. A recent study [46] revealed that high exercise capacity was associated independently with a reduced risk for premature mortality after (i.e., 28, 90, and 365 days) a first myocardial infarction (Fig. 3). It should be highlighted that many clinicians and exercise professionals avoid the usage of threshold-based messaging related to cardiorespiratory fitness and health indicators emphasizing the important health benefits that can be seen with relatively small changes in fitness (similar to the arguments put forward related to physical activity and health). This appears to be a prudent, evidence-based approach to the promotion of the importance of increasing aerobic capacity (and health-related physical fitness).

The relationship between cardiorespiratory fitness and health status extends well beyond diseases of the cardiovascular system. For instance, Kokkinos and colleagues [47] recently demonstrated an independent and graded relationship between cardiorespiratory fitness and the incidence of chronic kidney disease. The risk for chronic kidney disease was 22% lower for every 1 MET increase in cardiorespiratory fitness. The relative risk reductions were 13%, 45%, and 58%, respectively, for low-fit (6.5 ± 1.0 MET), moderate-fit (7.7 ± 0.9 MET), and high-fit (9.5 ± 1.0 MET) individuals in comparison with the least-fit (4.8 ± 0.9 MET) participants.

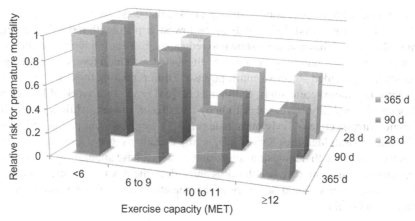

FIG. 3 Premature mortality was defined as all-cause mortality within 28, 90, or 365 days of a myocardial infarction. Referent group included individuals with an aerobic fitness below 6 METs. *Data from Shaya GE, Al-Mallah MH, Hung RK, Nasir K, Blumenthal RS, Ehrman JK, et al. High exercise capacity attenuates the risk of early mortality after a first myocardial infarction: the Henry Ford exercise testing (FIT) project. Mayo Clin Proc 2016;91(2):129–39.*

Changes in physical fitness over the life span are highly predictive of the risk for multiple chronic medical conditions and premature mortality [3]. An increase in fitness reduces the risk, while a decrease in fitness increases the risk [3,48–52]. The vast majority of this research relates to changes in aerobic fitness over a prolonged period of time. The greatest changes in health status occur in the least-fit individuals who become more physically fit. Blair and colleagues [50] revealed that previously inactive individuals that became physically fit (over a 5-year period) had a 44% risk reduction for premature mortality in comparison with those that remained unfit. Erikssen and colleagues also revealed that enhanced physical fitness was associated with a graded reduction in the risk for premature mortality [48,49]. Individuals that have a high aerobic fitness at baseline and are able to maintain or improve their fitness levels are at the lowest risk [49,50]. These authors further reported that very small changes in fitness were associated with a reduced risk for premature mortality. Every 13% increase in physical fitness was associated with a 30% reduction in the risk for all-cause mortality (over a 13-year period) [49]. These findings were supported by the findings of Kokkinos and colleagues [51] who examined male veterans ($n=5314$, aged 65–92 years). The authors revealed that there was an inverse and graded relationship between aerobic fitness and all-cause mortality in older men. Men who increased their aerobic fitness from the low fitness category had a 35% reduced risk for all-cause mortality (in comparison with those that maintained a low fitness level ($\leq$4 MET)). For each 1 MET increase in aerobic fitness, there was a 12% lower risk for premature mortality. Individuals that achieved a fitness level of 5.1–6.0 MET had a 38% lower risk, while those that had a fitness level of >9 MET had a 61% lower risk (compared with the least-fit individuals). Similarly, Brawner and colleagues [52] recently revealed that a change in maximal exercise capacity over time was inversely related to the risk of all-cause mortality in both men and women ($n=10,854$, 43% women). Each 1 MET increase in aerobic fitness was associated with an approximate 15% lower risk for all-cause mortality in both men and women. Increasing from low (i.e., <8 MET) to intermediate or high fitness levels was associated with a reduced risk for all-cause mortality in both men (37%) and women (44%). Collectively, these findings are consistent with our previous recommendations wherein we [53] stated that "it is important to have a high aerobic fitness early into life, and be able to maintain or increase this fitness across the lifespan for the optimal reduction in the risk for chronic disease and premature mortality."

It is also important to highlight the emerging evidence supporting the importance of achieving and maintaining a high level of musculoskeletal fitness (i.e., muscular strength, muscular endurance, muscular power, flexibility, and back fitness) [3,54,55]. In two systematic reviews of the literature [54,55], we found that enhanced musculoskeletal fitness was associated positively with functional status, glucose homeostasis, bone health, mobility, psychological well-being, and overall quality of life and negatively associated with fall risk, morbidity, and premature mortality. Similarly, enhancements in musculoskeletal fitness via exercise training had a positive effect on health status (particularly in individuals with a low musculoskeletal fitness and functional reserve) [3,7]. Other recent studies have shown that musculoskeletal fitness is a significant predictor of weight gain over a 20-year period [56], quality of life [57], mobility [58], and premature mortality [59]. Musculoskeletal fitness across the life span is very important for the maintenance of functional status in the frail elderly and those living with chronic medical conditions [3,54,55,60].

For clinicians, it is important to highlight that simple measures of musculoskeletal fitness (such as grip strength) are related to the risk for chronic disease, disability, and premature mortality [53,59,61–65]. For instance, Rantanen et al. [61] revealed that an increased rate of decline in muscular strength (>1.5% per year) or a very low grip strength (< 21 kg) was associated with a greater incidence of chronic conditions (e.g., type 2 diabetes, stroke, arthritis, coronary heart disease, and pulmonary disorders) over a 27-year period. Individuals in the lowest grip strength tertile had an eightfold increased risk for disability. High muscular strength has also been shown to reduce the risk for functional limitations over a 5-year period [66], and the risks associated with low muscular strength are attenuated with high levels of routine physical activity [67].

The InCHIANTI study [58] provided objective cutoff points for differentiating those with poor mobility (i.e., 30 kg for men and 20 kg for women). In a more recent study, Sallinen et al. [68] refined the work of Lauretani et al. [58] reporting that grip strength cut points for mobility increased with body mass index in men, while a single threshold value (21 kg) was appropriate for women (irrespective of body mass index) [68]. Using these cut points and Canadian data from York University [57] and Canadian Health Measures Survey [69], many Canadians would approach this functional threshold by the age of 69. Collectively, this research demonstrates a strong evidence base supporting the inclusion of daily activities that tax the musculoskeletal system to improve health status and reduce the risk for chronic disease, disability, and premature mortality [7]. Once again, experts should be aware of the potential perils of using simple cut-points (or thresholds) for musculoskeletal fitness and health.

INTERNATIONAL MESSAGING AND KNOWLEDGE TRANSLATION

The translation of scientific evidence relating physical activity and health outcomes has seen concerted efforts at the national and international levels. Physical activity guidelines have been published and widely disseminated throughout the world [70] including harmonized international recommendations from the World Health Organization [1]. How the actual scientific evidence has been translated to the general population has been very interesting providing an important case study in how knowledge translation errors can be introduced within health promotion strategies. For instance, in the United States, the Centers for Disease Control and Prevention (CDC; a highly influential organization) made a very clear statement regarding the minimal level of physical activity advocated for health benefits based on the 2008 Physical Activity Guidelines for Americans [71]. In a website dedicated to "Physical Activity Basics," the CDC posed the question "how much physical activity do adults need?" The CDC further went on to state that "according to the 2008 Physical Activity Guidelines for Americans, *you need to do* two types of physical activity each week *to improve* your health-aerobic and muscle-strengthening. For Important Health Benefits: Adults *need at least* 2 hours and 30 minutes (150 minutes) of moderate-intensity aerobic activity (i.e., brisk walking) every week and muscle-strengthening activities on 2 or more days a week that work all major muscle groups (legs, hips, back, abdomen, chest, shoulders, and arms) [71]."

The World Health Organization provides similar recommendations [72] stating "Adults aged 18–64 should do *at least* 150 minutes of moderate-intensity aerobic physical activity throughout the week or do at least 75 minutes of vigorous-intensity aerobic physical activity throughout the week or an equivalent combination of moderate- and vigorous-intensity activity. Aerobic activity should be performed in bouts of at least 10 minutes duration. For additional health benefits, adults should increase their moderate-intensity aerobic physical activity to 300 minutes per week, or engage in 150 minutes of vigorous-intensity aerobic physical activity per week, or an equivalent combination of moderate- and vigorous-intensity activity. Muscle-strengthening activities should be done involving major muscle groups on 2 or more days a week." Similar statements are promoted in other nations such as Canada and the United Kingdom.

These messages are often provided alongside contradictory statements highlighting how small the volumes of physical activity can lead to health benefits. For instance, the American College of Sports Medicine states "adults should get *at least* 150 minutes of moderate-intensity exercise per week. Gradual progression of exercise time, frequency and intensity is recommended for best adherence and least injury risk. People unable to meet these minimums can still benefit from some activity [73]."

Messaging that states that one "must" reach a certain level of physical activity (i.e., 150 min/week of MVPA) to accrue health benefits implies that there is a threshold or minimal level below which point limited health benefits are achieved. As we recently discussed [4], threshold-based messaging would indicate that the relationship between physical activity and health status would be "L" or "S" shaped (see Fig. 4) [4] with health benefits only occurring at the threshold of 150 min/week MVPA. However, as reviewed above, there is limited evidence to support a health benefit threshold. In fact, there is overwhelming and irrefutable evidence that the dose-response relationship is such that the largest health benefits are seen when moving away from the inactive state to a more active state, such that a small change in physical activity and/or fitness can lead to marked health benefits. There also appears to be an attenuation of benefit in extremely active/fit individuals [4].

Statements such as "you need to do," "must," or "*at least*" imply that there is a threshold below which health benefits are not accrued, whereas, the statements such as "should aim for" are more reflective of a continuum of health benefits arising from becoming more physically active. As we discussed previously, the simple turn of phrase from "should" to "must"

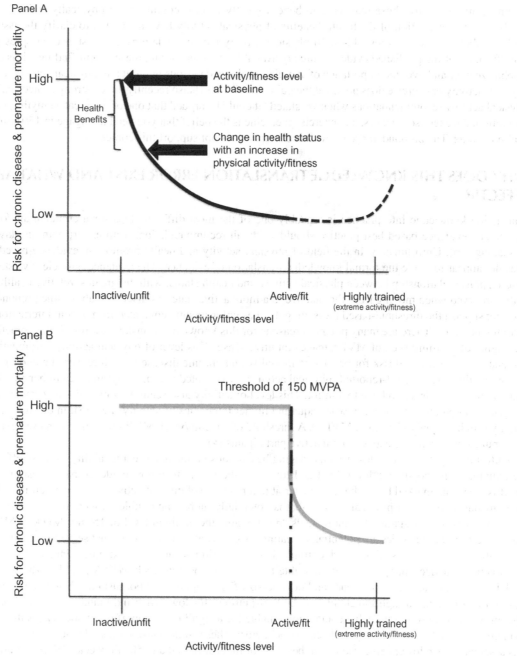

FIG. 4 (A) In individuals who are physically inactive/unfit, a small change in physical activity/fitness will lead to a significant improvement in health status including a reduction in the risk for chronic disease and premature mortality. Dashed line represents the potential attenuation in health status seen in highly (extremely) trained endurance athletes. (B) If current messaging regarding physical activity (i.e., individuals should engage in at least 150 min of weekly moderate to vigorous physical activity (MVPA) for health benefits) were evidence-based the shape of the dose response curve (*blue line;* *light gray in print versions*) would show a clear threshold at 150 min of MVPA wherein the benefits are accrued. Thus, the relationship would be "L-" or "S-" shaped. However, the overwhelming evidence indicates that this not the case. *(A) Modified with permission from Bredin SSD, Jamnik V, Gledhill N, Warburton DER. Effective pre-participation screening and risk stratification. In: Warburton DER, editor. Health-related exercise prescription for the qualified exercise professional. 6th ed. Vancouver: Health & Fitness Society of BC; 2016. p. 1–30.*

or "need to" can have a significant impact on the messaging and how it is interpreted and used by the general population [4]. The inconsistencies in messaging related to physical activity and health evidence have been raised in various countries (including Canada, the United Kingdom, and the United States) [4,37,74,75]. It has become apparent that recommendations containing "thresholds" for minimal levels of physical activity are not strictly evidence-based and are often the result of consensus interpretation of the evidence [4,37]. As discussed later, this failure in knowledge translation may present a significant barrier for those that serve to benefit greatly from becoming more physically active [4]. Recently, authors involved in the evaluation of the health benefits of physical activity have attempted to clarify the issue and highlight the evidence that contradicts threshold-based physical activity recommendations. For instance, Despres [37] stated recently that "from all of the published evidence already available and considering these recent findings, it is clear that we should emphasize over and over the importance of avoiding prolonged periods without any physical activity, even if the volume of physical activity or exercise does not meet the guidelines." We [4] also recently took a strong stance against threshold-based physical activity recommendations when we stated "it could be argued that one of the greatest myths perpetuated within PA promotion, the exercise sciences, and exercise medicine is the belief that you need to engage in 150 min/week of MVPA for health benefits. The preponderance of evidence simply does not support this contention."

WHY DOES THIS KNOWLEDGE TRANSLATION ERROR EXIST AND WHAT ARE THE EFFECTS?

Translating knowledge into practice is arguably one of the most difficult steps for a clinician and/or research scientist. Ultimately, evidence-based best practice should guide all recommendations with expert opinion playing a minor role in messaging [76]. Unfortunately, in the field of physical activity and health promotion, evidence-based best practice often does not appear to guide the formal knowledge translation of evidence. The evidence base clearly establishes that there is dose-response relationship between physical activity and health status, with the greatest relative health benefits being generally observed when moving from the inactive to a more active state. As discussed above, the preponderance of evidence does not support the threshold-based messaging that has unfortunately emanated from recent international physical activity recommendations. There are many potential reasons for this knowledge translation error. From a practical perspective, the messaging of 150 min/week of MVPA makes intuitive sense. This level of physical activity participation has been clearly associated with a reduced risk for premature mortality and chronic disease [7]. Moreover, the current evidence would indicate that this level of (self-reported) physical activity is associated with near optimal health benefits [26]. As such, many feel that this target is a good one to aim for. This level of activity also seemingly provides an easy target to understand for the general population particularly when equating higher intensities of activity (i.e., 75 min of vigorous-intensity physical activity roughly equals 150 min of MVPA). A threshold of 150 min/week MVPA has also been used extensively for surveillance purposes to separate active from inactive participants [4].

Unfortunately, this threshold-based messaging has various perils that affect its ability to cause effective change at both the individual and population level. First and foremost, the failure to follow evidence-based best practice is a critical error. It is clear that marked health benefits occur at much lower volumes of activity. However, current threshold messaging recommends a volume of physical activity that is potentially more than double on what is required for clinically relevant health benefits [4]. By stating that individuals "must attain" the minimal level of 150 min/week of MVPA for health benefits, experts and/or agencies are potentially creating a significant barrier to those that serve to benefit from becoming more physically active (i.e., those living with chronic medical conditions and/or the extremely inactive).

In actual practice, many clients are not able to reach the 150 min/week MVPA goal [77,78]. Also, arbitrary targets/guidelines often foster an "all-or-none" and/or "one-size-fits-all approach." For instance, Knox and colleagues [74] recently reported that creating a high threshold may be "off-putting" for low-active individuals. They stressed how the target of 150 min/week MVPA was unrealistic and unobtainable for a significant proportion of society (as this would translate into an increase of 100%–400% in physical activity behavior). These authors also indicated that this threshold-based messaging was associated with lower perceived health benefits for modest (less than 150 min/week MVPA) durations [74] despite the clear health benefits seen at this volume of activity.

There are various other issues surrounding generic physical activity messaging that affect greatly the ability to reach a critical mass and/or elicit significant health changes at the individual or population level. For instance, a growing body of evidence indicates that the general population is often unaware and/or does not effectively use generic physical activity guidelines [77,79,80]. This is further compounded by the fact that even key health and fitness organizations often report a relative lack of uptake despite concerted efforts to increase the dissemination of messaging from physical activity guidelines [81]. Generic physical activity guidelines have also increasingly been criticized for the inability to provide meaningful recommendations that can be used by qualified exercise professionals and/or other healthcare professionals with individual clients. Patient-centered care (or personalized medicine) requires the customization of healthcare with evidence-based best practice and communication strategies being tailored to the individual client. As such, generic physical activity guidelines with basic recommendations are seldom used in actual clinical practice [77,78]. For instance, qualified exercise professionals prescribe exercise prospectively and objectively [77,78]. It is clear that physical activity goals must be attainable and individualized to the client's needs [4]. The preponderance of the literature indicates that relatively small changes in physical activity/fitness can lead to clinically relevant changes in health status including improvements in functional status and reduced risk for premature mortality and chronic disease [4]. As such, the recommendation of the threshold of 150 min/week MVPA would not be consistent with evidence-based best practice or effective patient-centered care particularly for those that serve to benefit greatly from simply becoming more active.

It is also important to highlight the inherent errors associated with self-reported physical activity levels that are often not addressed in current messaging. The majority of data relating physical activity levels to health benefits have been derived from epidemiological trials using self-report. This includes the majority of studies that make up key systematic reviews in the field. A careful review of the literature and the accuracy of self-report data reveals that the dose of physical activity reported in survey-based physical activity studies could be overestimated by 50% or more [4,69,82,83]. Therefore, prescribing a volume of activity or exercise based on these data would not be advisable. It is not clear why current surveillance strategies still rely on physical activity dosages that were established via self-report particularly when other more robust technologies are available (such as accelerometry). The marked overestimation of self-reported physical activity is acknowledged in the scientific literature but unfortunately is seldom translated in current physical activity recommendations. This further reflects the need to promote the benefits of simply becoming more physically active.

Generic physical activity guidelines have also been criticized for the inability to address the key factors responsible for improving healthy lifestyle behaviors [84–86]. For instance, Segar and coworkers [85] recently stated "There is a need to identify new messages that can make physical activity more relevant and compelling to patients and the general population." Affective judgment constructs (e.g., affective attitude, enjoyment, and intrinsic motivation) have been associated positively with physical activity behavior [87–89]. There is extensive evidence indicating that enjoyment is a key factor in the pursuit and maintenance of healthy physical activity behaviors [90]. Researchers have accordingly emphasized the need to highlight well-being through enjoyable physical activities. The promotions of positive affective benefits (such as "feeling good") appear to resonate more with individuals and have more potential for promoting behavioral change than simple messages related to health (e.g., "you must engage in a 150 min of MVPA to achieve health benefits") [85]. It has been argued that current international recommendations (and related messaging and counseling) should progress to include the latest advancements in affective and behavioral science [85].

CONCLUSIONS

The evidence relating physical activity to health benefits clearly supports the potential for 150 min/week of MVPA to reduce markedly the risks for premature mortality and at least 25 chronic medical conditions. However, this evidence has been met with an unfortunate knowledge translation error such that key messages are often "lost in translation." The focus on health benefits occurring at a threshold of 150 min/week of MVPA is simply not evidence-based, but pervasive in global messaging related to international physical activity guidelines. Clinically, relevant health benefits can be accrued at very small volumes of activity (e.g., likely less than half of current recommendations). Simply moving more can lead to health benefits and improvements in well-being. The evidence is irrefutable; virtually, everyone can benefit from becoming more physically active [4]. Current knowledge translation strategies should move away from a traditional health-benefit-based approach emphasizing the positive effects of routine physical activity on enjoyment, well-being, and vitality [85].

ACKNOWLEDGMENTS

The authors have received funding from the Michael Smith Foundation for Health Research, the Canadian Institutes of Health Research, the Canada Foundation for Innovation, the BC Knowledge Development Fund, the Natural Sciences and Engineering Research Council of Canada, the BC Ministry of Health, and GE Healthcare.

DISCLOSURES

The authors were responsible for creating the systematic reviews (and related papers) that informed the 2011 Canadian Physical Activity Guidelines.

REFERENCES

[1] World Health Organization. Global recommendations on physical activity for health. Switzerland: World Health Organization; 2010.

[2] Brown WJ, Pavey T, Bauman AE. Comparing population attributable risks for heart disease across the adult lifespan in women. Br J Sports Med 2015;49(16):1069–76.

[3] Warburton DE, Nicol C, Bredin SS. Health benefits of physical activity: the evidence. Can Med Assoc J 2006;174(6):801–9.

[4] Warburton DE, Bredin SS. Reflections on physical activity and health: what should we recommend? Can J Cardiol 2016;32(4):495–504.

[5] Warburton DER, Taunton J, Bredin SSD, Isserow S. The risk-benefit paradox of exercise. BC Med Assoc J 2016;58(4):210–8.

[6] Warburton DE, Nicol C, Bredin SS. Prescribing exercise as preventive therapy. Can Med Assoc J 2006;174(7):961–74.

[7] Warburton DE, Charlesworth S, Ivey A, Nettlefold L, Bredin SS. A systematic review of the evidence for Canada's Physical Activity Guidelines for Adults. Int J Behav Nutr Phys Act 2010;7:39.

[8] Pedersen BK, Saltin B. Exercise as medicine - evidence for prescribing exercise as therapy in 26 different chronic diseases. Scand J Med Sci Sports 2015;25(Suppl. 3):1–72.

[9] Gledhill N, Jamnik V. Canadian Physical Activity, Fitness and Lifestyle Approach. 3rd ed. Ottawa: Canadian Society for Exercise Physiology; 2003.

[10] Warburton DER, Katzmarzyk PT, Rhodes RE, Shephard RJ. Evidence-informed physical activity guidelines for Canadian adults. Appl Physiol Nutr Metab 2007;32(2E):S16–68.

[11] Kushi LH, Fee RM, Folsom AR, Mink PJ, Anderson KE, Sellers TA. Physical activity and mortality in postmenopausal women. JAMA 1997;277(16):1287–92.

[12] Leon AS, Connett J, Jacobs Jr DR, Rauramaa R. Leisure-time physical activity levels and risk of coronary heart disease and death. The Multiple Risk Factor Intervention Trial. JAMA 1987;258(17):2388–95.

[13] Paffenbarger Jr. RS, Hyde RT, Wing AL, Lee IM, Jung DL, Kampert JB. The association of changes in physical-activity level and other lifestyle characteristics with mortality among men. N Engl J Med 1993;328(8):538–45.

[14] Lee IM, Skerrett PJ. Physical activity and all-cause mortality: what is the dose-response relation? Med Sci Sports Exerc 2001;33(6):S459–71. discussion S93-4.

[15] Lee IM, Rexrode KM, Cook NR, Manson JE, Buring JE. Physical activity and coronary heart disease in women: is "no pain, no gain" passe? JAMA 2001;285(11):1447–54.

[16] Myers J, Kaykha A, George S, Abella J, Zaheer N, Lear S, et al. Fitness versus physical activity patterns in predicting mortality in men. Am J Med 2004;117(12):912–8.

[17] Myers J, Prakash M, Froelicher V, Do D, Partington S, Atwood JE. Exercise capacity and mortality among men referred for exercise testing. N Engl J Med 2002;346(11):793–801.

[18] Villeneuve PJ, Morrison HI, Craig CL, Schaubel DE. Physical activity, physical fitness, and risk of dying. Epidemiology 1998;9(6):626–31.

[19] Wisloff U, Nilsen TI, Droyvold WB, Morkved S, Slordahl SA, Vatten LJ. A single weekly bout of exercise may reduce cardiovascular mortality: how little pain for cardiac gain? 'The HUNT study, Norway'. Eur J Cardiovasc Prev Rehabil 2006;13(5):798–804.

[20] Foulds HJ, Bredin SS, Charlesworth SA, Ivey AC, Warburton DE. Exercise volume and intensity: a dose-response relationship with health benefits. Eur J Appl Physiol 2014;114(8):1563–71.

[21] Paterson DH, Warburton DE. Physical activity and functional limitations in older adults: a systematic review related to Canada's Physical Activity Guidelines. Int J Behav Nutr Phys Act 2010;7:38.

[22] Lollgen H, Bockenhoff A, Knapp G. Physical activity and all-cause mortality: an updated meta-analysis with different intensity categories. Int J Sports Med 2009;30(3):213–24.

[23] Wen CP, Wai JP, Tsai MK, Yang YC, Cheng TY, Lee MC, et al. Minimum amount of physical activity for reduced mortality and extended life expectancy: a prospective cohort study. Lancet 2011;378(9798):1244–53.

[24] Lee DC, Pate RR, Lavie CJ, Sui X, Church TS, Blair SN. Leisure-time running reduces all-cause and cardiovascular mortality risk. J Am Coll Cardiol 2014;64(5):472–81.

[25] Moore SC, Patel AV, Matthews CE, Berrington de Gonzalez A, Park Y, Katki HA, et al. Leisure time physical activity of moderate to vigorous intensity and mortality: a large pooled cohort analysis. PLoS Med 2012;9(11):e1001335.

[26] Arem H, Moore SC, Patel A, Hartge P, Berrington de Gonzalez A, Visvanathan K, et al. Leisure time physical activity and mortality: a detailed pooled analysis of the dose-response relationship. JAMA Intern Med 2015;175(6):959–67.

[27] Sattelmair J, Pertman J, Ding EL, Kohl 3rd HW, Haskell W, Lee IM. Dose response between physical activity and risk of coronary heart disease: a meta-analysis. Circulation 2011;124(7):789–95.

[28] Kyu HH, Bachman VF, Alexander LT, Mumford JE, Afshin A, Estep K, et al. Physical activity and risk of breast cancer, colon cancer, diabetes, ischemic heart disease, and ischemic stroke events: systematic review and dose-response meta-analysis for the Global Burden of Disease Study 2013. BMJ 2016;354:i3857.

[29] Stone JA, Campbell NR, Genest J, Harris S, Pipe A, Warburton DER. Health behaviour interventions and cardiovascular disease risk factor modifications. In: Stone JA, editor. Canadian guidelines for cardiac rehabilitation and cardiovascular disease prevention. 3rd ed. Winnipeg, MB: Canadian Association of Cardiac Rehabilitation; 2009. p. 251–340.

[30] Ginis KA, Hicks AL, Latimer AE, Warburton DE, Bourne C, Ditor DS, et al. The development of evidence-informed physical activity guidelines for adults with spinal cord injury. Spinal Cord 2011;49(11):1088–96.

[31] Bredin SSD, Warburton DER, Lang DJ. The health benefits and challenges of exercise training in persons living with schizophrenia: a pilot study. Brain Sci 2013;3(2):821–48.

[32] Giacomantonio NB, Bredin SS, Foulds HJ, Warburton DE. A systematic review of the health benefits of exercise rehabilitation in persons living with atrial fibrillation. Can J Cardiol 2013;29(4):483–91.

[33] Hupin D, Roche F, Gremeaux V, Chatard JC, Oriol M, Gaspoz JM, et al. Even a low-dose of moderate-to-vigorous physical activity reduces mortality by 22% in adults aged >/=60 years: a systematic review and meta-analysis. Br J Sports Med 2015;49(19):1262–7.

[34] Bohn B, Herbst A, Pfeifer M, Krakow D, Zimny S, Kopp F, et al. Impact of physical activity on glycemic control and prevalence of cardiovascular risk factors in adults with type 1 diabetes: a cross-sectional multicenter study of 18,028 patients. Diabetes Care 2015;38(8):1536–43.

[35] Bao W, Tobias DK, Bowers K, Chavarro J, Vaag A, Grunnet LG, et al. Physical activity and sedentary behaviors associated with risk of progression from gestational diabetes mellitus to type 2 diabetes mellitus: a prospective cohort study. JAMA Intern Med 2014;174(7):1047–55.

[36] Bouchard C, Blair SN, Katzmarzyk PT. Less sitting, more physical activity, or higher fitness? Mayo Clin Proc 2015;90(11):1533–40.

[37] Despres JP. Physical activity, sedentary behaviours, and cardiovascular health: when will cardiorespiratory fitness become a vital sign? Can J Cardiol 2016;32(4):505–13.

[38] Blair SN, Cheng Y, Holder JS. Is physical activity or physical fitness more important in defining health benefits? Med Sci Sports Exerc 2001;33(6 Suppl.):S379–99. discussion S419-20.

[39] Williams PT. Physical fitness and activity as separate heart disease risk factors: a meta-analysis. Med Sci Sports Exerc 2001;33(5):754–61.

[40] Warburton DER, Katzmarzyk PT, Rhodes RE, Shephard RJ. Evidence-informed physical activity guidelines for Canadian adults. Can J Public Health 2007;98(2):S16–68.

[41] Eriksen L, Gronbaek M, Helge JW, Tolstrup JS. Cardiorespiratory fitness in 16025 adults aged 18–91 years and associations with physical activity and sitting time. Scand J Med Sci Sports 2015;26(12):1435–43.

[42] Bouchard C, Shephard RJ. Physical activity fitness and health: the model and key concepts. In: Bouchard C, Shephard RJ, Stephens T, editors. Physical activity fitness and health: international proceedings and consensus statement. Champaign, IL: Human Kinetics; 1994. p. 77–88.

[43] Myers J, McAuley P, Lavie CJ, Despres JP, Arena R, Kokkinos P. Physical activity and cardiorespiratory fitness as major markers of cardiovascular risk: their independent and interwoven importance to health status. Prog Cardiovasc Dis 2015;57(4):306–14.

[44] Kokkinos P, Faselis C, Myers J, Sui X, Zhang J, Blair SN. Age-specific exercise capacity threshold for mortality risk assessment in male veterans. Circulation 2014;130(8):653–8.

[45] Paterson DH, Cunningham DA, Koval JJ, St Croix CM. Aerobic fitness in a population of independently living men and women aged 55–86 years. Med Sci Sports Exerc 1999;31(12):1813–20.

[46] Shaya GE, Al-Mallah MH, Hung RK, Nasir K, Blumenthal RS, Ehrman JK, et al. High exercise capacity attenuates the risk of early mortality after a first myocardial infarction: the henry ford exercise testing (FIT) project. Mayo Clin Proc 2016;91(2):129–39.

[47] Kokkinos P, Faselis C, Myers J, Sui X, Zhang J, Tsimploulis A, et al. Exercise capacity and risk of chronic kidney disease in US veterans: a cohort study. Mayo Clin Proc 2015;90(4):461–8.

[48] Erikssen G. Physical fitness and changes in mortality: the survival of the fittest. Sports Med 2001;31(8):571–6.

[49] Erikssen G, Liestol K, Bjornholt J, Thaulow E, Sandvik L, Erikssen J. Changes in physical fitness and changes in mortality. Lancet 1998;352(9130):759–62.

[50] Blair SN, Kohl 3rd HW, Barlow CE, Paffenbarger Jr RS, Gibbons LW, Macera CA. Changes in physical fitness and all-cause mortality. A prospective study of healthy and unhealthy men. JAMA 1995;273(14):1093–8.

[51] Kokkinos P, Myers J, Faselis C, Panagiotakos DB, Doumas M, Pittaras A, et al. Exercise capacity and mortality in older men: a 20-year follow-up study. Circulation 2010;122(8):790–7.

[52] Brawner CA, Al-Mallah MH, Ehrman JK, Qureshi WT, Blaha MJ, Keteyian SJ. Change in maximal exercise capacity is associated with survival in men and women. Mayo Clin Proc 2017;92(3):383–90.

[53] Warburton DER. The health benefits of physical activity: a brief review. In: Warburton DER, editor. Health-related exercise prescription for the qualified exercise professional. 5th ed. Vancouver: Health & Fitness Society of BC; 2015. p. 1–17.

[54] Warburton DE, Gledhill N, Quinney A. Musculoskeletal fitness and health. Can J Appl Physiol 2001;26(2):217–37.

[55] Warburton DE, Gledhill N, Quinney A. The effects of changes in musculoskeletal fitness on health. Can J Appl Physiol 2001;26(2):161–216.

[56] Mason C, Brien SE, Craig CL, Gauvin L, Katzmarzyk PT. Musculoskeletal fitness and weight gain in Canada. Med Sci Sports Exerc 2007;39(1):38–43.

[57] Payne N, Gledhill N, Katzmarzyk PT, Jamnik V, Ferguson S. Health implications of musculoskeletal fitness. Can J Appl Physiol 2000;25(2):114–26.

[58] Lauretani F, Russo CR, Bandinelli S, Bartali B, Cavazzini C, Di Iorio A, et al. Age-associated changes in skeletal muscles and their effect on mobility: an operational diagnosis of sarcopenia. J Appl Physiol 2003;95(5):1851–60.

[59] Katzmarzyk PT, Craig CL. Musculoskeletal fitness and risk of mortality. Med Sci Sports Exerc 2002;34(5):740–4.

[60] American College of Sports Medicine. Position stand: exercise and physical activity for older adults. Med Sci Sports Exerc 1998;30(6):992–1008.

[61] Rantanen T, Masaki K, Foley D, Izmirlian G, White L, Guralnik JM. Grip strength changes over 27 yr in Japanese-American men. J Appl Physiol 1998;85(6):2047–53.

[62] Metter EJ, Talbot LA, Schrager M, Conwit R. Skeletal muscle strength as a predictor of all-cause mortality in healthy men. J Gerontol A Biol Sci Med Sci 2002;57(10):B359–65.

[63] Fujita Y, Nakamura Y, Hiraoka J, Kobayashi K, Sakata K, Nagai M, et al. Physical-strength tests and mortality among visitors to health-promotion centers in Japan. J Clin Epidemiol 1995;48(11):1349–59.

[64] Rantanen T, Harris T, Leveille SG, Visser M, Foley D, Masaki K, et al. Muscle strength and body mass index as long-term predictors of mortality in initially healthy men. J Gerontol A Biol Sci Med Sci 2000;55(3):M168–73.

[65] Rantanen T, Volpato S, Ferrucci L, Heikkinen E, Fried LP, Guralnik JM. Handgrip strength and cause-specific and total mortality in older disabled women: exploring the mechanism. J Am Geriatr Soc 2003;51(5):636–41.

[66] Brill PA, Macera CA, Davis DR, Blair SN, Gordon N. Muscular strength and physical function. Med Sci Sports Exerc 2000;32(2):412–6.

[67] Portegijs E, Rantanen T, Sipila S, Laukkanen P, Heikkinen E. Physical activity compensates for increased mortality risk among older people with poor muscle strength. Scand J Med Sci Sports 2007;17(5):473–9.

[68] Sallinen J, Stenholm S, Rantanen T, Heliovaara M, Sainio P, Koskinen S. Hand-grip strength cut points to screen older persons at risk for mobility limitation. J Am Geriatr Soc 2010;58(9):1721–6.

[69] Statistics Canada. Canadian Health Measures Survey Cycle 2. Ottawa: Statistics Canada; 2012. 2012-09-20.

[70] Tremblay MS, Warburton DE, Janssen I, Paterson DH, Latimer AE, Rhodes RE, et al. New Canadian physical activity guidelines. Appl Physiol Nutr Metab 2011;36(1):36–46.

[71] Centers for Disease Control and Prevention. How much physical activity do adults need?: Division of Nutrition, Physical Activity, and Obesity. National Center for Chronic Disease Prevention and Health Promotion; 2015.

[72] World Health Organization. Physical activity and adults. Geneva: World Health Organization; 2017.

[73] American College of Sports Medicine. ACSM issues new recommendations on quantity and quality of exercise. American College of Sports Medicine; 2011.

[74] Knox EC, Webb OJ, Esliger DW, Biddle SJ, Sherar LB. Using threshold messages to promote physical activity: implications for public perceptions of health effects. Eur J Public Health 2014;24(2):195–9.

[75] Lee IM. How much physical activity is good enough? National physical activity plan alliance commentaries on physical activity and health. Nat Phys Plan Alliance 2015;1(2).

[76] Tobe SW, Stone JA, Brouwers M, Bhattacharyya O, Walker KM, Dawes M, et al. Harmonization of guidelines for the prevention and treatment of cardiovascular disease: the C-CHANGE Initiative. Can Med Assoc J 2011;183(15):E1135–50.

[77] Bredin SS, Warburton DE. Physical activity line: effective knowledge translation of evidence-based best practice in the real-world setting. Can Fam Physician 2013;59(9):967–8.

[78] Bredin SS, Gledhill N, Jamnik VK, Warburton DE. PAR-Q+ and ePARmed-X+: new risk stratification and physical activity clearance strategy for physicians and patients alike. Can Fam Physician 2013;59(3):273–7.

[79] LeBlanc AG, Berry T, Deshpande S, Duggan M, Faulkner G, Latimer-Cheung AE, et al. Knowledge and awareness of Canadian Physical Activity and Sedentary Behaviour Guidelines: a synthesis of existing evidence. Appl Physiol Nutr Metab 2015;40(7):716–24.

[80] Dale LP, LeBlanc AG, Orr K, Berry T, Deshpande S, Latimer-Cheung AE, et al. Canadian physical activity guidelines for adults: are Canadians aware? Appl Physiol Nutr Metab 2016;41(9):1008–11.

[81] Gainforth HL, Berry T, Faulkner G, Rhodes RE, Spence JC, Tremblay MS, et al. Evaluating the uptake of Canada's new physical activity and sedentary behavior guidelines on service organizations' websites. Transl Behav Med 2013;3(2):172–9.

[82] Tucker JM, Welk GJ, Beyler NK. Physical activity in U.S.: adults compliance with the Physical Activity Guidelines for Americans. Am J Prev Med 2011;40(4):454–61.

[83] Prince SA, Adamo KB, Hamel ME, Hardt J, Gorber SC, Tremblay M. A comparison of direct versus self-report measures for assessing physical activity in adults: a systematic review. Int J Behav Nutr Phys Act 2008;5:56.

[84] Fortier M, Guerin E, Segar ML. Words matter: reframing exercise is medicine for the general population to optimize motivation and create sustainable behaviour change. Appl Physiol Nutr Metab 2016;41(11):1212–5.

[85] Segar ML, Guerin E, Phillips E, Fortier M. From a vital sign to vitality: selling exercise so patients want to buy it. Curr Sports Med Rep 2016;15(4):276–81.

[86] Segar ML, Richardson CR. Prescribing pleasure and meaning: cultivating walking motivation and maintenance. Am J Prev Med 2014;47(6):838–41.

[87] Rhodes RE, Fiala B. Building motivation and sustainability into the prescription and recommendations for physical activity and exercise therapy: the evidence. Physiother Theory Pract 2009;25(5–6):424–41.

[88] Rhodes RE, Fiala B, Conner M. A review and meta-analysis of affective judgments and physical activity in adult populations. Ann Behav Med 2009;38(3):180–204.

[89] Rhodes RE, Quinlan A. Predictors of physical activity change among adults using observational designs. Sports Med 2015;45(3):423–41.

[90] Rhodes RE, Warburton DE, Bredin SS. Predicting the effect of interactive video bikes on exercise adherence: an efficacy trial. Psychol Health Med 2009;14(6):631–40.

Chapter 14

Community-Based Maintenance Cardiac Rehabilitation

Sandra Mandic*, Anna Rolleston[†], Garrick Hately*, Stacey Reading[‡]

*University of Otago, Dunedin, New Zealand, [†]Faculty of Medicine and Health Science, University of Auckland, Auckland, New Zealand, [‡]Department of Exercise Sciences, University of Auckland, Auckland, New Zealand

CARDIAC REHABILITATION OVERVIEW

Unhealthy lifestyle practices such as physical inactivity, poor dietary habits, and cigarette smoking lead to development of risk factors, progression of atherosclerosis, and, ultimately, to adverse outcomes or clinical end points. Comprehensive multidisciplinary cardiac rehabilitation (CR) and risk reduction programs play an important role in primary and secondary prevention of cardiovascular disease [1]. The approach to CR and physical activity recommendations for cardiovascular disease patients have evolved considerably since the 1950s [2]. Modern CR is a multifaceted multidisciplinary and cost-effective intervention that improves functional capacity, recovery, and psychological well-being of individuals with cardiovascular disease [3].

Core Components

Modern CR focuses on the delivery of comprehensive, patient-focused interventions aimed at attenuating both disease progression and risk of event reoccurrence [3]. Core components of comprehensive modern CR include patient assessment, physical activity counseling, supervised exercise training, management of cardiovascular risk factors (including diabetes, dyslipidemia, hypertension, obesity, and smoking), nutrition counseling, and psychosocial management [1,3]. Therefore, CR programs require a multidisciplinary approach and involvement of a wide range of health professionals including cardiac nurses, clinical exercise physiologists, physiotherapists, physicians, cardiologists, dietitians, psychologists, and social workers [3].

Phases

The current model of CR involves three distinct phases: inpatient, outpatient, and maintenance CR [1]. Inpatient (phase I) CR involves patient education and early mobilization to minimize the detrimental effects of the bed rest following acute cardiac events. Outpatient (phase II) CR includes patient assessment, individualized and supervised exercise program, management of the patient's cardiovascular risk factors, patient education, and counseling about the importance of lifestyle modifications such as dietary changes, smoking cessation, nutrition counseling, psychosocial management, and patient education to self-monitor their symptoms. Maintenance (phase III) CR involves independent continuation of the rehabilitation process through either home-based or community-based exercise programs with a focus on lifelong maintenance of the healthy lifestyle habits and management of cardiovascular risk factors.

Benefits

Increased physical activity, structured exercise, and/or improved fitness have a number of potential cardioprotective effects including antiinflammatory, antithrombotic, antiarrhythmic, antiatherogenic, and antiischemic effects and lead to improvement in cardiovascular risk factors [1]. Participation in short-term outpatient CR programs improves exercise capacity and quality of life [4] and reduces all-cause mortality by 27% and cardiac mortality by 31% in patients with coronary artery disease (CAD) [5].

Lifestyle in Heart Health and Disease. https://doi.org/10.1016/B978-0-12-811279-3.00014-8

Challenges

Despite numerous benefits, CR remains an underutilized intervention in individuals with cardiovascular disease [6,7]. Main barriers to participation in CR include the lack of referral, cost, transportation, and inconvenient location and/or schedule [8]. The current model of CR delivery is unsustainable and needs significant improvement to provide cost-effective, patient-centered, and comprehensive secondary prevention of atherosclerotic cardiovascular disease [1]. Other models of CR such as mHealth [9] and community-based programs may need to be embraced.

The Need for Community-Based Maintenance Programs

The short duration of outpatient (phase II) CR programs (4–12 weeks) [10], high-dropout rates [11,12], and poor adherence to exercise after completion of the outpatient CR [13,14] diminishes the benefits that could ultimately be achieved through participation in those programs. Several studies report poor long-term adherence to exercise recommendations [15,16], a decline in exercise capacity [17], and worsening of cardiovascular risk factors [15,18,19] 12–18 months following outpatient CR. Therefore, setting up maintenance CR programs and offering them in local communities may be an effective intervention to increase CR uptake and preserve the physical and psychosocial benefits achieved during the earlier phases of CR.

COMMUNITY-BASED MAINTENANCE CARDIAC REHABILITATION

Optimal postcardiac event disease management includes adoption and continuation of the healthy behaviors and exercise routines that were established in the hospital-based phase II CR program. Unfortunately, most phase II CR programs are too short (e.g., 4–12 weeks) to effectively promote long-term adoption of new healthy behaviors. Community-based maintenance CR programs are an effective way to support patients by bridging the period between completion of a phase II CR program and full adoption of new healthy behaviors. Community-based programs are often more economical to operate, and they often remove many access barriers (e.g., the program is often housed close to their residence) that prevent patients from continuing to exercise regularly [20].

Participants

Patients who are highly motivated may be more likely to enroll and continue participation in long-term exercise programs including community-based CR [8]. A study from New Zealand found that community-based CR maintenance programs attract elderly, white, married, and retired males of moderate-to-high socioeconomic status who lived in close proximity to the program [21]. Men were more likely to attend the clubs for secondary prevention, while women were more likely to attend for primary prevention [21]. These findings are consistent with previous studies demonstrating that men are more likely to be referred to and participate in outpatient CR programs [22]. However, a high proportion of elderly individuals participating in these community-based CR programs [21] differs from the low participation rates of elderly individuals in outpatient CR programs [22,23]. The community location, schedule, low cost, peer support, and social activities afforded by these CR clubs may be particularly appealing for older individuals [21].

Outcomes

Only a few previous studies investigated community-based CR maintenance programs, despite suggestions that such programs could improve exercise capacity and physical function, facilitate management of cardiovascular risk factors, and have positive effects of psychosocial outcomes and quality of life.

Management of Cardiovascular Risk Factors

Previous studies reported both psychosocial and physical benefits gained from attending maintenance community-based CR programs by individuals with CAD [24,25]. Several studies reported improved management of cardiovascular risk factors [26] and favorable changes in body composition [12] and lipoprotein profile [12] in cardiac patients participating in long-term CR programs. Therefore, participation in maintenance CR programs may aid in the long-term management of cardiovascular risk factors.

Encouraging Physical Activity

Although active individuals are more likely to attend CR programs [27], CAD patients have poor long-term adherence to physical activity recommendations 1–1.5 years following outpatient CR program [15,16]. Approximately one-third of participants in community-based maintenance CR are sedentary (<1000 kcal per week) despite being enrolled in CR [21,28]. Considering an average energy expenditure of 230 kcal per CR exercise session [29], it is essential that maintenance CR programs encourage physical activity outside of scheduled supervised exercise sessions to meet the minimum physical activity guidelines ($\geq$150 min of moderate-to-vigorous physical activity per week). Potential use of the gentle reminders via the phone or web to promote physical activity on nonclub days should be explored in the future.

Improving Exercise Capacity

Several studies reported improved exercise capacity [12,26] and slowing down of age-related decline in exercise capacity [26] in cardiac patients participating in a long-term CR program. A recent study found that older adults with CAD who participated at the community-based maintenance CR program had lower measured peak oxygen consumption (VO_{2peak}) and predicted VO_{2peak} compared with age- and gender-matched "very active" healthy individuals but similar to their "less active" healthy peers [28]. In older adults with CAD who participated in the community-based CR program over 1.6-year period, exercise capacity (measured using 6-min walk test and 10 m shuttle walk) and physical function remained unchanged [30], lower body muscle strength improved, handgrip strength was reduced, and participants experienced unfavorable changes in body composition, regardless of attendance rates [30]. Further studies are needed to determine whether participation in these programs has favorable the long-term effects on improving and/or slowing down the aging-related decline in exercise capacity and physical function in individuals with CAD.

Psychosocial Outcomes and Quality of Life

CAD patients have two to three times greater risk of developing anxiety and depression compared with the general population [31]. CAD patients with symptoms of depression also experience a greater loss of physical function, fewer social interactions, worse health-related quality of life, and higher mortality rates compared with CAD patients who do not have depression [32]. Participation in CR improves psychological outcomes including anxiety, vitality, general mental health, and quality of life [33,34]. Adults with CAD with either a "good" or "poor" prognosis who participated in a long-term (12 months or more) CR program significantly improved psychological well-being by reducing tension, anxiety, and depression [35]. Significantly greater improvements in quality of life were observed in CAD patients who exercised regularly compared with nonexercising CAD patients [35,36]. Therefore, greater improvements in psychological well-being and quality of life were observed when CAD patients participate in exercise-based CR programs and simultaneously receive psychological support compared with nonexercising CAD patients. It is possible that psychological benefits of participation in CR are in part due to reducing patients' concerns about their physical function and safety of exercise and noticeable improvements in performing activities of daily living. These improvements lead to reduced stress and anxiety and improved confidence in the beneficial effects of participation in the CR programs [37]. CR programs are well positioned to provide comprehensive and long-term services that address the psychological issues after cardiac events. Therefore, participation in community-based CR programs, where social support and recognition of the emotional challenges associated with CAD, especially for elderly individuals, may be important for improving the long-term prognosis in CAD patients.

Taken together, these findings suggest that long-term participation in maintenance CR programs facilitates management of cardiovascular risk factors and improves exercise capacity and physical function in individuals with CAD. However, participation in such programs does not provide a sufficient amount of physical activity on its own and needs to be supplemented with additional physical activity outside of the supervised CR training sessions.

ADHERENCE

Program adherence is one of the major CR challenges, particularly in the maintenance phase. More than half of the patients enrolled in CR maintenance programs based in health facilities drop out within 6 months [11], and less than one-third remain in an extended CR program after 1 year [12]. Attendance at exercise sessions was highest in the first year (68%) and reduced on average by half over the following 5 years in middle-aged men with CAD [38]. Retrospective analysis of attendance to a community-based CR maintenance program showed a consistent average annual attendance rate of ~40% (average attendance of <1 session per week) for up to 5 years in older adults with CAD [21]. In the same study, session attendance was positively correlated with peak oxygen consumption and first-year attendance was strongly correlated with attendance in subsequent years [21]. Compared with nonattenders, older adults with CAD participating in community-based CR programs and in particular high attenders perceived greater benefits from attending CR, had fewer barriers, and perceived stronger cues to action [39]. High attenders also perceived a larger gain of social benefits compared with the low attenders [39]. In the same study, CR program newsletters, personal health concerns, and others having heart problems were stronger cues to action for high attenders compared with nonattenders [39].

Individuals who do not enroll in CR maintenance programs likely perceive no need to participate and/or do not like the format of such programs [39,40]. However, those individuals should also have readily available access to a medical support and advice regarding secondary prevention of cardiac events [41]. This could be achieved by providing an online community with advice on cardiac health maintenance.

Taken together, to promote participation and encourage attendance, community-based maintenance CR programs should emphasize both physical and social benefits of participating in CR, address modifiable barriers and provide encouragement. CR program newsletters could be used to advocate the physical and social benefits of participation in the program [39].

SOCIAL SUPPORT

Social support is an integral part of the community-based maintenance CR programs. Creating a sense of community through peer support encourages attendance to community-based CR programs [42]. CR program newsletters may be used as tool for creating a sense of community and promoting adherence to CR [39]. Such newsletters could be distributed to patients in in-hospital and outpatient CR programs to facilitate transition into the maintenance phase of CR [39].

SPECIAL CONSIDERATIONS

The proper tailoring of exercise and lifestyle interventions to a diverse range of populations as part of CR is an important consideration for a health system that is often western centric. Tailoring programs appropriately requires an understanding of the unique beliefs, values, and potential barriers to participation for different groups of people. This section will discuss aspects of culture that is often poorly integrated into traditional CR programs and also the challenges presented when CR needs to be provided in rural communities.

CULTURAL ASPECTS

The ability to develop a CR program that is culturally appropriate and tailored to a specific population may be one of the most important, yet most often overlooked, requirements. For example, indigenous Australians are underrepresented in CR participation and improving services to better meet the needs of this population has been identified as a priority [43]. A common assumption made by health-care providers is that translation of materials into another language is sufficient cultural integration. Unfortunately, translation of materials, although important, is a superficial acknowledgment of a person's cultural context. Including greater cultural considerations was identified in New Zealand as a way of improving attendance at CR by indigenous Māori [44]. Developing a cultural-specific CR program requires strong cultural understanding. Program goals and the methods used to achieve them must align with central cultural themes and values if the program is to be sustainable and efficacious within the community. Seeking design input from community leaders and other community members will improve cultural sensitivity and compatibility and therefore uptake within the target community.

Cultural Aspects of CR: Insights From New Zealand

In New Zealand, Māori are the indigenous people making up 15% of the total population. As with other indigenous cultures around the world, there are disparities in many areas of health for Māori compared with non-Māori New Zealanders including access to health services and overall health outcomes.

The difference in life expectancy between Māori (for female, 77.1 years and, for males, 73.0 years) and non-Māori (83.2 years for females and 79.5 years for males) is primarily due to cardiovascular disease [45]. The total cardiovascular disease mortality rate among Māori is more than twice as high as that among non-Māori, and in 2012–14, Māori were more than 1.5 times as likely as non-Māori to be hospitalized for cardiovascular disease [46]. Innovative approaches, using principles associated with a Māori worldview, are required to improve cardiovascular health outcomes and life expectancy for Māori.

In New Zealand, persistent disparity in cardiovascular health has led to the utilization of kaupapa Māori health services and viewing health from a Māori worldview in an attempt to change the way a traditional health system delivers to Māori.

An understanding that western medicine and science is disparate in many ways compared with a Māori holistic view of health has meant that health research and health services are being thought about differently. The Treaty of Waitangi, the founding document of New Zealand, dictates a basis from which western society ensures that Māori are acknowledged in societal processes and that the importance of a Māori way of life is upheld. A focus on Māori health equity is therefore an essential part of health service provision in New Zealand, guaranteed under the Treaty of Waitangi.

A Māori view of health is holistic and takes into account more than the physical and physiological components of the body that is the norm for western medicine. A number of models of Māori health exist, and a commonly referenced model is Sir Professor Mason Durie's Te Whare Tapa Wha (the four-walled house) [47]. Te Whare Tapa Wha describes a house that is strong due to the integrity of all four of its walls. The walls represent four important areas of consideration for Māori in terms of health and well-being:

- Taha tinana (physical)
- Taha hinengaro (mental and emotional)
- Taha whanau (family)
- Taha wairua (spirituality)

In addition, the model states that the house must also be built on solid ground and therefore acknowledgment that connection to whenua (land) in health is also of importance. Therefore, for a person to be well, all four aspects of health need to be considered and a person's connection to nature and land.

There are still only limited examples of Māori-focused programs for CR. Internationally, there is also a paucity of studies involving indigenous peoples although some work has been done with aboriginal communities in Australia, including a technology-enabled CR program delivered through smartphones to indigenous populations [48]. Best practice for CR recommends that all patients who experience a cardiac event have access to phase I and phase II programs and that rehabilitation be tailored to patients' needs, which should include cultural context [49]. The Heart Guide Aotearoa program was developed in New Zealand to increase participation and completion in CR and improve the management of cardiovascular disease, particularly among Māori [50].

Programs that focus on equity for Māori in health are sometimes developed within the "interface space" [51]. The interface space is ideal, which recognizes that two disparate worlds, in this case western medicine and Māori health, should be able to find some common ground without compromising the foundations upon which either is situated. To design a program from the interface space, one must draw important aspects from both worlds and then weave them together to make something new and innovative. For this approach to work, there needs to be strong medical leadership and strong Māori leadership, and there needs to be equity in that leadership relationship. The use of an interface space approach to design programs to reduce health disparities for Māori is distinct from kaupapa Māori. Kaupapa Māori methodologies can be defined as "by Māori, for Māori" and fully developed beneath the cloak of Te Ao Māori (the Māori world). Kaupapa Māori is an accepted framework for health research in New Zealand and is underpinned by critical theory. Māori beliefs and values are placed at the center of the research process, and findings are applied based on Māori views of the world. Kaupapa Māori informs the development of programs and services that use the interface space ideal.

Currently in New Zealand, a cluster controlled trial entitled "The effect of a 12-week exercise and lifestyle management program on cardiac risk reduction: A controlled trial using a kaupapa Māori philosophy" is being performed to test the use of the interface space methodology for improving heart health. Although not strictly a CR project, the research is the first in heart health to trial an intervention for Māori developed within the interface space. The project is the continuation of a pilot program that successfully integrated a kaupapa Māori methodology within a clinical setting [52,53]. The pilot program identified that the acknowledgment of culture within a health setting and allowing aspects of culture to be embedded into a clinical program was important for Māori. Providing programs that allow self-determination for Māori is a significant and positive step forward for equity.

CR IN RURAL COMMUNITIES

The cultural aspect of CR services, outlined above, also needs to be considered within rural programs. For example, in New Zealand, a higher proportion of Māori than non-Māori New Zealanders lives in rural areas. Rurality plays an important role in determining the nature, level of access to, and provision of health services. When considering issues of equity, rurality per se does not necessarily lead to a rural-urban disparity, but living in a rural community may exacerbate the effects of socioeconomic disadvantage, poorer service availability, higher levels of personal risk, and more hazardous environmental, occupational, and transportation conditions [54]. In addition, in New Zealand, Māori make up a higher proportion of the population living in rural and remote areas of the country [55], and therefore, rurality has an impact on equity for Māori due to location of health services, cost of transport, and availability of services.

When considering establishing services for CR in rural communities, the lack of infrastructure and health and medical expertise needs to be taken into account. In New Zealand, the delivery of CR services varies significantly by provider [56], and therefore, the rural communities serviced by those providers experience different barriers.

Ideally, CR services would be delivered directly to patients either at home or within their local community. The venue for the delivery of a CR program would need to be provided by the community and facilitated by an individual or group who were based in that community. The cost of sending health professionals, usually nursing staff, into rural and remote areas is often the reason why services are not provided in these locations. Some kind of "transfer of care" needs to be established where an urban-based CR provider oversees the delivery of the rural program, but local "champions" deliver the program and provide for its sustainability. In rural communities, finding local champions to promote and run CR programs means that urban resources do not have to be distributed out into rural areas, and the rural community is empowered through a process of upskilling champions and providing expertise, to some extent, within that community. Finally, population demographics and cultural aspects need to be considered when setting up CR services in rural communities.

EXAMPLES OF COMMUNITY-BASED MAINTENANCE CR PROGRAMS

Community-Led Cardiac Rehabilitation "Clubs"

In New Zealand, there are over 40 community-based CR maintenance "clubs." One of these clubs, the Otago Phoenix Club, founded by Dr. Edwin (Ted) R. Nye, is one of the earliest CR programs in the world established in Dunedin, New Zealand, in 1968. The nature of two community-based CR maintenance "clubs" in New Zealand, including the Otago Phoenix Club, has been described previously [21,30] and is summarized below.

Exercise Sessions

Both of these community-based CR "clubs" offered two supervised 60 min group exercise sessions per week, led by a physiotherapist or an exercise specialist. The group exercise sessions consisted of a combination of aerobic, strength, flexibility, balance, and coordination exercises. One of the clubs also offered the members access to an indoor pool in the same building immediately after the gym-based exercise sessions. Both clubs trained their members in cardiopulmonary resuscitation and safety awareness.

Social Support

The social support network was an integral part of the club's function. Members supported one another by offering transport, hospital visiting, and helping with lawn mowing and heavy chores. Both clubs had regular social events (such as dinners/lunches, potlucks, and golf tournaments) and set up their own websites and newsletters focused on healthy lifestyle and issues related to heart disease. The members are also active in supporting and raising funds for the local community.

Program Governance

The clubs were self-governed by the club members. The clubs formed a committee of interested members and elected executive officers and subgroups to organize walking and tramping events, track clearing activities, and social events. The club treasurer managed the club funds and payment of club's expenses. The clubs applied for grants as a charitable organization to purchase additional equipment and pay their exercise facilitators.

Cost

One club had an annual membership fee, while the other worked on a user pays basis. It is important to note that since the members self-govern the club, these programs may have evolved over time to suit the needs of the most common members [21].

An example of the Otago Phoenix Club newsletter heading is presented in Fig. 1. The Otago Phoenix Club website is http://www.otagophoenix.hostzi.com/index.html.

Take H.E.A.R.T. Program New Brunswick, Canada

The Take Healthy Eating Active Recreation Today (Take H.E.A.R.T.) program was developed as a community-based CAD prevention and rehabilitation program by Drs. S Reading and B Miedema of the University of New Brunswick and

FIG. 1   The Otago Phoenix Club.

Dalhousie University in Fredericton New Brunswick, Canada. The program was designed to address some of the difficulties and barriers present in New Brunswick that limit patient access to exercise-based rehabilitation services. The population of New Brunswick (approximately 750,000) is distributed among three urban cities (i.e., between 50,000 and 75,000 each) and a number of small rural communities (i.e., <15,000). Exercise rehabilitation services offered by the provincial Department of Health are available only within the major urban centers. Those living outside urban centers are required to drive, often through inclement weather to access service (e.g., New Brunswick is located in the northern hemisphere and snow covered for 4 or more months each year). As a result of this barrier, many New Brunswick citizens are unable to access exercise programing to prevent CAD or pursue rehabilitation following an acute coronary event. The Take H.E.A.R.T. program was conceived and developed to address this barrier by implementing a program that was capable of delivering safe and effective exercise rehabilitation and nutritional counseling services to adults living in small communities across New Brunswick. Key design features of the program are the program's use of existing community infrastructure and professional resources (i.e., avoided duplication or adding cost burden); its minimal reliance on equipment (i.e., reduced cost burden); its portability (i.e., easily and consistently implemented across several communities); and a built in robust assessment to monitor and evaluate program efficacy (i.e., demonstrate value for money).

The Take H.E.A.R.T. program consisted of pre/postprogram health-related fitness assessments (i.e., tests of aerobic, anthropometric, and musculoskeletal fitness), 36 group exercise sessions supervised by the Canadian Society for Exercise Physiology Certified Exercise Physiologist (CSEP-CEP), and 10 nutritional education seminars delivered by a registered dietitian (RD). An indoor facility is required to run the program in northern climates, and in most cases, the host community provided space in their community recreation center or a school or church basement was used. The required professional staff (e.g., a CEP and RD) often lived in the community or in a nearby community providing a local champion to lead the program. The exercise program consists of brisk walking, calisthenic-type exercise, resistance training using the body mass as the resistance element, and balance and flexibility training [57]. The nutritional education seminars focused on helping patients understand Canada's Food Guide and how to make healthy food choices. The sessions included a field trip to a local supermarket with the dietitian to put the course theory into practice.

Early in program development, it was observed that charging a small fee improved program attrition and attendance rates presumably because payment of a fee represented an investment by the patient and therefore increased the likelihood of completing the program. A program of 20 participants charged a modest fee of $150 CAN per year per participant is typically sufficient to offset all program costs (e.g., program booklet printing, service fees for CEP and RD, facility rental if necessary) without presenting a new program access barrier.

The efficacy of the Take H.E.A.R.T. program was assessed by measures of attendance and attrition and by objective field testing of aerobic capacity, body composition, and musculoskeletal fitness. Questionnaires were used to assess quality of life (e.g., $SF36_{v2}$) and nutritional knowledge [58]. Analysis of three program sites that offered a total of 10- and 12-week program cycles over a 4-year period revealed that participation in a Take H.E.A.R.T. program typically improved an individual's aerobic and musculoskeletal fitness in addition to positively modifying their body composition. The analysis involved a total of $n = 392$ adults (69% female) of which 66% completed the program defined as providing pre- and postprogram assessment data. Patients attended $62.4\% \pm 24\%$ (mean $\pm$ SD) of the program sessions. Aerobic fitness was assessed using a submaximal effort walking test to estimate the maximal aerobic capacity (i.e., Rockport 1-mile test [59]). Postprogram participants were able to complete the 1 mi distance faster (postminus preprogram 95% CI, -1.1 ± 0.3 min; $P < 0.05$) for a similar heart rate (pre- vs. postmean $\pm$ SD, 118 ± 20 vs. 121 ± 7 bpm) that translated into a 1.3 ± 0.3 MET (95% CI; $P < 0.05$) increase in aerobic capacity. Body mass and waist circumference decreased by 1.1 ± 0.4 kg and 3.0 ± 1.0 cm (95% CI; $P < 0.05$), respectively. The combined left plus right grip strength score was significantly ($P < .05$) increased by 1.3 ± 0.8 kg.

SF36v$_2$ physical and mental component scores were (mean ± SD) 49.0 ± 7.8 and 49.6 ± 10.8, respectively, prior to beginning the program, and these scores increased by 2.0 ± 1.3 and 2.9 ± 1.7 (95% CI; $P < 0.05$) postprogram.

The strength of the Take H.E.A.R.T. program is its emphasis on providing a cost-effective, safe, and effective program delivered by professionally trained staff in the local community. The program utilizes infrastructure and professional resources that already exist in the community. Most communities in New Brunswick have access to a community dietitian through the Department of Health. The Canadian Society for Exercise Physiology has a long-established accreditation program for exercise professionals, and provincial tertiary institutions provide the academic preparation in exercise science that is required to achieve accreditation. The Take H.E.A.R.T. program employed a robust and standardized assessment of patient outcomes. Far too often, community-based programs rely on weak metrics such as attendance or body mass changes to characterize program efficacy. Limitations of the program include the need for a local community member (i.e., a CEP or RD) to champion the program and obtain buy-in from community leaders (i.e., through provision of space) and community members themselves.

Heart Guide Aotearoa in New Zealand

The Heart Guide Aotearoa is "an individualized, menu-based, cognitive behavioral, and chronic disease management program for people with coronary heart disease" [50]. The program is Māori-centric but based on an initiative from the United Kingdom [60] and was developed in consultation with the author of the original United Kingdom program, Māori clinical staff, and consumers.

The 6-week program involves four contacts (two face-to-face and two by telephone) with the participant. The first is a home visit of 45–60 min duration for the purpose of assessment. During week 1, the facilitator performs a brief follow up and review (by telephone) that occurs again during week 3. At week 6, there is a final face-to-face meeting, review, and design of a maintenance action plan. The interventions that form the Heart Guide Aotearoa program are based on cognitive behavioral principles. The program includes personalized assessment, goal setting, problem solving, self-monitoring, action planning, relaxation training, and correction of unhelpful health beliefs. The general approach taken by facilitators of the program is one of individual support, responsiveness to client needs, and empowerment of clients to self-manage and take control of their own health.

The whānau (family) and home-based nature of the program was intended to overcome access issues due to rehabilitation programs typically being hospital based and only available during working hours. It was suggested that this made them inaccessible to many people under 65 years due to work commitments and to those without transport or living in rural areas. As Māori are far more likely to be in the under 65 years age group when they experience a cardiac event and are overrepresented in the lower socioeconomic groups compared with non-Māori, the place and timing of hospital-based programs probably contributed to low uptake by Māori [50]. The Heart Guide Aotearoa is therefore an attempt to redress inequities and make CR more broadly available by providing the service in the patient's own home.

The University of Auckland Health and Performance Clinic, New Zealand

The Department of Exercise Sciences within the University of Auckland has operated a community-based CR program for more than 15 years in the Auckland suburb of Glen Innes. The program provides phase II and III CR exercise programming to urban city residents that have been treated for acute or chronic cardiovascular disease within one of the local district health board services. The program applies current best practice, and it serves as a "living laboratory" for students completing postgraduate degrees in clinical exercise physiology.

The program's operational model mimics that of a typical allied health practice. Following referral from a medical practitioner, the patient is assigned to a rehabilitation management team consisting of a staff clinical exercise physiologist and several MSc and postgraduate diploma/fourth year honors undergraduate level students. The team is responsible for assessing the patient's current functional capacity and developing an exercise prescription that is goal orientated and compatible with the client's ability, their medical condition, and current medical management. Each case is regularly reviewed to ensure the prescription, and patient progress remains on track.

Each patient assessment typically includes a movement and mobility analysis, cardiopulmonary exercise testing with metabolic gas analysis, muscle strength, power and endurance testing, and measurement of mental well-being by questionnaire (i.e., SF36v$_2$ or HADs). The exercise prescription is derived from information obtained in the assessment. The first phase of a prescription (i.e., typically 8–12 weeks) aims to help patients become accustomed to routine participation in exercise. Movement abnormalities and/or muscle strength imbalances and deficits are corrected where possible to reduce biomechanical strain and injury risk. Aerobic and resistance training are used to ensure the patient possesses sufficient aerobic capacity, muscular strength, power and endurance to support activities of daily living, and any preferred recreational physical activities. Once sufficient base conditioning is established, the exercise prescription enters a conditioning phase to maximize functional capacity over a 12–24 week window. High-intensity interval training and muscular power training techniques may be employed if appropriate during this phase. Patients that wish to continue in the program beyond 6 months enter a maintenance phase and receive periodic reassessment and program adjustment.

Patients attend the Health and Performance Clinic two or three times each week for 60 min sessions. Although prescriptions are individualized, patients exercise in a large common area with 15–20 other patients attending the clinic at the same time each week. During each exercise session, there is significant interaction between patients, staff, and student trainees to help clients develop peer relationships and a sense of community and connectedness to the clinic.

The program is resource heavy in terms of equipment (i.e., use of state-of-the-art metabolic gas analyzer, muscle function, and biomechanical measurement equipment) and personnel (i.e., staff possess MSc degrees in exercise physiology or related discipline and registration with a clinical exercise science professional body such as the American College of Sports Medicine or equivalent) and is thus expensive. The cost of delivering the program is offset by the University of Auckland so that patients attending the initial 12-week program do so free of charge and those attending beyond the initial 12 weeks pay a modest fee to partially offset the cost of service. In 2016, $n = 246$ patients aged 64 ± 13 years (i.e., mean $\pm$ SD; 38% female) enrolled in the initial 12-week CR program. Successful program completion defined as patients providing pre- and post-12-week assessment data was 82% of the enrolled patients. On average, patients attended 26 ± 5 (i.e., mean $\pm$ SD) of 36 possible sessions for a 72% attendance rate. Maximal cardiopulmonary exercise tests with metabolic gas analysis were completed for 98 patients. Peak cycling workload increased by 18.5 W (postminus preprogram 95% CI, 18.5 ± 4.2 W; $P < 0.05$) that was supported by a measured increase in peak aerobic capacity of 0.4 METs (95% CI, 0.4 ± 0.3 METs; $P < 0.05$). Patient body mass was stable (mean $\pm$ SD, 81.5 ± 17.3 vs. 81.1 ± 17.2 kg) despite a small reduction in waist circumference (95% CI, -1.0 ± 0.7 cm; $P < 0.05$). An increase in patient muscular strength was noted through increases in both combined left- and right-hand grip strength score (95% CI, 2.3 ± 1.8 kg; $P < 0.05$) and maximal load lifted by leg press exercise (23.4 ± 10.1 kg; $P < 0.05$). The SF36$_{v2}$-derived physical component summary (mean $\pm$ SD, 42.8 ± 10.7) and mental component summary scores (41.6 ± 15.4) were significantly less than the general US population mean score (i.e., 50 ± 10) at intake. Following 12 weeks of exercise rehabilitation, the physical and mental component scores increased by 5.3 ± 2.0 and 6.1 ± 3.1 (95% CI), respectively, and were no longer different from the general US population score for both components.

The strength of the University of Auckland Health and Performance CR program reside in the staff and equipment resources that are available to the program. This program is able to undertake sophisticated patient assessment and utilize the information gained to direct appropriate individualized exercise prescription design. The clinic has dedicated operational space and it possesses a vast array of ergometers and other exercise rehabilitation equipment. Each staff member is an accredited professional supported by academic researchers and medical professionals. The clinic is able to actively seek out and accommodate patients with difficult and complex disease (i.e., heart failure, postaortic dissection, aortic stenosis, and multiple comorbidities) providing those with increased risk to be able to undertake CR in a community-based setting. The primary limitation to University of Auckland program is that it is prohibitively expensive to operate in a fee for service model.

RECOMMENDATIONS FOR SETTING UP COMMUNITY-BASED CR PROGRAMS

Based on the current evidence and examples of the CR programs described above, the recommendations for setting up community-based CR programs are outlined in Table 1.

CHALLENGES

Community-based CR maintenance programs face similar challenges to the outpatient CR programs including issues associated with the lack of referrals, cost, transportation, inconvenient location, and/or schedule [8]. In addition, limited availability of community-based CR programs, need for community buy-in, poor adherence to long-term exercise programs, and the lack of funding (and therefore limited opportunity to attract qualified personnel and adequate equipment, venue, and/or resources) represent particular challenges for establishing and maintaining such programs in the long term. Ultimately, community-based CR maintenance programs should aspire to attract a diverse group of individuals with cardiovascular disease and particularly currently underrepresented populations such as women, elderly, and ethnic minorities. The location of the community can be a challenge, especially if it is particularly remote. Consideration of how to provide access to CR services for remote communities is important for equity, and the solutions are likely to be different in different locations.

TABLE 1 Recommendations for Setting up Community-Based CR Programs.

Recommendations

- Obtain a buy-in from the community (get a community champion involved)
- Set up a program that meets the needs of the local community
- Program supervision should include accredited exercise professionals with additional input from medical and allied health providers
- Incorporate robust assessment to measure several aspects of program efficacy
- Incorporate regular social events including dinners, lunches, or potlucks
- Emphasize physical, functional, and social benefits of participation
- Encourage physical activity outside of the regular exercise sessions to ensure participants are achieving at least minimum physical activity recommendations for adults
- Create a program newsletter to emphasize the benefits of participation, create a sense of a community, encourage regular attendance, and promote physical activity outside of scheduled supervised exercise sessions
- Consider setting up program website
- Acknowledge a persons' cultural context and worldview within the program
- Consider where people live and problem-solve potential transport issues or deliver the program locally

Community-based maintenance CR programs may not be a suitable solution for all patients. To promote a greater reach of maintenance CR programs, other delivery modalities such as the Internet or mobile phones and peer-led group classes should also be considered.

SUMMARY

Community-based CR programs promote and support long-term adherence to exercise and healthy lifestyle behavior and play an important role in a continuum of care for individuals with cardiovascular disease. Although the format of those programs varies greatly, most programs offer exercise sessions, and some also provide health-related education, nutritional guidance, and behavioral counseling. This chapter discussed outcomes of participation in community-based CR programs, program adherence, special considerations including cultural aspects and program implementation in rural communities, and several examples of CR programs. Future design of such programs needs to involve the community and community champions, take into account social and cultural needs of the program participants, consider setting up program newsletter and website to reinforce the healthy lifestyle messages and facilitate social support, and, whenever possible, offer such programs locally. The most successful programs tend to be those that are based on best practice guidelines and take into consideration cultural and social sensitivity of the program participants.

REFERENCES

[1] Sandesara PB, Lambert CT, Gordon NF, Fletcher GF, Franklin BA, Wenger NK, et al. Cardiac rehabilitation and risk reduction: time to "rebrand and reinvigorate". J Am Coll Cardiol 2015;65(4):389–95.
[2] Redfern J. The evolution of physical activity and recommendations for patients with coronary heart disease. Heart Lung Circ 2016;25(8):759–64.
[3] Piepoli MF, Corra U, Benzer W, Bjarnason-Wehrens B, Dendale P, Gaita D, et al. Secondary prevention through cardiac rehabilitation: from knowledge to implementation. A position paper from the cardiac rehabilitation section of the European association of cardiovascular prevention and rehabilitation. Eur J Cardiovasc Prev Rehabil 2010;17(1):1–17.
[4] Ades PA. Cardiac rehabilitation and secondary prevention of coronary heart disease. N Engl J Med 2001;345(12):892–902.
[5] Jolliffe J, Rees K, Taylor RS, Thompson D, Oldridge N, Ebrahim S. Exercise-based rehabilitation for coronary heart disease. Cochrane Database Syst Rev 2001;1. CD001800.
[6] Lavie CJ, Thomas RJ, Squires RW, Allison TG, Milani RV. Exercise training and cardiac rehabilitation in primary and secondary prevention of coronary heart disease. Mayo Clin Proc 2009;84(4):373–83.
[7] Mensah GA. Rehabilitation referral revisited: rhyme, reason, and response. J Cardiopulm Rehabil 2004;24(3):175–7.
[8] Bock BC, Carmona-Barros RE, Esler JL, Tilkemeier PL. Program participation and physical activity maintenance after cardiac rehabilitation. Behav Modif 2003;27(1):37–53.

[9] Rawstorn JC, Gant N, Meads A, Warren I, Maddison R. Remotely delivered exercise-based cardiac rehabilitation: design and content development of a novel mHealth platform. JMIR Mhealth Uhealth 2016;4(2). e57.

[10] Taylor RS, Brown A, Ebrahim S, Jolliffe J, Noorani H, Rees K, et al. Exercise-based rehabilitation for patients with coronary heart disease: systematic review and meta-analysis of randomized controlled trials. Am J Med 2004;116(10):682–92.

[11] Oldridge NB, Streiner DL. The health belief model: predicting compliance and dropout in cardiac rehabilitation. Med Sci Sports Exerc 1990;22(5):678–83.

[12] Brubaker PH, Warner Jr JG, Rejeski WJ, Edwards DG, Matrazzo BA, Ribisl PM, et al. Comparison of standard- and extended-length participation in cardiac rehabilitation on body composition, functional capacity, and blood lipids. Am J Cardiol 1996;78(7):769–73.

[13] Balady GJ, Jette D, Scheer J, Downing J. Changes in exercise capacity following cardiac rehabilitation in patients stratified according to age and gender. Results of the Massachusetts association of cardiovascular and pulmonary rehabilitation multicenter database. J Cardiopulm Rehabil 1996;16(1):38–46.

[14] Oldridge NB, Stoedefalke KG. Compliance and motivation in cardiac exercise programs. Clin Sports Med 1984;3(2):443–54.

[15] Hansen D, Dendale P, Raskin A, Schoonis A, Berger J, Vlassak I, et al. Long-term effect of rehabilitation in coronary artery disease patients: randomized clinical trial of the impact of exercise volume. Clin Rehabil 2010;24(4):319–27.

[16] Reid RD, Morrin LI, Pipe AL, Dafoe WA, Higginson LA, Wielgosz AT, et al. Determinants of physical activity after hospitalization for coronary artery disease: the tracking exercise after cardiac hospitalization (TEACH) study. Eur J Cardiovasc Prev Rehabil 2006;13(4):529–37.

[17] Oerkild B, Frederiksen M, Hansen JF, Simonsen L, Skovgaard LT, Prescott E. Home-based cardiac rehabilitation is as effective as centre-based cardiac rehabilitation among elderly with coronary heart disease: results from a randomised clinical trial. Age Ageing 2011;40(1):78–85.

[18] Willich SN, Muller-Nordhorn J, Kulig M, Binting S, Gohlke H, Hahmann H, et al. Cardiac risk factors, medication, and recurrent clinical events after acute coronary disease; a prospective cohort study. Eur Heart J 2001;22(4):307–13.

[19] Lear SA, Spinelli JJ, Linden W, Brozic A, Kiess M, Frohlich JJ, et al. The extensive lifestyle management intervention (ELMI) after cardiac rehabilitation: a 4-year randomized controlled trial. Am Heart J 2006;152(2):333–9.

[20] Mosleh SM, Bond CM, Lee AJ, Kiger A, Campbell NC. Effects of community based cardiac rehabilitation: comparison with a hospital-based programme. Eur J Cardiovasc Nurs 2015;14(2):108–16.

[21] Mandic S, Body D, Barclay L, Walker R, Nye ER, Grace SL, et al. Community-based cardiac rehabilitation maintenance programs: use and effects. Heart Lung Circ 2015;24(7):710–8.

[22] Witt BJ, Jacobsen SJ, Weston SA, Killian JM, Meverden RA, Allison TG, et al. Cardiac rehabilitation after myocardial infarction in the community. J Am Coll Cardiol 2004;44(5):988–96.

[23] Suaya JA, Stason WB, Ades PA, Normand SL, Shepard DS. Cardiac rehabilitation and survival in older coronary patients. J Am Coll Cardiol 2009;54(1):25–33.

[24] Clark AM, Barbour RS, White M, MacIntyre PD. Promoting participation in cardiac rehabilitation: patient choices and experiences. J Adv Nurs 2004;47(1):5–14.

[25] Pâquet M, Bolduc N, Xhignesse M, Vanasse A. Re-engineering cardiac rehabilitation programmes: considering the patient's point of view. J Adv Nurs 2005;51(6):567–76.

[26] Gayda M, Juneau M, Levesque S, Guertin MC, Nigam A. Effects of long-term and ongoing cardiac rehabilitation in elderly patients with coronary heart disease. Am J Geriatr Cardiol 2006;15(6):345–51.

[27] Daly J, Sindone AP, Thompson DR, Hancock K, et al. Barriers to participation in and adherence to cardiac rehabilitation programs: a critical literature review. Prog Cardiovasc Nurs 2002;17(1):8–17.

[28] Mandic S, Stevens E, Hodge C, Brown C, Walker R, Body D, et al. Long-term effects of cardiac rehabilitation in elderly individuals with stable coronary artery disease. Disabil Rehabil 2016;38(9):837–43.

[29] Schairer JR, Kostelnik T, Proffitt SM, Faitel KI, Windeler S, Rickman LB, et al. Caloric expenditure during cardiac rehabilitation. J Cardiopulm Rehabil 1998;18(4):290–4.

[30] Mandic S, Hodge C, Stevens E, Walker R, Nye ER, Body D, et al. Effects of community-based cardiac rehabilitation on body composition and physical function in individuals with stable coronary artery disease: 1.6-year follow-up. Biomed Res Int 2013;2013.

[31] Lane D, Carroll D, Ring C, Beevers DG, Lip GY. The prevalence and persistence of depression and anxiety following myocardial infarction. Br J Health Psychol 2002;7(Pt 1):11–21.

[32] Alla F, Briancon S, Guillemin F, Juilliere Y, Mertes PM, Villemot JP, et al. Self-rating of quality of life provides additional prognostic information in heart failure. Insights into the EPICAL study. Eur J Heart Fail 2002;4(3):337–43.

[33] Seki E, Watanabe Y, Sunayama S, Iwama Y, Shimada K, Kawakami K, et al. Effects of phase III cardiac rehabilitation programs on health-related quality of life in elderly patients with coronary artery disease: Juntendo Cardiac Rehabilitation Program (J-CARP). Circ J 2003;67(1):73–7.

[34] Herman R, Liebergall M, Rott D. Correlation between participation in a cardiac rehabilitation program and quality of life of patients with coronary artery disease. Rehabil Nurs 2014;39(4):192–7.

[35] Dugmore LD, Tipson RJ, Phillips MH, Flint EJ, Stentiford NH, Bone MF, et al. Changes in cardiorespiratory fitness, psychological wellbeing, quality of life, and vocational status following a 12 month cardiac exercise rehabilitation programme. Heart 1999;81(4):359–66.

[36] Izawa KP, Yamada S, Oka K, Watanabe S, Omiya K, Iijima S, et al. Long-term exercise maintenance, physical activity, and health-related quality of life after cardiac rehabilitation. Am J Phys Med Rehabil 2004;83(12):884–92.

[37] Shepherd CW, While AE. Cardiac rehabilitation and quality of life: a systematic review. Int J Nurs Stud 2012;49(6):755–71.

[38] Niebauer J, Hambrecht R, Velich T, Hauer K, Marburger C, Kalberer B, et al. Attenuated progression of coronary artery disease after 6 years of multifactorial risk intervention: role of physical exercise. Circulation 1997;96(8):2534–41.

[39] Horwood H, Williams MJ, Mandic S. Examining motivations and barriers for attending maintenance community-based cardiac rehabilitation using the health-belief model. Heart Lung Circ 2015;24(10):980–7.

[40] Farley RL, Wade TD, Birchmore L. Factors influencing attendance at cardiac rehabilitation among coronary heart disease patients. Eur J Cardiovasc Nurs 2003;2(3):205–12.

[41] Guiraud T, Granger R, Gremeaux V, Bousquet M, Richard L, Soukarié L, et al. Telephone support oriented by accelerometric measurements enhances adherence to physical activity recommendations in noncompliant patients after a cardiac rehabilitation program. Arch Phys Med Rehabil 2012;93(12):2141–7.

[42] Clark AM, Whelan HK, Barbour R, MacIntyre PD. A realist study of the mechanisms of cardiac rehabilitation. J Adv Nurs 2005;52(4):362–71.

[43] Hamilton S, Mills B, McRae S, Thompson S. Cardiac rehabilitation for aboriginal and Torres strait islander people in Western Australia. BMC Cardiovasc Disord 2016;16:150.

[44] Hutchinson P, Meyer A, Marshall B. Factors influencing outpatient cardiac rehabilitation attendance. Rehabil Nurs 2015;40(6):360–7.

[45] MacPherson L. New Zealand period life tables: 2012–14. Wellington: Statistics New Zealand; 2015.

[46] Ministry of Health. Tatau Kahukura: Māori health chart book. Wellington: Ministry of Health; 2015.

[47] Durie M. Whaiora: Māori health development. Auckland: Oxford University Press; 1998.

[48] Bradford D, Hansen D, Karunanithi M. Making an APPropriate care program for indigenous cardiac disease: customization of an existing cardiac rehabilitation program. Stud Health Technol Inform 2015;216:343–7.

[49] New Zealand Guidelines Group. Cardiac rehabilitation: evidence-based best practice guideline. Wellington: New Zealand Guidelines Group; 2002.

[50] Henwood W, Barnes HM. Heart guide aotearoa evaluation: final report. Auckland: Te Ropu Whariki, Massey University; 2008.

[51] Durie M. Exploring the interface between science and indigenous knowledge. In: 5th APEC research and development leaders Forum; Christchurch; 2004.

[52] Rolleston AK, Doughty R, Poppe K. Pounamu: integration of kaupapa Māori concepts in health research: a way forward for Māori cardiovascular health? J Prim Health Care 2016;8(1):60–6.

[53] Rolleston A, Doughty RN, Poppe K. The effect of a 12-week exercise and lifestyle management programme on cardiac risk reduction: a pilot using a Kaupapa Māori philosophy. Int J Indig Health 2017;11(2):116–30.

[54] Smith KB, Humphreys JS, Wilson MG. Addressing the health disadvantage of rural populations: how does epidemiological evidence inform rural health policies and research? Aust J Rural Health 2008;16(2):56–66.

[55] Zealand SN. New Zealand: an urban/rural profile. Wellington: Statistics New Zealand; 2004.

[56] Kira G, Doolan-Noble F, Humphreys G, Williams G, O'Shaughnessy H, Devlin G. A national survey of cardiac rehabilitation services in New Zealand: 2015. N Z Med J 2016;129(1435):50–8.

[57] Miedema B, Reading SA, Hamilton RA, Morrison KS, Thompson AE. Can certified health professionals treat obesity in a community-based programme? A quasi-experimental study. BMJ Open 2014;5. e006650.

[58] Miedema BP, Bowes AP, Hamilton RP, Reading SP. Assessing the efficacy of a group mediated nutritional knowledge intervention for individuals with obesity. Can J Diet Pract Res 2016;77(4):206–9.

[59] Kline GM, Porcari JP, Hintermeister R, Freedson PS, Ward A, McCarron RF, et al. Estimation of VO_{2max} from a one-mile track walk, gender, age, and body weight. Med Sci Sports Exerc 1987;19(3):253–9.

[60] Lewin B, Robertson IH, Cay EL, Irving JB, Campbell M. Effects of self-help post-myocardial-infarction rehabilitation on psychological adjustment and use of health services. Lancet 1992;339(8800):1036–40.

Chapter 15

Determinants of Exercise Ventilatory Inefficiency in Heart Failure With Reduced or Preserved Ejection Fraction: Application of Classical and Emerging Integrative Physiology Concepts

Erik H. Van Iterson*, Thomas P. Olson*

Department of Cardiovascular Medicine, Mayo Clinic, Rochester, MN, United States

INTRODUCTION

Chronic heart failure (HF) is a debilitating and life-threatening syndrome to which there is no known cure. Whether diagnosis of HF derives from a classification of reduced (HFrEF) or preserved ejection fraction (HFpEF), a hallmark feature universal to HF is exercise intolerance [1–11]. Importantly, it is suggested that the magnitude of exercise intolerance is inversely related to both short- and long-term prognosis [1,2,6–10,12–15]. An improved pathophysiological understanding of exercise intolerance is urgently needed as HF is a public health epidemic. While there are > 5.5 million adults with HF in the United States and ~ 700,000 new diagnoses annually, equally important, 1 year mortality in both men and women is > 20% [1–3,16,17].

Irrespective of definition, a substantial contributor to exercise intolerance in HF is an abnormally elevated rate of breathing leading to what has been characterized as inefficient ventilation, which is exacerbated at even modest levels of exercise [6–10,12–15,18–20]. With this, a key variable measured during cardiopulmonary exercise testing (CPET) is the slope or ratio of the ventilatory equivalent for carbon dioxide ($\dot{V}_E/\dot{V}CO_2$), which is suggested to appropriately phenotype the extent of exercise ventilatory inefficiency [6–10,12–15,18–23]. An increasing magnitude of ventilatory inefficiency quantified using $\dot{V}_E/\dot{V}CO_2$ slope (>34) or other metrics is strongly related to increased hospitalizations and mortality in HF [6–10,12,13,15,18].

While it is currently recognized that symptoms such as dyspnea in addition to ventilatory inefficiency play a disparaging role in the ability of HF patients to adequately perform routine tasks of daily living, it is a steepened $\dot{V}_E/\dot{V}CO_2$ slope demonstrated during CPET that unequivocally suggests there are immediate clinical implications associated with objectively quantifying the level of exercise ventilatory inefficiency [6–10,12,13,15]. However, despite the breadth of accumulating evidence supporting the use of CPET indices (including $\dot{V}_E/\dot{V}CO_2$ slope) to quantify exercise intolerance [6–10,12–15,18–24], the field as an entirety has not established a clear and concise understanding of the underlying integrative pathophysiological origins of exercise intolerance and the linked mechanisms underpinning inefficient ventilation during exercise in HF patients. Thus, this chapter is dedicated to bridging the multitude of separate lines of study in order to provide an integrative examination of factors that are foremost consequential to exercise ventilatory inefficiency in HF [6–10,12–15,18–33]. To accomplish this, the following major points of discussion are limited to studies of human HF, which have examined ventilatory responses to exercise while including assessment of factors directly influential to and/or a major consequence of abnormal $\dot{V}_E/\dot{V}CO_2$ (slope and/or ratio) and closely related derivatives.

FUNDAMENTALS OF VENTILATION: TOWARD AN IMPROVED UNDERSTANDING OF VENTILATORY (IN) EFFICIENCY IN HEART FAILURE

To fully appreciate the following discussion of exercise ventilatory function, it is important to briefly review classical theory of human ventilation relating in large part to the physiological constructs of "Ideal" alveolar gas and air equation(s) [34–39]. While alveolar equations and underlying individual parts used to study human ventilation are less useful for acutely identifying pathological mechanisms underpinning abnormal breathing patterns at the cellular level, it is evident these equations and related elements indeed represent more than pulmonary-centric factors in describing changes in respiration. This integrated view of human ventilatory function is keenly underscored when we examine one of several arrangements of variables in the alveolar gas equation at sea level (while also considering and meeting the assumptions of Haldane transformation equations) [34–41]:

$$PAO_2 = PIO_2 + \frac{PACO_2 \times FIO_2 \times (1 - R)}{R} - \frac{PACO_2}{R}$$

where PAO_2 may also be derived using the alveolar ventilation equation equal to

$$PAO_2 = PIO_2 - \frac{0.863 \times \dot{V}O_2 \times (1 - FIO_2)}{\dot{V}_A} - FIO_2 \times PACO_2$$

when alveolar ventilation ($\dot{V}_A$) is known as

$$\dot{V}_A = \frac{0.863 \times \dot{V}O_2 \times R}{PACO_2}$$

and, with knowledge of $\dot{V}_E$, derived, for example, as

$$f_B \times V_T = \dot{V}_E = \dot{V}_I \times \frac{\%N_2I}{\%N_2E}$$

and physiological dead space to tidal volume ratio (V_D/V_T) may be computed as

$$\frac{V_D}{V_T} = 1 - \frac{863 \times \dot{V}CO_2}{\dot{V}_E \times PaCO_2}$$

where PAO_2 is alveolar tension of O_2; PIO_2 is inspired tension of O_2; $PACO_2$ is alveolar tension of CO_2; FIO_2 is inspired fraction of O_2 (room air equal to 0.209) where inspired fraction of CO_2 is considered equal to zero; R is respiratory exchange ratio at the lung; $\dot{V}O_2$ is pulmonary oxygen uptake at standard temperature and pressure, dry (STPD); f_B is breathing frequency; $\%N_2E$ is percentage nitrogen in expired air; $\%N_2I$ is percentage nitrogen in inspired air (typically 79.04); $\dot{V}_I$ is inspired air volume/min; $PaCO_2$ is arterial blood tension of CO_2 (as used in V_D/V_T where it is assumed that $PACO_2$ and $PaCO_2$ are equivalent, otherwise $PACO_2$ should be used); and 0.863 is the correction factor needed when computing partial pressure from fractional concentration involving both volumes/gas (STPD) and volumes/flows (body temperature and pressure, saturated (BTPS)) standards of measurement (when gas exchange indexes are expressed in mL/min).

Thus, it is perhaps apparent to the trained human integrative physiologist while understandably more subtle to others that there are a complexity of interactions involving an infinite network of central and peripheral factors (i.e., physiology, anatomy, nervous system, and blood biochemistry), which determine the dynamics of human ventilation. It follows, therefore, that solely reporting changes in $\dot{V}_E$ relative to that of $\dot{V}CO_2$ is not a sound means to advance the integrative physiological understanding of exercise intolerance to which ventilatory inefficiency plays a major role in HF.

PHYSIOLOGIC DEAD SPACE OR ARTERIAL CARBON DIOXIDE? THE VENTILATORY EQUIVALENT FOR CARBON DIOXIDE IN HEART FAILURE

Given the breadth of studies that have quantified while also linking increased $\dot{V}_E/\dot{V}CO_2$ (slope and/or ratio) to worsened HF prognosis, it is clear there is little need or novelty in comprehensively debating the clinical relevance of this index [1,2,6–10,12–15]. By comparison, and to the best of our knowledge, what appears to be largely understood about the more remarkable integrative physiological features of an elevated $\dot{V}_E/\dot{V}CO_2$ (slope and/or ratio) in HF comes from no more than several studies that have been individually well conducted with properly reasoned interpretations in HFrEF (none in HFpEF). Nonetheless, these studies do not unequivocally arrive at similar conclusions regarding the major contributors to increased $\dot{V}_E/\dot{V}CO_2$ and, hence, ventilatory inefficiency (summary of studies where $PaCO_2$ was directly measured for the calculation of V_D/V_T with respect to quantifying ventilatory responses to exercise in HF [9,19–24,42], Table 1).

For example, Sullivan et al. [21] observed that while performing cardiac catheterization during CPET in 64 HFrEF patients, elevated V_D/V_T contributed a marked proportion to ventilatory inefficiency. More specifically, Sullivan et al. [21] suggested that the origin of the observed relationship between increased V_D/V_T (due to ↑ V_D, but not ↓ V_T) and $\dot{V}_E/\dot{V}CO_2$ is likely attributable to impaired pulmonary perfusion (i.e., ↑ V_D/V_T through an inverse relationship with cardiac output ($\dot{Q}$)) leading to a high ventilation-perfusion ratio ($\dot{V}_A/\dot{Q} > 1.0$). However, in the process of drawing their conclusions, Sullivan et al. [21] unfortunately does not present direct analyses between $PaCO_2$ and $\dot{V}_E/\dot{V}CO_2$ despite dedicating a substantial proportion of the study discussion to highlight support for strong relationships involving increased V_D and reduced $\dot{Q}$ with elevated $\dot{V}_E/\dot{V}CO_2$. Thus, while Sullivan et al. [21] astutely demonstrates there may be a compelling case for the importance of $\dot{V}_A/\dot{Q}$ mismatch provoked by decreased pulmonary perfusion, which may contribute to increased V_D and $\dot{V}_E/\dot{V}CO_2$ in HFrEF, the omission of relationship tests involving $PaCO_2$ and the absence of specific testing of ventilatory control mechanisms do not provide robust evidence-based support that ventilatory control mechanisms are indeed intact in HFrEF.

By contrast, both Guazzi et al. [9] and Woods et al. [22], for example, demonstrate intriguing evidence suggesting that in addition to provocations from high V_D/V_T, decreasing $PaCO_2$ is appreciably related to a rising $\dot{V}_E/\dot{V}CO_2$ (slope and ratio, respectively) during exercise in HFrEF. Indeed, Guazzi et al. [9] observed that HFrEF patients demonstrating lower levels of peak exercise $PaCO_2$ (<35 mm Hg) have exacerbated $\dot{V}_E/\dot{V}CO_2$ slope (38 ± 9 vs. 30 ± 5, $P < 0.01$, respectively) compared with HFrEF patients with higher peak exercise $PaCO_2$ (≥ 35 mm Hg). Likewise, the difference in $\dot{V}_E/\dot{V}CO_2$ slope between high and low peak exercise $PaCO_2$ thresholds across HFrEF appears similar when separately stratifying patients by peak exercise V_D/V_T between high (≥ 0.22) and low (<0.22) cut points (37 ± 10 vs. 32 ± 6, $P < 0.01$, respectively) [9]. In a similar manner, Woods et al. [22] demonstrated during submaximal exercise testing in HFrEF that like $\dot{V}_E/\dot{V}CO_2$ slope observed in Guazzi et al. [9], both $PaCO_2$ and V_D/V_T play important roles in increased $\dot{V}_E/\dot{V}CO_2$ ratio. Thus, given observations of Guazzi et al. [9] and Woods et al. [22], it is possible that HFrEF who regulate at an abnormally low $PaCO_2$ set point (e.g., apneic threshold) accompanied by an augmented ventilatory gain coupled with modest fluctuations in $PaCO_2$ may be more likely to demonstrate exercise tachypnea, which, given the denominator position of $PaCO_2$ in the V_D/V_T equation (*see discussion on alveolar equations*), would be consistent with high-low $\dot{V}_E/\dot{V}CO_2$ slope responders observed by Guazzi et al. [9].

Nevertheless, in addition to equally well-reasoned interpretations of Guazzi et al. [9], Woods et al. [22], and others [42] who suggest pulmonary-centric factors cannot entirely explain increased $\dot{V}_E/\dot{V}CO_2$ (slope or ratio), there are others, similar to Sullivan et al. [21], who place a stronger emphasis on the pathological role that high V_D driven by poor pulmonary perfusion (i.e., ↑ $\dot{V}_A/\dot{Q}$) plays in ventilatory inefficiency in HFrEF [20,23,24]. Whether the pulmonary-centric-only model or the integrative central and peripheral paradigm can wholly explain the genesis of ventilatory inefficiency in HFrEF, it is perhaps more of an enigma when attempting to answer the same question in HFpEF for whom there is a paucity of human integrative exercise physiology studies dedicated to phenotyping the origins of abnormal $\dot{V}_E/\dot{V}CO_2$ (slope or ratio) with respect to alveolar equations.

(RE)EMERGING CONCEPTS AND MECHANISMS OF VENTILATORY INEFFICIENCY: APPLICATION TO REDUCED OR PRESERVED EJECTION FRACTION HEART FAILURE

While the accumulating evidence to date suggests there is an evolving but broad understanding of what anatomical, physiological, and biochemical factors likely compose the HFpEF phenotype, there is a dearth of human-based research examining fundamental physiologically driven questions seeking to clarify the pathological origins of exercise intolerance unique to these patients. To this end and consistent with the aims of this chapter, we suggest that it is critical to continually advance our understanding of the underlying mechanisms contributing to the pathophysiology of HF with particular emphasis on studying those mechanisms directly translating to morbidity and mortality, which can only be accomplished through properly dedicated peer-reviewed scientific efforts.

In this context, by conducting patient-centered studies aimed at improving our understanding of the pathophysiology of exercise intolerance, recent cross-sectional observations (*unpublished data from our laboratory*) provide initial physiologically guided evidence suggesting exercise intolerance linked to ventilatory inefficiency (i.e., increased $\dot{V}_E/\dot{V}CO_2$ slope) is not mirrored across HFrEF and HFpEF. For example, while acknowledging causation cannot be implied when interpreting these data (consisting of a sample of HFrEF and HFpEF patients both meeting respective European Society of Cardiology (1) criteria), the use of alveolar equations and related parameters support the hypothesis that an elevated $\dot{V}_E/\dot{V}CO_2$ slope during CPET is differentially linked to decreasing $PaCO_2$ and increasing V_D/V_T when comparing HFrEF to HFpEF (Fig. 1).

TABLE 1 Studies of Exercise Ventilatory Function Relating to Physiological Dead Space to Tidal Volume Ratio Derived From Measured Arterial Carbon Dioxide Tension in Patients With Heart Failure

Study	Sample	Included Variable(s) During Exercise (mean±SD, Where Applicable When Reported)					$\dot{V}_A/\dot{Q}$	Ventilatory Control Test
		$\dot{V}O_2$ (mL/kg/min)	$\dot{V}_E/\dot{V}CO_2$	$\dot{V}_E/\dot{V}CO_2$ Slope	V_D/V_T	$PaCO_2$		
Sullivan et al. [21]	HFrEF, Age, 55±10 years, N=62	Peak (bike), 15.1±4.1	Yes, NR	No	Yes, NR	Yes	No	No
Clark et al. [42]	HFrEF, Age, 52 years[c], N=29	Peak (treadmill), 14.6±4.2	Yes, 40±10	No	Yes[a], 0.28±0.12	Yes[a], 36.9±4.7	No	No
Wasserman et al. [24]	HFrEF, Age, 60±3 years, N=130	Peak (bike), 17.0±4.7	No	No	Yes, NR	Yes, NR	No	No
Tanabe et al. [23]	HFrEF, Age, 52±10 years, N=105	Peak (bike), NR	Yes, NR	Yes[c]	Yes	Yes[b], NR	No	No
Wensel et al. [19]	HFrEF, Age, 65±11 years, N=30	Peak (treadmill), 15.2±4.2	Yes[c], 40.7	Yes[c], 32.5	Yes, 0.44±0.10	Yes[a], 36.5±5.3	No	No
Wensel et al. [20]	HFrEF, Age, 62±11 yrs, N=18	Peak (treadmill), 15.1±4.1	Yes, 45±12	Yes, 40±11	Yes, NR	Yes, 36±5	No	No
Guazzi et al. [9]	HFrEF, Age, 60±9 years, N=128	Peak (bike), 16.5±4.4	No	Yes, 34±8	Yes, 0.22±0.05	Yes, 38±5	No	No
Woods et al. [22]	HFrEF, Age, 54±8 years, N=33	Peak (bike), 13.0±3.8[d]	Yes, 45±9	No	Yes, 0.37±0.08	Yes, 33±5	No	No

[a]Reported from 9 out of 29 participants.
[b]Measured $PaCO_2$ but studied relationships involving end-tidal CO_2.
[c]Mean only reported.
[d]Analyses performed at 65% peak $\dot{V}O_2$. NR = mean±SD of total group not reported. See main text for definition of variables.

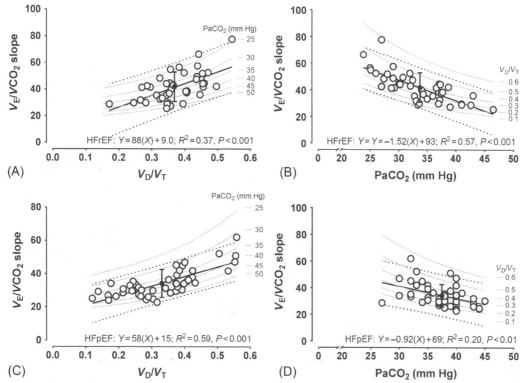

FIG. 1 Univariate ordinary least-squares regressions between peak exercise physiological dead space to tidal volume ratio ($V_D V_T$) or arterial carbon dioxide tension ($PaCO_2$) as independent predictors (abscissa) of the slope of the ventilatory equivalent for carbon dioxide ($\dot{V}_E/\dot{V}CO_2$, from rest to peak exercise) (ordinate) in patients with reduced (HFrEF, $N=42$; mean ± SD, left ventricular ejection fraction (LVEF) = 21 ± 6; weight = 86 ± 16 kg; body surface area (BSA) = 2.03 ± 0.20 m^2; and peak pulmonary oxygen uptake ($\dot{V}O_2$) = 8.7 ± 2.7 mL/kg/min) or preserved (HFpEF, $N=47$; mean ± SD, LVEF = 62 ± 8; weight = 98 ± 22 kg; BSA = 2.14 ± 0.28 m^2; and peak $\dot{V}O_2 = 8.1 \pm 2.4$ mL/kg/min; only LVEF differed vs. HFrEF at $P < 0.05$) ejection fraction. Respective patients with HFrEF or HFpEF (*unpublished data from our laboratory*) met European Society of Cardiology (1) criteria. For panels A–D: solid line is model goodness-of-fit line, dotted lines are 95% prediction bands of the model goodness-of-fit equation, gray bands are isopleths, and filled circle is mean ± SD of $\dot{V}_E/\dot{V}CO_2$ slope at the respective mean of the variable set on the abscissa.

More specifically, in addition to augmented $\dot{V}_E/\dot{V}CO_2$ slope (normal threshold, <30), it is noteworthy that while at similar levels of strength, the robust inverse relationship illustrated in Fig. 1 between $PaCO_2$ and $\dot{V}_E/\dot{V}CO_2$ slope in HFrEF is contrasted by direct linearity between V_D/V_T and $\dot{V}_E/\dot{V}CO_2$ slope in HFpEF. These data would appear to harmonize with an abnormally high V_D/V_T (normal range, ~0.15–0.29) at peak exercise accompanied by consistent levels of arterial pH (pHa, near the midpoint of normal range, 7.35–7.45) concurrent with reduced $PaCO_2$ (lower limits of the normal range, 35–45 mm Hg) across HF (Fig. 2) [43–46].

Despite V_D/V_T explaining >2x the variance associated with $\dot{V}_E/\dot{V}CO_2$ slope compared with $PaCO_2$ in HFpEF (Fig. 1, panels C and D, respectively), these patients also demonstrated a significant inverse relationship between $PaCO_2$ and $\dot{V}_E/\dot{V}CO_2$ slope that was less than half the strength of association observed in HFrEF presented in Fig. 1B. Likewise, although close to twofold less compared with HFpEF (Fig. 1C), increasing V_D/V_T also explained a large proportion of the variance in $\dot{V}_E/\dot{V}CO_2$ slope in HFrEF illustrated in Fig. 1A. Thus, it is apparent that although peak exercise pHa remained within normal limits across HFrEF and HFpEF (Fig. 2C), based on observations of lower peak exercise V_D/V_T (albeit greater than normal levels) but higher $PaCO_2$ in HFpEF compared with HFrEF (due to reduced $PaCO_2$ in HFrEF < 35 mm Hg) (Fig. 2, panels B and D, respectively), these data may be used to hypothesize that compared with HFrEF, different integrative pathophysiological mechanisms may compromise efficient ventilation in order to properly maintain acid-base equilibrium in HFpEF.

While no concurrent chemosensitivity or other muscle afferent testing was performed to objectively assess function of ventilatory control pathways in HF patients (similar to our critiques of Sullivan et al. [21] noted above), these data support the conclusions of Guazzi et al. [9] and Woods et al. [22] suggesting pathological mechanisms native to regulating $PaCO_2$ and anatomical/physiological constraints affecting V_D/V_T bear relevance to increased $\dot{V}_E/\dot{V}CO_2$ slope in HFrEF. Perhaps equally important, these data also provide novel perspectives to further our understanding of how the rise in $\dot{V}_E/\dot{V}CO_2$ slope may be affected by unique pathophysiology encompassing influential factors of both $PaCO_2$ and V_D/V_T in HFpEF.

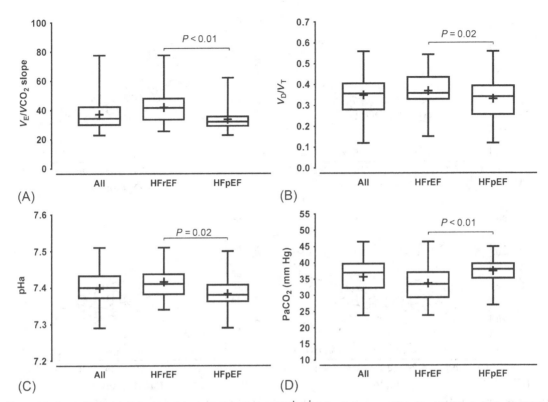

FIG. 2 Peak exercise slope of the ventilatory equivalent for carbon dioxide ($\dot{V}_E/\dot{V}CO_2$, from rest to peak exercise), physiological dead space to tidal volume ratio (V_DV_T), arterial pH (pHa), or arterial carbon dioxide tension (PaCO$_2$). Data are presented as interquartile range with mean equal to (+) in patients with reduced (HFrEF) or preserved (HFpEF) ejection fraction (see basic patient characteristics in caption for Fig. 1).

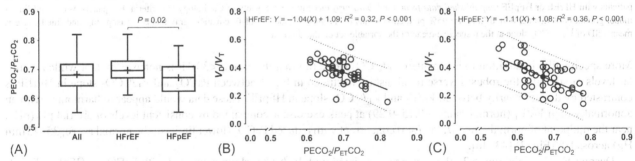

FIG. 3 Quotient of mixed-expired carbon dioxide to end-tidal carbon dioxide tension (PECO$_2$/P$_{ET}$CO$_2$) as a surrogate for global pulmonary ventilation-perfusion matching as described by Hansen et al. [47]. Panel A: Peak exercise PECO$_2$/P$_{ET}$CO$_2$, data presented as interquartile range with mean equal to (+) in patients with reduced (HFrEF) or preserved (HFpEF) ejection fraction (see basic patient characteristics in caption for Fig. 1). Panels B or C: Univariate ordinary least-squares regressions between peak exercise PECO$_2$/P$_{ET}$CO$_2$ (abscissa and independent) and peak exercise physiological dead space to tidal volume ratio (V_DV_T) (ordinate and dependent) in HFrEF or HFpEF, respectively. The filled circle is mean $\pm$ SD of V_DV_T at the respective mean of PECO$_2$/P$_{ET}$CO$_2$ set on the abscissa.

The extent to which reduced pulmonary perfusion impacts regional $\dot{V}_A/\dot{Q}$ inhomogeneity while subsequently contributing to increased V_D/V_T across HF remains unclear in the absence of accompanying cardiac catheterization data. However, the eloquent descriptions of Hansen et al. [47] illustrate that the quotient of mixed-expired CO$_2$ and end-tidal CO$_2$ (PECO$_2$/P$_{ET}$CO$_2$) may be used to estimate global airway versus perfusion limitation contributions to $\dot{V}_A/\dot{Q}$ mismatching. In this light, Fig. 3 illustrates PECO$_2$/P$_{ET}$CO$_2$ ratios (i.e., <0.70), which are not consistent with the presence of marked pulmonary perfusion limitation. Accordingly, because alveolar dead space (i.e., $V_{DAlv} = V_D$—anatomic dead space) constitutes the volume of gas entering unperfused alveoli, these data could be interpreted as $\dot{V}_A/\dot{Q}$ mismatch occurs primarily due to airway (ventilation) defects as opposed to physiological limitations (i.e., not ↓ pulmonary perfusion). Thus, though the above statements are currently hypothesis-generating suggestions, contributions from anatomical (as opposed to physiological) limitations to increased V_D/V_T may be considered consistent with pathology of the lungs (e.g., ↓ $\dot{V}_A/\dot{Q}$ and long-time-constant lung units contributed to by increased radial traction associated with thickened/fibrosed parenchyma and regional variations in lung compliance) resembling a restrictive pattern of airway disease, which has been demonstrated to occur secondary to HF [48–50].

On-Transient Kinetics of Ventilation and Gas Exchange: Emerging Concepts in Preserved Ejection Fraction Heart Failure

Although exercise studies of on-transient kinetics of ventilation and gas exchange associated with subanaerobic threshold intensities may not be traditionally used in a manner to support links between alveolar equations and $\dot{V}_E/\dot{V}CO_2$ slope, we suggest that there is certain relevance to such a discussion in HF. For example, while we acknowledge that the $\dot{V}_E$ to $\dot{V}CO_2$ relationship assumes a hyperbolic shape as exercise workload increases beyond the anaerobic threshold and that even at a hypothetical $\dot{V}CO_2$ equal to zero $\dot{V}_E$ would not be expected to intercept at zero [51,52], it could also be suggested that reporting the relationship between $\dot{V}_E$ and $\dot{V}CO_2$ as a linear function (i.e., first-degree polynomial) is accepted across most exercise paradigms including both incremental and square-wave exercise testing [53–55].

In this context, unpublished data from our laboratory in HFpEF examining on-transient kinetics of ventilation and $\dot{V}O_2$ during acute (6 min) fixed-load (20 W) square-wave cycle ergometry further support the hypothesis that pulmonary perfusion limitations may not be the single principal factor involved in driving the rise in V_D/V_T and $\dot{V}_E/\dot{V}CO_2$ slope likened to augmented V_{DAlv} (Fig. 4). As such, Fig. 4A describes on-transient kinetics of $\dot{V}O_2$ superimposed over on-kinetics of $\dot{V}CO_2$, demonstrating that both on-kinetics of $\dot{V}O_2$ and $\dot{V}CO_2$ are prolonged in HFpEF compared with what is expected in either younger (not shown) or age-matched healthy adults [56–63]. Compared with $\dot{V}O_2$, it is evident that on-kinetics of $\dot{V}CO_2$ are nearly twofold slower in HFpEF and age-matched controls. Following, it appears that although the asymptote for $\dot{V}_E$ is expectedly higher than that of $\dot{V}_A$ in Fig. 4B, occurring in a manner that temporally aligns with slowed on-kinetics of $\dot{V}CO_2$ in Fig. 4A, Fig. 4B illustrates that similar to age-matched controls, HFpEF demonstrated a similarly lengthened phase II (primary component) time constant (τ) for the on-kinetics of $\dot{V}_E$ and $\dot{V}_A$ (albeit modestly slower on-kinetics of $\dot{V}_E$ and $\dot{V}_A$ compared with $\dot{V}CO_2$). This discernable overlapping of τ across panels A and B (Fig. 4) is consistent with the contention that ventilation adjusts more as a function of $\dot{V}CO_2$ than $\dot{V}O_2$ [55], even in patients with HFpEF.

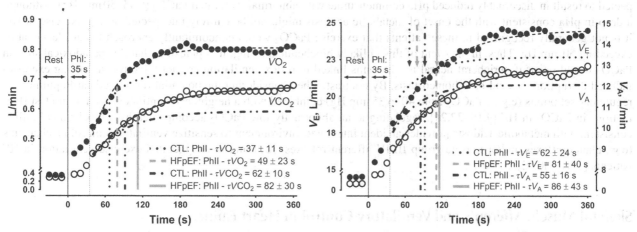

FIG. 4 On-transient kinetics of ventilation or gas exchange transitioning from rest to fixed-load (20 W at 65 rpm) upright cycle ergometry for 6 min in heart failure patients with preserved ejection fraction (male Caucasian HFpEF, $N=15$, met European Society of Cardiology (1) criteria; mean ± SD, age = 69 ± 7 years; body surface area = 2.17 ± 0.16 m²; and peak pulmonary oxygen uptake ($\dot{V}O_2$) = 17.2 ± 4.2 mL/kg/min) or age-matched healthy controls (male Caucasian CTL, $N=6$; mean ± SD, age = 70 ± 8 years; body surface area = 1.98 ± 0.13 m²; and peak $\dot{V}O_2$ = 27.8 ± 5.5 mL/kg/min). Breath-by-breath $\dot{V}O_2$, carbon dioxide output ($\dot{V}CO_2$), and minute ventilation ($\dot{V}_E$) were measured continuously via open-circuit spirometry interfaced with gas mass spectrometry. Alveolar ventilation ($\dot{V}_A$) was derived using alveolar equations (see text for detailed discussion of equations). Raw data from two repeated transients were each linear interpolated to 1 s intervals and then time aligned for ensemble averaging across transients into 10 s bins for analyses. Excluding the cardiodynamic phase (PhI = 35 s), phase II (PhII) on-transient ventilation or gas exchange kinetics were assessed via single exponential models as $Y_B + A_1(1 - e^{-(\tau-TD)/\tau})$, where Y_B is rest, A_1 is steady-state increase in ventilation or gas exchange above rest, TD is time delay, and τ is time constant. In panels A and B, large closed or open circles overlying dashed or solid model curve-fit lines represent data points for a patient with HFpEF representative of the study sample, whereas both small dotted and interspersed double-dotted lines without overlying large open or closed circles represent model curve-fit lines for a healthy CTL individual representative of the study sample. Downward facing arrows in panel B align vertically with respective PhII τ for on-kinetics of $\dot{V}O_2$ or $\dot{V}CO_2$ demonstrated in panel A.

In addition and perhaps more intriguing, because $\dot{V}_A$ is a direct consequence not only of changes in f_B and V_T but also due to adjustments in capillary perfusion of ventilated alveoli (*see discussion on alveolar equations*), if decreased pulmonary perfusion (i.e., $\uparrow \dot{V}_A/\dot{Q}$ as suggested in HFrEF [20,21]) is indeed foremost responsible for provoking increased V_D whereby a precipitous rise in $\dot{V}_E/\dot{V}CO_2$ slope follows, we hypothesize that the phase II τ for the on-kinetics of $\dot{V}_A$ would in fact be prominently shortened to the extent where the $\tau\dot{V}CO_2$ to $\tau\dot{V}_A$ ratio would be displaced to a value ≥ 1.0. However, this proposed convergence of $\dot{V}_A$ toward $\dot{V}CO_2$, based on the results of other populations [64], would theoretically be underpinned by impaired alveolar-capillary membrane gas transfer driven primarily by decreased pulmonary perfusion [20,21] (not airway or peripheral factors) that is not demonstrated as part of these observations (Fig. 4 in HFpEF: mean $\pm$ SD, $\tau\dot{V}CO_2/\tau\dot{V}_A = 0.75 \pm 0.27$ or $\tau\dot{V}CO_2/\tau\dot{V}_E = 0.73 \pm 0.27$). Accordingly, with respect to these data, the hypothesis that there is an absence of an independent driving effect of impaired pulmonary perfusion on the increased ventilatory response to exercise in HFpEF is consistent with selected observations of Olson et al. [28]. They demonstrated that although alveolar-capillary membrane conductance accompanied by pulmonary capillary blood volume was reduced in HFpEF compared with matched healthy controls at rest, at similar 20 W exercise, pulmonary capillary blood volume increased markedly in HFpEF leading to no differences between groups, whereas blunted alveolar-capillary membrane conductance at rest persisted at 20 W in HFpEF. Thus, while it is likely that $\dot{V}_A/\dot{Q}$ mismatch was also present in those data [28], a reduction in pulmonary perfusion does not appear to be the foremost driving factor.

Moreover, based on our above hypothesized model of dissociated on-kinetics between the unique parts of ventilation and gas exchange in HFpEF, a displacement of phase II on-kinetics of $\tau\dot{V}_A$ toward $\tau\dot{V}CO_2$ largely attributable to pulmonary arteriovenous shunting (i.e., due to $\downarrow$ pulmonary perfusion) would also reasonably be accompanied by a similar time course of transient arterial hypoxemia. However, HFpEF patients did not demonstrate obvious reductions in transcutaneous arterial saturation levels during exercise to suggest the presence of arterial O_2 desaturation (e.g., <93%–95% [65]) and arterial hypoxemia.

Lastly, impaired pulmonary perfusion could be expected to result in appreciable downstream reductions in pHa paralleled by abnormally elevated $PaCO_2$ relative to that of mixed alveolar CO_2. This would be the case in the absence of a calibrated rise in $\dot{V}_A$ commensurate with venous return and, hence, increased pulmonary CO_2 flow (i.e., $\downarrow \dot{V}_A/\dot{Q}$); or, conversely, if there is in indeed an augmented rise in $\dot{V}_A$ relative to pulmonary CO_2 flow (i.e., $\uparrow \dot{V}_A/\dot{Q}$ as suggested by others in HFrEF [20,21]). In the latter paradigm, because of direct attenuating effects on alveolar-capillary membrane gas transfer (i.e., increased gas diffusion is dependent on perfusion, whereas increased perfusion is not dependent on gas diffusion), while reiterating our suggestions from above, the presence of impaired pulmonary perfusion could be expected to result in discernably reduced pHa commensurate with abnormally elevated $PaCO_2$ (>45–50 mm Hg). Although a drop in pHa consistent with the onset of metabolic acidosis might not be entirely unexpected during exercise in HF, it is not generally recognized in these patients that exercise $PaCO_2$ would concomitantly encroach toward levels at or exceeding 50 mm Hg. Thus, in the event of this unlikely biochemical swing, it is plausible that this paradoxical shift in $PaCO_2$ would occur if peripheral neural mechanisms linked to reflex ventilation are not effective in sensing exercise-induced disruptions in acid-base homeostasis. By contrast, when considering the potential for altered cardiopulmonary regulatory set points (e.g., $\downarrow PaCO_2$ from 40 to 35 mm Hg) combined with a heightened ventilatory gain to modest fluctuations in $PaCO_2$ in HF [9,19–22,24], regulating at an abnormally low $PaCO_2$ accompanied by a biochemical milieu consistent with metabolic acidosis provides an ideal integrated environment to sensitize ventilatory control mechanisms (e.g., chemoreflexes and/or muscle group III/IV afferent reflexes) resulting in exaggerated exercise ventilation in HF patients [15,18,30,32,33,66,67].

Skeletal Muscle Afferents and Ventilatory Control in Heart Failure

While there is a clear interaction between neural afferents associated with both central and peripheral chemoreceptors and certain aspects of ventilatory control in adults across the health spectrum, there is an encouraging but still inexact understanding of the neurophysiological control of ventilation that originates at the skeletal muscle level in HF patients [15,18,30,32,33,66–68]. Therefore, because of distinct interactions between neural afferents and ventilatory control, this topic certainly warrants further discussion as it is related to the regulation of $\dot{V}_E/\dot{V}CO_2$.

A number of groups have soundly demonstrated the relationship between abnormal feedback from polymodal thin-fiber group III/IV muscle afferents and poor HF patient prognosis [15,18]. However, it is with advanced patient-based mechanistic studies [30,32,33,66–68] linking increased exercise ventilation (e.g., $\dot{V}_E/\dot{V}CO_2$ slope and/or ratio) to pathophysiology of networked peripheral (both somatic and autonomic) and central nervous system function that help to clarify this line of study in the context of alveolar equations. For example, Scott et al. [67] confirmed that peripheral neurophysiological mechanisms originating within skeletal muscle provoke abnormal rate-mediated elevations in ventilation in HF. In a novel manner, they demonstrated that pharmacological intervention specific to restoring acid-base blood homeostasis favorably influenced ventilatory responses to upper-limb muscle afferent stimulation (handgrip exercise). More specifically, Scott et al. [67] showed that acute infusion of sodium bicarbonate (HCO_3^-), a known hydrogen ion (H^+) buffer, prior to and continuing throughout muscle afferent stimulation, led to a pronounced attenuation in the reflex rise in $\dot{V}_E$ compared with a crossover condition of saline infusion in HFrEF. Those data further suggest that with reduced H^+ accumulation followed closely by lower $\dot{V}_E$, it could be expected that both pHa and $PaCO_2$ would have risen to levels consistent with normal acid-base equilibrium in the setting of the sodium HCO_3^- experimental arm.

Likewise, in a separate but complimentary experimental paradigm, Olson et al. [30] tested the hypothesis that by inhibiting the communication between muscle afferents and brainstem centers of ventilatory control via lower-lumbar intrathecal injection of the μ-opioid agonist fentanyl, the sympathetic efferent rate-mediated rise in $\dot{V}_E$ would be blunted during submaximal locomotor ergometry in HFrEF. Indeed, consistent with observations of Scott et al. [67], Olson et al. [30] demonstrated that commensurate with the normalized rise in $PaCO_2$ observed during the exercise plus fentanyl experimental transient, f_B, $\dot{V}_E$, and $\dot{V}_E/\dot{V}CO_2$ (both slope and ratio) decreased compared with a placebo crossover exercise condition (i.e., sham lower-lumbar injection) in HFrEF. Thus, while those data of Scott et al. [67] and Olson et al. [30] are encouraging and supportive of the discussions of Guazzi et al. [9] and Woods et al. [22] suggesting that mechanisms of ventilatory control are not fully intact and thereby play a major role in provoking increased $\dot{V}_E/\dot{V}CO_2$ in HF, it should also be considered that those observations have been isolated to patients with HFrEF in the absence of separate studies properly performed in HFpEF. Nevertheless, while it is clear that specific studies focusing on testing potential links between muscle afferents and ventilatory control are needed in HFpEF, there is precedence for such a discussion as pathology of skeletal muscle proposed in HFpEF (e.g., decreased blood acid-base equilibrium contributed to by increased energy contributions from anaerobic metabolism due to an increased proportion of type II vs type I skeletal muscle fibers and decreased enzymatic activity/function at mitochondrion levels [69–74]) may prime the intramuscular environment for an altered group III/IV afferent pathway affecting increased $\dot{V}_E$ to a similar extent as has been observed in HFrEF [30,32,33,66–68].

SUMMARY AND CONCLUDING REMARKS

Heart failure is a major public health crisis. The population of adults with HF is expected to exceed 8 million by 2020 [17]. Exercise intolerance is closely related to increased morbidity, reduced quality of life, and worse prognosis [1,2,6–10,12–15]. While there is no unifying support for a specific pathophysiological mechanism explaining exercise intolerance in HF, there is a large body of evidence suggesting ventilatory inefficiency is a major contributor to this hallmark feature [6–10,12–15,18–22,24]. Although ample evidence has evolved highlighting the clinical implications linked to exercise ventilatory inefficiency, paradoxically, there is not an equally concerted dedication toward conducting integrative physiological driven studies in human HF aimed at dissecting where variables such as $\dot{V}_E/\dot{V}CO_2$ slope manifest from.

Therefore, this chapter sought to bridge the discussion between the clinical relevance of exercise ventilatory inefficiency (i.e., $\dot{V}_E/\dot{V}CO_2$ slope and/or ratio) specific to both HFrEF and HFpEF with what is currently understood about the classical constructs of human ventilation as articulated by "Ideal" alveolar gas and air equations [34–41]. While we acknowledge that solving the conundrum of exercise ventilatory inefficiency and exercise intolerance cannot be solved by a sudden growth of independent human integrative physiological studies, the noted disparity in agreement between works in this field regarding the fundamental contributions of V_D/V_T and $PaCO_2$ to increased $\dot{V}_E/\dot{V}CO_2$ in HFrEF should be resolved [9,14,19–24]. Additional areas for advancement in this field should focus on the paucity of data unique to patients with HFpEF as relating to the pathophysiology of increased $\dot{V}_E/\dot{V}CO_2$ (slope and/or ratio) and the "gold" standard principles of human respiratory physiology [34–41]. These gaps in knowledge are clinically impactful given the direct prognostic translation of ventilatory inefficiency in patients across the HF spectrum [1,2,6–10,12–15].

In conclusion, our detailed discussion of what is currently understood about links between alveolar equations and $\dot{V}_E/\dot{V}CO_2$ combined with new and noteworthy hypothesis-generating data presented in both HFrEF and HFpEF patients provides merit for the suggestion that ventilatory inefficiency is not a consequence of independent pulmonary-centric pathophysiology. Moreover, while it is likely that pathology of mechanical and physiological factors contributing to increased V_D/V_T and impaired mechanisms of ventilatory control affecting regulation of $PaCO_2$ play profound roles in leading to increased $\dot{V}_E/\dot{V}CO_2$ slope across HF, it is equally apparent that interpretations of the current paradigms intrinsic to HFrEF cannot at this point in time be applied to HFpEF. Evolving our human integrative physiological understanding of the underlying mechanisms associated with regulation of common outcome variables such as $\dot{V}_E/\dot{V}CO_2$ slope is a high-priority clinical and research need as exercise intolerance is a hallmark feature universal to HF [1,2,6–11,16].

ACKNOWLEDGMENTS

Funding for this work was supported by National Institutes of Health Grant RO1-HL126638 (TPO) and American Heart Association Grant #16POST30260021 (EHV).

REFERENCES

[1] McMurray JJ, Adamopoulos S, Anker SD, Auricchio A, Böhm M, Dickstein K, et al. ESC Guidelines for the diagnosis and treatment of acute and chronic heart failure 2012. Eur J Heart Fail 2012;14(8):803–69.

[2] Paulus WJ, Tschöpe C, Sanderson JE, Rusconi C, Flachskampf FA, Rademakers FE, et al. How to diagnose diastolic heart failure: a consensus statement on the diagnosis of heart failure with normal left ventricular ejection fraction by the Heart Failure and Echocardiography Associations of the European Society of Cardiology. Eur Heart J 2007;28(20):2539–50.

[3] Bhatia RS, Tu JV, Lee DS, Austin PC, Fang J, Haouzi A, et al. Outcome of heart failure with preserved ejection fraction in a population-based study. N Engl J Med 2006;355(3):260–9.

[4] Zile MR, Brutsaert DL. New concepts in diastolic dysfunction and diastolic heart failure. Part II. Causal mechanisms and treatment. Circulation 2002;105(12):1503–8.

[5] Zile MR, Brutsaert DL. New concepts in diastolic dysfunction and diastolic heart failure. Part I. Diagnosis, prognosis, and measurements of diastolic function. Circulation 2002;105(11):1387–93.

[6] Shafiq A, Brawner CA, Aldred HA, Lewis B, Williams CT, Tita C, et al. Prognostic value of cardiopulmonary exercise testing in heart failure with preserved ejection fraction. The Henry Ford HospITal CardioPulmonary EXercise Testing (FIT-CPX) project. Am Heart J 2016;174:167–72.

[7] Guazzi M, Myers J, Arena R. Cardiopulmonary exercise testing in the clinical and prognostic assessment of diastolic heart failure. J Am Coll Cardiol 2005;46(10):1883–90.

[8] Guazzi M, Myers J, Peberdy MA, Bensimhon D, Chase P, Arena R. Cardiopulmonary exercise testing variables reflect the degree of diastolic dysfunction in patients with heart failure-normal ejection fraction. J Cardiopulm Rehabil Prev 2010;30(3):165–72.

[9] Guazzi M, Reina G, Tumminello G, Guazzi MD. Exercise ventilation inefficiency and cardiovascular mortality in heart failure: the critical independent prognostic value of the arterial CO_2 partial pressure. Eur Heart J 2005;26(5):472–80.

[10] Arena R, Myers J, Aslam SS, Varughese EB, Peberdy MA. Peak VO_2 and VE/VCO_2 slope in patients with heart failure: a prognostic comparison. Am Heart J 2004;147(2):354–60.

[11] Arena R, Myers J, Aslam SS, Varughese EB, Peberdy MA. Technical considerations related to the minute ventilation/carbon dioxide output slope in patients with heart failure. Chest 2003;124(2):720–7.

[12] Arena R, Guazzi M, Myers J, Peberdy MA. Prognostic characteristics of cardiopulmonary exercise testing in heart failure: comparing American and European models. Eur J Cardiovasc Prev Rehabil 2005;12(6):562–7.

[13] Arena R, Myers J, Hsu L, Peberdy MA, Pinkstaff S, Bensimhon D, et al. The minute ventilation/carbon dioxide production slope is prognostically superior to the oxygen uptake efficiency slope. J Card Fail 2007;13(6):462–9.

[14] Clark AL, Chua TP, Coats AJ. Anatomical dead space, ventilatory pattern, and exercise capacity in chronic heart failure. Br Heart J 1995;74(4):377–80.

[15] Ponikowski P, Francis DP, Piepoli MF, Davies LC, Chua TP, Davos CH, et al. Enhanced ventilatory response to exercise in patients with chronic heart failure and preserved exercise tolerance: marker of abnormal cardiorespiratory reflex control and predictor of poor prognosis. Circulation 2001;103(7):967–72.

[16] Heidenreich PA, Albert NM, Allen LA, Bluemke DA, Butler J, Fonarow GC, et al. Forecasting the impact of heart failure in the united states a policy statement from the american heart association. Circulation: Heart Failure 2013;6(3):606–19.

[17] Benjamin EJ, Blaha MJ, Chiuve SE, Cushman M, Das SR, Deo R, et al. Heart disease and stroke statistics—2017 update: a report from the American Heart Association. Circulation 2017;135(10):e146–603. https://doi.org/10.1161/CIR.0000000000000485.

[18] Ponikowski PP, Chua TP, Francis DP, Capucci A, Coats AJ, Piepoli MF. Muscle ergoreceptor overactivity reflects deterioration in clinical status and cardiorespiratory reflex control in chronic heart failure. Circulation 2001;104(19):2324–30.

[19] Wensel R, Georgiadou P, Francis DP, Bayne S, Scott AC, Genth-Zotz S, et al. Differential contribution of dead space ventilation and low arterial pCO_2 to exercise hyperpnea in patients with chronic heart failure secondary to ischemic or idiopathic dilated cardiomyopathy. Am J Cardiol 2004;93(3):318–23.

[20] Wensel R, Francis DP, Georgiadou P, Scott A, Genth-Zotz S, Anker SD, et al. Exercise hyperventilation in chronic heart failure is not caused by systemic lactic acidosis. Eur J Heart Fail 2005;7(7):1105–11.

[21] Sullivan M, Higginbotham M, Cobb F. Increased exercise ventilation in patients with chronic heart failure: intact ventilatory control despite hemodynamic and pulmonary abnormalities. Circulation 1988;77(3):552–9.

[22] Woods PR, Olson TP, Frantz RP, Johnson BD. Causes of breathing inefficiency during exercise in heart failure. J Card Fail 2010;16(10):835–42.

[23] Tanabe Y, Hosaka Y, Ito M, Ito E, Suzuki K. Significance of end-tidal P(CO(2)) response to exercise and its relation to functional capacity in patients with chronic heart failure. Chest 2001;119(3):811–7.

[24] Wasserman K, Zhang YY, Gitt A, Belardinelli R, Koike A, Lubarsky L, et al. Lung function and exercise gas exchange in chronic heart failure. Circulation 1997;96(7):2221–7.

[25] Clark AL, Piepoli M, Coats AJ. Skeletal muscle and the control of ventilation on exercise: evidence for metabolic receptors. Eur J Clin Invest 1995;25(5):299–305.

[26] Olson LJ, Snyder EM, Beck KC, Johnson BD. Reduced rate of alveolar-capillary recruitment and fall of pulmonary diffusing capacity during exercise in patients with heart failure. J Card Fail 2006;12(4):299–306.

[27] Olson TP, Johnson BD. Influence of cardiomegaly on disordered breathing during exercise in chronic heart failure. Eur J Heart Fail 2011;13(3).

[28] Olson TP, Johnson BD, Borlaug BA. Impaired pulmonary diffusion in heart failure with preserved ejection fraction. JACC Heart Fail 2016;4(6):490–8.

[29] Olson TP, Joyner MJ, Dietz NM, Eisenach JH, Curry TB, Johnson BD. Effects of respiratory muscle work on blood flow distribution during exercise in heart failure. J Physiol 2010;588(Pt 13):2487–501.

[30] Olson TP, Joyner MJ, Eisenach JH, Curry TB, Johnson BD. Influence of locomotor muscle afferent inhibition on the ventilatory response to exercise in heart failure. Exp Physiol 2014;99(2):414–26.

[31] Piepoli M, Clark AL, Volterrani M, Adamopoulos S, Sleight P, Coats AJ. Contribution of muscle afferents to the hemodynamic, autonomic, and ventilatory responses to exercise in patients with chronic heart failure: effects of physical training. Circulation 1996;93(5):940–52.

[32] Scott AC, Davies LC, Coats AJ, Piepoli M. Relationship of skeletal muscle metaboreceptors in the upper and lower limbs with the respiratory control in patients with heart failure. Clin Sci (Lond) 2002;102(1):23–30.

[33] Scott AC, Francis DP, Davies LC, Ponikowski P, Coats AJ, Piepoli MF. Contribution of skeletal muscle 'ergoreceptors' in the human leg to respiratory control in chronic heart failure. J Physiol 2000;529(3):863–70.

[34] Whipp BJ, Ward SA. Determinants and control of breathing during muscular exercise. Br J Sports Med 1998;32(3):199–211.

[35] Cropp GJ, Comroe JH. Role of mixed venous blood PCO_2 in respiratory control. J Appl Physiol 1961;16(6):1029–33.

[36] Comroe JH. Lung. Chicago: Year Book Publishers; 1955.

[37] Comroe JH, Dripps RD. The oxygen tension of arterial blood and alveolar air in normal human subjects. Am J Physiol—Legacy Content 1944;142(5):700–7.

[38] Comroe JH. The hyperpnea of muscular exercise. Physiol Rev 1944;24(3):319–39.

[39] Fenn WO, Rahn H, Otis AB. A theoretical study of the composition of the alveolar air at altitude. Am J Physiol—Legacy Content 1946;146(5):637–53.

[40] Wilmore JH, Costill DL. Adequacy of the Haldane transformation in the computation of exercise VO_2 in man. J Appl Physiol 1973;35(1):85–9.

[41] Wagner JA, Horvath S, Dahms TE, Reed S. Validation of open-circuit method for the determination of oxygen consumption. J Appl Physiol 1973;34(6):859–63.

[42] Clark AL, Poole-Wilson PA, Coats AJS. Relation between ventilation and carbon dioxide production in patients with chronic heart failure. J Am Coll Cardiol 1992;20(6):1326–32.

[43] Whipp BJ, Wasserman K. Alveolar-arterial gas tension differences during graded exercise. J Appl Physiol 1969;27(3):361–5.

[44] Furuike AN, Sue DY, Hansen JE, Wasserman K. Comparison of physiologic dead space/tidal volume ratio and alveolar-arterial PO_2 difference during incremental and constant work exercise 1, 2. Am Rev Respir Dis 1982;126(3):579–83.

[45] Hansen JE, Sue DY, Wasserman K. Predicted values for clinical exercise testing. Am Rev Respir Dis 1984;129(2 Pt 2):S49–55.

[46] Sun X-G, Hansen JE, Garatachea N, Storer TW, Wasserman K. Ventilatory efficiency during exercise in healthy subjects. Am J Respir Crit Care Med 2002;166(11):1443–8.

[47] Hansen JE, Ulubay G, Chow BF, Sun X-G, Wasserman K. Mixed-expired and end-tidal CO_2 distinguish between ventilation and perfusion defects during exercise testing in patients with lung and heart diseases. CHEST J 2007;132(3):977–83.

[48] Agostoni P, Pellegrino R, Conca C, Rodarte JR, Brusasco V. Exercise hyperpnea in chronic heart failure: relationships to lung stiffness and expiratory flow limitation. J Appl Physiol 2002;92(4):1409–16.

[49] Johnson BD, Beck KC, Olson LJ, O'Malley KA, Allison TG, Squires RW, et al. Ventilatory constraints during exercise in patients with chronic heart failure. CHEST J 2000;117(2):321–32.

[50] Mancini DM. Pulmonary factors limiting exercise capacity in patients with heart failure. Prog Cardiovasc Dis 1995;37(6):347–70.

[51] Ward SA, Whipp BJ. Ventilatory control during exercise with increased external dead space. J Appl Physiol Respir Environ Exerc Physiol 1980;48(2):225–31.

[52] Gargiulo P, Apostolo A, Perrone-Filardi P, Sciomer S, Palange P, Agostoni P. A non invasive estimate of dead space ventilation from exercise measurements. PLoS ONE 2014;9(1):e87395.

[53] Wasserman K, Whipp BJ, Casaburi R. Respiratory control during exercise. Comprehensive Physiol 1986;.

[54] Whipp BJ, Ward SA. Cardiopulmonary coupling during exercise. J Exp Biol 1982;100(1):175–93.

[55] Wasserman K, Van Kessel AL, Burton GG. Interaction of physiological mechanisms during exercise. J Appl Physiol 1967;22(1):71–85.

[56] Mezzani A, Grassi B, Giordano A, Corrà U, Colombo S, Giannuzzi P. Age-related prolongation of phase I of $\dot{V}O_2$ on-kinetics in healthy humans. Am J Phys Regul Integr Comp Phys 2010;299(3):R968–76.

[57] Murias JM, Spencer MD, Kowalchuk JM, Paterson DH. Influence of phase I duration on phase II VO_2 kinetics parameter estimates in older and young adults. Am J Physiol Regul Integr Comp Physiol 2011;301(1):R218–24.

[58] Babcock MA, Paterson DH, Cunningham DA, Dickinson JR. Exercise on-transient gas exchange kinetics are slowed as a function of age. Med Sci Sports Exerc 1994;26(4):440–6.

[59] Bell C, Paterson D, Kowalchuk J, Cunningham D. Oxygen uptake kinetics of older humans are slowed with age but are unaffected by hyperoxia. Exp Physiol 1999;84(4):747–59.

[60] Cunningham DA, Himann JE, Paterson DH, Dickinson JR. Gas exchange dynamics with sinusoidal work in young and elderly women. Respir Physiol 1993;91(1):43–56.

[61] DeLorey DS, Kowalchuk JM, Paterson DH. Effect of age on O(2) uptake kinetics and the adaptation of muscle deoxygenation at the onset of moderate-intensity cycling exercise. J Appl Physiol (1985) 2004;97(1):165–72.

[62] DeLorey DS, Paterson DH, Kowalchuk JM. Effects of ageing on muscle O_2 utilization and muscle oxygenation during the transition to moderate-intensity exercise. Appl Physiol Nutr Metab 2007;32(6):1251–62.

[63] Chilibeck P, Paterson D, Smith W, Cunningham D. Cardiorespiratory kinetics during exercise of different muscle groups and mass in old and young. J Appl Physiol 1996;81(3):1388–94.

[64] Oren A, Whipp BJ, Wasserman K. Effect of acid-base status on the kinetics of the ventilatory response to moderate exercise. J Appl Physiol 1982;52(4):1013–7.

[65] Dempsey JA, Wagner PD. Exercise-induced arterial hypoxemia. J Appl Physiol 1999;87(6):1997–2006.

[66] Olson TP, Joyner MJ, Johnson BD. Influence of locomotor muscle metaboreceptor stimulation on the ventilatory response to exercise in heart failure. Circ Heart Fail 2010;3(2):212–9.

[67] Scott AC, Wensel R, Davos CH, Georgiadou P, Kemp M, Hooper J, et al. Skeletal muscle reflex in heart failure patients: role of hydrogen. Circulation 2003;107(2):300–6.

[68] Scott AC, Wensel R, Davos CH, Kemp M, Kaczmarek A, Hooper J, et al. Chemical mediators of the muscle ergoreflex in chronic heart failure: a putative role for prostaglandins in reflex ventilatory control. Circulation 2002;106(2):214–20.

[69] Kitzman DW, Nicklas B, Kraus WE, Lyles MF, Eggebeen J, Morgan TM, et al. Skeletal muscle abnormalities and exercise intolerance in older patients with heart failure and preserved ejection fraction. Am J Physiol Heart Circ Physiol 2014;306(9):H1364–70.

[70] Haykowsky MJ, Brubaker PH, Stewart KP, Morgan TM, Eggebeen J, Kitzman DW. Effect of endurance training on the determinants of peak exercise oxygen consumption in elderly patients with stable compensated heart failure and preserved ejection fraction. J Am Coll Cardiol 2012;60(2):120–8.

[71] Haykowsky MJ, Brubaker PH, John JM, Stewart KP, Morgan TM, Kitzman DW. Determinants of exercise intolerance in elderly heart failure patients with preserved ejection fraction. J Am Coll Cardiol 2011;58(3):265–74.

[72] Zamani P, Rawat D, Shiva-Kumar P, Geraci S, Bhuva R, Konda P, et al. Effect of inorganic nitrate on exercise capacity in heart failure with preserved ejection fraction. Circulation 2015;131(4):371–80 [discussion 80].

[73] Dhakal BP, Malhotra R, Murphy RM, Pappagianopoulos PP, Baggish AL, Weiner RB, et al. Mechanisms of exercise intolerance in heart failure with preserved ejection fraction: the role of abnormal peripheral oxygen extraction. Circ Heart Fail 2015;8(2):286–94.

[74] Lee JF, Barrett-O'Keefe Z, Nelson AD, Garten RS, Ryan JJ, Nativi-Nicolau JN, et al. Impaired skeletal muscle vasodilation during exercise in heart failure with preserved ejection fraction. Int J Cardiol 2016;211:14–21.

Part III

Alcohol, Tobacco, and Other Drugs of Abuse

Chapter 16

Relationships of Alcohol Consumption With Risks for Type 2 Diabetes Mellitus and Cardiovascular Disease in Men and Women

Ichiro Wakabayashi

Department of Environmental and Preventive Medicine, Hyogo College of Medicine, Nishinomiya, Japan

INTRODUCTION

The risk of cardiovascular disease is increased in patients with diabetes. Unfavorable lifestyles, including excessive calorie intake and the lack of physical activity, cause obesity, which is a major risk factor for type 2 diabetes. Alcohol drinking and cigarette smoking are popular lifestyles. In the past 100 years, there has been an accumulation of information on the relationship between alcohol drinking and health hazards. There are various alcohol-related health problems including malignant tumors (e.g., hepatoma, esophageal cancer, colon cancer, and breast cancer), liver dysfunction, accidents following acute alcohol consumption, and alcohol dependency. On the other hand, light-to-moderate alcohol consumption has been known to be associated with a decrease in the risk of thrombotic diseases including coronary artery disease, ischemic type of stroke, and peripheral arterial disease. A J-shaped relationship between amount of alcohol consumption and total mortality has been shown in a meta-analysis study using results of 34 previous prospective studies [1]. This means that a small amount of alcohol drinking, such as less than two drinks (24 g ethanol) for men and one drink (12 g ethanol) for women, acts favorably for life extension. This is consistent with the famous statement "it has long been recognized that the problems with alcohol relate not to the use of a bad thing but to the abuse of a good thing" by Abraham Lincoln [2].

In this chapter, the relations of habitual alcohol drinking with type 2 diabetes and with cardiovascular disease are introduced, and it is discussed whether and how alcohol influences cardiovascular health.

ALCOHOL AND CEREBRO- AND CARDIOVASCULAR DISEASES

In many developed countries, heart disease is the leading cause of death, although it is the second death cause in Japan (percentage of total deaths in 2015: malignant tumor, 28.7%; heart disease, 15.2%; pneumonia, 9.4%; and cerebrovascular disease, 8.7%) [3]. The relationship between alcohol and the risk of ischemic heart disease depends on the amount of alcohol consumption. According to a meta-analysis study [4], habitual light-to-moderate alcohol intake, such as 20 g ethanol per day, resulted in maximum reduction of the risk of ischemic heart disease, and there was a significant reduction with intake of 72 g or less of ethanol per day, whereas there was a significant increase in the risk with intake of 89 g or more of ethanol per day. Heavy alcohol drinking also causes cardiomyopathy [5] and cardiac arrhythmia [6].

Atherosclerosis is deeply involved in the pathogenesis of ischemic heart disease, and its major risk factors are known to be dyslipidemia, hypertension, smoking, and diabetes. The most important mechanism for the inverse association between alcohol drinking and risk of ischemic heart disease is alcohol-induced elevation of HDL cholesterol level in the blood. HDL is involved in reverse transport of cholesterol from blood vessels to the liver. The HDL-elevating action of alcohol has been shown to explain about 50% of the reason for the beneficial effect of alcohol on coronary risk [7]. In addition, LDL cholesterol level has been shown to be lower in drinkers than in nondrinkers [8], resulting in a much lower atherogenic index (ratio of LDL cholesterol to HDL cholesterol), a classical predictor of coronary artery disease, in drinkers. Another mechanism for the above beneficial action of alcohol is alcohol-induced favorable change in the blood coagulation-fibrinolysis balance toward atherogenesis. Alcohol is known to inhibit platelet aggregation [9]. Fibrinogen level and coagulation factor VII level have been shown to be lower in drinkers than in nondrinkers [10], while plasminogen activator level was reported to be higher in drinkers than in nondrinkers [11]. Thus, alcohol is thought to inhibit thrombus formation, which is an important process of atherosclerotic progression and triggers a cardiovascular event. Moreover, the risks of diabetes and metabolic syndrome have been shown to be lower in light-to-moderate drinkers than in nondrinkers as described below. In addition to the effects of alcohol itself, polyphenol-containing beverages such as red wine are thought to have protective effects on cardiovascular health through their antioxidant free radical-scavenging actions including inhibition of LDL oxidation and increase in synthesis of nitric oxide [12]. However, it is still debatable whether red wine is superior to other alcoholic beverages, because there are possibilities of confounding by socioeconomic factors, e.g., a healthier lifestyle of wine drinkers than that of beer and spirit drinkers [13]. On the other hand, alcohol drinking induces elevation of blood pressure, a major risk factor for atherosclerotic disease, although the exact mechanism for this action of alcohol remains unknown. Triglyceride level is increased by excessive alcohol drinking. Furthermore, excessive alcohol drinking causes an increase in oxidative stress, resulting in dysfunction of the vascular endothelial cells and acceleration of atherosclerosis [14]. Thus, in heavy drinkers, these harmful actions of alcohol on cardiovascular health are thought to overcome the abovementioned beneficial actions.

Stroke can be divided into two types, ischemic type (cerebral infarction) and hemorrhagic type (cerebral hemorrhage and subarachnoid hemorrhage). As a risk factor, hypertension is more important in the pathogenesis of stroke, especially in hemorrhagic type, while dyslipidemia is more important in the pathogenesis of ischemic heart disease. Similar to the relationship between alcohol intake and risk of ischemic heart disease, there is a J-shaped relationship between the amount of alcohol intake and risk of ischemic type of stroke [15], and the favorable effect of light drinking is explained by the abovementioned beneficial actions of alcohol on cholesterol levels and blood coagulation-fibrinolysis balance, while the unfavorable effect of heavy drinking is explained by alcohol-induced hypertension and hypertriglyceridemia. In addition, dehydration and arrhythmia, especially atrial fibrillation, induced by heavy drinking have been shown to be partly involved in the increased risk of cerebral infarction in drinkers. A linear positive association between amount of alcohol intake and risk of the hemorrhagic type of stroke has been reported, and this unfavorable effect of alcohol is mainly explained by alcohol-induced elevation of blood pressure [16]. Alcohol-induced inhibition of blood coagulation and increase in blood fibrinolysis are thought to exacerbate the hemorrhage. The pattern of drinking alcohol also influences the risk of cerebro- and cardiovascular diseases. Incidental heavy or binge drinking (five or more drinks on an occasion) is known to increase the risk of hemorrhagic stroke and ischemic heart disease. A high frequency of light-to-moderate drinking during the week has been shown to be more beneficial for cardiovascular health because the action of alcohol on HDL cholesterol level is more prolonged [17]. The relationships of habitual alcohol drinking with cardiovascular risk factors are summarized in Fig. 1.

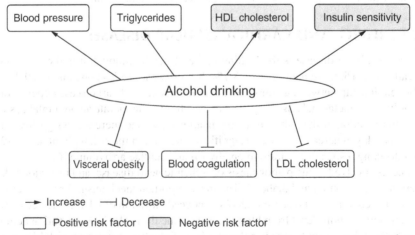

FIG. 1 Associations of alcohol drinking with cardiovascular risk factors. Beneficial effects of alcohol on cardiovascular health are mediated by increases of HDL cholesterol and insulin sensitivity and by decreases of blood coagulation activity, LDL cholesterol level, and risk of visceral obesity. Harmful effects of alcohol on cardiovascular health are mediated by an elevation of blood pressure and an increase of triglyceride level.

ALCOHOL AND RISK OF DIABETES

In previous meta-analysis studies, light-to-moderate alcohol drinking was shown to be inversely associated with incidence of type 2 diabetes, and there was a U-shaped relationship between amount of alcohol intake and risk of diabetes [18–20]. In a study by Baliunas et al. using the results of 20 cohort studies, the relative risk (RR) for type 2 diabetes was lowest with consumption of 22 g ethanol per day (RR = 0.87) in men and 24 g ethanol per day (RR = 0.60) in women, while there was an increased risk with consumption of >60 g and >50 g ethanol per day in men and women, respectively [21]. The risk of coronary artery disease in patients with type 2 diabetes has also been shown to be lower in light-to-moderate alcohol drinkers than in nondrinkers [18,22], and this finding is similar to that in the general population [4].

In previous studies on the relationship between alcohol and type 2 diabetes, confounding for the relationship by adiposity was suggested: A prospective study in the United States showed that there was a U-shaped association between alcohol consumption and risk of diabetes in women who were overweight or obese, while there was an inverse association between them in women with a body mass index (BMI) of <25 kg/m^2 [23]. In Japanese men, there was a positive relationship between moderate-to-heavy alcohol intake and incidental diabetes in men with low BMI, while an inverse relationship between them was shown in men with high BMI [24–26]. There was reportedly no significant association between alcohol intake and risk of type 2 diabetes in Japanese women [26]. In a recent study in which data for Europeans in eight countries were analyzed, there was an inverse association between moderate alcohol drinking and risk of diabetes in women but not in men, and the association was more prominent in obese women than in nonobese women [27]. Thus, there is a possibility that the relationship between alcohol consumption and risk of diabetes is modified by sex, adiposity, and race and/or ethnicity. Further studies with adjustment for these confounding factors are needed to clarify the relationship between alcohol consumption and risk of diabetes.

ALCOHOL AND GLUCOSE METABOLISM

The inverse association between alcohol consumption and the risk of type 2 diabetes is explained by an increase in insulin sensitivity by alcohol [28]. As a mechanism for this, alcohol increases gene transcription in adipocytes for adiponectin [29], which is known to promote insulin sensitivity and have an anti-inflammatory action [30]. By a meta-analysis, it has been shown that blood adiponectin level was consistently increased after drinking alcohol [31]. Leptin, another adipocyte-secreted hormone, also increases insulin sensitivity [32]. Chronic alcohol administration has been demonstrated to increase mRNA expression of adiponectin and leptin in mouse white adipose tissue [33]. On the other hand, an inverse relationship between alcohol drinking and blood adiponectin level has been reported in Japanese men [34], and this relationship depended on polymorphism of the alcohol dehydrogenase 1B (ADH1B) gene [35]. Thus, there is a possibility that the relationship between alcohol and adiponectin varies among different races and/or ethnicities and is influenced by gene polymorphisms of alcohol-metabolizing enzymes.

Metabolic syndrome, a cluster of risk factors including visceral obesity, diabetes, dyslipidemia, and hypertension, is known to be a potent predictor of cerebro- and cardiovascular diseases. There are two major sets of international criteria for diagnosing metabolic syndrome. Metabolic syndrome is defined according to the criteria of the National Cholesterol Education Program (NCEP-ATP) III as the presence of three or more risk factors [36], and it is defined according to the criteria of the International Diabetes Federation (IDF) as the presence of two or more risk factors in addition to visceral obesity diagnosed as a large waist circumference [37]. Risk factors included in the criteria are visceral obesity, high blood pressure, dyslipidemia (low HDL cholesterol and/or high triglycerides), and hyperglycemia. There is a discrepancy in the findings of previous studies regarding the relationship between habitual alcohol drinking and risk of metabolic syndrome: it has been shown that the risk is lower, not different, and higher in drinkers than in nondrinkers. However, an inverse association between light drinking and risk of metabolic syndrome in the general population has been shown by meta-analysis studies: A meta-analysis study using results of seven prospective studies showed that the risk of metabolic syndrome was decreased by light-to-moderate drinking such as 40 g or less of ethanol per day in men and 20 g or less of ethanol per day in women [38]. Elevation of HDL cholesterol and increase in insulin sensitivity are thought to contribute to the lower risk of metabolic syndrome in light-to-moderate drinkers. Another meta-analysis study using results of six prospective studies showed that the risk of metabolic syndrome was significantly lower in very light drinkers (0.1–5 g ethanol per day) and significantly higher in heavy drinkers (more than 35 g ethanol per day) than in nondrinkers [39]. The prevalence of metabolic syndrome in persons with alcohol use disorders (AUDs) has been shown to be high by a recent meta-analysis study using data of cross-sectional analysis in five studies: more than one in five persons with AUDs has metabolic syndrome [40]. This increased risk of metabolic syndrome may be due to alcohol abuse, resulting in an increase in prevalence of high blood pressure and hypertriglyceridemia, which are components of metabolic syndrome. In Japanese men, light drinking was shown to be inversely associated with metabolic syndrome in the general population [41] but not in men with diabetes

[42]. On the other hand, heavy drinking was shown to be positively associated with metabolic syndrome both in the general population [41] and in men with diabetes [42]. These findings suggest that the decreased risk of diabetes by light drinking is at least partly involved in the inverse association between light drinking and metabolic syndrome in the general population.

ALCOHOL AND GLYCEMIC STATUS

Glycemic status is strongly affected by diet and thus cannot be correctly evaluated by a single measurement of blood glucose level. Hemoglobin A1c is a conventional measure used for evaluating long-term glycemic status that does not require fasting, although it has been pointed out that one-third fewer cases of undiagnosed diabetes are identified by a hemoglobin A1c cutoff point of $\geq 6.5\%$ (48 mmol/mol) than by a fasting glucose cutoff point of ≥ 126 mg/dL (7.0 mmol/L) in the US population [43]. An inverse association between alcohol consumption and hemoglobin A1c level in the general population has been shown in previous cross-sectional and case-control studies [44,45]. Fig. 2 shows a comparison of hemoglobin A1c levels in different categories of drinkers in Japanese men and women. Since the amount of alcohol consumption was much smaller in women than in men, different categories of drinkers were used for men and women: female subjects were divided into four groups of nondrinkers, occasional drinkers, regular light drinkers (< 22 g ethanol per day), and regular heavy drinkers ($\geq$ 22 g ethanol per day), while male subjects were divided into five groups of nondrinkers, regular light drinkers (< 22 g ethanol per day), regular moderate drinkers ($\geq$22 and <44 g ethanol per day), regular heavy drinkers ($\geq$44 and <66 g ethanol per day), and regular very heavy drinkers ($\geq$66 g ethanol per day). Male occasional drinkers were excluded from analysis because their alcohol consumption levels were too varied. Age, BMI, and histories of smoking and regular exercise were adjusted in the analysis. The results showed that in men, hemoglobin A1c was significantly lower in the regular light, moderate, heavy, and very heavy drinker groups than in the nondrinker group, and there was no significant difference in hemoglobin A1c among the light, moderate, heavy, and very heavy drinker groups [46]. In women, hemoglobin A1c was significantly lower in the occasional, regular light, and regular heavy drinker groups than in the nondrinker group and was significantly lower in the regular light and regular heavy drinker groups than in the occasional drinker group, while there was no significant difference in hemoglobin A1c between the regular light and heavy drinker groups [47]. Therefore, both in men and in women, alcohol drinking was inversely associated with hemoglobin A1c level, but there was no dose dependency in the relationship of regular drinking with hemoglobin A1c in men and women. In addition, the inverse association between alcohol and hemoglobin A1c has been reported to be independent of BMI and waist circumference [46,47]. This finding is reasonable because alcohol has been shown to promote insulin sensitivity independently of body weight and body fat [33,48]. In women, hemoglobin A1c was significantly lower in regular light drinkers than in occasional light drinkers (Fig. 2), and this agrees with the findings of the previous study that drinking

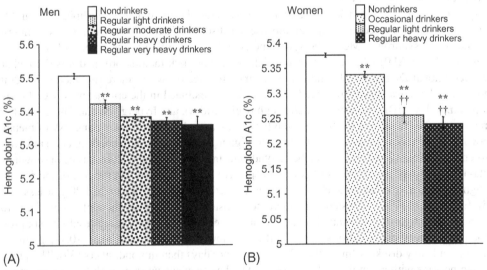

FIG. 2 Relationships between alcohol consumption and hemoglobin A1c levels in men (A) and women (B) at 35–60 years of age. Those receiving medication therapy for diabetes were excluded from the subjects for analysis. Male subjects ($n=32,111$) were divided into five groups: nondrinkers, regular light drinkers (<22 g ethanol/day), regular moderate drinkers ($\geq$22 g and <44 g ethanol/day), regular heavy drinkers ($\geq$44 g and <66 g ethanol/day), and regular very heavy drinkers ($\geq$66 g ethanol/day). Female subjects ($n=18,352$) were divided into four groups of nondrinkers, occasional drinkers, regular light drinkers (<22 g ethanol/day), and regular heavy drinkers ($\geq$22 g ethanol/day). Symbols denote significant differences from nondrinkers (**, $p<0.01$) and occasional drinkers (††, $p<0.01$).

frequency was inversely associated with the risk of myocardial infarction and that this association was in part attributable to the relationship of alcohol with hemoglobin A1c [17]. There seems to be no gender-related difference in the relationship between alcohol and glycemic status [44–47].

ALCOHOL AND OBESITY

Obesity deteriorates insulin sensitivity and increases the risk of type 2 diabetes. Although there have been a number of studies on the relationship between alcohol and body weight, their results are inconsistent, showing positive, inverse, and no associations between them [49]. Alcohol has bilateral effects on body weight and obesity: Alcohol drinking results in energy intake, enhances appetite, and reduces fat oxidation, while alcohol increases energy expenditure, interferes with digestion and absorption of nutrients, and increases sympathetic activity [49–51]. The former and latter effects of alcohol may explain the positive and inverse associations, respectively, between alcohol consumption and body weight. Their relationship was shown to be influenced by age in a study in which data for 36,121 Japanese men aged 20–70 years were analyzed [52]. The subjects in that study were divided by age into four quartile groups. The odds ratio of each drinker subgroup versus the nondrinker subgroup for abdominal obesity evaluated by high waist-to-height ratio (WHtR ≥ 0.5) was estimated in each quartile group for age. As shown in Table 1, in the younger groups (first and second quartiles for age), the odds ratios of light, moderate, and heavy drinker subgroups versus the nondrinker subgroup were significantly lower than the reference level of 1.00. In the third quartile group for age, the odds ratio of the light drinker subgroups versus the nondrinker subgroup was significantly lower than the reference level. However, in the oldest group (fourth quartile for age), the odds ratios of the light, moderate, and heavy drinker groups versus the nondrinker group were not significantly different from the reference level (Table 1). Therefore, the relationship between alcohol and obesity was confounded by age, and the inverse association between them tended to become weaker with advance of age [52]. A gender difference in the relationship between alcohol consumption and obesity has also been pointed out: there may be positive and inverse associations between alcohol intake and obesity in men and women, respectively [53]. Thus, age and gender should be taken into account when the relationship between alcohol and obesity is discussed. In addition, the relationship may be affected by beverage type, drinking frequency, and drinking pattern. The frequency of drinking was shown to be inversely related to obesity [54], while binge drinking was shown to be positively associated with overweight and obesity [55].

The association between obesity and hyperglycemia in the general population has been shown to be modified by alcohol consumption [56]: The odds ratio for hyperglycemia of men with versus men without high WHtR was significantly lower in light (<22 g ethanol per day) and moderate (≥ 22 and <44 g ethanol per day) drinkers than in nondrinkers but was not significantly different in nondrinkers and heavy drinkers (≥ 44 g ethanol per day), and significant interactions of light-to-moderate drinking and obesity with hyperglycemia were found. Thus, light-to-moderate alcohol drinking is thought to reduce the impact of obesity on the risk of diabetes, and this agrees with the U-shaped relationship between amount of alcohol intake and risk of diabetes [18,19].

TABLE 1 Odds Ratios for Visceral Obesity of Each Drinker Subgroup Versus the Nondrinker Subgroup in Four Quartiles for Age of Japanese Men

	Q$_1$ of Age (20–36 Years)	Q$_2$ of Age (37–46 Years)	Q$_3$ of Age (47–55 Years)	Q$_4$ of Age (56–70 Years)
Nondrinkers	1	1	1	1
Light drinkers (<22 g/day)	0.59** (0.50–0.71)	0.68** (0.59–0.79)	0.84* (0.72–0.96)	1.05 (0.91–1.21)
Moderate drinkers (≥ 22, <44 g/day)	0.60** (0.52–0.68)	0.75** (0.67–0.83)	0.93 (0.84–1.03)	1.03 (0.93–1.14)
Heavy drinkers (≥ 44 g/day)	0.81** (0.69–0.94)	0.82** (0.72–0.92)	1.00 (0.89–1.12)	1.02 (0.90–1.15)

Odds ratios with their 95% confidence intervals in parentheses for visceral obesity (WHtR ≥ 0.5) are shown. Age and history of smoking were adjusted for estimation of odds ratios. Q$_1$, first quartile for age; Q$_2$, second quartile for age; Q$_3$, third quartile for age; Q$_4$, fourth quartile for age. Asterisks denote significant differences from the reference level of 1.00 (*, $p<0.05$; **, $p<0.01$).

ADIPOSITY-RELATED INDICES AND DIABETES

Obesity is a major risk factor of type 2 diabetes. In addition, adiposity-related indexes have been shown to be useful for discriminating the risk of diabetes. Lipid accumulation product (LAP), a continuous marker of lipid overaccumulation, has been proposed to be a predictor of diabetes and cardiovascular disease [57–59]. LAP is determined by triglyceride level (TG) and waist circumference (WC) as follows: LAP=TG (mmol/l)×(WC (cm)−65) for men and LAP=TG (mmol/l)×(WC (cm)−58) for women. LAP has been shown to be better than BMI, a common index for obesity, for identifying diabetes [60]. More recently, we have proposed cardiometabolic index (CMI) as a discriminator of diabetes [61]. CMI is defined as the product of WHtR and triglycerides-to-HDL-cholesterol ratio (TG/HDL ratio). WHtR, an index of waist circumference corrected by height, is known to be a more reasonable marker for abdominal obesity than waist circumference alone [62] and has been shown to be a better discriminator for cardiovascular disease than waist circumference or BMI [63–65]. TG/HDL ratio has been shown to be a better predictor of coronary artery disease than lipid alone or a classical lipid-related index such as LDL-cholesterol-to-HDL-cholesterol ratio [66,67]. Moreover, TG/HDL ratio has been demonstrated to reflect small dense LDL particles [68], which are known to be more atherogenic than larger buoyant particles [69–71]. Therefore, CMI is thought to be an ideal adiposity and blood lipid-related index for evaluating the risk of cardiovascular disease. In our recent study comparing LAP and CMI in patients with peripheral arterial disease, CMI but not LAP showed significant associations with intima-media thickness of the common carotid artery and exercise-induced decrease in ankle-brachial systolic pressure index [72]. Further prospective studies are needed to confirm the usefulness of CMI for predicting cardiovascular disease.

MODIFICATIONS BY ALCOHOL OF ADIPOSITY-RELATED INDICES AND OF THEIR RELATIONS WITH DIABETES

LAP and CMI are composed of adiposity and lipid measures, which are influenced by alcohol consumption. Thus, these indexes are also modified by alcohol drinking. Alcohol intake has been reported to show U-shaped and J-shaped relationships with WHtR and triglycerides, respectively [73,74]. A linear positive relationship has been shown between alcohol intake and HDL cholesterol. Putting these relationships together, there are J-shaped and U-shaped relationships between alcohol and LAP and between alcohol and CMI, respectively, as shown in Fig. 3 [73,74]. The inverse associations of light drinking with LAP and CMI are in line with its association with the risk of diabetes [18,19], which is reasonable since LAP and CMI are discriminators of the risk of diabetes [60,61]. The odds ratios for hyperglycemia of men with versus men without

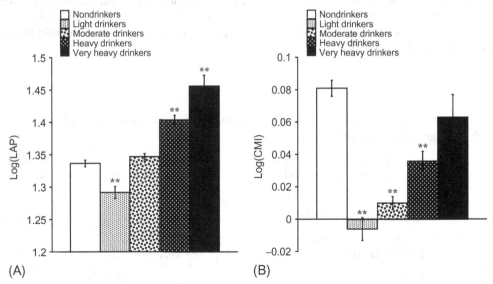

FIG. 3   Relationships of alcohol consumption with LAP (A) and CMI (B) in men at 35–60 years of age (n=21,572). Men receiving medication therapy for dyslipidemia were excluded from subjects. Subjects were divided into five groups of nondrinkers, light drinkers (<22 g ethanol/day), moderate drinkers (≥22 g and <44 g ethanol/day), heavy drinkers (≥44 g and <66 g ethanol/day), and very heavy drinkers (≥66 g ethanol/day). Since LAP and CMI did not show normal distributions, they were used for analysis after logarithmic transformation. Subjects showing LAP levels of zero or lower (n=194) were excluded from analysis because their log-transformed values could not be calculated. Asterisks denote significant differences from nondrinkers (**, p<0.01).

high CMI were significantly higher than the reference level of 1.00 in nondrinker, light (<22 g ethanol per day), moderate (≥22 and <44 g ethanol per day), heavy (≥44 and <66 g ethanol per day), and very heavy (≥66 g ethanol per day) drinkers. The odds ratio was significantly lower in light, moderate, and heavy drinkers than in nondrinkers and was not significantly different in nondrinkers and very heavy drinkers, and significant interactions of high CMI and light and moderate drinking with hyperglycemia were found [74]. Thus, the positive association between CMI and hyperglycemia is attenuated by light-to-moderate alcohol drinking. This finding suggests that the status of alcohol drinking should be taken into account when CMI is used for evaluation of the risk of diabetes.

CONCLUSION

There is an inverse association between alcohol consumption and the risk of type 2 diabetes. The risk of coronary artery disease is also inversely associated with light-to-moderate drinking in patients with diabetes and in the general population. Thus, habitual alcohol drinking is thought to prevent the onset of type 2 diabetes and its cardiovascular complications, which determine the prognosis of patients. On the other hand, there are various not only health-related problems but also social problems caused by alcohol. Nobody knows whether they will suffer from alcohol dependency in the future. Although light drinking has been shown to be preventive for type 2 diabetes and coronary artery disease in epidemiological studies, a pattern of daily light drinking is not common: Most light drinkers do not drink daily, and most daily drinkers are not light drinkers [75]. In fact, it was reported that only 2% of the US population had a light daily drinking pattern [76]. Moreover, the risk of hypoglycemia is increased by alcohol drinking in patients receiving medication therapy for diabetes, especially in patients taking insulin or insulin secretagogues [77]. Therefore, a habit of alcohol drinking, even light-to-moderate drinking, should never be recommended to persons with diabetes or persons without diabetes, despite of the cardioprotective action of alcohol at a low dose.

REFERENCES

[1] Di Castelnuovo A, Costanzo S, Bagnardi V, Donati MB, Iacoviello L, de Gaetano G. Alcohol dosing and total mortality in men and women: an updated meta-analysis of 34 prospective studies. Arch Intern Med 2006;166:2437–45.

[2] Ellison RC. Continuing reluctance to accept emerging scientific data on alcohol and health. AIM Digest 2002;11:6–7.

[3] Anonymous. Health, Labour and Welfare Statistics Association. Annual statistical report of national health conditions. J Health Welfare Stat 2016;63(9):65 [in Japanese].

[4] Corrao G, Rubbiati L, Bagnardi V, Zambon A, Poikolainen K. Alcohol and coronary heart disease: a meta-analysis. Addiction 2000;95:1505–23.

[5] Laonigro I, Correale M, Di Biase M, Altomare E. Alcohol abuse and heart failure. Eur J Heart Fail 2009;11:453–62.

[6] Tonelo D, Providência R, Gonçalves L. Holiday heart syndrome revisited after 34 years. Arq Bras Cardiol 2013;101:183–9.

[7] Gaziano JM, Buring JE, Breslow JL, Goldhaber SZ, Rosner B, VanDenburgh M, et al. Moderate alcohol intake, increased levels of high-density lipoprotein and its subfractions, and decreased risk of myocardial infarction. N Engl J Med 1993;329:1829–34.

[8] Castelli WP, Doyle JT, Gordon T, Hames CG, Hjortland MC, Hulley SB, et al. Alcohol and blood lipids. The cooperative lipoprotein phenotyping study. Lancet 1977;2:153–5.

[9] Rubin R. Effect of ethanol on platelet function. Alcohol Clin Exp Res 1999;23:1114–8.

[10] Salem RO, Laposata M. Effects of alcohol on hemostasis. Am J Clin Pathol 2005;123(Suppl):S96–S105.

[11] Ridker PM, Vaughan DE, Stampfer MJ, Glynn RJ, Hennekens CH. Association of moderate alcohol consumption and plasma concentration of endogenous tissue-type plasminogen activator. JAMA 1994;272:929–33.

[12] Caimi G, Carollo C, Lo Presti R. Wine and endothelial function. Drugs Exp Clin Res 2003;29:235–42.

[13] van de Wiel A, de Lange DW. Cardiovascular risk is more related to drinking pattern than to the type of alcoholic drinks. Neth J Med 2008;66:467–73.

[14] Lucas DL, Brown RA, Wassef M, Giles TD. Alcohol and the cardiovascular system: research challenges and opportunities. J Am Coll Cardiol 2005;45:1916–24.

[15] Mazzaglia G, Britton AR, Altmann DR, Chenet L. Exploring the relationship between alcohol consumption and non-fatal or fatal stroke: a systematic review. Addiction 2001;96:1743–56.

[16] Reynolds K, Lewis B, Nolen JD, Kinney GL, Sathya B, He J. Alcohol consumption and risk of stroke: a meta-analysis. JAMA 2003;289:579–88.

[17] Mukamal KJ, Jensen MK, Grønbaek M, Stampfer MJ, Manson JE, Pischon T, et al. Drinking frequency, mediating biomarkers, and risk of myocardial infarction in women and men. Circulation 2005;112:1406–13.

[18] Howard AA, Arnsten JH, Gourevitch MN. Effect of alcohol consumption on diabetes mellitus: a systematic review. Ann Intern Med 2004;140:211–9.

[19] Koppes LL, Dekker JM, Hendriks HF, Bouter LM, Heine RJ. Moderate alcohol consumption lowers the risk of type 2 diabetes: a meta-analysis of prospective observational studies. Diabetes Care 2005;28:719–25.

[20] Li XH, Yu FF, Zhou YH, He J. Association between alcohol consumption and the risk of incident type 2 diabetes: a systematic review and dose-response meta-analysis. Am J Clin Nutr 2016;103:818–29.

[21] Baliunas DO, Taylor BJ, Irving H, Roerecke M, Patra J, Mohapatra S, et al. Alcohol as a risk factor for type 2 diabetes: a systematic review and meta-analysis. Diabetes Care 2009;32:2123–32.

[22] Koppes LL, Dekker JM, Hendriks HF, Bouter LM, Heine RJ. Meta-analysis of the relationship between alcohol consumption and coronary heart disease and mortality in type 2 diabetic patients. Diabetologia 2006;49:648–52.

[23] Hu FB, Manson JE, Stampfer MJ, Colditz G, Liu S, Solomon CG, et al. Diet, lifestyle, and the risk of type 2 diabetes mellitus in women. N Engl J Med 2001;345:790–7.

[24] Tsumura K, Hayashi T, Suematsu C, Endo G, Fujii S, Okada K. Daily alcohol consumption and the risk of type 2 diabetes in Japanese men: the Osaka Health Survey. Diabetes Care 1999;22:1432–7.

[25] Watanabe M, Barzi F, Neal B, Ueshima H, Miyoshi Y, Okayama A, et al. Alcohol consumption and the risk of diabetes by body mass index levels in a cohort of 5,636 Japanese. Diabetes Res Clin Pract 2002;57:191–7.

[26] Waki K, Noda M, Sasaki S, Matsumura Y, Takahashi Y, Isogawa A, et al. Alcohol consumption and other risk factors for self-reported diabetes among middle-aged Japanese: a population-based prospective study in the JPHC study cohort I. Diabet Med 2005;22:323–31.

[27] Beulens JW, Van Der Schouw YT, Bergmann MM, Rohrmann S, Schulze MB, Buijsse B, et al. Alcohol consumption and risk of type 2 diabetes in European men and women: influence of beverage type and body size: The EPIC-InterAct study. J Intern Med 2012;272:358–70.

[28] Hulthe J, Fagerberg B. Alcohol consumption and insulin sensitivity: a review. Metab Syndr Relat Disord 2005;3:45–50.

[29] Joosten MM, Beulens JW, Kersten S, Hendriks HF. Moderate alcohol consumption increases insulin sensitivity and ADIPOQ expression in post-menopausal women: a randomised, crossover trial. Diabetologia 2008;51:1375–81.

[30] Aldhahi W, Hamdy O. Adipokines, inflammation, and the endothelium in diabetes. Curr Diab Rep 2003;3:293–8.

[31] Brien SE, Ronksley PE, Turner BJ, Mukamal KJ, Ghali WA. Effect of alcohol consumption on biological markers associated with risk of coronary heart disease: systematic review and meta-analysis of interventional studies. BMJ 2011;342:d636.

[32] Mantzoros CS, Magkos F, Brinkoetter M, Sienkiewicz E, Dardeno TA, Kim SY, et al. Leptin in human physiology and pathophysiology. Am J Physiol Endocrinol Metab 2011;301:E567–84.

[33] Paulson QX, Hong J, Holcomb VB, Nunez NP. Effects of body weight and alcohol consumption on insulin sensitivity. Nutr J 2010;9:14.

[34] Nishise Y, Saito T, Makino N, Okumoto K, Ito JI, Watanabe H, et al. Relationship between alcohol consumption and serum adiponectin levels: the Takahata study—a cross-sectional study of a healthy Japanese population. J Clin Endocrinol Metab 2010;95:3828–35.

[35] Maeda S, Mure K, Mugitani K, Watanabe Y, Iwane M, Mohara O, et al. Roles of the ALDH2 and ADH1B genotypes on the association between alcohol intake and serum adiponectin levels among Japanese male workers. Alcohol Clin Exp Res 2014;38:1559–66.

[36] Anonymous. Expert Panel on Detection, Evaluation, and Treatment of High Blood Cholesterol in Adults. Executive Summary of The Third Report of The National Cholesterol Education Program (NCEP) Expert Panel on Detection, Evaluation, And Treatment of High Blood Cholesterol In Adults (Adult Treatment Panel III). JAMA 2001;285:2486–97.

[37] Alberti KG, Zimmet P, Shaw J. IDF Epidemiology Task Force Consensus Group. The metabolic syndrome—a new worldwide definition. Lancet 2005;366:1059–62.

[38] Alkerwi A, Boutsen M, Vaillant M, Barre J, Lair ML, Albert A, et al. Alcohol consumption and the prevalence of metabolic syndrome: a meta-analysis of observational studies. Atherosclerosis 2009;204:624–35.

[39] Sun K, Ren M, Liu D, Wang C, Yang C, Yan L. Alcohol consumption and risk of metabolic syndrome: a meta-analysis of prospective studies. Clin Nutr 2014;33:596–602.

[40] Vancampfort D, Hallgren M, Mugisha J, De Hert M, Probst M, Monsieur D, et al. The prevalence of metabolic syndrome in alcohol use disorders: a systematic review and meta-analysis. Alcohol Alcohol 2016;51:515–21.

[41] Wakabayashi I. Cross-sectional relationship between alcohol consumption and prevalence of metabolic syndrome in Japanese men and women. J Atheroscler Thromb 2010;17:695–704.

[42] Wakabayashi I. Association between alcohol drinking and metabolic syndrome in Japanese male workers with diabetes mellitus. J Atheroscler Thromb 2011;18:684–92.

[43] Cowie CC, Rust KF, Byrd-Holt DD, Gregg EW, Ford ES, Geiss LS, et al. Prevalence of diabetes and high risk for diabetes using A1C criteria in the U.S. population in 1988-2006. Diabetes Care 2010;33:562–8.

[44] Gulliford MC, Ukoumunne OC. Determinants of glycated haemoglobin in the general population: associations with diet, alcohol and cigarette smoking. Eur J Clin Nutr 2001;55:615–23.

[45] Harding AH, Sargeant LA, Khaw KT, Welch A, Oakes S, Luben RN, et al. Cross-sectional association between total level and type of alcohol consumption and glycosylated haemoglobin level: the EPIC-Norfolk Study. Eur J Clin Nutr 2002;56:882–90.

[46] Wakabayashi I. Obesity-independent inverse association between regular alcohol consumption and hemoglobin A1c. Obes Facts 2012;5:60–7.

[47] Shimomura T, Wakabayashi I. Association between alcohol consumption and glycemic status in middle-aged women. Can J Diabetes 2015;39:502–6.

[48] Hong J, Smith RR, Harvey AE, Núñez NP. Alcohol consumption promotes insulin sensitivity without affecting body fat levels. Int J Obes (Lond) 2009;33:197–203.

[49] Suter PM. Is alcohol consumption a risk factor for weight gain and obesity? Crit Rev Clin Lab Sci 2005;42:197–227.

[50] Lieber CS. Perspectives: do alcohol calories count? Am J Clin Nutr 1991;54:976–82.

[51] Lieber CS. ALCOHOL: its metabolism and interaction with nutrients. Annu Rev Nutr 2000;20:395–430.

[52] Wakabayashi I. Age-dependent inverse association between alcohol consumption and obesity in Japanese men. Obesity (Silver Spring) 2011;19:1881–6.

[53] Yeomans MR. Alcohol, appetite and energy balance: is alcohol intake a risk factor for obesity? Physiol Behav 2010;100:82–9.

[54] Tolstrup JS, Heitmann BL, Tjønneland AM, Overvad OK, Sørensen TI, Grønbaek MN. The relation between drinking pattern and body mass index and waist and hip circumference. Int J Obes (Lond) 2005;29:490–7.

[55] Arif AA, Rohrer JE. Patterns of alcohol drinking and its association with obesity: data from the Third National Health and Nutrition Examination Survey, 1988-1994. BMC Public Health 2005;5:126.

[56] Wakabayashi I. Light-to-moderate alcohol drinking reduces the impact of obesity on the risk of diabetes mellitus. J Stud Alcohol Drugs 2014;75:1032–8.

[57] Kahn HS. The "lipid accumulation product" performs better than the body mass index for recognizing cardiovascular risk: a population-based comparison. BMC Cardiovasc Disord 2005;5:26.

[58] Wehr E, Pilz S, Boehm BO, März W, Obermayer-Pietsch B. The lipid accumulation product is associated with increased mortality in normal weight postmenopausal women. Obesity (Silver Spring) 2011;19:1873–80.

[59] Bozorgmanesh M, Hadaegh F, Azizi F. Diabetes prediction, lipid accumulation product, and adiposity measures; 6-year follow-up: Tehran lipid and glucose study. Lipids Health Dis 2010;9:45.

[60] Kahn HS. The lipid accumulation product is better than BMI for identifying diabetes: a population-based comparison. Diabetes Care 2006;29:151–3.

[61] Wakabayashi I, Daimon T. The "cardiometabolic index" as a new marker determined by adiposity and blood lipids for discrimination of diabetes mellitus. Clin Chim Acta 2015;438:274–8.

[62] Hsieh SD, Muto T. Metabolic syndrome in Japanese men and women with special reference to the anthropometric criteria for the assessment of obesity: Proposal to use the waist-to-height ratio. Prev Med 2006;42:135–9.

[63] Lee CM, Huxley RR, Wildman RP, Woodward M. Indices of abdominal obesity are better discriminators of cardiovascular risk factors than BMI: a meta-analysis. J Clin Epidemiol 2008;61:646–53.

[64] Gruson E, Montaye M, Kee F, Wagner A, Bingham A, Ruidavets JB, et al. Anthropometric assessment of abdominal obesity and coronary heart disease risk in men: the PRIME study. Heart 2010;96:136–40.

[65] Ashwell M, Gunn P, Gibson S. Waist-to-height ratio is a better screening tool than waist circumference and BMI for adult cardiometabolic risk factors: systematic review and meta-analysis. Obes Rev 2012;13:275–86.

[66] Gaziano JM, Hennekens CH, O'Donnell CJ, Breslow JL, Buring JE. Fasting triglycerides, high-density lipoprotein, and risk of myocardial infarction. Circulation 1997;96:2520–5.

[67] Jeppesen J, Hein HO, Suadicani P, Gyntelberg F. Relation of high TG-low HDL cholesterol and LDL cholesterol to the incidence of ischemic heart disease. An 8-year follow-up in the Copenhagen Male Study. Arterioscler Thromb Vasc Biol 1997;17:1114–20.

[68] Dobiášová M, Frohlich J. The plasma parameter log (TG/HDL-C) as an atherogenic index: correlation with lipoprotein particle size and esterification rate in apoB-lipoprotein-depleted plasma (FER$_{HDL}$). Clin Biochem 2001;34:583–8.

[69] de Graaf J, Hak-Lemmers HL, Hectors MP, Demacker PN, Hendriks JC, Stalenhoef AF. Enhanced susceptibility to in vitro oxidation of the dense low density lipoprotein subfraction in healthy subjects. Arterioscler Thromb 1991;11:298–306.

[70] Tribble DL, Holl LG, Wood PD, Krauss RM. Variations in oxidative susceptibility among six low density lipoprotein subfractions of differing density and particle size. Atherosclerosis 1992;93:189–99.

[71] Galeano NF, Milne R, Marcel YL, Walsh MT, Levy E, Ngu'yen TD, et al. Apoprotein B structure and receptor recognition of triglyceride-rich low density lipoprotein (LDL) is modified in small LDL but not in triglyceride-rich LDL of normal size. J Biol Chem 1994;269:511–9.

[72] Wakabayashi I, Sotoda Y, Hirooka S, Orita H. Association between cardiometabolic index and atherosclerotic progression in patients with peripheral arterial disease. Clin Chim Acta 2015;446:231–6.

[73] Wakabayashi I. Relationship between alcohol intake and lipid accumulation product in middle-aged men. Alcohol Alcohol 2013;48:535–42.

[74] Wakabayashi I. A U-shaped relationship between alcohol consumption and cardiometabolic index in middle-aged men. Lipids Health Dis 2016;15:50.

[75] Knupfer G. Drinking for health: the daily light drinker fiction. Br J Addict 1987;82:547–55.

[76] Russell M, Cooper ML, Frone MR, Welte JW. Alcohol drinking patterns and blood pressure. Am J Public Health 1991;81:452–7.

[77] Anonymous. American Diabetes Association. 4. Lifestyle management. Diabetes Care 2017;40(1):S33–43.

Chapter 17

Lifestyle Features and Heart Disease

Vijay Singh, Ronald Ross Watson

University of Arizona, Mel and Enid Zuckerman College of Public Health, Tucson, AZ, United States

INTRODUCTION

Cardiovascular disease (CVD) is one of the most pressing issues in the world of health today as it affects a majority of populations not only in the United States but also in the world as a whole. CVD is the single largest cause of death in the United States, killing more than 800,000 people a year [1]. CVD consists of a wide range of diseases that affect the heart or blood vessels [1]. CVD also involves coronary artery disease (CAD), which entails health issues such as angina and myocardial infarction. Other forms of CVDs include stroke, heart failure, venous thrombosis, and peripheral artery disease.

The primary causes for CVD are dependent on the specific type of disease being studied. The causes that are noted with patients suffering from CVD can range from the following: hypertension, smoking, the lack of physical activity, hyperlipidemia, poor diet, and excessive alcohol consumption [1]. Overall, the progression of CVD is dependent on how well patients manage their lifestyle.

This review will investigate the different lifestyle choices that patients can make in order to alleviate CVD complications, such as exercise, smoking, dieting, and consumption of omega-3 fatty acids.

RISK FACTORS

There are a variety of risk factors involved for patients being affected by CVD, including age, gender, tobacco use, physical inactivity, alcohol consumption, unhealthy diet, obesity, genetics, hypertension, diabetes, hyperlipidemia, poverty, educational status, and air pollution [2]. The aforementioned risk factors vary depending on ethnic group, demographics, and socioeconomic status and gradient [2].

Family history of CVD can increase individual risk by threefold, especially if this is found with parents [3]. The risk of CVD is tripled by each decade of life accumulated by an individual [4]. The chance of stroke doubles every decade of life after the age of 55 [5]. A key indicator of age being a risk factor for CVD is its correlation with serum cholesterol levels [6]. In most demographics, serum cholesterol levels increase as age increases. This trend tends to happen to men and women between 45 and 65 years of age. As patients age, the vascular wall will begin to reduce and also cause the loss of arterial elasticity and compliance, which can also subsequently lead to CAD [7].

Men are at a greater risk of heart disease than women [7]. Once women patients are past menopause, it has also been discussed that a woman's risk of CAD is similar to a man's [7]. If a woman has diabetes, she is more prone to developing CAD than a man with diabetes. In general, CAD is more common in middle-aged men than women. A suggested explanation as to why CAD is different among genders is due to hormonal differences [6]. Estrogen is the principal sex hormone for women, as it has many defensive effects for glucose metabolism and the hemostatic system. Estrogen also helps develop endothelial cell function for women. The production of estrogen decreases after the postmenopause period, and this will subsequently change female lipid metabolism by decreasing high-density lipoprotein (HDL) cholesterol levels while increasing low-density lipoprotein (LDL) and total cholesterol levels [6]. Between men and women, there are differences in body composition, weight, height, body fat, heart rate, stroke volume, and heart compliance as their age progresses [7].

Nonmodifiable risk factors include age, genetics, and type 1 diabetes. Age is an influencing factor for patients as it can lead to many chronic diseases to occur as the body begins to deteriorate over time [8]. As patients begin to age more, the body becomes more susceptible of stress, and free radicals begin to freely generate, causing a breakdown of organ structure and function. Past epidemiological research has shown that patients who have a family history of heart disease are more prone to CVD complications [8].

Lifestyle in Heart Health and Disease. https://doi.org/10.1016/B978-0-12-811279-3.00017-3

EXERCISE

Consistent physical activity using large muscle groups, such as through walking, running, or swimming, causes cardiovascular adaptations that progressively increase exercise capacity, endurance, and skeletal muscle strength [9]. Maintaining a consistent regiment for physical activity is also able to prevent development of other cardiovascular issues such as CAD and reduce symptoms with patients that have established CVD. Evidence from other previous research studies has shown that exercise can reduce the risk of other various chronic diseases, such as type 2 diabetes, osteoporosis, obesity, and depression, along with breast and colon cancer [9]. Many organizations throughout the United States have encouraged health-care professionals to introduce exercise-based therapies to patients suffering from CVD in order to find new preventative measures.

Physical activity is described as bodily movement by skeletal muscles that cause energy expenditure beyond that of resting. Exercise is a subclass of physical activity that is organized, consistent, and purposeful for the common goal of having proper facilitation of physical fitness [10]. Exercise can consist of cardio-based fitness activities, weight lifting, flexibility, and other various forms in order to ensure proper physical activity maintenance.

Physical activity is a part of everyone's life. Several studies have shown a positive correlation between physical activity and good health [8].

When looking at positive factors of exercise on the cardiovascular system, blood pressure and heart rate measurements are very important to analyze. Blood pressure is directly correlated to blood volume and vascular resistance [8]. Vascular resistance is controlled throughout the nervous and endocrine systems via maintenance of hormones. Hormones can trigger physiological mechanisms to take place throughout the cardiovascular system, including vasoconstriction and vasodilation of the blood vessels and salt and water retention. These modifications can take place if the hormones target a nephrons glomerular filtration rate [8]. The more constricted a blood vessel becomes, the greater the resistance produced on the blood flow, causing a higher blood pressure measurement.

In order to understand the effects of exercise on blood pressure, measurements should be analyzed for both systolic and diastolic blood pressure. Systolic blood pressure is measured by the arterial pressured exerted when the heart is contracting. Diastolic blood pressure is determined when pressure is exerted on the arterial walls when the heart is relaxing [8]. When a patient is going through very rigorous exercise, oxygen demand increases for the heart, causing cardiac output and stroke volume to increase. In general, the lower a patient's blood pressure is, the less break down of their arteries. High blood pressure is a major risk factor for CVD, as it can increase the risk of ischemia and renal-related diseases [11].

Electrolytes, lipoproteins, and total lipid count play an essential role in cardiovascular health maintenance [12]. Lipids are most ideally measured as high-density lipoprotein (HDL), low-density lipoprotein (LDL), and total cholesterol levels. HDL and LDL levels differ through function and structural configuration. LDLs have a much higher triglyceride count than HDLs [12].

Although there are many advantages to daily exercise activity, there are also many consequences that can also occur [12]. In order to optimize proper exercise and cardiovascular health, it is recommended that physical activity is maintained at moderate levels. Moderate activities can include basic exercises such as walking, bike riding, rowing, and running on treadmills [13]. The most optimal way to control exercise is through having a schedule that can maintain various frequencies and intensity. If a patient has a history of CVD, he or she must report to a physician before beginning a routine exercise regiment [13].

Atherosclerosis is when the arteries become constricted and less flexible. This is especially prevalent with fat, cholesterol, and other various substances in the vessels form plaque that accumulates in the walls of the arteries [14]. As plaque builds up, the arteries become more constricted, and more blood needs to diffuse throughout the body in order to accommodate for this change. Risk factors that are present with atherogenesis become more apparent as childhood progresses. There is growing evidence that lifestyle modifications can occur to help alleviate any risk factors that are present for atherosclerosis and other similar diseases [14]. The effort that is needed toward alleviating risk factors should begin early in life.

SMOKING

Exposure to secondhand smoke and other similar environmental factors can increase risks of having CVD for nonsmokers. The risk of CVD increases with every cigarette that an individual uses per day, especially when smoking continues for years [15]. Studies have also shown that smoking cigarettes with low concentrations of tar and nicotine does not reduce the risks of CVD [15]. Thirty percent of coronary heart disease (CHD) deaths in the United States are attributable to smoking [16]. Smoking also doubles the risk of ischemic stroke [17]. Health-care settings can provide an important precedent for patients that want to begin smoking cessation. When patients are at a physicians' office for a checkup, they are very receptive to learning about how to reduce risk factors and attempt to begin healthy lifestyles [18]. There has been tremendous evidence given that can show how smoking cessation can greatly reduce the risks involved with CVD. Evidence greatly suggests that smoke cessation is the most effective means of reducing CVD [18].

Smoking has been found to kill more people through heart disease than lung cancer [19]. Smoking is the most important factor to prevent in order to alleviate symptoms associated with CVD. This is noted especially with nonsmokers being exposed to environmental tobacco smoke [20]. Environment tobacco smoke adversely affects platelet function and impairs arterial endothelium that slowly progresses the risk of CVD. Overall, CVD is a significant consequence of environmental tobacco smoke. With data ranging from epidemiological, physiological, and biochemical studies, environmental tobacco smoke has major adverse effects that can affect patient cardiovascular health.

BODY WEIGHT

Body weight is correlated to a variety of different risk factors for CVD, but it may also contain an independent consequence with the risk of CHD [21]. The key factors associated with this are the correlation between body mass index (BMI), smoking, cholesterol levels, and blood pressure [21]. There are also significant studies associated with how patient BMI and the aforementioned risk factors are linked to CHD mortality rates. BMI is the most used indicator in order to evaluate population obesity studies. One issue with using BMI as a measure is that it is unable to account body fat patterns, specifically for the waist-hip ratio and skinfold quantities, which both have been correlated to cardiovascular mortality [22]. Additionally, body weight correlates with morbidity and mortality, whereas height is linked with good health. BMI can then show negative effects on patients that are both obese and short [22].

Obesity is an independent risk factor for male patients and also has an increased risk of CHD among female patients [21]. By preventing the overweight BMI category, the risk of CHD can be drastically reduced. Weight control should be an important part of patients' lifestyle in order to reduce CVD. In order to assess whether or not obesity is a risk factor of CVD, studies must focus on factors such as blood pressure, lipid levels, reduced glucose metabolism, or other similar mechanisms that revolve around vitals and basic metabolic function [22].

DIET

Improving diet is an imperative factor for facilitating CVD. A variety of recommendations to better cardiovascular health include having an overall healthy diet; aiming for a healthy body weight, normal blood pressure, and normal blood glucose level; and being physically active [23]. In general, diet and lifestyle changes are necessary in order to ensure proper growth and development. It has been well noted through past clinical findings that there are multiple dietary factors involved with the risk of CVD development and its aforementioned risk factors [23]. An emphasis is needed on the whole diet rather than on just individual nutrient intake in order to ensure proper alleviation of CVD [24]. Having a healthy overall diet consists of the consumption of fruits, vegetables, grains, dairy products, and lean meats.

Obesity is an independent risk factor of CVD [25]. Having excess body weight based on your height and age can adversely affect the risk of CVD. This is especially prevalent through an increase in LDL cholesterol levels, triglyceride levels, blood pressure, and blood glucose levels. Excess body weight can also increase the risk of developing CHD, heart failure, stroke, and cardiac arrhythmias [25]. The overall factors of this increase in obesity are ubiquitous throughout the world. The major implications include increased portion sizes, high-calorie-density foods, convenient accessibility to inexpensive food, and commercial influences. A multifactorial initiative needs to be taken in order to reduce overall body weight, obesity, and CVD demographics throughout the world [23]. This can be attained through having proper dieting and healthy weight based on height and age.

Consumption of omega-3 fatty acids (fish oil) is correlated with reduced risk of sudden death and CAD [26]. Consistent fish consumption can maintain the displacement of other foods higher in saturated and trans-fatty acids from a patient's diet. This is especially prevalent in fatty meats and full-fat dairy products. Omega-3 fatty acids occur in certain fish, such as, salmon, tuna, and halibut. Additionally, other seafood is shown to contain omega-3 fatty acids, such as algae, some plants, and nut oils. Omega-3 fatty acids play a very important role in brain development and also lowering the risk of cardiovascular disease [26]. Moreover, omega-3 fatty acids lower blood pressure for people with hypertension [26].

CONCLUSION

Important lifestyle choices should be made and administered for patients in order to better reduce risk factors of cardiovascular health [23]. In order to effectively avoid CVD, patients must take note of lifestyle benefits and aim for a healthy body weight, diet, proper exercise, and avoid exposure to negative environmental influences such as tobacco and alcohol consumption. Successfully completing these objectives will require individuals to make daily lifestyle and environmental changes. A difficulty that may be faced for researchers and health-care providers is to find effective strategies that can lead to optimal lifestyle changes among patients and also throughout specific demographics [23]. Adjusting to new lifestyle changes will drastically benefit patients through all aspects of health, especially alleviating the symptoms that coincide with CVD.

REFERENCES

[1] Villareal MR. Global atlas on cardiovascular prevention and control. United Nations - World Health Organization; 2011.

[2] Howard BV, Wylie-Rosett J. Sugar and cardiovascular disease: a statement for healthcare professionals from the Committee on Nutrition of the Council on Nutrition, Physical Activity, and Metabolism of the American Heart Association. Circulation 2002;106(4).

[3] Kathiresan S, Srivastava D. Genetics of human cardiovascular disease. Cell 2012;148(6).

[4] Nikpay M, Goel A, Won HW, Hall LM, Willenborg C, Kanoni S, et al. A comprehensive 1,000 Genomes-based genome-wide association meta-analysis of coronary artery disease. Nat Genet 2015;47(10).

[5] Nikpay M, Goel A, Won HW, Hall LM, Willenborg C, Kanoni S, et al. Atherogenic dyslipidemia and cardiovascular risk factors in obese children. Int J Endocrinol 2015;2015:1–9.

[6] American Heart Association. Understand Your Risks to Prevent a Heart Attack. heart.org [accessed 06.03.10].

[7] Jousilahti P, Vartiainen E, Tuomilehto J, Puska P. Sex, age, cardiovascular risk factors, and coronary heart disease: a prospective follow-up study of 14 786 middle-aged men and women in Finland. Circulation 2004;99.

[8] Buttar HS, Li T, Ravi N. Prevention of cardiovascular diseases: role of exercise, dietary interventions, obesity and smoking cessation. Pulsus: Exp Clin Cardiol 2005;10(4).

[9] Thompson PD, Buchner D, Piña IL, Balady GJ, Williams MA, Marcus BH, et al. Exercise and physical activity in the prevention and treatment of atherosclerotic cardiovascular disease: a statement from the Council on Clinical Cardiology (Subcommittee on Exercise, Rehabilitation, and Prevention) and the Council on Nutrition, Physical Activity, and Metabolism (Subcommittee on Physical Activity). Circulation 2003;107(24).

[10] Fletcher GF, Balady GJ, Amsterdam EA. Exercise standards for testing and training: a statement for healthcare professionals from the American Heart Association. Circulation 2001;104.

[11] Puffer JC. Exercise and heart disease. Clin Cornerstone 2001;3.

[12] Thomson JA. Human nutrition course notes. Faculty of Applied Health Sciences; 2004.

[13] Manson JE, Hu FB, Rich-Edwards JW. A prospective study of walking as compared with vigorous exercise in the prevention of coronary heart disease in women. N Engl J Med 1999;341.

[14] Massin MM, Hövels-Gürich H, Seghaye MC. Atherosclerosis lifestyle risk factors in children with congenital heart disease. Eur J Prev Cardiol 1999;14.

[15] Surgeon General Report, Smoking and Cardiovascular Disease. CDC 2014.

[16] US Dept of Health and Human Services. Reducing the health consequences of smoking: 25 years of progress. A report of the Surgeon General. US Department of Health and Human Services, Public Health Service, Centers for Disease Control, Center for Chronic Disease Prevention and Health Promotion, Office on Smoking and Health. vol. 298. DHHS Publication (CDC); 1989.

[17] Shinton R, Beevers G. Meta-analysis of relation between cigarette smoking and stroke. BMJ 1989;298.

[18] Ockene JK. Smoking intervention: the expanding role of the physician. Am J Public Health 1987;77.

[19] Glantz SA, Parmley WW. Passive smoking and heart disease: epidemiology, physiology, and biochemistry. Circulation 1991;83(1).

[20] Wells A. An estimate of adult mortality in the United States from passive smoking. Environ Int 1988;14(1).

[21] Jousilahti P, Tuomilehto K, Vartiainen E, Pekkanen J, Puska P. Body weight, cardiovascular risk factors, and coronary mortality 15-year: follow-up of middle-aged men and women in eastern Finland. Circulation 1996;93(1).

[22] Gillum RF. The association of body fat distribution with hypertension, hypertensive heart disease, coronary heart disease, diabetes and cardiovascular risk factors in men and women aged 18-79 years. J Chronic Dis 1987;40(1).

[23] Lichtenstein AH, Appel LJ, Brands M, Carnethon M, Daniels S, Franch HA, et al. Diet and lifestyle recommendations revision 2006: a scientific statement from the American Heart Association Nutrition Committee. Circulation 2006;114(1).

[24] US Department of Health and Human Services, US Department of Agriculture. Dietary Guidelines for Americans. US Government Printing Office; 2005.

[25] Rashid MN, Fuentes F, Touchon RC, Wehner PS. Obesity and the risk for cardiovascular disease. Prev Cardiol 2003;6(1).

[26] Kris-Etherton PM, Harris WS, Appel LJ. Fish consumption, fish oil, omega-3 fatty acids, and cardiovascular disease. Circulation 2002;106(1).

Chapter 18

Prevention of Cardiovascular Disease Among People Living With HIV: A Tailored Smoking Cessation Program Treating Depression

Stephanie Wiebe, Lousie Balfour, Paul A. McPherson

Department of Psychology, The Ottawa Hospital, University of Ottawa, Ottawa, ON, Canada

The risk of suffering a myocardial infarction is two to four times greater among people living with HIV/AIDS (PHAs) as compared with the general population, and it is more likely to occur at a younger age [1–3]. As in the general population, among PHAs, cigarette smoking is a significant predictor of CVD [4]. Explanations for the elevated risk of CVD among PHAs include chronic inflammation due to the virus itself, inflammation due to antiretroviral medications, and cigarette smoking among PHAs [5]. Approximately 40–70% of PHAs smoke cigarettes [6], a rate that is 2.3–4.4 times higher than smoking rates among the general population [7]. Indeed, although the advent of antiretroviral therapy in the mid-1990s has led to decreased mortality due to HIV/AIDS, the number of deaths due to CVD has increased substantially [8,9]. Furthermore, PHAs now lose more life years due to smoking-related illnesses than to HIV/AIDS [10]. Given both the high rate of CVD among PHAs and the detrimental role of cigarette smoking and that smoking is the only modifiable risk factor that could be eliminated from PHAs, smoking cessation interventions are of highest priority to decrease the risk of CVD among PHAs [11].

UNIQUE FACTORS IMPACTING SMOKING CESSATION AMONG PHAS

Given the high rates of smoking and the negative impact of smoking in the HIV population, the need to foster smoking cessation among people living with HIV is clear. However, this population faces a number of unique issues related to quitting as compared with the general population that may impact the success of smoking cessation attempts.

Depression

The rates of depression among people living with HIV are high, ranging from 40% to 60%, which is much higher than the rates of depression in the general population [12]. Depression is an independent risk factor for CVD [13], a link that may be explained by inflammation [13] or adverse health behaviors associated with depression such as the lack of exercise and smoking [14]. Depression and smoking are both independently associated with increased inflammatory markers [14], resulting in a cumulative effect among those who are depressed and smoke cigarettes [15]. Furthermore, compounding the problem, depression is a well-known barrier to successful smoking cessation [16]. Depression is itself associated with higher rates of cigarette smoking and higher relapse rates from smoking cessation programs [16].

Social Support

Having a supportive social network significantly predicts successful smoking cessation among PHAs [17]. However, research has shown that on average, 43% of the people in a PHA's social support network are made up of other people who smoke [18]. Furthermore, a lack of social support is associated with depression among PHAs, which has a negative impact on quitting [19]. This may be a significant challenge for those who are attempting to quit because paradoxically the very source of social support that may foster a successful quit attempt may also be a trigger for relapse [20]. A recent study examining the influence of social support on smoking cessation outcomes among PHA smokers found that frequency of contact with social support people was significantly associated with 7-day point prevalence smoking abstinence at 6-month follow-up [21]. This effect was mediated by adherence to the smoking cessation intervention (i.e., nicotine replacement patch) [21]. Given the stigma associated with HIV, many PHAs experience the loss of social support following diagnosis [22]. This highlights the importance of examining the role of social support in smoking cessation among PHAs.

SMOKING CESSATION AMONG PHAs

The need to support people living with HIV to quit smoking has been identified, and attempts have been made to support smoking cessation attempts in this population. Counseling interventions in addition to smoking cessation medication have generally yielded the best results. Quit rates found at 3-month follow-up ranged from 19.2% to 36.8% for counseling and varenicline or NRT. This is much higher than the rates of 9.7%–10.3% found with medication and usual care [23,24]. In terms of longer term follow-up, Humfleet et al. [25] compared individual counseling, internet counseling, and self-help in addition to NRT in a 52-week follow-up study. They found that the counseling groups showed higher quit rates than self-help. In a smaller RCT study comparing the NRT patch plus a self-guided reading intervention with the NRT patch plus an intensive group therapy intervention for PHAs, Ingersoll et al. [26] found a 22% abstinence rate overall with no significant difference between groups; however, the small sample size limits interpretation of results. Two large RCT studies demonstrated quit rates ranging from 3.3% to 9% with NRT and usual care versus 10%–11.9% for counseling and medication at 3- and 6-month follow-up [27,28]. Although none of the above studies excluded patients with depression, two of the more successful studies addressed depressive symptoms briefly in the context of their counseling sessions [23,25]. We took this as preliminary evidence that treating depression among PHAs attempting to quit smoking may have a beneficial impact on their quit attempt.

THE HIV-TAILORED SMOKING CESSATION PROGRAM

Given the imperative for fostering smoking cessation among PHAs to lower the prevalence of CVD in this population and the unique issues related to quitting in this population, we developed a tailored smoking cessation counseling to meet the specific needs of PHAs. Our counseling program involves five face-to-face sessions including educational, motivational interviewing, and cognitive-behavioral interventions to help PHAs quit smoking. The HIV-tailored smoking cessation counseling program is based on an empirically supported smoking cessation program developed at the University of Ottawa Heart Institute (UOHI). This program is known as the "Ottawa Model for Smoking Cessation" and is a state-of-the-art program for people with cardiovascular disease recognized for its national standards of excellence [29]. In the remainder of this chapter, we will proceed to describe the HIV-tailored smoking cessation counseling program and outline preliminary research supporting the effectiveness of this approach.

Theoretical Framework

The HIV-tailored smoking cessation program draws on the investment model of commitment [30]. The premise of the investment model of commitment is that individuals have a vested interest in successfully performing a behavior and that commitment is enhanced by promoting active participation on the part of the patient in decisions about their own health care [31]. The HIV-tailored smoking cessation program is therefore designed to enhance participants' commitment to their own health care by promoting active participation in making informed decisions about their health such as changing smoking behavior. The counseling approach also draws strongly from motivational interviewing [32]. Motivational interviewing is useful in the health-care setting with patients in terms of the prospect of making a health behavior change and adhering to treatment. The approach focuses on building autonomy and active participation in making a health behavior change. A key principle is to encourage change from within. Motivational interviewing is always a patient-centered, nonjudgmental, and nonconfrontational approach. It relies on four main interventions of the counselor: asking open-ended questions, providing affirmations, reflective listening, and periodically providing summary statements to the patient.

Given the high prevalence of depression among PHAs and the finding that depression is a barrier to a successful smoking cessation attempt, we incorporated evidence-based interventions for treating depression into our smoking cessation counseling program.

Description of the HIV-Tailored Smoking Cessation Counselling Program

Our program consists of five face-to-face counseling sessions of approximately 45–60 min in length at important intervals in the smoking cessation process (prior to quitting, on the chosen quit date, week 4, week 12, and week 24). Educational interventions included discussing the possible impact of smoking and quitting smoking on mood, outlining symptoms of stress, and other possible side effects of quitting smoking and using smoking cessation medications. Motivational interviewing included discussing pros and cons of quitting smoking including positive motivational aspects of quitting on HIV-related health factors (i.e., preventing CVD and immune functioning). Cognitive-behavioral interventions included identifying depressive cognitions, cognitive restructuring, developing positive coping strategies, relaxation training, and behavioral activation [33]. Assertiveness skills are also taught as part of this program in order to help patients address interpersonal challenges related to their smoking cessation attempt effectively and increase the benefits of their social support network while decreasing negative impacts (i.e., pressure to join friends for a cigarette). We will now turn to a description of the counseling sessions.

Session One: Preparing to Quit

The initial session focuses on issues related to preparation to quit smoking. The goals of the counseling intervention and the role of the counselor are provided. The main focus of the session is about the patient's full range of feelings about quitting, from worries and fears to hope and pride. The patient's goals for quitting are discussed and validated. Together, the patient and counselor identify possible barriers to quitting smoking. The possible HIV-specific barriers are explored (i.e., stigma and social isolation). HIV-specific motivating factors are also explored (i.e., improved immune functioning). The patient is provided education about smoking cessation including withdrawal symptoms and use of smoking cessation medication. Patients are encouraged to explore ways of preparing themselves for their quit date and are given home exercises to complete prior to their quit date such as getting rid of cigarettes, discussing their planned quit attempt with their social support network, and possibly starting to taper down their cigarette usage. The counselor and patient collaboratively decide on a quit date, reviewing pros and cons to the date selected. The second counseling session is scheduled for the quit date.

Session Two: Quit Date

The second counseling session occurs on the planned quit date and focuses on preparing for the first few weeks of quitting. The use of smoking cessation medication is reviewed including dosing and possible side effects. Education is reviewed regarding withdrawal symptoms, coping with cravings, anticipating possible changes in mood, and possible weight changes. Information about stress and depression is given, and ways of coping with stress and low mood are discussed. The possibility of lapses occurring is explicitly addressed, and ways to manage lapses are discussed. The difference between a "lapse" and a "relapse" is emphasized to normalize lapses that may occur and validate the difficulty of quitting smoking with the aim of reducing guilt and shame that may be associated with lapses. The third session is scheduled for week 4.

Session Three: Week Four of Quit Attempt

This is the first session following the patient's quit attempt. In this session, the counselor reviews the patient's experience encouraging sharing about both successes and challenges. The patient is encouraged to discuss possible barriers to staying smoke-free: difficulties with coping with cravings, withdrawal symptoms, side effects of smoking cessation medications, adhering to smoking cessation medications, social factors that make staying smoke-free difficult (i.e., smoking behavior of friends and family), aspects of their routine that may make staying smoke-free difficult, changes in mood, and coping with stress. Problem solving around possible barriers is facilitated, and strategies are recommended to facilitate remaining smoke-free. The difference between a "lapse" and a "relapse" is reviewed and their occurrence normalized. The circumstances surrounding any lapses that have occurred are explored with the aim of formulating strategies to prevent future lapses.

Session Four: Week Twelve of Quit Attempt

This is the second session following the patient's quit date. The counselor invites the patient to share the potential barriers to remaining smoke-free they may have experienced over the past 12 weeks. The patient is encouraged to share about successes and challenges they have faced. The counselor asks the patient about their experiences with smoking cessation medication and their feelings about stopping the medication if relevant. The links between smoking, stress, mood, and HIV health care are discussed and related to the patient's own life experiences. Coping strategies are discussed for dealing with stress and low mood. Relaxation strategies are provided including progressive muscle relaxation [34]. The impact of smoking cessation on social relationships is explored, and assertiveness strategies are taught if needed. Additionally, the link between thoughts, feelings, and behaviors/coping is discussed; information is provided on cognitive distortions, negative self-talk, and their impact on mood.

Session Five: Six-Month Follow-up

This is the final session. The session focuses on a review of the patient's progress including successes and challenges and thoughts and feelings about completing the smoking cessation counseling sessions and remaining smoke-free without the support of the counselor. The patient is invited to share any concerns of fears they may have—these are explored and validated. Relevant coping strategies are reviewed. Anticipated barriers to remaining smoke-free are discussed. Coping strategies are reviewed. Referral resources are provided as needed. The counselor provides a summary of the successes and ongoing challenges and provides encouragement in coping with ongoing struggles moving forward.

PILOT STUDY OUTCOME RESULTS

We examined the effectiveness of the HIV-tailored counseling program in a pilot study in which participants received the counseling program in addition to nicotine replacement therapy (NRT) [35]. We recruited 50 patients with HIV being followed by the HIV program at a hospital in Ottawa, Ontario, who wanted to quit smoking. We provided all 50 participants with the five-session smoking cessation counseling in addition to NRT. Counseling sessions were provided at baseline, at quit date, and at 4, 12, and 24 weeks post quit date. We assessed patient smoking status at the 4th, 12th, and 24th week sessions through self-report and biochemical testing using exhaled carbon monoxide. Participants were classified as abstinent from smoking if they reported abstinence and also demonstrated exhaled CO at <10 ppm. We administered questionnaire packages at baseline, quit date, and 4th, 12th, and 24th week follow-up. Telephone-based support was also provided as needed for the duration of the study.

We found a successful biometrically confirmed quit rate of 28% at a 24th week follow-up appointment among participants from our program. This quit rate is significantly higher than results of similar studies in the literature on a program that utilized a motivational enhancement counseling intervention [27,35].

In contrast with previous studies in the literature, we specifically targeted depression, and in contrast with these studies (i.e., Humfleet et al., 2013), we did not find that initial depression scores predicted relapse. Furthermore, although we found that 52% of our participants scored above a clinical cutoff for depression at baseline, these depressed participants demonstrated lower depression at the 24th week follow-up appointment with a large effect size ($d = 1$). Furthermore, those with clinically significant depression at baseline were no more likely to relapse at week 24 than their nondepressed counterparts, suggesting that treating depression as part of our program may have led to increased quit rates among depressed participants. Therefore, our findings suggest that PHAs with depression were equally likely to successfully quit smoking as their nondepressed counterparts. These results suggest that patients with depression may be as likely to be able to engage in a successful smoking cessation attempt if they have the coping strategies to cope with the impact of quitting on mood and find more healthy ways to regulate their mood. This is good news for the significant proportion of PHAs who are depressed that may need to quit smoking.

CHANGE IN PSYCHOLOGICAL CORRELATES OF SUCCESSFUL QUITTING

In order to gain a better understanding of the changes that occurred among our participants in the pilot study who successfully remained smoke-free, we examined change in key psychological variables across counseling session thought to be related to a successful quit attempt. Specifically, we examined change in self-efficacy, depression, and attachment insecurity across the five counseling sessions. We identified these variables for their relevance to our key hypothesized mechanisms of change including improved mood, increased self-efficacy to stay quit, and interpersonal functioning (i.e., attachment style). Attachment style was also selected due to the correlation found in past research between attachment insecurity and depression [36]. From our original sample of 50 participants, 14 (28%) successfully remained smoke-free. These 14 participants made up the sample of this study. Using hierarchical linear modeling [37], we tested whether participants' trajectory of change in our key variables resulted in significant linear growth curves (i.e., increased in a linear direction across counseling sessions). We found that participants demonstrated linear improvements across counseling sessions in internal self-efficacy (i.e., self-efficacy to refrain from smoking when presented with internal cues to smoke such as low mood or stress) and external self-efficacy (i.e., self-efficacy to refrain from smoking when confronted with external cues to smoke such as friends smoking around them). We also found linear reductions across counseling sessions in depression scores. We did not, however, find significant reductions in self-reported attachment security. This is perhaps not surprising as attachment style may take longer to change and may be more likely to occur in the context of specific relationship changes or in the context of interpersonal or couple psychotherapy [38–40]. Overall, these results provide evidence that our counseling intervention helped to improve self-efficacy and reduce symptoms of depression among successful quitters, suggesting that these may be key factors involved in the process of change.

CONCLUSIONS

We developed an HIV-tailored smoking cessation program focusing on factors specific to people living with HIV. Our program incorporates education, motivational interviewing, and CBT interventions to promote motivation to quit, treat depression, and foster improved social support with the goal of improving quit rates in this population. The results of our pilot study provided evidence for the effectiveness of our intervention suggesting that tailoring interventions for specific populations and treating symptoms of depression may lead to improved quit rates and thus reduced risk of CVD. We recommend the development of tailored approaches for other high-risk populations (i.e., patients with COPD) in order to reduce CVD risk for these patients as well. Our results also point to the utility of psychological interventions for smoking cessation, especially given the link between smoking and depression and the barrier that depressive symptoms present to a successful quit attempt.

REFERENCES

[1] Boccara F. Cardiovascular complications and atherosclerotic manifestations in the HIV-infected population: type, incidence and associated risk factors. Aids 2008;22(Suppl 3):S19–26. https://doi.org/10.1097/01.aids.0000327512.76126.6e.

[2] Triant VA, Lee H, Hadigan C, Grinspoon SK. Increased acute myocardial infarction rates and cardiovascular risk factors among patients with Human Immunodeficiency Virus disease. J Clin Endocrinol Metab 2007;92(7):2506–12. https://doi.org/10.1210/jc.2006-2190.Increased.

[3] Triant VA, Meigs JB, Grinspoon SK. Association of C-reactive protein and HIV infection with acute myocardial infarction. J Acquir Immune Defic Syndr 2009;51(3):268–73. https://doi.org/10.1097/QAI.0b013e3181a9992c.Association.

[4] Furber AS, Maheswaran R, Newell JN, Carroll C. Is smoking tobacco an independent risk factor for HIV infection and progression to AIDS? A systemic review. Sex Transm Infect 2007;83(1):41–6. https://doi.org/10.1136/sti.2005.019505.

[5] Cropsey KL, Hendricks PS, Jardin B, et al. A pilot study of Screening, Brief Intervention, and Referral for Treatment (SBIRT) in non-treatment seeking smokers with HIV. Addict Behav 2013. https://doi.org/10.1016/j.addbeh.2013.05.003.

[6] Cioe PA, Crawford SL, Stein MD. Cardiovascular risk-factor knowledge and risk perception among HIV-infected adults. J Assoc Nurses AIDS Care 2014;25(1):60–9. https://doi.org/10.1016/j.jana.2013.07.006.

[7] Health Canada. Canadian tobacco use monitoring survey. 2012:1–3. https://doi.org/10.1016/B978-01-2-380920-9.00009-2.

[8] Palella FJ, Baker RK, Moorman AC, et al. Mortality in the highly active antiretroviral therapy era: changing causes of death and disease in the HIV outpatient study. J Acquir Immune Defic Syndr 2006;43(1):27–34. https://doi.org/10.1097/01.qai.0000233310.90484.16.

[9] Petoumenos K, Worm S, Reiss P, et al. Rates of cardiovascular disease following smoking cessation in patients with HIV infection: results from the D:A:D study. HIV Med 2011;12(7):412–21. https://doi.org/10.1111/j.1468-1293.2010.00901.x.

[10] Helleberg M, Afzal S, Kronborg G, et al. Mortality attributable to smoking among HIV-1-infected individuals: a nationwide, population-based cohort study. Clin Infect Dis 2013;56(5):727–34. https://doi.org/10.1093/cid/cis933.

[11] Triant VA. HIV infection and coronary heart disease: an intersection of epidemics. J Infect Dis 2012;205(Suppl. 3):355–61. https://doi.org/10.1093/infdis/jis195.

[12] Tsao JCI, Dobalian A, Moreau C, Dobalian K. Stability of anxiety and depression in a national sample of adults with human immunodeficiency virus. J Nerv Ment Dis 2004. https://doi.org/10.1097/01.nmd.0000110282.61088.cc.

[13] Gouweleeuw L, Naudé PJW, Rots M, DeJongste MJL, Eisel ULM, Schoemaker RG. The role of neutrophil gelatinase associated lipocalin (NGAL) as biological constituent linking depression and cardiovascular disease. Brain Behav Immun 2015;46:23–32. https://doi.org/10.1016/j.bbi.2014.12.026.

[14] Nunes SOV, Vargas HO, Brum J, et al. A comparison of inflammatory markers in depressed and nondepressed smokers. Nicotine Tob Res 2012;14(5):540–6. https://doi.org/10.1093/ntr/ntr247.

[15] Nunes SOV, Piccoli de Melo LG, Pizzode Castro MR, et al. Atherogenic index of plasma and atherogenic coefficient are increased in major depression and bipolar disorder, especially when comorbid with tobacco use disorder. J Affect Disord 2015;172:55–62. https://doi.org/10.1016/j.jad.2014.09.038.

[16] Malpass D, Higgs S. Acute psychomotor, subjective and physiological responses to smoking in depressed outpatient smokers and matched controls. Psychopharmacology (Berlin) 2007;190(3):363–72. https://doi.org/10.1007/s00213-006-0612-z.

[17] de Dios MA, Stanton CA, Caviness CM, Niaura R, Stein M. The social support and social network characteristics of smokers in methadone maintenance treatment. Am J Drug Alcohol Abuse 2013;39(1):50–6. https://doi.org/10.3109/00952990.2011.653424.

[18] Humfleet GL, Delucchi K, Kelley K, Hall SM, Dilley J, Harrison G. Characteristics of HIV-positive cigarette smokers: a sample of smokers facing multiple challenges. AIDS Educ Prev 2009. https://doi.org/10.1521/aeap.2009.21.3_supp.54.

[19] Lyketsos CG, Hoover DR, Guccione M, et al. Changes in depressive symptoms as AIDS develops. Am J Psychiatry 1996;153(11):1430–7.

[20] Fowler J, Christakis N. Quitting in droves: collective dynamics of smoking behavior in a large social network. N Engl J Med 2010;358(21):2249–58. https://doi.org/10.1056/NEJMsa0706154.

[21] MA De Dios, Stanton CA, Cano MÁ, Lloyd-richardson E, Niaura R. The influence of social support on smoking cessation treatment adherence among HIV + smokers. Nicotine Tob Res 2016;1126–33. https://doi.org/10.1093/ntr/ntv144.

[22] Gostin L, Webber D. The AIDS litigation project: HIV/AIDS in the courts in the 1990s, Part 2. AIDS Public Policy J 1998;13(1):3–19.

[23] Vidrine DJ, Arduino RC, Lazev AB, Gritz ER. A randomized trial of a proactive cellular telephone intervention for smokers living with HIV/AIDS. AIDS 2006;20(2):253–60. https://doi.org/10.1097/01.aids.0000198094.23691.58.

[24] Moadel AB, Bernstein SL, Mermelstein RJ, Arnsten JH, Dolce EH, Shuter J. A randomized controlled trial of a tailored group smoking cessation intervention for HIV-infected smokers. J Acquir Immune Defic Syndr 2012;61(2):208–15. https://doi.org/10.1097/QAI.0b013e3182645679.

[25] Humfleet GL, Hall SM, Delucchi KL, Dilley JW. A randomized clinical trial of smoking cessation treatments provided in HIV clinical care settings. Nicotine Tob Res 2013;15(8):1436–45. https://doi.org/10.1093/ntr/ntt005.

[26] Ingersoll KS, Cropsey KL, Heckman CJ. A test of motivational plus nicotine replacement interventions for HIV positive smokers. AIDS Behav 2009;13(3):545–54. https://doi.org/10.1007/s10461-007-9334-4.

[27] Lloyd-Richardson EE, Stanton CA, Papandonatos GD, et al. Motivation and patch treatment for HIV+ smokers: a randomized controlled trial. Addiction 2009;104(11):1891–900. https://doi.org/10.1111/j.1360-0443.2009.02623.x.

[28] Vidrine DJ, Marks RM, Arduino RC, Gritz ER. Efficacy of cell phone-delivered smoking cessation counseling for persons living with HIV/AIDS: 3-month outcomes. Nicotine Tob Res 2012;14(1):106–10. https://doi.org/10.1093/ntr/ntr121.

[29] Reid RD, Mullen KA, D'Angelo MES, et al. Smoking cessation for hospitalized smokers: an evaluation of the "Ottawa Model.". Nicotine Tob Res 2009;12(1):11–8. https://doi.org/10.1093/ntr/ntp165.

[30] Rusbult CE. Commitment and satisfaction in romantic associations: a test of the investment model. J Exp Soc Psychol 1980;186:172–86.

[31] Putnam DE, Finney JW, Barkley PL, Bonner MJ. Enhancing commitment improves adherence to a medical regimen. J Consult Clin Psychol 1994;62(1):191–4.

[32] Rollnick S, Miller WR, Butler CC. Motivational interviewing in health care: helping patients change behavior (applications of motivational interviewing). New York: Guilford Press; 2008. 2563. https://doi.org/10.1080/15412550802093108.

[33] Beck JS. Cognitive behavior therapy. 2nd ed. New York: Guilford Press; 2011.

[34] Jacobson E. Progressive relaxation. Chicago: University of Chicago Press; 1974.

[35] Balfour L, Wiebe SA, Cameron DW, Sandre D, Pipe A, Cooper C, et al. An HIV tailored quit-smoking counselling intervention targeting depressive symptoms plus Nicotine Replacement Therapy (NRT). AIDS Care 2017;29(1):24–31. https://doi.org/10.1080/09540121.2016.1201195.

[36] Marganska A, Gallagher M, Miranda R. Adult attachment, emotion dysregulation, and symptoms of depression and generalized anxiety disorder. Am J Orthopsychiatry 2013;83(1):131–41. https://doi.org/10.1111/ajop.12001.

[37] Singer JD, Willett JB. Applied longitudinal data analysis modeling change and event occurrence. New York: Oxford University Press; 2003.

[38] Moser MB, Johnson SM. Changes in relationship-specific attachment in emotionally focused couple therapy. J Marital Fam Ther 2016;42 (April):231–45. https://doi.org/10.1111/jmft.12139.

[39] Scharfe E, Cole V. Stability and change of attachment representations during emerging adulthood: an examination of mediators and moderators of change. Pers Relat 2006;13(3):363–74. https://doi.org/10.1111/j.1475-6811.2006.00123.x.

[40] Maxwell H, Tasca GA, Ritchie K, Balfour L, Bissada H. Change in attachment insecurity is related to improved outcomes 1-year post group therapy in women with binge eating disorder. Psychotherapy (Chic) 2014;51(1):https://doi.org/10.1037/a0031100.

Chapter 19

Factors Associated With Tobacco Use Among Patients With MCC: Multidisciplinary Visions about the Lifestyle on Health and Cardiovascular Disease

Arise G. de Siqueira Galil*, Arthur da Silva Gomes*,†, Bárbara A.B.B. de Andrade*,‡, Mariana M. Gusmão*,‡, Tatiane da Silva Campos*,§, Marcela M. de Melo*, Eliane F.C. Banhato*,¶

*IMEPEN Foundation, Faculty of Medicine, Federal University of Juiz de Fora, Juiz de Fora, Minas Gerais, Brazil, †Department of Biomedical Sciences, Federal University of Fluminense, Niterói, Rio de Janeiro, Brazil, ‡Institute of Human Sciences, Department of Psychology, Federal University of Juiz de Fora, Juiz de Fora, Minas Gerais, Brazil, §Faculty of Nursing, State University of Rio de Janeiro, Rio de Janeiro, Brazil, ¶Department of Psychology, Higher Education Center of Juiz de Fora, Juiz de Fora, Minas Gerais, Brazil

INTRODUCTION

The increase in life expectancy and aging of the population has led to changes in the epidemiological profile, with transition from acute and infectious diseases to chronic noncommunicable diseases (CNCDs). Chronic disease is defined as "conditions occurring in the last year or more, which require continued medical attention and/or cause limitations in daily living activities" [1,2]. The CNCDs include a wide variety of physical diseases such as arthritis, acquired immunodeficiency syndrome (AIDS), chronic respiratory conditions, cancer, diabetes mellitus (DM) and its complications, heart disease, and systemic arterial hypertension (SAH), frequently associated with behavioral conditions, such as the use of psychoactive substances (e.g., tobacco and alcohol) [3]. Multimorbidity or the co-occurrence of two or more chronic conditions, known as multiple chronic conditions (MCCs), is defined as a set of comorbidities and their implications that stand out as a relevant public health problem, which adequate management is a challenge for both professionals and health services around the world [4]. MCCs are more frequently observed in the elderly population, being present in more than 50% of them [4,5]. Moreover, they increase the limitations and injuries of their patients, reduce the quality of life, and increase hospitalization, morbidity, and mortality, compared with those who have an isolated chronic condition [6,7].

The answer to the multimorbidity problem is complex and requires a task force that goes through a better organized health management, including an organized health-care system, interdependent care networks, multidisciplinary team, and integral health care, leading to improvement in quality of life and reduction of costs and of the secondary damages to these diseases [6,7]. Preventive measures are critical as well as health promotion, aiming at eliminating or controlling risk factors for MCCs by encouraging healthy eating, physical activity, stress reduction, anxiety and depression control, alcohol abuse control, and smoking cessation [8]. An interesting fact is that the interventions made to manage a risk factor interpose, complement, and potentiate the action on that specific factor. Among these measures, smoking cessation has the greater cost-benefit impact and reduction of cardiovascular and all-cause mortality [9,10].

Smoking is the CNCD responsible for the higher prevalence of premature diseases and deaths worldwide, and tobacco use predisposes to the appearance of other CNCDs [11]. Smoking is currently defined both as a chronic disease and as a risk factor for cardiovascular disease. In addition, chronic obstructive pulmonary disease (COPD), tuberculosis, bronchial asthma, reduced fertility and sexual impotence, cataract, glaucoma, periodontitis, and ectopic pregnancy, among other diseases, have been associated with smoking [12–14].

Lifestyle in Heart Health and Disease. https://doi.org/10.1016/B978-0-12-811279-3.00019-7

The aim of this chapter was to identify factors associated with smoking in patients with MCC and the impact of their cessation on cardiovascular and global health. It was also sought to identify models of multidisciplinary intervention as a priority goal in the smoking cessation process.

SMOKING, CARDIOVASCULAR DISEASES, AND MULTIPLE CHRONIC CONDITIONS

Cerebrovascular and cardiovascular diseases (CVD) are the leading causes of morbimortality worldwide. Tobacco use, sedentary lifestyle, alcohol abuse, and high-saturated-fat and high-carbohydrate diet are known and modifiable risk factors responsible for the incidence of these diseases [4,8,15]. The occurrence of acute myocardial infarction (AMI) worldwide, in both sexes and at all ages, is a clear example of the repercussion of these factors [15,16]. This population profile usually mirrors the population with MCC, which represents approximately 72% of world mortality [4,8].

Smoking causes coronary heart disease and AMI due to multifactorial mechanisms, confirmed by large studies such as INTERHEART [16,17]. In addition to coronary disease, smoking causes stroke, aortic diseases and peripheral vascular disease, premature cardiovascular disease, and sudden death [13]. Since the 1980s, studies have been developed in the search for the association of systemic atherosclerosis and smoking, and the results confirm this relationship [14,18–22]. Smoking induces endothelial injury, formation of atheromas, and overlapping of prothrombotic influences [14,21,22].

Smoking cessation represents a central goal for cardiovascular diseases' primary and secondary prevention due to the diverse benefits that promote the physical health of all smokers of all ages, despite the duration and intensity of the addiction [14,23]. In addition to the rapid reduction of cardiovascular risk, it also improves the cost-benefit ratio for prevention programs (Table 1). Smoking cessation is also a recommended measure in treatment of several chronic diseases of high morbidity and mortality, such as hypertension, DM, COPD, and chronic kidney disease (CKD) [24–29].

Studies such as Bakhru et al. (2005) have shown that the inflammatory component for atherosclerosis progression in smokers may be reversible in the face of cessation of addiction and exposure to tobacco [30]. Also, smoking cessation reduces the risk of disease, even though they can remain high for a decade or more after cessation [31].

However, tobacco use has been neglected as a cardiovascular risk factor, receiving less attention than hypertension, dyslipidemia, or DM [14,32]. In the medical clinic, there is a need to intensify the identification of smoking patients, to evaluate their smoking profile, and to offer and/or recommend treatment periodically, depending on their motivation stage to stop smoking. In Table 1, we show the benefits of smoking cessation according to the duration of cessation and its repercussion in the various organs and systems.

Studies have shown that the deleterious effects of tobacco on health go beyond active smoking. Currently, passive smoking has been considered per se a new cardiovascular risk factor [14,23]. There is consistent evidence for the effects associated with passive smoking, particularly of vascular vasoconstriction and hypoxia [14,16]. More recently, new concepts emerge about the so-called tertiary smoking, where deposition of smoke-derived particles or recirculation after this deposit is detected in individuals who do not smoke or who have not had direct contact with people who smoke. Thus, it is essential to ensure that there is no safe level for tobacco use [33].

TABLE 1 Smoking Cessation Benefits

	Repercussion of Tobacco Cessation Over Time
20 min	Blood and pulse return to normal
8 h	Oxygen levels in the blood return to normal
24 h	Carbon monoxide is eliminated from the body
72 h	Breathing becomes easier, and the general disposition increases
2–12 weeks	Circulation improves throughout the body
3–9 months	Pulmonary function improves between 5% and 10%
5 years	The risk of heart attack is reduced by half
10 years	The risk of lung cancer falls by half, and the risk of heart attack is the same as of those who never smoked

Source: Adapted from Eriksen M, Mackay J, Schluger N, Gomeshtaph FI, Drope J. The Tobacco Atlas. Fifth edition revised, expanded, and update. Captured in: 19/03/2015; 2015. Disponible in: http://www.TobaccoAtlas.org.pdf; Reichert J, de Araújo AJ, Gonçalves CMC, de Godoy I, Chatkin JM, Sales MDPU, et al. Diretrizes para cessação do tabagismo-2008. J Bras Pneumol 2008;34(10):845–80.

In hypertensive patients, smoking causes an acute increase in blood pressure and heart rate, probably due to actions secondary to nicotine, which acts as an adrenergic agonist, promoting local and systemic release of catecholamines. The perpetuation of exposure can rapidly develop nicotinic dependence, but hemodynamic effects tend to stabilize or decline [33]. The chronic effects of smoking on SAH are also characterized by phenomena directed to sympathetic hyperreactivity. Hypertensive smokers have a worse cardiovascular risk profile than nonsmokers despite treatment optimization [34]. Furthermore, cardiovascular aggression and increased morbimortality generated by tobacco use have multifactorial origins, with higher precipitation and occurrence of vascular events independent of the affected area [13]. These factors alone or combined may lead to uncontrolled pressure, both acute and chronic [33,35]. According to Vlachopoulos et al. (2010), smoking was associated with increased arterial stiffness and increased left ventricular mass in previously hypertensive individuals. Such an effect can also be observed in those who have ceased addiction [31].

Diabetes mellitus is not a single disease, but a heterogeneous group of metabolic disorders that have hyperglycemia in common resulting from defects in insulin action, insulin secretion, or both [36]. Smoking has a positive contribution to glycotoxicity and increased insulin resistance, common components in DM. Nicotine and tobacco accelerate the free radicals and apoptosis of beta-pancreatic cells as well as the mobilization of the insulin-sensitive transporter (GLUT4), which promotes the uptake of glucose into adipose and skeletal muscle tissues. This mobilization of GLUT4 may explain the frequent association of increased cardiovascular mortality in diabetic smokers [37]. The European Association for the Study of Diabetes (EASD) and the American Diabetes Association (ADA) recommend smoking cessation as an integral component of DM management [36,38]. Some studies, however, identify weight gain and worsening of postcessation blood glucose levels as unfavorable points for smoking cessation in the diabetic [37,39,40]. On the other hand, more striking than the weight gain and the possible glycemic control will be the positive impact in the reduction of endothelial dysfunction, in morbidity and mortality secondary to cessation, improving the lipid profile and contributing to the respective improvement of the glycemic profile [34,41,42].

CKD is considered an emerging cardiovascular risk factor, with growing prevalence in the world. It is associated with the exponential increase of users in late stages of the disease, being aggressive and having high cost [43,44]. Within the population with MCC, presenting CKD increases per se the cardiovascular risk in 20–40 times, when compared with those without the disease [45,46]. In recent years, smoking has emerged as an independent risk factor for injury and progression of renal disease, particularly in hypertensive and diabetic patients [43,47–52]. Smokers with CKD have a higher prevalence of CVD and its recurrences, as well as premature CVD, respiratory diseases, and cancer [19,43,44,46,51–53]. They also present greater progression to advanced stages of disease, higher rejection and mortality in renal transplanted patients [52,54]. The benefit of smoking cessation includes preservation of renal function even in previously existing CKD patients. Dialysis units could provide a conducive environment to help users quit smoking [44]. Smoking cessation is recommended in preoperative renal transplant patients, improving graft and recipient survival [43,44].

Besides being a lung disease, COPD is also a systemic inflammatory disease. The literature shows that COPD is more prevalent among smokers and ex-smokers than among nonsmokers. There is evidence that quitting smoking is the most effective and cost-effective intervention in reducing and stopping COPD progression [29]. A study by Galil et al. (2016) found that smoking in a population with MCC increases the chance to develop COPD in all age groups and that cessation reduces its prevalence [34].

Among these modifiable risk factors for MCCs, many cohort studies and meta-analyses have shown the benefits of physical activity. However, this relationship is still not well characterized, because studies that focused on physical activity and chronic diseases usually showed a single domain of activity. According to the WHO, a minimum total physical activity level (regardless of domains, including leisure time, housekeeping, occupation, and/or transport) of 150 min/week is recommended. It is known that patients who perform the minimum recommended amount of physical activity on a regular basis have a significant reduction in the risk of developing chronic diseases [55]. Regular physical activity has a positive effect on anxiety reduction due to the release of endorphins that may counterbalance the dopaminergic need, normally released by nicotinic receptors in smokers, in the cessation process [56]. Also, regular moderate-intensity physical exercises may reduce craving occurrence during smoking cessation [57]. Encouraging the practice of physical activity in the period prior to the higher consumption of cigarettes has also been a positive strategy [58,59]. Supervised exercises in smoker women with signs of depression may facilitate short smoking cessation in this population [60]. A recent study by O'Donovan et al. (2017), in addition to reinforcing the association of exercise to the reduction of CVD risk and cancer mortality, demonstrated that vigorous physical exercise could reduce the risk of all-cause mortality, CVD, and cancer in about 30% among smokers and former smokers [61].

Regarding the distribution of body fat, several biological mechanisms demonstrate its direct relation with smoking. High levels of cortisol increase lipogenesis and promote differentiation of adipocytes and deposition of abdominal fat, which alters lipoprotein lipase enzyme's activity, contributing to the accumulation of fat in the gluteal and abdominal regions [62]. The action of smoking on body weight seems to be mediated by nicotine. Experimental and human studies have demonstrated the impact of nicotine on suppressing appetite and reducing body weight. Patients submitted to nicotine replacement therapy during the process of smoking cessation present delay in weight gain. When the medication is discontinued, patients usually gain weight [63]. Therefore, fear of weight gain is a barrier to smoking cessation and an important cause of relapse for many smokers [64]. The main risk factors for greater weight gain in the abstinence process are being female, being African American descent, being younger than 55 years, consuming more than 20 cigarettes per day, and having low socioeconomic status [63].

As for eating habits, another important modifiable risk factor, smokers deserve special attention, since they are known to be vulnerable and known for adopting behaviors that favor weight gain, sedentary lifestyle, and inadequate diet. This group is characterized by the choice of foods with low nutritional quality and high energy density, when compared with nonsmokers [58,59]. During periods of smoking abstinence, there are periods of weight gain that has magnitude variable, with an average gain of 5–6 kg in 2 years. It should be emphasized that the highest gain coincides with the immediate period of cessation [58,59]. Smokers, in general, have a lower body mass index (BMI) than nonsmokers. However, recent studies in the population with MCC revealed a high prevalence of obesity. Smokers with co-occurrence of obesity present a higher risk of weight gain after cessation, which may lead to numerous health consequences, especially the increase in CVD [34,64–66].

In addition to factors related to nutritional aspects and dietary preferences that directly interfere with weight gain during abstinence, the literature has pointed to a possible relation between appetite regulating hormones such as leptin in the genesis of relapse processes among individuals who long for abstinence. This hormone has the main effect of being anorexigenic, promoting a sensation of satiety. Studies have shown that in addition to leptin, cortisol has also influenced cravings during short-term smoking cessation [67,68]. It is emphasized that encouraging the ingestion of dried fruit and oilseeds may help reduce these phenomena during the cessation process, particularly in the brief one [68–70].

In general, an inverse relationship can be observed between the consumption of fruits and vegetables and the use of cigarettes. Smokers who consume more fruits and vegetables smoke fewer cigarettes a day. Haibach et al. (2012) studied the difference between nutrient intake in smokers and nonsmokers, noting that smokers, when compared with nonsmokers, consume larger amounts of saturated fat, cholesterol, and alcohol and a smaller amount of monounsaturated fat, fiber, vitamins C and E, and beta-carotene, commonly found in fruits and vegetables [71]. Aguiar-Nemer et al. (2013) analyzed the influence of abstinence and craving on the dietary and nutritional status of smokers in a multidisciplinary intervention program and identified that the majority of foods consumed by smokers to reduce cigarette cravings were fruit followed by coffee, water, and sweets [65]. Gomes et al. (2014) observed the reduction of cravings with the consumption of dried fruits and oilseeds during smoking cessation process [69]. Furthermore, dark chocolate (70% or more of cocoa content) can also be used to reduce stress and may contribute to smoking cessation, since in the cessation period, patients usually present a greater change in mood and stress load [72].

Alcohol use is related to a variety of diseases, disorders, and injuries, as well as legal and social problems. Alcohol is related to mouth, esophagus, and larynx cancer; hepatic cirrhosis; pancreatitis; SAH; DM; some types of stroke; and depression, among others. High-risk drinking can bring countless losses to the user, as well as alcohol dependence per se [73]. Therefore, it is important to identify and quantify alcohol consumption in clinical practice [74,75].

It is observed that the smoking habit coexists with the drinking habit. The prevalence of alcoholism in smokers is 10–14 times higher than in nonsmokers. Most users report that alcohol use predates tobacco use [76]. Studies by Campos et al. (2014) and Galil et al. (2016) have shown a prevalence of abuse of 5.2% in alcohol use in the MCC population, whereas the smokers in the study had a prevalence of 13.4%. Allied, being a smoker increased the chance of alcohol abuse in 3.82 times (95% CI, 1.89–7.59), comparing with ex-smokers and nonsmokers [34,77]. Heavy smokers are also heavy drinkers, a habit that can aggravate weight gain, elevate triglyceride levels, and increase blood pressure [78]. Moderate drinkers have better profiles than abstainers, while heavy drinkers have worse profiles compared with drinkers of low and moderate intensity or even with people who do not drink. Thus, moderate alcohol consumption is generally associated with a healthier lifestyle [76]. A review by Dorie et al. (2016) suggests that smoking cessation interventions targeting smokers in treatment and recovery of alcohol and other dependencies increase abstinence from tobacco [78].

Studies have shown that the association between psychiatric disorders and smoking habits is bidirectionally and more often represented by the development of stress, anxiety, and/or depression [79,80]. In the case of anxiety and its different types, there has been a high prevalence of smoking in panic disorder, with or without agoraphobia, and in post-traumatic stress disorder [79]. Anxiety sufferers are those with higher smoking loads and higher prevalences of relapses during the smoking cessation process [81,82].

The association between depression and smoking is also well documented, with a high prevalence of depressive symptoms and/or depression among those who smoke [80,83–85]. Farinha et al. (2013) also detected the association between nicotine dependence and severity of anxiety and depression symptoms [84]. An international study involving tobacco and alcohol users showed that any of these groups had a three times greater risk of depression than nonsmokers and alcohol users [83].

Currently, stress is used as an indicator of morbidity, since a person with a high level of stress can develop physical exhaustion, characterized by muscular tension, hypertension, nausea, and emotional exhaustion whose symptomatology is anxiety and tension, implications for the development and worsening of CVD [86].

Tobacco use is known as a way to reduce stress, constituting a coping strategy. Cardoso (2009) identified that smokers show a higher level of stress than nonsmokers [87]. Ismael (2007) identified that stress is one of the most frequent reasons for relapse in 61.9% of smokers [88]. The implementation of stress management programs would be useful to emphasize the importance of health-promoting habits, such as practice of physical exercises, adoption of a balanced diet, and discouragement of harmful practices such as smoking.

MULTIDISCIPLINARY VISIONS ABOUT THE LIFESTYLE ON HEALTH AND CARDIOVASCULAR DISEASE

There is no justification to consider chronic conditions as a single isolated problem or to include them in the traditional categories of communicable and noncommunicable diseases anymore. New ideas, methods, and innovative programs in the prevention and management of chronic conditions become important. An innovative treatment would be based on the demands that a problem imposes on the health system. Chronic conditions present similar requirements regarding management strategies, and therefore, a characteristic health care has been developing. According to the new concepts of chronic conditions, it is important to focus on the quality of life of the patient and his/her family. It is necessary to understand that the patient is not a passive participant in the treatment and should be considered a health producer instead [89].

NCDs have a great impact on morbimortality, and in 80% of the cases, they are present in low- and middle-income countries, with a population composed primarily of vulnerable and low schooling. In order to cope with each isolated chronic condition and MCC, it is necessary to evaluate and systematically monitor these conditions. Also, we need to endorse health promotion, in a continuous way, encouraging the creation of physical spaces suitable for physical activity and developing food guides with regional and cultural suitability and tobacco-free environments. Similarly and simultaneously, it is fundamental to promote real integral care, starting from a correct risk stratification and registration of priority groups, elaborating plans of care and self-care, organization of network flows, and pharmacological assistance, among other actions [90].

This whole context fits and can be inserted in the actions to combat smoking. Thus, the guiding principle of smoking cessation guidelines would be for all smokers to receive treatment or some kind of guidance on their use of substances in view of the various health problems associated with smoking. Users motivated for cessation would be referred for treatment. However, it should not be forgotten that treatment for nicotine dependence and substance abuse commonly involves repeated cycles of attempts to stop smoking and relapse. Smokers with many years of addiction can spend up to seven times for such cycles before achieving cessation [58,59]. Despite the relapse process, many smokers may present incremental gains when addiction is treated as a chronic disease, such as reducing the amount of cigarettes smoked per day and increasing periods of abstinence [91,92]. Nonmotivated users, for example, should receive brief interventions with the purpose of increasing the readiness to change addictive behaviors [58,91]. Often, even in precontemplative periods, those who are unmotivated to stop smoking may respond positively to the motivational interview.

The development of new tools to facilitate interventions is a key task. Data show that decisions that lead to quitting attempts are different from those that lead to quitting. Brief interventions work primarily on motivating quitting attempts. Attempts at cessation may be frequent, but cessation success is rare. Working on cessation attempts would be an excellent rate of intervention effectiveness. Providing cessation care is highly sensitive to increase confidence in success by being more effective in generating cessation attempts than by counseling alone [58,59].

From this context, the health professionals, especially physicians, should intervene in a more proactive manner in relation to stimulating smoking cessation [93,94]. At this point, what is relevant is the use of the five A's: ask, advise, assess, assist, and arrange [59]. In addition, clinical guidelines recommend that all users should receive pharmacotherapy to achieve cessation, except in contraindicated cases such as pregnant women, light smokers, and adolescents [47,58,59].

Monitoring users undergoing treatment should be done longitudinally, in an attempt to help them cope with risk situations, preventing lapses or relapses [91,92].

MULTIDISCIPLINARY TEAM VERSUS CLINICAL PRACTICE

In clinical practice, the use of drug treatment is available, and more recently, nonpharmacological strategies have been gaining strength, with good indications of effectiveness. In order for a nonpharmacological intervention to bring positive results, it is important that it aims to act on the risk factors of the disease in order to contribute to change in lifestyle. Such a change should involve the stimulus for developing healthier habits and controlling emotional variables that may be hindering adherence to the treatment and favoring the maintenance of the disease.

Based on the principle that CVD is associated with a deficient lifestyle, it is important to find ways to minimize the negative effects of stress, which are among the main risk factors for these diseases [95–97]. Studies that propose nondrug interventions that may contribute, together with medical treatment, to the control of the disease are very useful.

Among nonpharmacological practices, some effective actions to stimulate smoking cessation are written information through self-administered material, motivational interviewing, cognitive-behavioral intervention, stress monitoring, and nonconventional access [58,59,98]. Proactive phone support, automated text messaging programs, and printed self-help materials can help smokers who seek help and are motivated to stop smoking and are easily accessible. Individual or group sessions of behavioral support can also increase success for cessation, especially in middle- and high-income countries.

Motivational interviewing is a user-centered, directive approach in which your feelings, ideas, and beliefs about tobacco are explored to help you overcome substance-related ambivalence [99]. The cognitive-behavioral approach (CBA) techniques help the smoker to modify the pattern of smoking behavior, avoiding situations related to relapse. This is reflected both in learning to resist the urge to smoke and in adopting strategies that counteract the act of smoking.

The smoker should learn to recognize withdrawal symptoms and duration and prepare to cope with them, especially in the first few days without smoking. The main one, the craving (imperious desire to smoke), usually lasts between 1 and 5 min, and it is important to develop a substitutive strategy until the symptom wears off [58,59,100]. Social support consists of reinforcing motivations to stop, emphasizing the benefits of cessation, increasing self-efficacy, combating beliefs and rationalizations around consumption, preventing residual problems of cessation (weight gain, irritability, and negative mood), and supporting the smoker to settle ambivalence if motivation declines. Social support through friends and family is critical in tobacco resistance. Smoking-free environments at work and at home and encouragement for other smokers to seek cessation assistance contribute positively to strengthening recovery [98,100]. Strategies to support smoking cessation can be performed by any member of the multidisciplinary health team who has been adequately trained to approach the smoker [32,58,59]. The task of cessation rests with health professionals, who often accept training but distance themselves from the approach of the five A's [58,59].

During the period of treatment for smoking cessation, a raise in stress level is common. Stress control training (SCT) is an intervention that aims to perform a functional analysis of stressors through the objective performance of four pillars of stress control: (1) relaxation, (2) feeding, (3) physical exercise, and (4) modifications in the cognitive sphere, which includes techniques for problem solving, assertiveness training, visualization exercises, identification, and modification of thinking. Studies have shown that SCT is capable of promoting lasting lifestyle changes that favor stress management and the maintenance of a better quality of life [101–106]. Other forms of therapy can be offered, such as music therapy and relaxation sessions and self-perception of stress-inducing situations [107].

However, the most effective strategy in smoking cessation is the combination of approaches, especially cognitive-behavioral and pharmacological approaches [59]. Studies show that this combination results in higher abstinence rates (assist) that increased from 40% to 60% of smoking cessation [58,59,98]. In addition, the existence of guidelines for the cessation process makes the intervention more incisive.

Regarding the pharmacological approach to smoking cessation, we ask what does it mean to offer treatment to quit smoking? And we answer that it is relevant to help the patient overcome the symptoms of withdrawal and overcome relapses, to enable the ex-smoker to feel happy and rewarded for having stopped smoking, and to manage more effectively the use of medications that should aid in the control of withdrawal symptoms and the compulsive desire to smoke [92]. It is worth remembering that the CCA should be included regardless of the treatment instituted. Currently advocated drugs are nicotine replacement therapy (long-acting, adhesive, and short-acting, chewing gum), bupropion, and varenicline as first-line drugs [58,59,98]. Bupropion is one of the most indicated drugs because it has few important side effects. Other studies have shown that varenicline, bupropion, and nicotine replacement therapy have a two- to threefold higher rate of cessation when compared with placebo [59,98,108,109].

Among all pharmacotherapy options, varenicline leads the first-line drugs. Approved in 2006 by the US FDA, this drug reduces the reward associated with smoking and alleviates nicotine craving and withdrawal [110]. It is particularly suitable for the light smoker due to the action of the substance's active principles, which promote nicotine receptor saturation, reduce the reinforcing effects of use, and reduce craving and abstinence crisis, as well as cigarette reward effects. In addition, the highest abstinence rates in clinical trials evaluated after 6 months of cessation are due to the use of varenicline [59].

In general, monotherapy is sufficient for most patients. For patients who cannot achieve cessation with only one medication allied with CBA, combined pharmacotherapy is indicated. This process consists in using the nicotine patch with other short-acting nicotine replacement products, such as chewing gum or tablets or even a combination of bupropion and nicotine replacement. These combinations show even more effective results than single drug therapy, including varenicline [58,59,98]. Users with cardiovascular complications can use drug-cessation therapies because they are less harmful to health than complications that the disease can cause [14,111]. Inpatients with cardiovascular complications (catheterization, angioplasty, or myocardial revascularization) should begin the treatment with NRT in order to reduce withdrawal symptoms and the risk of a new cardiovascular event. In heavy smokers with COPD, the use of varenicline for longer than 7–12 traditional weeks of treatment has been shown to be a good strategy to increase cessation rates [112,113].

Drug treatment helps a lot in improving adherence to smoking cessation. Nicotine replacement therapy, bupropion, nortriptyline, varenicline, and cytisine may offer help in stopping smoking when administered in combination with some behavioral support. Of these, cytisine and nortriptyline (second-line medications) are globally accessible [58,59,114]. West et al. (2015) conducted a review aiming to assess the efficacy, effectiveness, and accessibility of health-care interventions to promote and assist smoking cessation [114]. The paper was prepared by searching the Cochrane randomized clinical trial database for major tobacco cessation interventions. The accessibility of each intervention was assessed for exemplary countries in each category of World Bank income (low, medium low, medium high, and high) [114]. As a result, it was demonstrated that brief counseling offered by a health professional, with a good timing and addressed to smokers who attend health services, can promote smoking cessation and is accessible to countries in all income categories.

Evidence on smoking cessation suggests that CBA and the use of varenicline may promote cessation [114]. Negative predictors for cessation could influence the course of interventions. Shimadu et al. (2016), through logistic regression, assessed that younger age and higher scores on the Beck Depression Inventory-II questionnaire for depression screening were inversely correlated with smoking cessation success rates [115,116]. In conclusion, pharmacological therapies play an important role in the success of interventions for smoking cessation, along with the search for facilitating factors and individual barriers to the maintenance of smoking cessation, which still require more long-term follow-up studies.

EXPERIENCE OF A MULTIDISCIPLINARY TEAM IN THE MANAGEMENT OF SMOKERS WITH MULTIPLE CHRONIC CONDITIONS

In Brazil, data on smoking status in the population with MCCs are still poorly spread. Therefore, there is a need to prioritize the detection and handling of MCC, which may result in the development of strategies that reduce both the prevalence and the morbimortality resulting from them [6,8].

The health system in Brazil is universal and consists of primary, secondary, and tertiary (hospital) facilities, some of which are linked to research programs of public academic institutions, offering a unique opportunity to test interventions and develop programs known as "Centers of Excellence." An example is the Centro Hiperdia Minas de Juiz de Fora (CHDM-JF), which was designed to attend patients with MCC, specifically hypertensive, diabetic, and CKD patients, with high and very high cardiovascular risk. These patients come to the CHDM-JF referenced by the city's and region's basic health units or the centers of medical specialties of the Federal University of Juiz de Fora and are accompanied by a multidisciplinary team in an attempt to reduce injuries and improve their quality of life.

The Integral Tobacco Assistance Unit (UAI-T) is a multidisciplinary project developed since 2012 at CHDM-JF to supporting smokers with MCC. It operates on two fronts: educational and care. At the educational level, it promotes training and continued education for health professionals regarding the management of tobacco cessation and, in the care plan, identifies, evaluates, treats, and follows up smokers with MCC motivated to quit smoking at CHDM-JF.

Following the norms of the Brazilian National Cancer Institute (INCA)/Health Ministry, the sessions with the groups of smokers are developed as described. Initially, a sensitization meeting is held that lasts approximately 30 min. In this session, it is intended to show the participants the negative repercussions of smoking on health in a global way and the exponential increase of these consequences for high-risk cardiovascular individuals. We also present the viability of the treatment and the great impact that cessation has in improving health, reducing morbimortality, and reducing expenses. We then scheduled for these patients continued care, through CBA, over a period of four consecutive weeks. Each meeting lasts approximately 45 min, addressing issues such as the damages that the cigarette represents for each participant in the group, the need and ways to react to reducing the number of cigarettes, coping with abstinence and cravings, the possibility of relapses and their prevention, relaxation training, and strengthening of cessation skills.

Specifically, in the first meeting, users are clarified about the characteristics of the dependency, which may be physical, psychological, and/or behavioral. In the second session, associated with the cognitive approach, participants who need pharmacological treatment are identified, based on the specificity of each case (stage of action). The forms of tobacco cessation (abrupt or sequenced and with or without medication) are explained. In this meeting, participants are invited to share their perceptions regarding their experience of "learning to quit smoking" such as major difficulties encountered during the process (barriers), the main positive advantages of cessation (individual strengths), tips, and suggestions. At the third meeting, conflicting situations or hypotheses of fissures, lapses, and/or relapses are reinforced. The importance of practicing physical exercises and the concepts of healthy eating are also highlighted in this session, aiming to encourage the increased consumption of dried fruits and oleaginous and good food choices. For those who initiated drug treatment, the strengthening factors of not smoking are reinforced. At the fourth session, the group is moved to those users who are already in the maintenance phase, and their testimonials can be a stimulus for participants to maintain abstinence and prevent relapse. After the four consecutive sessions, the follow-up period begins, where those users who have managed to stop the addiction will receive medical and psychological support every 15 days in the first 2 months and then a monthly follow-up from the third to the twelfth month. The program is finished when the users reach a smoking cessation period of 12 months or longer [110,117]. These participants, now considered ex-smokers, receive a smoking cessation certificate from UAI-T.

Regarding the profile of the MCC patients seen in CHDM-JF, a study by Campos et al. (2014) identified that 12% were current smokers and the mean age was 55.24 ± 11.69 years old. Most of them were female, were elderly (≥ 60 years old), and had low educational status. Sedentary lifestyle and obesity were highly prevalent at 71.4% and 45.3%, respectively. The prevalences of hypertension, diabetes, chronic kidney disease, and declared atherosclerotic disease were highly prevalent with frequencies of 92.2%, 59.4%, 55.9%, and 23.5%, respectively. Lower percentages of participants had asthma (2.5%), cancer (2.3%), and COPD (1.8%). Regarding MCC, 10.8% of patients were with two chronic disease and 4.8% of patients with three chronic diseases. Screening for depression was positive in 45% of the participants. Anxiety and stress level were not assessed. Alcohol abuse was found in 13.4% of smokers [77].

UAI-T invited 273 users to participate in the smoking cessation program, and only 13.9% of them attended to start the treatment. According to the contemplation level for cessation, 78.9% were motivated to quit smoking and attended the first session. Adherence to all CCA sessions reached more than 65%, but only 10.5% ceased tobacco use after all four sessions [77,118]. Among the elderly participants, it was identified that they had predominantly low schooling, were female, were married and sedentary, had an age of 64.85 ± 4.04 years, and had an average time of addiction of 37.34 ± 11.54 years. High nicotine dependence was found in 35.6% and previous attempts to stop smoking in 71.2% of them [119]. The presence of CKD, coronary disease, and COPD was significantly associated with the elderly, whereas depression was less prevalent in this age group [119]. Andrade et al. (2016) have identified that smoking cessation is widely reported, but achieving and maintaining abstinence is a difficult process. Of the 98 participants who attended in a 12-month period, 66.1% were highly dependent on nicotine, 27.4% were reported being abstinent for 25.1 ± 7.7 months, and of these, two were reported having lapses. Of those who were current smoking, 44.4% remained 6.4 months without smoking, and 93.3% stated that they want to stop smoking [120]. Despite maintenance of the addiction, these current smokers consumed 14.7 ± 4.4 cigarette per day versus 21.8 ± 1.33 cigarettes per day when they started the treatment. These data combined encourage the systematic intervention for cessation of smokers in this age group, associated with the possibility of reducing the diseases [120]. The study concludes that the higher rates of smoking cessation coincides with the period of greater performance of the multidisciplinary team (first 6 months). This observation reinforces the importance of multidisciplinary team work in smoking cessation process and its maintenance period [120,121].

FUTURE PERSPECTIVES

Most smoking programs target adult smokers. Encouraging children and young adults not to start the addiction is a cost-effective strategy. Similarly, the implementation of programs to stimulate the cessation of young smokers with less years of addiction would be an intervention of great relevance. Encouraging greater engagement of this population in physical activity and nutritional programs would reduce cravings and relapses and incorporate healthy habits into these young people that will reflect on their cardiovascular health in the coming decades.

The introduction of electronic media, interactive online games, text messages, and telephone contact, respecting the literacy of each population group and their biopsychosocial characteristics, are strategies being improved. By multidisciplinary team action, it would be possible to extend patients' reach to the modification of their stage of motivation using these kinds of new technology, providing a greater search for programs aiming to promote smoking cessation.

Another important point is that, in addition to increasing the access to prevention and treatment programs for smoking cessation, it is essential to prepare health professionals. It is also important to promote continued education at all levels of training and clinical practice regarding the modifiable risk factors that most impact cardiovascular and all-cause morbimortality. The implementation of such measures will promote cost-effective interventions in the smoking population with MCC.

FINAL CONSIDERATIONS

As we can see, reaching abstinence is an arduous and complex process involving many neurobiological and environmental factors. Developing different therapeutic options is of great relevance for the treatment of an intensive approach to smoking. Therefore, it is important to study possible nutritional and nondrug options that can act as adjunctive therapy in smoking cessation treatment. In addition, acting on the anti-inflammatory and antioxidant modulation of these individuals could reduce possible comorbidities associated with prolonged smoking.

Currently, there is a precariousness of smoking assistance programs that perform full and multiprofessional actions in the city of Juiz de Fora, MG, Brazil, and possibly in several countries with similar socioeconomic profiles. The implementation of multiprofessional teams engaged in a comprehensive approach to smokers with MCC could help increase and maintain the abstinence rates of this specific population. The training of professionals working in these sectors and the dissemination of benefits achieved through multiprofessional therapeutic approaches could be effective change tools in the complex management of complications related to smoking.

REFERENCES

[1] Warshaw G. Introduction: advances and challenges in care of older people with chronic illness. Generations 2006;30(3):5–10.

[2] Hwang W, Weller W, Ireys H, Anderson G. Out-of-pocket medical spending for care of chronic conditions. Health Aff 2001;20(6):267–78.

[3] Carvalho MD, Morais NOL, Silva JJB. Presentation of the strategic action plan for coping with chronic diseases in Brazil from 2011 to 2022. Epidemiol Serv Saúde 2011;20(4):425–38.

[4] Goodman RA, Boyd C, Tinetti ME, Von Kohorn I, Parekh AK, McGinnis JM. IOM and DHHS meeting on making clinical practice guidelines appropriate for patients with multiple chronic conditions. Ann Family Med 2014;12(3):256–9.

[5] Ward BW. Prevalence of multiple chronic conditions among US adults: estimates from the National Health Interview Survey, 2010. Prev Chronic Dis 2013;10.

[6] World Health Organization. Global status report on noncommunicable diseases. Geneva, Switzerland: World Health Organization; 2010.

[7] CDC—Center for Disease Control and Prevention. Request for Proposal Department of Public Health-Public Health Initiatives BranchTobacco Use Prevention and Control Program. Disponible: http://www.das.state.ct.us/rfpdoc/DPH06/bids/2009-0919.pdf.

[8] Malta DC, Morais Neto OLD, Silva Junior JBD. Apresentação do plano de ações estratégicas para o enfrentamento das doenças crônicas não transmissíveis no Brasil, 2011 a 2022. Epidemiol Serviços Saúde 2011;20(4):425–38.

[9] Unal B, Critchley JA, Capewell S. Modelling the decline in coronary heart disease deaths in England and Wales, 1981–2000: comparing contributions from primary prevention and secondary prevention. BMJ 2005;331. https://doi.org/10.1136/bmj.38561.633345.8.

[10] Malta DC, Silva Jr JBD. O Plano de Ações Estratégicas para o Enfrentamento das Doenças Crônicas Não Transmissíveis no Brasil e a definição das metas globais para o enfrentamento dessas doenças até 2025: uma revisão. Epidemiol Serviços Saúde 2013;22(1):151–64.

[11] Pearce N, Ebrahim S, McKee M, Lamptey P, Barreto ML, Matheson D, et al. Global prevention and control of NCDs: limitations of the standard approach. J Public Health Policy 2015;36(4):408–25.

[12] Araújo A, et al. Abordagem de populações especiais: tabagismo e mulher- razões para abordagem específica de gênero. In: Gigliotti A, Presman S, editors. Atualização no tratamento do tabagismo. Rio de Janeiro: ABP, Saúþ; 2006. p. 107–28.

[13] US Department of Health and Human Services. The health consequences of smoking—50 years of progress: a report of the Surgeon General. Atlanta, GA: US Department of Health and Human Services, Centers for Disease Control and Prevention, National Center for Chronic Disease Prevention and Health Promotion, Office on Smoking and Health; 2014. p. 17.

[14] Rigotti NA, Clair C. Managing tobacco use: the neglected cardiovascular disease risk factor. Eur Heart J 2013;34(42):3259–67.

[15] O'Donnell MJ, Xavier D, Liu L, Zhang H, Chin SL, Rao-Melacini P, et al. Risk factors for ischaemic and intracerebral haemorrhagic stroke in 22 countries (the INTERSTROKE study): a case-control study. Lancet 2010;376(9735):112–23.

[16] Lanas F, Avezum A, Bautista LE, Diaz R, Luna M, Islam S, et al. Risk factors for acute myocardial infarction in Latin America. Circulation 2007;115(9):1067–74.

[17] Yusuf S, Hawken S, Ôunpuu S, Dans T, Avezum A, Lanas F, et al. Effect of potentially modifiable risk factors associated with myocardial infarction in 52 countries (the INTERHEART study): case-control study. Lancet 2004;364(9438):937–52.

[18] Tiengo A, Fadini GP, Avogaro A. The metabolic syndrome, diabetes and lung dysfunction. Diabetes Metabol 2008;34(5):447–54.

[19] Waters D, Lespe J, Gladstone P, Boccuzzi SJ, Cook T, Hudgin R, et al. Effects of cigarette smoking on the angiographic evolution of coronary atherosclerosis. Circulation 1996;94(4):614–21.

[20] Li S, Yun M, Fernandez C, Xu J, Srinivasan SR, Chen W, et al. Cigarette smoking exacerbates the adverse effects of age and metabolic syndrome on subclinical atherosclerosis: the Bogalusa Heart Study. PLoS ONE 2014;9(5). e96368.

[21] Siasos G, Tsigkou V, Kokkou E, Oikonomou E, Vavuranakis M, Vlachopoulos C, et al. Smoking and atherosclerosis: mechanisms of disease and new therapeutic approaches. Curr Med Chem 2014;21(34):3936–48.

[22] Johnson HM, Gossett LK, Piper ME, Aeschlimann SE, Korcarz CE, Baker TB, et al. Effects of smoking and smoking cessation on endothelial function: 1-year outcomes from a randomized clinical trial. J Am Coll Cardiol 2010;55(18):1988–95.

[23] Carnevale R, Sciarretta S, Violi F, Nocella C, Loffredo L, Perri L, et al. Acute impact of tobacco vs electronic cigarette smoking on oxidative stress and vascular function. CHEST J 2016;150(3):606–12.

[24] Diretrizes. VI Diretrizes Brasileiras de Hipertensão. Arq Bras Cardiol 2010;95(1).

[25] Council ES, Redon J, Narkiewicz K, Nilsson PM, Burnier M, Viigimaa M, et al. 2013 ESH/ESC Guidelines for the management of arterial hypertension. Eur Heart J 2013;34:2159–219.

[26] Algoritmo para o tratamento do diabetes tipo 2—Atualização 2011. Posicionamento Oficial SBD n°2; 2011.

[27] Kursztajn GM, Souza EE, Romão J, Bastos MG, Meyer F, Andrada NC. Projeto Diretrizes. Doença renal Crônica (Pré-terapia renal substitutiva). Associação Médica Brasileira/Conselho Federal de Medicina; 2011.

[28] Infância AN. I Diretriz de Prevenção da Aterosclerose na Infância e na Adolescência. Arq Bras Cardiol 2005;85(VI).

[29] Gold. Global initiative for chronic obstructive lung. Global Strategy for the diagnosis, management, and prevention of chronic obstructive pulmonary disease; Update 2016. Disponible: http://goldcopd.org/global-strategy-diagnosis-management-prevention-copd-2016/.

[30]   Bakhru A, Erlinger TP. Smoking cessation and cardiovascular disease risk factors: results from the Third National Health and Nutrition Examination Survey. PLoS Med 2005;2(6). e160.

[31]   Vlachopoulos C, Aznaouridis K, Stefanadis C. Prediction of cardiovascular events and all-cause mortality with arterial stiffness: a systematic review and meta-analysis. J Am Coll Cardiol 2010;55(13):1318–27.

[32]   Jemal A, Vineis P, Bray F, Torre L, Forman D. The Cancer Atlas. 2nd ed. Atlanta, GA: American Cancer Society; 2014.

[33]   Malachias MVB, Souza WKSB, Plavnik FL, Rodrigues CIS, Brandão AA, Neves MFT, et al. 7ª Diretriz Brasileira de Hipertensão Arterial. Arq Bras Cardiol 2016;107(3):1–104.

[34]   de Siqueira Galil AG, Cupertino AP, Banhato EF, Campos TS, Colugnati FA, Richter KP, et al. Factors associated with tobacco use among patients with multiple chronic conditions. Int J Cardiol 2016;221:1004–7.

[35]   Sousa MG. Tabagismo e Hipertensão arterial: como o tabaco eleva a pressão. Rev Bras Hipertens 2015;22(3):78–83.

[36]   Diretrizes da Sociedade Brasileira de Diabetes (2015-2016)/Adolfo Milech...[et al.]; Organização José Egidio Paulo de Oliv0065ira, Sérgio Vencio—São Paulo: A.C. Farmacêutica; 2016.

[37]   Aveyard P, Lycett D, Farley A. Managing smoking cessation-related weight gain. Pol Arch Med Wewn 2012;122(10):494–8.

[38]   ESC Guidelines on diabetes, pre-diabetes, and cardiovascular diseases developed in collaboration with the EASD—Summary. The Task Force on diabetes, pre-diabetes, and cardiovascular diseases of the European Society of Cardiology (ESC) and developed in collaboration with the European Association for the Study of Diabetes (EASD). Diabetes Vascul Dis Res 2014;11((3):133–73.

[39]   Tonstad S. Cigarette smoking, smoking cessation, and diabetes. Diabetes Res Clin Pract 2009;85(1):4–13.

[40]   Ussher MH, Taylor A, Faulkner G. Exercise interventions for smoking cessation. Cochrane Database Syst Rev 2014;8.

[41]   Karthikeyan VJ, Bakris G, MacFadyen RJ. The ADVANCE trial: further PROGRESS with HOPE. J Hum Hypertens 2007;21(12):911.

[42]   Khan HA, Sobki SH, Khan SA. Association between glycaemic control and serum lipids profile in type 2 diabetic patients: HbA1c predicts dys-lipidaemia. Clin Exp Med 2007;7(1):24–9.

[43]   Júnior E, Fernando U, Elihimas HCDS, Lemos VM, Leão MDA, Sá MPBDO, et al. Smoking as risk factor for chronic kidney disease: systematic review. J Bras Nefrol 2014;36(4):519–28.

[44]   Coresh J, Turin TC, Matsushita K, Sang Y, Ballew SH, Appel LJ, et al. Decline in estimated glomerular filtration rate and subsequent risk of end-stage renal disease and mortality. JAMA 2014;311(24):2518–31.

[45]   Nazar CMJ, Kindratt TB, Ahmad SMA, Ahmed M, Anderson J. Barriers to the successful practice of chronic kidney diseases at the primary health care level; a systematic review. J Renal Inj Prevent 2014;3(3):61.

[46]   Brasil. Ministério da Saúde. Secretaria de Atenção à Saúde. Departamento de Atenção Especializada e Temática. Diretrizes Clínicas para o Cuidado ao paciente com Doença Renal Crônica—DRC no Sistema Único de Saúde/Ministério da Saúde. Secretaria de Atenção à Saúde. Departamento de Atenção Especializada e Temática. Brasília: Ministério da Saúde; 2014. p. 37.

[47]   Eriksen M, Mackay J, Schluger N, Gomeshtaph FI, Drope J. The Tobacco Atlas. Fifth edition revised, expanded, and update. Captured in: 19/03/2015; 2015. Disponible in: http://www.TobaccoAtlas.org.pdf.

[48]   Hallan SI, Orth SR. Smoking is a risk factor in the progression to kidney failure. Kidney Int 2011;80(5):516–23.

[49]   Bucharles SGE, Pecoits-filho R. Doença Renal Crônica: Mecanismo da Progressão e Abordagem Terapêutica. J Bras Nefrol 2009;31(1):6–12.

[50]   Rezonzew G, Chumley P, Feng W, Hua P, Siegal GP, Jaimes EA. Nicotine exposure and the progression of chronic kidney disease: role of the α7-nicotinic acetylcholine receptor. Am J Physiol Renal Physiol 2012;303(2):F304–12.

[51]   McCullough PA, Li S, Jurkovitz CT, Stevens L, Collins AJ, Chen SC, et al. Chronic kidney disease, prevalence of premature cardiovascular disease, and relationship to short-term mortality. Am Heart J 2008;156(2):277–83.

[52]   Puddu PE, Menotti A. The impact of basic lifestyle behaviour on health: how to lower the risk of coronary heart disease, other cardiovascular diseases, cancer and all-cause mortality. Lifestyle adaptation: a global approach. ESC Council Cardiol Pract 2015;13(32).

[53]   Rostron BL, Chang CM, Pechacek TF. Estimation of cigarette smoking-attributable morbidity in the United States. JAMA Intern Med 2014;174(12):1922–8.

[54]   RRaymond CB, Naylor HK. Smoking cessation in patients with chronic kidney diseases. CANNT J 2010;20:24–9.

[55]   Kyu HH, Bachman VF, Alexander LT, Mumford JE, Afshin A, Estep K, et al. Physical activity and risk of breast cancer, colon cancer, diabetes, ischemic heart disease, and ischemic stroke events: systematic review and dose-response meta-analysis for the Global Burden of Disease Study. BMJ 2013;354:i3857.

[56]   Edwards MK, Loprinzi PD. Experimentally increasing sedentary behavior results in increased anxiety in an active young adult population. J Affect Disord 2016;204:166–73.

[57]   Daniel JZ, Cropley M, Fife-Schaw C. The effect of exercise in reducing desire to smoke and cigarette withdrawal symptoms is not caused by distraction. Addiction 2006;101(8):1187–92.

[58]   Reichert J, de Araújo AJ, Gonçalves CMC, de Godoy I, Chatkin JM, Sales MDPU, et al. Diretrizes para cessação do tabagismo-2008. J Bras Pneumol 2008;34(10):845–80.

[59]   Fiore MC, Jaen CR, Baker T, Bailey WC, Benowitz NL, Curry SEEA, et al. Treating tobacco use and dependence: 2008 update. Rockville, MD: US Department of Health and Human Services; 2008.

[60]   American Psychiatric Association. DSM-5: Manual diagnóstico e estatístico de transtornos mentais. Porto Alegre: Artmed Editora; 2014.

[61]   O'Donovan G, Hamer M, Stamatakis E. Relationships between exercise, smoking habit and mortality in more than 100,000 adults. Int J Cancer 2017;140(8):1819–27.

[62]   Canoy D, Wareham N, Luben R, Welch A, Bingham S, Day N, et al. Cigarette Smoking and Fat Distribution in 21, 828 British Men and Women: a Population-based study. Obesity 2005;13(8):1466–75.

[63]   Chatkin R, Chatkin JM. Tabagismo e variação ponderal: a fisiopatologia e genética podem explicar esta associação? J Bras Pneumol 2007;33(6).

[64] Leslie WS, Koshy PR, Mackenzie M, Murray HM, Boyle S, Lean ME, et al. Changes in body weight and food choice in those attempting smoking cessation: a cluster randomised controlled trial. BMC Public Health 2012;12(1):389.

[65] de Aguiar Nemer AS, de Melo MM, Luquetti SCPD, de Almeida Vargas AL, Rodrigues GRS, Leite MA, et al. Craving and food choices in patients under treatment for smoking cessation. Int J Food Sci Nutr Diet 2013;2(3):35–9.

[66] Roubos EW, Dahmen M, Kozicz T, Xu L. Leptin and the hypothalamo-pituitary-adrenal stress axis. Gen Comp Endocrinol 2012;177(1):28–36.

[67] von der Goltz C, Koopmann A, Dinter C, Richter A, Rockenbach C, Grosshans M, et al. Orexin and leptin are associated with nicotine craving: a link between smoking, appetite and reward. Psychoneuroendocrinology 2010;35(4):570–7.

[68] al'Absi M, Hatsukami D, Davis GL, Wittmers LE. Prospective examination of effects of smoking abstinence on cortisol and withdrawal symptoms as predictors of early smoking relapse. Drug Alcohol Depend 2004;73(3):267–78.

[69] Da Silva Gomes A, Toffolo MCF, Van Keulen HV, E Silva FMC, Ferreira AP, Luquetti SCPD, et al. Influence of the leptin and cortisol levels on craving and smoking cessation. Psychiatr Res 2015;229((1):126–32.

[70] Potretzke S, Nakajima M, Cragin T, al'Absi M. Changes in circulating leptin levels during acute stress and associations with craving in abstinent smokers: a preliminary investigation. Psychoneuroendocrinology 2014;47:232–40.

[71] Haibach JP, Homish GG, Giovino GA. A longitudinal evaluation of fruit and vegetable consumption and cigarette smoking. Nicotine Tob Res 2013;15(2):355–63.

[72] Al Sunni A, Latif R. Effects of chocolate intake on perceived stress; a controlled clinical study. Int J Health Sci 2014;8(4):393.

[73] Chaieb JA, Castellarin C. Associação tabagismo-alcoolismo: introdução às grandes dependências humanas. Rev Saude Publica 1998;32(3):246–54.

[74] Pednekar MS, Vasa J, Narake SS, Sinha DN, Gupta PC. Tobacco and Alcohol Associated Mortality among Men by Socioeconomic Status in In-dia. Epidemiol Open J 2016;1(1):2–15.

[75] Gowing LR, Ali RL, Allsop S, Marsden J, Turf EE, West R, et al. Global statistics on addictive behaviours: 2014 status report. Addiction 2015;110(6):904–19.

[76] Babor TF, Higgins-biddle JC, Saunders JB, Monteiro MG. AUDIT—Teste para Identificação de Problemas Relacionados ao Uso de Álcool. Roteiro para uso em atenção primária. Organização Mundial de Saúde—Departamento de Saúde Mental e Dependências Químicas; 2006.

[77] Campos TDS, Richter KP, Cupertino AP, Galil AG, Banhato EF, Colugnati FA, et al. Cigarette smoking among patients with chronic diseases. Int J Cardiol 2014;174(3):808.

[78] Apollonio D, Philipps R, Bero L. Interventions for tobacco use cessation in people in treatment for or recovery from substance use disorders. The Cochrane Library; 2016. Disponible: http://onlinelibrary.wiley.com/doi/10.1002/14651858.CD010274.pub2/abstract;jsessionid=A22553F5F73C9 9B9D42F8EF84131480A.f04t01.

[79] Morissette SB, Tull MT, Gulliver SB, Kamholz BW, Zimering RT. Anxiety, anxiety disorders, tobacco use, and nicotine: a critical review of inter-relationships. Psychol Bull 2007;133(2):245.

[80] Nunes SAOV, Vargas HO, de Souza Lanssoni MMB, de Castro MRP, Nunes MVA, Barbosa LR, et al. Avaliação das características clínicas dos fumantes que buscaram tratamento em Centro de Referência do Sistema Único de Saúde (SUS). Biosaúde 2016;8(1):3–24.

[81] Aguiar M, Todo-Bom F, Felizardo M, Macedo R, Caeiro F, Sotto-Mayor R, et al. Four years' follow up at a smoking cessation clinic. Rev Portuguesa Pneumol (Engl Ed) 2009;15(2):179–97.

[82] Collins BN, Lepore SJ. Association between anxiety and smoking in a sample of urban black men. J Immigr Minor Health 2009;11(1):29–34.

[83] Epstein JF, Induni M, Wilson T. Patterns of clinically significant symptoms of depression among heavy users of alcohol and cigarettes. Prev Chronic Dis 2009;6(1):A09.

[84] Farinha H, Almeida JR, Aleixo AR, Oliveira H, Xavier F, Santos AI. Relação do Tabagismo com ansiedade e depressão nos Cuidados de Saúde Primários. Acta Medica Port 2013;26(5):523–30.

[85] Banhato EFC, Galil AGDS, Campos TDS, Colugnati FA, Richter KP, Bastos MG, et al. Depression symptoms among patients with multiple chronic conditions. J Depress Anxiety 2016;5:230.

[86] Pafaro RC, De Martino MMF. Estudo do estresse do enfermeiro com dupla jornada de trabalho em um hospital de oncologia pediátrica de Campinas. Revista da Escola de Enfermagem da USP; 2004.

[87] Cardoso BAP, Santos MLSCD, Berardinelli LMM. A relação estilo de vida e tabagismo entre acadêmicos de enfermagem. Rev Eletrôn Enfermagem 2009;11(2):368.

[88] Ismael SMC. Efetividade da terapia cognitivo-comportamental na terapêutica do tabagista. Doctoral Dissertation, Universidade de São Paulo; 2007.

[89] Holman H, Lorig K. Patients as partners in managing chronic disease; 2000.

[90] Malta DC, Oliveira TP, Santos MAS, Andrade SSCA, Silva MMA. Avanços do Plano de Ações Estratégicas para o Enfrentamento das Doenças Crônicas não Transmissíveis no Brasil, 2011–2015. Epidemiol Serviços Saúde 2016;25(2):373–90.

[91] McLellan AT, Lewis DC, O'brien CP, Kleber HD. Drug dependence, a chronic medical illness: implications for treatment, insurance, and outcomes evaluation. JAMA 2000;284(13):1689–95.

[92] Joseph AM, Fu SS, Lindgren B, Rothman AJ, Kodl M, Lando H, et al. Chronic disease management for tobacco dependence: a randomized, controlled trial. Arch Intern Med 2011;171(21):1894–900.

[93] Richter KP, Shireman TI, Ellerbeck EF, Cupertino AP, Catley D, Cox LS, et al. Comparative and cost effectiveness of telemedicine versus telephone counseling for smoking cessation. J Med Internet Res 2015;17(5).

[94] Park ER, Gareen IF, Japuntich S, Lennes I, Hyland K, DeMello S, et al. Primary care provider-delivered smoking cessation interventions and smoking cessation among participants in the National Lung Screening Trial. JAMA Intern Med 2015;175(9):1509–16.

[95] Muniz LC, Schneider BC, Silva ICM, Matijasevich A, Santos IS. Accumulated behavioral risk factors for cardiovascular diseases. Southern Brazil Rev Saúde Pública [Internet] 2012;46(3):534–42. Available from: http://www.scielo.br/pdf/rsp/v46n3/en_3690.pdf.

[96] Whittaker R, McRobbie H, Bullen C, Borland R, Rodgers A, Gu Y. Mobile phone-based interventions for smoking cessation; 2016. Available from: http://www.thehealthwell.info/node/115301 [accessed 05.03.17].

[97] Lopes MAS. Os estilos de vida nos doentes com cardiopatia isquêmica. Dissertação de Mestrado apresentada no Curso de Mestrado em Enfermagem Médico-Cirúrgica. Repositório Científico de Acesso Aberto de Portugal, Instituto Politécnico de Viseu, Escola Superior de Saúde de Viseu; 2012.

[98] Estratégias para o cuidado da pessoa com doença crônica: o cuidado da pessoa tabagista. In: Básica MdSSDdA, ed2015:1-156.

[99] Miller W, Rollnick S. Motivational interviewing: preparing people for change. J Healthc Qual 2003;25(3):46.

[100] Ministerio de Salud. Argentina Nos Incluye. Guioa de práctica Clínica Nacional de Tratamiento de la Adicción al Tabaco. Recomendaciones basadas em la evidencia científica. Disponível: www.msal.gov.ar. Capturado em:05052016.

[101] Lipp MN, Bignotto MM, Alcino AB. Efeitos do treino de controle dos stress social na reatividade cardiovascular de pacientes hipertensos. Teoria Invest Prática 1997;2(1):137–46.

[102] Alcino AB. Stress Social e Reatividade Cardiovascular infantil: Um estudo psicofisiológico. In: Lipp M (Org.), Pesquisas sobre stress no Brasil. SP: Papirus, vol. I(2); 1996. p. 35–46.

[103] Lipp MEN, Alcino AB, Bignotto MM, Malagris LE. O treino de controle do stress para hipertensos: uma contribuição para a medicina comportamental. Estud Psicol (Campinas) 1998;15(3):59–66.

[104] Lipp MEN. O Tratamento Psicológico do Stress. In: Lipp MEN (Org.), Mecanismos Neuropsicofisiológicos do Stress: Teoria e Aplicações Clínicas. São Paulo: Casa do Psicólogo; 2003. p. 187–92.

[105] Lipp M, Malagris L. Estresse: aspectos históricos, teóricos e clínicos. In Rangé B (Org.), Psicoterapias cognitivo-comportamentais: um diálogo com a psiquiatria (2a ed, Cap. 39). Porto Alegre: Artmed; 2011. p. 617–32.

[106] Malagris LEN, Brunini TMC, Moss MB, Silva PJA, Esposito BR, Ribeiro ACM. Evidências biológicas do treino de controle do stress em pacientes com hipertensão. Psicol Reflex Crit 2009;22(1):1–9.

[107] Malagris LEN, Lipp MEN, Chicayban LM. Hipertensão arterial sistêmica: contribuição de fatores emocionais e possibilidades de atuação da Psicologia. In: Seidl EMF, Miyazaki MCOS (Coords.) ,Psicologia da saúde: pesquisa e atuação profissional no contexto de enfermidades crônicas. Curitiba, Juruá Editora; 2014. p. 73–102.

[108] Oncken C, Gonzales D, Nides M, Rennard S, Watsky E, Billing CB, et al. Efficacy and safety of the novel selective nicotinic acetylcholine receptor partial agonist, varenicline, for smoking cessation. Arch Intern Med 2006;166(15):1571–7.

[109] Nides M, Oncken C, Gonzales D, Rennard S, Watsky EJ, Anziano R, et al. Smoking cessation with varenicline, a selective α4β2 nicotinic receptor partial agonist: results from a 7-week, randomized, placebo- and bupropion-controlled trial with 1-year follow-up. Arch Intern Med 2006;166(15):1561–8.

[110] Monteiro CA, Cavalcante TM, Moura EC, Claro RM, Szwarcwald CL. Population-based evidence of a strong decline in the prevalence of smokers in Brazil (1989–2003). Bull World Health Organ 2007;85(7):527–34.

[111] Reid RD, Mullen KA, Pipe AL. Systematic approaches to smoking cessation in the cardiac setting. Curr Opin Cardiol 2011;26(5):443–8.

[112] Wu P, Wilson K, Dimoulas P, Mills EJ. Effectiveness of smoking cessation therapies: a systematic review and meta-analysis. BMC Public Health 2006;6(1):300.

[113] Sansores RH, Ramírez-Venegas A, Arellano-Rocha R, Noé-Díaz V, García-Gómez L, Pérez Bautista O, et al. Use of varenicline for more than 12 months for smoking cessation in heavy chronic obstructive pulmonary disease smokers unmotivated to quit: a pilot study. Ther Adv Respir Dis 2016;10(5):383–90.

[114] West R, Raw M, McNeill A, Stead L, Aveyard P, Bitton J, et al. Health-care interventions to promote and assist tobacco cessation: a review of efficacy, effectiveness and affordability for use in national guideline development. Addiction 2015;110(9):1388–403.

[115] Tulloch HE, Pipe AL, Els C, Clyde MJ, Reid RD. Flexible, dual-form nicotine replacement therapy or varenicline in comparison with nicotine patch for smoking cessation: a randomized controlled trial. BMC Med 2016;14(1):80.

[116] Shimadu S, Hamajima N, Okada Y, Oguri T, Murohara T, Ban N, et al. Factors influencing sustainable efficacy of smoking cessation treatment with varenicline beyond nine months. Nagoya J Med Sci 2016;78(2):205.

[117] World Health Organization—WHO. Noncommunicable diseases country profiles. Geneva: World Health Organization—WHO; 2011 209.

[118] Campos TS, Galil AGS, Ferreira MA, Banhato EFC, Cupertino AP, Bastos MG. Perfil do Uso de Tabaco Industrializado em uma População de Alto Risco Cardiovascular. Arq Bras Cardiol 2016;103(3 (Suppl 1)):112.

[119] Banhato EFC, Galil AGS, Andrade BABB, Miranda KFF, Ferreira MA, Valente RGC, et al. Elderly profile with multiple chronic conditions treated in the intervention group for smoking cessation. In: Society for Research on Nicotine & Tobacco. 2016 annual meeting abstracts; 2016. p. 134–5 [2015;85(Suppl 2):59].

[120] Andrade BABB, Galil AGS, Cruz MV, Campos TS, Valente RGC, Banhato EFC, et al. Multidisciplinary care unit for smoking users with multiple chronic conditions: tobacco status in a follow-up of 12 months. In: Society for Research on Nicotine & Tobacco. 2016 annual meeting abstracts 2016; 2016. p. 151.

[121] Galil AGS, Cruz MV, Andrade BABB, Miranda KFF, Ferreira MA, Fantini LS, et al. Conociendo el seguimento de fumadores de alto riesgo cardiovascular em uma unidad multidisciplinaria de apoyo a la cesácio del tabaquismo em Brasil. Arch Cardiol Mex 2015;85(2):59.

FURTHER READING

[1] Chiolero A, Faeh D, Paccaud F, Cornuz J. Consequences of smoking for body weight, body fat distribution, and insulin resistance. Am J Clin Nutr 2008;87(4):801–9.

[2] Erhardt L. Cigarette smoking: an undertreated risk factor for cardiovascular disease. Atherosclerosis 2009;205(1):23–32.

Part IV

Social, Population, and Family Effects on the Heart and Arteries

Chapter 20

Lifestyle Interventions in Patients With Serious Mental Illness

Aaron Gluth*, DeJuan White[†,‡], Martha Ward[†,‡]

*Division of Hospital Medicine, Department of Psychiatry and Behavioral Sciences, Emory University, Atlanta, GA, United States, †Department of Psychiatry and Behavioral Sciences, Emory University, Atlanta, GA, United States, ‡Department of General Medicine, Emory University, Atlanta, GA, United States

BACKGROUND

The burden of morbidity and mortality due to cardiovascular disease (CVD) within the general population is well established [1]. Mounting evidence suggests that people with serious mental illness (SMI) suffer from disproportionately high rates of CVD and conditions known to be risk factors for CVD [2]. Studies have also consistently found that patients with SMI have significantly shorter life spans than the general population. A large 16-state study from 2003 found that people with SMI treated in the public sector were dying an average of 25 years earlier than people without SMI [3]. A recent meta-analysis by Walker et al. found a median of 10.1 years of potential life lost and a pooled relative mortality risk of 2.22 (95% CI, 2.12–2.33) in patients with mental illness [4].

The factors contributing to the mortality gap are numerous, complex, and often interconnected. Commonly cited examples of such factors include suicide, violence, accidents, medication side effects, poverty, access to care, health-related behaviors, and general medical comorbidity. It is important to recognize, however, that medical illnesses contribute more to the mortality gap than suicides or accidents [2,5,6]. Cardiovascular disease (CVD) and the associated risk factors are particularly important contributors to the mortality gap, and CVD is the leading cause of death in those with SMI [6–8]. There are elevated rates of obesity, metabolic syndrome, diabetes, and tobacco use in people with SMI. Those with SMI are also more likely to be sedentary and tend to consume less fruit, vegetables, and fiber compared with the general population [2,6,9,10]. Many psychotropic medications, particularly the atypical antipsychotics, are known to cause or exacerbate weight gain, dyslipidemia, and glucose dysregulation [11].

Nonpharmacological interventions (diet, exercise, and behavior modifications) are crucial tools in managing metabolic syndrome, obesity, and related conditions in the general population. Five percent or greater weight loss has been shown to significantly decrease risk of cardiovascular disease. Additionally, improved fitness (regardless of weight change) decreases mortality related to CVD [12]. Decades of accumulated evidence suggests that lifestyle interventions can be effective in preventing CVD and in managing metabolic risk factors in the general population. In fact, many physicians consider lifestyle interventions to be the cornerstone of treatment for CVD risk factors. A smaller but rapidly growing body of research has explored the benefits of lifestyle interventions in the SMI population.

EFFICACY OF LIFESTYLE INTERVENTIONS IN THE SMI POPULATION

Key Reviews—Trials and publications examining lifestyle interventions in the SMI population have generally been small and heterogeneous, so reviews and meta-analyses are indispensable in interpreting the evidence. Disappointingly, reviews suggest that lifestyle interventions are only modestly effective in the SMI population. A 2010 review by Cabassa et al. examined seven randomized controlled trials that demonstrated that nonpharmacological interventions resulted in a mean weight loss of 1.6 kg in subjects with SMI. This is significantly less than the 3.6–5 kg weight loss seen in meta-analyses of lifestyle interventions in the general population [13].

A number of reviews demonstrate study interventions that produce varying amounts of statistically significant weight loss but fail to show the clinically significant (5%–10%) weight loss deemed necessary to reduce cardiovascular risk. A systematic review of 14 studies by Verhaeghe et al. revealed a mean weight loss of only 1.96 kg (95% CI, 0.12 to −3.80 kg). Weight loss was statistically significant in 9 of the 14 reviewed studies, but none reached 5% weight loss. The authors were also unable to find studies that examined cost-effectiveness of the interventions in question [14].

A 2008 systematic review and meta-analysis of 10 RCTs reported a 2.56 kg (95% CI, −3.20–1.92 kg, $P < 0.001$) decrease, which corresponded to a 3.2% mean weight loss. The studies included were limited by short follow-up (18.2 week mean follow-up) and inconsistently reported methods [15].

Bartels and Desilets included a larger number of trials in their review, which showed statistically significant weight loss in 22 of 24 studies, yet the median total body weight loss was only 2.6% [16].

A 2012 systematic review and meta-analysis by Bonfioli et al. included 13 RCTs that studied weight loss interventions in patients with psychotic illnesses. They noted a statistically significant mean of 3.12% weight loss but observed that this falls short of the 5% goal. The authors pointed out that the mean follow-up of 18 weeks was relatively short; they also raised the possibility that "in individuals taking atypical antipsychotics, outcomes associated with metabolic risk factors may have greater health implications than weight changes alone," suggesting that we may be underestimating the potential benefits of interventions in this particular population [17].

Reviews on lifestyle interventions in SMI are not universally underwhelming. Gabriele et al. reviewed 16 studies that addressed behavioral interventions in patients taking antipsychotics and found a bell curve-like relationship between study duration and amount of weight lost. The mean weight lost was 2.63 kg for 12th–16th-week interventions, 4.24 kg for 6th-month interventions, and 3.05 kg for 12th–18th-month interventions. Importantly, the 12th–18th-month intervention category only consisted of two studies. The authors also noted evidence of improved hemoglobin A1c and insulin regulation in the studies examined. This review raises interesting questions about the relationship between intervention duration and effectiveness. It may be helpful to have future studies that test long-term (greater than 12 months) lifestyle modifications [18].

Key Positive Studies—The systematic reviews and meta-analyses of nonpharmacological interventions on weight and cardiovascular risk factors in SMI have largely yielded negative to modest results. This may be due to small sample sizes, short study/follow-up duration, and considerable study heterogeneity. Also, despite growing interest and an expanding body of literature pertaining to lifestyle interventions in SMI, the quantity of publications on the subject is still small compared with that of lifestyle interventions in the general population. Nonetheless, given the modest results generally seen in these reviews and the unclear clinical significance, some authors assert that there is currently insignificant (or very limited) evidence to recommend lifestyle modifications to manage obesity and cardiovascular risk in patients with SMI, particularly given the lack of data on cost-effectiveness [14,19,20]. Despite this somewhat nihilistic conclusion, there are a few randomized controlled trials yielding clinically significant results. It may prove useful to examine these studies closer to try to understand how particular elements of study design contribute to more robust results.

A small ($n = 14$) crossover style randomized controlled trial (RCT) by Jean-Baptiste et al. examined intensive lifestyle interventions versus no intervention in obese (BMI > 30) patients with psychotic disorders who were on antipsychotics. During the active treatment phase, the mean weight difference between the experimental group and the control group was 5.6 kg. It was also noted that weight loss continued beyond the duration of the intervention, to as far as 6 months out, without any booster treatments. The small size of the trial is an obvious limitation. The intensity of the intervention is also noteworthy and calls into question the feasibility of the intervention in relatively resource-poor settings. Subjects received lifestyle counseling, dietary education, physical activity enhancement, pedometers, tailored nutrition support, a grocery store tour, and healthy cooking classes [21].

A 2006 RCT by McKibbin et al. randomized 57 subjects with schizophrenia and diabetes to a diet and exercise intervention or usual care plus information. The intervention was 24 weeks of Diabetes Awareness and Rehabilitation Training (DART), which consisted of teaching sessions pertaining to physical activity and nutrition. The subjects were all 40 years or older and were recruited from day treatment programs, board-and-care facilities, and community clubrooms. The intervention arm showed a mean weight change of −5.4 kg compared with the usual treatment group [22].

Wu et al. included 53 schizophrenic patients in an RCT comparing lifestyle changes with usual treatment in an inpatient setting in Taiwan. All of the subjects were on clozapine and had BMIs > 27 (the threshold for obesity in Taiwan). The intervention consisted of a balanced and calorically restricted diet and a supervised exercise program. The study lasted 6 months and resulted in a mean weight difference of −5.2 kg in the intervention arm compared with the control arm. This translated to a 5.4% reduction in BMI. The intervention arm also enjoyed significantly decreased waist circumference, triglycerides, insulin, and IGFBP-3 levels. These results suggest that intensive supervised interventions may be beneficial for inpatients with significant risk factors for metabolic disease (clozapine use and preexisting obesity) [23].

A 2016 intent-to-treat randomized, controlled, parallel superiority study completed in an outpatient VA setting randomized 122 veterans to a lifestyle balance (LB) intervention or to a usual care intervention. The subjects were adults who had experienced either a 7% gain in body weight or had a BMI > 25 while on a second-generation antipsychotic. The LB group received education about diet and exercise, dietary monitoring, individual counseling regarding lifestyle choices, and recommended (not mandatory) exercise routine. The LB group also took quizzes pertaining to lifestyle changes, and they received modest rewards (e.g., gift certificates) for completing goals. The usual care group received recommendations from clinicians to improve their nutrition and physical activity, and they had access to publically available educational materials about lifestyle modifications. The average weight difference at a year was 5.2 kg, favoring the intervention group. Despite the difference in weight-based outcomes, both groups completed the study with similar knowledge about lifestyle interventions, suggesting that education alone is probably not sufficient to achieve favorable outcomes [24].

OTHER IMPORTANT STUDIES

Although not all studies demonstrated robust clinically significant weight loss, a number of studies yield other notable findings. The In SHAPE RCT randomized 133 overweight or obese patients with SMI to receive either a supervised fitness and nutrition program (the In SHAPE program) or free fitness club membership and education. The intervention program consisted of weekly exercise and education sessions with a personal trainer who had received additional training in nutrition. The trial lasted 12 months and revealed a significant increase in the fitness of the experimental group, as measured by serial 6 min walk tests. The intervention group enjoyed a 97.3 ft increase in their 6 min walk distance, while the control group had a 20 ft decrease in their distance (a 117.3 ft between-group difference). No significant changes were seen in the BMI or mean weight changes between the intervention and control groups, but the authors reference an observation study by Lee et al., which suggests that level of fitness may be a more important marker of all-cause and CVD-related mortality than BMI and weight changes [25,26]. Bartels et al. published a slightly larger pragmatic replication of the In SHAPE study. The replication included a more heterogeneous population spread out across multiple outpatient community clinics. Once again, the control group had an improvement in the 6 min walk test, but they also had a statistically significant improvement in weight and BMI. The authors conclude that at 12 months, about half of the In SHAPE group (51%) achieved clinically significant cardiovascular risk reduction, which they defined as a weight loss of 5% or an increase of >50 m (164 ft) on the 6 min walk test [27].

The STRIDE trial was an intent-to-treat RCT that randomized 200 obese adults on antipsychotic medication to a nonpharmacological lifestyle intervention or to usual care. The experimental condition was based on the PREMIER multi-component lifestyle intervention. The sixth-month intervention included moderate physical activity, calorie restriction, and dietary changes based on the DASH diet [28]. The treatment arm had a 4.4 kg weight decrease compared with the control arm at 6 months, but at 12th-month follow-up, the difference had attenuated to 2.6 kg. There were some remarkable findings noted in this trial, aside from the quantitative weight differences. The authors note that the 4.4 kg weight difference seen in the STRIDE trial was similar to the 4.7 kg weight difference seen in the PREMIER trial [28], which was a study conducted in the general population. It is noteworthy that similar results were achieved by a group that was heavier at baseline, on antipsychotics, and faced with the constellation of challenges that are specific to the SMI population. The intervention group also had a significant improvement in fasting glucose levels and a significant decrease in medical hospitalizations (6.7 vs 18.8%) compared with the control group [29].

The Randomized Trial of Achieving Healthy Lifestyles in Psychiatric Rehabilitation (ACHIEVE) examined 291 overweight or obese patients in outpatient community psychiatric rehabilitation centers. The subjects recruited included patients with schizophrenia, schizoaffective disorder, major depression, and bipolar disorder. The intervention group received group weight management sessions, individual weight management sessions, and group exercise sessions. The sessions were also tailored to be easily digestible for patients with potential deficits in memory and executive function. The control group received standard nutrition and exercise information along with the option to attend quarterly nonweight-related health classes (e.g., cancer screening). Throughout the duration of the study (18 months), the experimental group had a progressive decrease in weight compared with controls. At the 18th-month conclusion, mean weight change was −3.2 kg, and 37.8% of the intervention group achieved 5% or more of their initial weight compared with 22.7% of controls. ACHIEVE is noteworthy due to the diversity of psychiatric diagnoses included and also due to the progressive and sustained nature of weight lost. [30]

KEYS TO SUCCESS

There is significant heterogeneity in the design and the results of the studies examining lifestyle modifications in SMI. While this can present challenges in reviewing and synthesizing the available data, the heterogeneity can also provide useful hints about which factors are associated with favorable outcomes. In their 2012 systematic review, Bartels and Desilets attempted to identify the useful and successful aspects of health promotion programs for SMI populations. They reviewed

30 publications (6 reviews and 24 trials) pertinent to nonpharmacological weight management in patients with SMI and found that interventions with longer duration (3 months or more) were more likely to be successful. Programs that addressed both nutrition and physical activity were more likely to produce positive results. Treatments that were structured, manualized, and monitored were also more effective [16]. As previously discussed, interventions that utilize active engagement, supervision, and incentives are probably more effective than education alone [24]. In the general population, multiple studies suggest that tailoring and personalization of lifestyle interventions yield greater results compared with generic interventions; the same principle may apply to the SMI population [31].

UNIQUE BARRIERS

Addressing lifestyle choices to modify cardiovascular risk is challenging in any demographic, but those with SMI must contend with a unique set of barriers above and beyond those of the general population. Ward et al. suggest that in order to maximize effectiveness, interventions should be tailored to address the barriers commonly encountered by the SMI population such as poverty, medication side effects, and psychiatric symptoms. [32]

SMI is associated with disproportionate rates of poverty [33]. People of lower socioeconomic status (SES) may have difficulty affording athletic clothing, appropriate footwear, gym memberships, and exercise equipment. Low-SES communities may be less conducive to safe outdoor activities and exercise. Utilizing the resources available in the immediate environment can help to circumvent socioeconomic challenges. Walking on level ground and climbing up- and downstairs can be meaningful components of a weight management program [20]. Providing instructional exercise handouts, books, and videos could be a relatively cost-effective intervention (although, as noted above, probably not sufficient as a lone intervention). Providing free or subsidized gym memberships is another way to help mitigate socioeconomic disparities [34]. Other free or public resources such as schools, senior centers, and parks can also be potential sites for fitness-related activities [32]. Exercise sessions and equipment made available within the clinic may help to overcome perceived cost-related and time-related barriers [35,36]. Some of our psychotherapist colleagues conduct "walk and talk" sessions, which combine therapy encounters with brisk walks. Wu et al. instituted a successful supervised exercise program within a Taiwanese long-term inpatient psychiatric unit [23], though these results may not be generalizable to countries and locales where long-term hospitalization is less common. Minority ethnic groups, people of low SES, and people with SMI often perceive that their access to health-care services is poor, largely due to transportation issues. Troubleshooting transportation issues with shuttles, public transportation vouchers, and bus or train fare could improve patient participation [36–38].

Low SES is also associated with unhealthy eating habits. Persons receiving food stamps have been shown to consume fewer whole grains, more sugar-containing drinks, more potatoes, and more red meat compared with the rest of the population [39]. Those living in poverty often suffer from limited access to fresh fruits and vegetables, in part due to the presence of food desserts [40]. Providing patients with healthier groceries and meals is a simple intervention in the setting of well-funded and short-lived trials but may not be fiscally and logistically sustainable on a large scale. Other potential solutions include escorting patients to restaurants and grocery stores in order to educate them about healthier choices, helping patients to identify food banks with fruit and vegetable options, and providing instruction on nutritional cooking techniques. [13,35] An increasing number of farmer's markets accept food stamps and could be valuable sources of fresh food for patients. Some studies yielded favorable results by focusing on total caloric intake or portion control, as opposed to complex diets and detailed nutritional education [23,35].

Psychiatric symptoms present a common barrier to effective health-care delivery. Amotivation is common in psychotic and depressive disorders and can contribute to poor adherence with health promotion programs and a lack of engagement with lifestyle interventions. Interventions done in group settings may improve engagement, bolster social connections, and reduce attrition and can make the program more enjoyable [36]. Incentivizing participants with modest rewards (e.g., gift certificates and token systems) can improve outcomes [24,34]. Provider/facilitator factors can also have a positive effect on patient motivation, and staff can effectively model healthy behaviors alongside patients in order to improve adherence and solidify therapeutic alliance [41].

Positive symptoms of psychosis, including paranoia and hallucinations, can pose barriers to adherence and to therapeutic alliance [32,42]. This underscores the importance of optimized management of the underlying psychiatric illness. Incorporating nonpharmacological lifestyle interventions into visits with established mental health providers may promote engagement and therapeutic alliance [43,44].

Patients with SMI frequently suffer from cognitive impairment. Neuropsychiatric studies in the SMI population reveal deficits in memory, attention, processing speed, and executive functions [45]. Some studies included interventions that were structured deliberately to mitigate cognitive deficits. Authors provided simplified handout materials, they printed materials in large font sizes, and they gave frequent quizzes to reinforce lessons. They also took advantage of mnemonic aides and visual learning materials [13]. The ACHIEVE group tailored their intervention specifically to address cognitive issues by dividing information into smaller sections and by utilizing repetition to solidify important concepts [30].

Additionally, certain side effects of psychotropic medications can present substantial barriers to lifestyle interventions. Psychotropic medications (particularly antipsychotics and mood stabilizers) can be sedating. Sedation can reduce motivation to exercise and can impair attention during health education sessions. In order to address sedation, exercise and education sessions can be timed in the early morning before medications are administered or later in the day when the side effects have had time to wear off. Dry mouth is a common side effect, which can sometimes prompt patients to increase consumption of sugary drinks. Providers can encourage patients to use sugar-free hard candies or water to replace sugar-containing beverages. Extrapyramidal side effects (EPS) such as tremor, bradykinesia, and rigidity can make exercise challenging and uncomfortable. EPS can often be managed effectively by adding anticholinergic medications, adjusting the dose of the offending medication, or switching medications. Finally, weight gain is a common side effect, particularly with use of second-generation antipsychotics (SGAs). This underscores the importance of utilizing lifestyle interventions when these medications are prescribed [32]. There is also evidence to support switching to a less metabolically active medication in order to manage SGA-induced weight gain [46].

FUTURE DIRECTIONS

The body of data on lifestyle interventions in serious mental illness (SMI) reveals the difficulty of achieving substantial weight loss in this population. Improving metabolic outcomes in this complex population may require novel approaches and techniques. A number of studies have recruited subjects with chronic mental illness, established obesity, and long-standing use of antipsychotics. Relatively, fewer studies have looked at primary prevention measures in patients with recently diagnosed SMI (before weight gain begins), but the available evidence suggests that such early interventions may attenuate weight gain and improve dietary measures in adolescents and young adults with first-episode psychosis [47,48]. Technology has changed the landscape of health and medicine tremendously. Mobile health technologies have become increasingly popular within the general population. Small pilot studies suggest that mobile health technologies such as smartphone applications and wearable activity trackers (e.g., Fitbits) could be useful additions to lifestyle intervention programs within the SMI population [49,50].

REFERENCES

[1] Mozaffarian D, Benjamin EJ, Go AS, et al. Heart disease and stroke statistics-2016 update a report from the American Heart Association. Circulation 2016;133(4):e38–48.

[2] De Hert M, Correll CU, Bobes J, et al. Physical illness in patients with severe mental disorders. I. Prevalence, impact of medications and disparities in health care. World Psychiatry 2011;10(1):52–77.

[3] Parks J, Svendsen D, Singer P, Foti ME. Morbidity and mortality in people with serious mental illness. Alexandria, VA: Council NAoSMHPDMD; 2006.

[4] Walker ER, McGee RE, Druss BG. Mortality in mental disorders and global disease burden implications. JAMA Psychiatry 2015;72(4):334–41.

[5] Druss BG, Zhao L, Von Esenwein S, Morrato EH, Marcus SC. Understanding excess mortality in persons with mental illness: 17-year follow up of a nationally representative US survey. Med Care 2011;49(6):599–604.

[6] McCreadie RG. Diet, smoking and cardiovascular risk in people with schizophrenia: descriptive study. Br J Psychiatry 2003;183:534–9.

[7] Compton MT, Daumit GL, Druss BG. Cigarette smoking and overweight/obesity among individuals with serious mental illnesses: a preventive perspective. Harv Rev Psychiatry 2006;14(4):212–22.

[8] Osborn DPJ, Nazareth I, King MB. Physical activity, dietary habits and Coronary Heart Disease risk factor knowledge amongst people with severe mental illness. Soc Psychiatry Psychiatr Epidemiol 2007;42(10):787–93.

[9] Brown S, Birtwistle J, Roe L, Thompson C. The unhealthy lifestyle of people with schizophrenia. Psychol Med 1999;29(3):697–701.

[10] Strassnig M, Brar JS, Ganguli R. Nutritional assessment of patients with schizophrenia : a preliminary study. Schizophr Bull 2003;29(2):393–7.

[11] De Hert M, Detraux J, van Winkel R, et al. Metabolic and cardiovascular adverse effects associated with treatment with antipsychotic drugs. Nat Rev Endocrinol 2012;8:114–26.

[12] Klein S, Burke LE, Bray GA, et al. Clinical implications of obesity with specific focus on cardiovascular disease: a statement for professionals from the American Heart Association Council on Nutrition, Physical Activity, and Metabolism. Circulation 2004;110(18):2952–67.

[13] Cabassa LJ, Ezell JM, Lewis-Fernandez R. Lifestyle interventions for adults with serious mental illness: a systematic literature review. Psychiatr Serv 2010;61(8):774–82.

[14] Verhaeghe N, De Maeseneer J, Maes L, et al. Effectiveness and cost-effectiveness of lifestyle interventions on physical activity and eating habits in persons with severe mental disorders: a systematic review. Int J Behav Nutr Phys Act 2011;8:28.

[15] Álvarez-Jiménez M, Hetrick SE, González-Blanch C, Gleeson JF, McGorry PD. Non-pharmacological management of antipsychotic-induced weight gain: systematic review and meta-analysis of randomised controlled trials. Br J Psychiatry 2008;193(2):101–7.

[16] Bartels S, Desilets R. Health promotion programs for persons with serious mental illness: what works? Washington, DC: Solutions S-HCfIH; 2012 [Prepared by the Dartmouth Health Promotion Research Team].

[17] Bonfioli E, Berti L, Goss C, Muraro F, Burti L. Health promotion lifestyle interventions for weight management in psychosis: a systematic review and meta-analysis of randomised controlled trials. BMC Psychiatry 2012;12(1):78.

[18] Gabriele JM, Dubbert PM, Reeves RR. Efficacy of behavioural interventions in managing atypical antipsychotic weight gain. Obes Rev 2009;10(4):442–55.

[19] Faulkner G, Cohn T, Remington G. Interventions to reduce weight gain in schizophrenia. Cochrane Database Syst Rev 2007;33(1):654–6.

[20] Lowe T, Lubos E. Effectiveness of weight management interventions for people with serious mental illness who receive treatment with atypical antipsychotic medications. A literature review. J Psychiatr Ment Health Nurs 2008;15(10):857–63.

[21] Jean-Baptiste M, Tek C, Liskov E, et al. A pilot study of a weight management program with food provision in schizophrenia. Schizophr Res 2007;96(1-3):198–205.

[22] McKibbin CL, Patterson TL, Norman G, et al. A lifestyle intervention for older schizophrenia patients with diabetes mellitus: a randomized controlled trial. Schizophr Res 2006;86(1–3):36–44.

[23] Wu MK, Wang CK, Bai YM, Huang CY, Lee SD. Outcomes of obese, clozapine-treated inpatients with schizophrenia placed on a six-month diet and physical activity program. Psychiatr Serv 2007;58(4):544–50.

[24] Erickson ZD, Mena SJ, Pierre JM, et al. Behavioral interventions for antipsychotic medication associated obesity: a randomized, controlled clinical trial. J Clin Psychiatry 2016;77(2):e183–9.

[25] Bartels SJ, Pratt SI, Aschbrenner KA, et al. Clinically significant improved fitness and weight loss among overweight persons with serious mental illness. Psychiatr Serv 2013;64(8):729–36.

[26] Lee D, Sui X, Artero EG, et al. Long-term effects of changes in cardiorespiratory fitness and body mass index on all-cause and cardiovascular disease mortality in men: the Aerobics Center Longitudinal Study. Circulation 2011;124(23):2483–90.

[27] Bartels SJ, Pratt SI, Aschbrenner KA, et al. Pragmatic replication trial of health promotion coaching for obesity in serious mental illness and maintenance of outcomes. Am J Psychiatry 2015;172(4):344–52.

[28] Funk KL, Elmer PJ, Stevens VJ, et al. PREMIER—a trial of lifestyle interventions for blood pressure control: intervention design and rationale. Health Promot Pract 2008;9(3):271–80.

[29] Green CA, Yarborough BJH, Leo MC, et al. The STRIDE weight loss and lifestyle intervention for individuals taking antipsychotic medications: a randomized trial. Am J Psychiatry 2015;172(1):71–81.

[30] Daumit GL, Dickerson FB, Wang N-Y, et al. A behavioral weight-loss intervention in persons with serious mental illness. N Engl J Med 2013;368(17):1594–602.

[31] Eyles HC, Mhurchu CN. Does tailoring make a difference? A systematic review of the long-term effectiveness of tailored nutrition education for adults. Nutr Rev 2009;67(8):464–80.

[32] Ward MC, White DT, Druss BG. A meta-review of lifestyle interventions for cardiovascular risk factors in the general medical population. J Clin Psychiatry 2015;e477–86.

[33] Saraceno B, Levav I, Kohn R. The public mental health significance of research on socio-economic factors in schizophrenia and major depression. World Psychiatry 2005;4(3):181–5.

[34] Loh C, Meyer JM, Leckband SG. A comprehensive review of behavioral interventions for weight management in schizophrenia. Ann Clin Psychiatry 2006;18(1):23–31.

[35] Bradshaw T, Lovell K, Harris N. Healthy living interventions and schizophrenia: a systematic review. J Adv Nurs 2005;49(6):634–54.

[36] Roberts SH, Bailey JE. Incentives and barriers to lifestyle interventions for people with severe mental illness: a narrative synthesis of quantitative, qualitative and mixed methods studies. J Adv Nurs 2011;67(4):690–708.

[37] Johnson M, Everson-Hock E, Jones R, Woods HB, Payne N, Goyder E. What are the barriers to primary prevention of type 2 diabetes in black and minority ethnic groups in the UK? A qualitative evidence synthesis. Diabetes Res Clin Pract 2011;93(2):150–8.

[38] Bauer MS, Williford WO, McBride L, McBride K, Shea NM. Perceived barriers to health care access in a treated population. Int J Psychiatry Med 2005;35(1):13–26.

[39] Leung CW, Ding EL, Catalano PJ, Villamor E, Rimm EB, Willett WC. Dietary intake and dietary quality of low-income adults in the Supplemental Nutrition Assistance Program. Am J Clin Nutr 2012;96(4):977–88.

[40] Baker EA, Schootman M, Barnidge E, Kelly C. The role of race and poverty in access to foods that enable individuals to adhere to dietary guidelines. Prev Chronic Dis 2006;3(3):A76.

[41] Roberts SH, Bailey JE. An ethnographic study of the incentives and barriers to lifestyle interventions for people with severe mental illness. J Adv Nurs 2013;69(11):2514–24.

[42] Kilbourne AM, McCarthy JF, Post EP, et al. Access to and satisfaction with care comparing patients with and without serious mental illness. Int J Psychiatry Med 2006;36(4):383–99.

[43] Faulkner G, Soundy A, Lloyd K. Schizophrenia and weight management: a systematic review of interventions to control weight. Acta Psychiatr Scand 2003;324–32.

[44] Kemp V, Bates A, Isaac M. Behavioural interventions to reduce the risk of physical illness in persons living with mental illness. Curr Opin Psychiatry 2009;22(2):194–9.

[45] Reichenberg A, Harvey PD, Bowie CR, et al. Neuropsychological function and dysfunction in schizophrenia and psychotic affective disorders. Schizophr Bull 2009;35(5):1022–9.

[46] Hasnain M, Vieweg WVR. Weight considerations in psychotropic drug prescribing and switching. Postgrad Med 2013;125(5):117–29.

[47] Teasdale SB, Ward PB, Rosenbaum S, et al. A nutrition intervention is effective in improving dietary components linked to cardiometabolic risk in youth with first-episode psychosis. Br J Nutr 2016;115(11):1–7.

[48] Curtis J, Watkins A, Rosenbaum S, et al. Evaluating an individualized lifestyle and life skills intervention to prevent antipsychotic-induced weight gain in first-episode psychosis. Early Interv Psychiatry 2016;10(3):267–76.

[49] Aschbrenner KA, Naslund JA, Shevenell M, Mueser KT, Bartels SJ. Feasibility of behavioral weight loss treatment enhanced with peer support and mobile health technology for individuals with serious mental illness. Psychiatry Q 2015;87(3):1–10.

[50] Naslund JA, Aschbrenner KA, Bartels SJ. Wearable devices and smartphones for activity tracking among people with serious mental illness. Ment Health and Phys Act 2016;10:10–7.

Chapter 21

Chocolate and Its Component's Effect on Cardiovascular Disease

Katha Patel, Ronald Ross Watson

University of Arizona, Mel and Enid Zuckerman College of Public Health, Tucson, AZ, United States

INTRODUCTION

Today, cardiovascular disease (CVD) is an immense public health issue that largely impacts patients, health systems, and society in general. This great pandemic victimizes individuals in both industrialized and developing countries [1]. According to the World Health Organization (WHO), cardiovascular disease remains the primary cause of global death; yearly, more individuals die from cardiovascular disease than from any other disease [2]. The Center for Disease Control and Prevention (CDC) reports that in the United States alone, roughly 610,000 people die each year from some form of cardiovascular disease; that accounts for one in every four deaths [3].

Cardiovascular disease, which is often interchanged with heart disease, can be generally characterized as a group of disorders relating to the human heart and blood vessels [2]. Some major diseases under this umbrella include cerebrovascular disease, peripheral arterial disease, coronary heart disease, rheumatic heart disease, congenital heart disease, and deep vein thrombosis [2]. Along with these chronic diseases, cardiac health problems can also occur in acute events such as heart attacks and strokes [2,3]. These episodes are primarily triggered by intense blockages of vessels that hinder blood flow [2]. Often, these acute events manifest through symptoms such as chest pain, shortness of breath, nausea, cold sweats, and upper body pain [3].

The development of cardiovascular disease can be significantly preventable [2]. This is as many, but not all, of the risk factors for cardiovascular disease are influenced by lifestyle choices [3]. Lifestyle factors such as nutrition, physical activity, alcohol use, tobacco use, and drug use portray strong correlations to the risk of developing cardiovascular disease [2,3]. Diverging from these lifestyle factors, certain health conditions such as high blood pressure, high blood lipid levels, and high blood glucose levels are also noteworthy risk factors [2,4,5]. With this, endothelial dysfunction, modified lipoproteins and lipids, activated platelet aggregation, and inflammatory response can also play a large role in progressing the development of thrombosis, atherosclerosis, and other cardiovascular diseases [6–11].

In efforts to reduce this development of cardiovascular disease, there has been a large interest in finding foods with properties that can either decrease the risk factors of CVD or increase the manageability of any present risk factors for CVD [4]. Currently, studies support that plant-derived foods containing flavonoids possess cardioprotective properties to a certain extent; some of these plant sources of flavonoids include tea, olives, coffee, red wine, grape juice, berries, and cocoa [4,9]. More specifically, one concentration of interest focuses on chocolate, cocoa, and their many components.

Based on observational and epidemiological conclusions, chocolate consumption is correlated with reducing the risk of developing cardiovascular disease and hypertension significantly. Supporting this, recent studies and experiments convey that chocolate and cocoa have a positive impact on heart health due to their antiinflammatory, antioxidant, and antihypertensive qualities. Cocoa and chocolate contribute to the vascular environment, impacting endothelial function, platelet aggregation, and the bioavailability and activation of nitric oxide [12]. Compiling many of these findings, the goal of this review is to summarize the current epidemiological studies and human intervention trials on the role of chocolate and its components on cardiovascular disease.

BACKGROUND

History of Chocolate in Heart Health

Despite its primary consumption for pleasure, cocoa and chocolate have also been consumed and used for a wide range of medicinal purposes for many years; historically, many of these medicinal uses of chocolate relate to cardiovascular health.

Lifestyle in Heart Health and Disease. https://doi.org/10.1016/B978-0-12-811279-3.00021-5

For example, the 17th-century Aztecs, who revered cocoa beans in their culture, believed in a strong correlation between the heart-shaped cocoa bean and the human heart [13]. Often, the word chocolate was synonymous with the words yollotzal and eztili, which, respectively, translate to heart and blood in the Aztec culture [13]. This relationship between chocolate and the heart leads the Aztecs to make bitter cocoa infused beverages to treat various heart disorders [9,13]. The Old World explorers that traveled to the New World also discovered cocoa beans to have a positive effect on cardiovascular health; William Hughes, a new-world physician, stated that chocolate and cocoa tended to reduce inflammation and open and smoothen the arteries [13]. The medicinal use of chocolate soon spread from the New World to the Old World due to the introduction of chocolate in Spain in 1528 by Hernan Cortes [14,15]. As chocolate spread throughout Europe, many physicians began prescribing it as a cure for various ailments, including diseases related to the heart and vessels [6]. In this time, chocolate was a sign of wealth and class, limiting its availability to the average citizen [13]. However, the introduction of mechanical mills in the 1800s allowed for an efficient extraction of cocoa butter, making chocolate more readily available to the average consumer [13,15]. As a result of this increased availability, sweet chocolate-flavored products and medicines such as headache chocolate for heart disease patients were commonly produced in America and Europe; this was especially seen during the 20th century [13].

History portrays the commonality of chocolate's use for medicinal purposes related to heart health. In line with this, despite certain gaps in the medicinal use of chocolate, today, chocolate and its various components are studied extensively to analyze chocolate's physiological impact on cardiovascular health.

Origins

Cacao beans originate from the seedpods found on the tropical tree, *Theobroma cacao* L., which is native to South and Central America; the tree translates to "food of the gods" [9,13,14]. Grown and harvested for their seedpods, this tree's cacao beans are used to produce chocolate and other cocoa products [13,16] .

There are three distinctive groups of cacao trees: forastero, trinitario, and criollo [13,16]. Each of these groups of cacao is unique. For example, accounting for most chocolate, the forastero tree's cacao beans are most commonly used in chocolate production [17]. They are highly efficient for bulk production of chocolate [18]. Unlike the forastero tree, the criollo's highly aromatic cacao beans are significantly rare in the chocolate market [17]. The chocolate produced from criollo cacao beans has a much milder and weaker flavor compared with the forastero's cacao beans; however, this bean is known for its exceptional quality [16,18]. Lastly, the trinitario tree is a hybrid between the other two categories of cacao [13,16]. It originated in the upper amazon region and has characteristics of both the other cacao beans [13]. This bean is often found in high-quality dark chocolate [13].

Native to many South American and Central American countries, these cacao trees primarily grow in tropical evergreen forests; most often, these trees thrive about 10° latitude north and south of the equator, where the climate is appropriate for their growth [14]. The climate in these tropical areas is characterized by warm temperatures (21–23°C), high humidity (70%–100%), and high annual rainfall (1500–2000 mm) [15]. The trees thrive under the shade from other taller trees, as direct sunlight and high winds hinder their survival [16,17].

The cacao tree grows cacao seedpods that contain the cacao beans used to produce chocolate products. Grown on plantations, the cacao tree is grown to about 4–8 m; typically, in nature, the maximum height of a cacao tree is about 7.5 m [13,16]. The cacao beans inside the seedpods often vary in size and mass due to the moisture content of where the tree is located [19]. The cacao beans typically have a length of between 20.0 and 26.00 mm and a width of between 10.0 and 14.00 mm that increases with moisture content [19]. In the same respect, the mass of the beans vary between 1.11 and 1.31 g [19]. Roughly, 20–50 cacao beans are present in one seedpod [13,19]. These cacao beans are harvested twice a year and used to produce chocolate products. The largest producers and growers of cacao beans are found in West Africa, Brazil, Ecuador, Malaysia, and Indonesia [15,16].

Chocolate Production

The processing and production of chocolate requires several steps. The general overview of processing cocoa bean is composed of opening the pods, removing any pulp, fermentation, drying, alkalization, roasting, and milling [20]. Most of these steps are standardized for mass production.

To begin this process, cacao beans and the surrounding creamish pulp are first removed from the seedpods during harvest [13,21]. The cacao beans are then dried and fermented for 2–8 days; this fermentation changes the cacao beans' name to cocoa beans [13,17]. The dried roasted cocoa beans are then used to make cocoa liquor, which contains nonfat cocoa solids and cocoa butter [13,17,20]. The amount of cocoa liquor present in chocolate is known as the percent of cacao; this is often found on chocolate packages [11].

This cocoa liquor is used in various combinations with other ingredients to create different chocolate products. For example, removing some cocoa butter from the cocoa liquor results in cocoa powder [21,22]. Combining cocoa liquor, cocoa butter, and sugar creates solid chocolate [14]. Among the various types of solid chocolate, milk chocolate is made when powdered or dried milk is added to cocoa liquor, cocoa butter, and sugar [11,14,21]. Furthermore, the potency and the darkness of the chocolate are a result of its quantity of cocoa liquor [21]. Milk chocolate contains 10%–12% cocoa liquor, while semisweet, bittersweet, or dark chocolate contains 35% or more cocoa liquor [11].

The process in which chocolate is produced, grown, and harvested is imperative as it influences chocolate's chemical composition. For example, altering the amount of chocolate liquor in a chocolate product can change the product's chemical composition. This is seen as chocolate liquor contains important health bioactive compounds such as lipids, vitamins, minerals, fiber, and polyphenols [11]. The liquor is especially rich in flavonoids. Because milk chocolate tends to have less chocolate liquor compared with dark chocolate, it has a smaller concentration of these flavonoids [6]. With this, the addition of alkaline and the drying process during chocolate production also decrease the flavonoid concentration in chocolate; sometimes, this decrease occurs by 60% [6,9,22]. The differences in chocolate's chemical composition may influence how the consumption of chocolate affects the body physiologically, especially as flavonoids play an important role in cocoa's cardioprotective effects.

As a sidenote, it is important to distinguish between manufactured chocolate and natural cacao; chocolate contains other components such as milk, sugar, and cocoa [11,14]. Because of this discrepancy, the results regarding cacao in certain studies may not be as applicable to chocolate.

CHOCOLATE AND CARDIOVASCULAR DISEASE: AN EPIDEMIOLOGICAL FOCUS

Currently, there are several noteworthy epidemiological cohort studies that provide significant insight into the budding relationship between chocolate, cocoa, and heart health.

One notable observational study examined the Kuna Indians that live on San Blas islands near Panama [23]. These Kuna Indians consume great amounts of cocoa in their daily diet; many consume about 4 cups or 30–40 oz of a cocoa-based drink daily [23,24]. Because the Kuna Indians do not roast their cacao beans, the beans contain 2000 mg of flavanols/100 g of cocoa; comparatively, processed cocoa contains 150 mg of flavanols/100 g of cocoa [24]. Interestingly, these Kuna Indians are unlikely to experience the development of arterial hypertension [25]. They also tend to be protected against age-dependent increase in blood pressure [25]. Clinical studies examining this health phenomena support that the Kuna Indians have lower blood pressure values; the blood pressure was as low in participants aged 60 compared with the participants aged 20–30 [25]. This population also has lower rates of stroke, myocardial infarction, and cancer compared with mainland Panamanians [25]. With this, the Kuna Indians have a lower incidence of cardiovascular mortality (9.21 ± 3.1 age-adjusted deaths per 100,000) compared with the mainland Pan-American citizens (83.4 ± 0.7 age-adjusted deaths per 100,000) [26]. It is observed that the Kuna Indians on the islands consume 10 times as much cocoa compared with Kuna Indians living on mainland [27]. As a result, it is hypothesized that the cause of this low incidence of cardiovascular disease is due to the Kuna Indians' high intake of cocoa beverage. Due to this, the epidemiological study of the Kuna Indians not only provides insight into the potential cardioprotective effects of cocoa but also inspires and sparks further study on the topic.

Published in 2006, the first definitive cohort study was called the Dutch Zutphen Elderly study. This study examined a group of 470 elderly Dutch men with no baseline chronic diseases. Individuals in the highest-quartile cocoa-intake group consumed 2.30 g of cocoa daily, while the lowest-quartile group consumed less than 0.36 g of cocoa daily. The study concluded that cocoa consumption is correlated to a reduction in cardiovascular disease and blood pressure. This is evident as those in the highest quartile of cocoa intake compared with those in the lowest quartile of cocoa intake had 50% lower rates for cardiovascular disease ($P = .004$). The highest-quartile cocoa-intake participants also reported a 3.7 mm Hg systolic and a 2.1 mm Hg diastolic decrease in blood pressure compared with the lowest-quartile cocoa-intake participants. After adjusting for various factors, the study concluded that cocoa intake is inversely associated with blood pressure; this is almost statistically significant for systolic blood pressure ($P = .06$) and is statistically significant for diastolic blood pressure ($P = .03$). Overall, this study highlighted the protective effects of cocoa, showing that cocoa intake has an inverse relationship with blood pressure, cardiovascular disease, and all-cause mortality [28].

In one Swedish study, chocolate consumption is connected with a decreased risk of ischemic heart disease; both a prospective study of the Swedish adults and a meta-analysis of prospective data were conducted. The study involved 67, 640 women and men from the Cohort of Swedish Men and the Swedish Mammography Cohort, who did not have cardiovascular disease at the baseline measurement. In the follow-up, about 4417 cases of myocardial ischemia occurred. The results found that the multivariable relative risk for developing cardiovascular disease was 0.87 (95% CI, .77–.98) ($P = .04$) for the participants that consumed three to four servings of chocolate per week. The meta-analysis found that the overall relative risk for the highest compared with the lowest chocolate intake was .90 (95% CI .82–.97). In the six studies analyzed in the meta-analysis, a total of 6851 cases of ischemic heart disease occurred. As a result of these data, the study concludes that chocolate consumption is associated with a reduced risk of ischemic heart disease [29].

Many other epidemiological studies reveal that chocolate consumption is associated with reduced rates not only for cardiovascular disease and acute events but also for cardiovascular and all-cause mortality in humans. For example, the Iowa Women's Health study, a prospective study on 34,489 postmenopausal women over a 16-year period, found that the consumption of foods rich in flavonoids such as cocoa reduces cardiovascular mortality ($P = .06$) [12,30]. Similar to this, a study using 19,357 German participants aged 35–65 found that increased chocolate consumption had smaller risk of myocardial infarction or stroke; reduction in blood pressure in part influenced this [31]. In an NHLBI family heart study involving 2117 participants, a cross-sectional analysis of data reveals that there is a converse relationship between chocolate consumption and the development of hardened atherosclerosis plaque in the coronary arteries [32]. Individuals consuming chocolate two to three times a week were about a third less likely to have plaque buildup in their coronary arteries compared with those who did not consume chocolate [32]. Meta-analysis of many experimental studies shows that high levels of chocolate consumption are associated with a reduction in a third of the risk of developing cardiometabolic disorders [33]. More specifically, chocolate consumption correlates with a 37% reduction in cardiovascular disease and 29% reduction in stroke [12,33].

This positive relationship between chocolate and cardiovascular disease is largely evident among several major epidemiological studies. It is significant to note that because these studies are observational in nature, cause and effect conclusions cannot be directly supported; however, these studies do provide significant insight into the superficial cardioprotective properties of chocolate [4]. Although chocolate may provide benefits toward heart health, other aspects of chocolate consumption must also be considered. For example, the high energy density and sugar and fat content of chocolate may result in weight gain, which is a significant risk factor for cardiometabolic disease [8]. As a result, the other factors of chocolate cannot be ignored, even if it does seem to provide large heart health benefits.

CHOCOLATE CONSUMPTION AND BLOOD PRESSURE

High blood pressure is a significant risk factor for the development of cardiovascular disease; hypertension affects 20%–30% of the world's adult population [34]. As a result, uncovering that flavanol-rich cocoa consumption can reduce blood pressure about 3–5 mm Hg is a large public health implication [4].The first evidence for chocolate's ability to lower blood pressure came from the Zutphen Elderly Study. This cohort described previously conveys that cocoa intake is inversely related to blood pressure; the systolic blood pressure was 3.8 mm Hg lower in the highest-quartile chocolate consumer compared with the lowest-quartile chocolate consumers [27]. Many other studies, small and large, have been conducted after this first one.

Some smaller independent studies also denote that the consumption of cocoa-based chocolate reduces blood pressure. One study involving 13 elderly subjects reports reductions in both systolic and diastolic blood pressures; systolic blood pressure decreased by 5.1 mm Hg, and diastolic blood pressure decreased by 1.8 mm Hg [8,35]. Another study displayed that systolic blood pressure decreases by 12 mm Hg after the consumption of dark chocolate, but not after consuming white chocolate [8,36]. This study of 20 subjects revealed reduction in diastolic daytime and nighttime blood pressures after consuming 100 g of flavonoid-rich (88 mg of flavanols) chocolate compared with 0 mg flavanol-containing white chocolate for 2 weeks [36]. The dark chocolate also improved serum LDL levels [36].

Published in 2007, the results of a large meta-analysis of randomized controlled studies (173 subjects) conveyed that cocoa ingestion can significantly reduce blood pressure [37]. In the metaanalysis, the systolic and diastolic blood pressures were reduced by 4.7 and 2.8 mm Hg, respectively [37]. This is a very significant finding as blood pressure is reduced in the same scope of antihypertensive drugs [8,37]. Another study found that the consumption of cocoa containing flavanols resulted in decreases in systolic blood pressure (SBP) and diastolic blood pressure (DBP); after 4 weeks of consuming chocolate products, systolic blood pressure and diastolic blood pressure decreased by 8.2%. ($P < .001$) [4]. This study is especially significant as this population was not hypertensive [4]. Additionally, a different meta-analysis that consisted of 13 trial arms found that cocoa-chocolate consumption has a significant reducing effect on blood pressure. The mean blood pressure change was SBP -3.2 ± 1.9 mm Hg ($P = .001$) and DBP -2.0 ± 1.3 mm Hg (.003) [38]. The meta-analysis further displayed that the blood pressure change was noteworthy for only the hypertensive or prehypertensive groups, while the normotensive groups did not see significant changes in blood pressure. In this meta-analysis, the flavanol doses of the chocolate and cocoa ranged from 30 to 1000 mg. The study also suggests that dark chocolate is more effective in causing blood pressure changes [38].

Another 18-week study showed that the ingestion of 6.3 g of dark chocolate containing 30 mg of polyphenols resulted in decreases in systolic and diastolic blood pressures; the decreases were 2.9 and 1.9 mm Hg, respectively. These values are compared with the consumption of white chocolate, which lacks polyphenols. The study also found that the dark-chocolate-consuming group had a higher NO concentration, which was determined by measuring S-nitrosoglutathione [39].

This relationship between chocolate and blood pressure is beneficial and provides large implications to public health. The exact mechanism for this blood pressure decrease has not been determined; however, the plant-derived flavanols and their role in lowering blood pressure are being researched [8].

Despite the many studies supporting cocoa's blood-pressure-reducing effect, it is important to realize that some studies claim that chocolate has no effect on blood pressure. This can possibly be due to the small nature of a particular study or the use of participants with regular healthy blood pressures. These factors can cause any observations to be hard to detect [8,34].

CHEMICAL COMPOSITION OF CHOCOLATE'S COCOA BEANS

To understand the effect of chocolate and cocoa consumption on heart health, it is first vital to understand the significant structures and components of cocoa and chocolate. Cocoa contains various components, most notably polyphenols, minerals, and lipids [8,11]. Fiber and theobromine are also significant components; 200 mg of theobromine, which is physiologically similar to caffeine, is found in one cup of cocoa [14]. Flavonoids are an important subclass polyphenols found in cocoa as they contribute heavily to its cardioprotective properties [4].

Lipids

The primary source of lipids in chocolate is cocoa butter; this accounts for about 50%–57% of the dry weight of cocoa beans [6]. Cocoa butter consists of a blend of monounsaturated and saturated fatty acids [11]. The monounsaturated fats include oleic acid (18:1, 35%), while the saturated fats include palmitic acid (16:0, 25%) and stearic acid (18:0, 35%) [6]. Research has shown that saturated fatty acids and trans-fatty acids are related to the increased risk of coronary heart disease [40]. Historically, saturated fatty acids elevate low-density lipoprotein (LDL) and cholesterol, increasing the risk of developing coronary heart disease [6]. Supporting this, palmitic fatty acid is commonly associated with increased LDL. Unlike most saturated fats, stearic acid is peculiar in that it doesn't lower high-density lipoprotein (HDL) or increase low-density lipoprotein (LDL) and total cholesterol [11]. It is supported to have a neutral cholesterolemic effect in humans; this is evident in chocolate consumption as a high proportion of chocolate's lipids are stearic acids [6,11]. This may be due to chain length, metabolism kinetics, or inefficient absorption [6,41]. As a result, it is currently unclear what the effects of stearic acid are on the risks of heart disease such as platelet aggregation and coagulant activity [6]; the in vivo and in vitro studies differ in results and require further research [17].

Other studies on the general effect of cocoa lipids on heart health are much clearer. In one study, the daily ingestion of 100 g of chocolate over a 2-week period led to a 12% reduction of serum and LDL cholesterol levels in hypertensive patients despite chocolate's high fat content [36]. Another study found that LDL cholesterol decreased by 7.5% after 15-day consumption of dark chocolate containing polyphenols [42]. When replacing a high-carbohydrate snack with one 1.6 oz chocolate bar a day, LDL cholesterol levels were not negatively affected [6]. Furthermore, a recent study found that cocoa may reduce LDL oxidation and may regulate the rate of cholesterol absorption [17]. One particular study found that chocolate inhibited LDL oxidation by 75% compared with red wine's 37%–65% [9]. However, more research is needed to determine the precise mechanisms of chocolate's lipids effect on health.

Minerals

Chocolate contains a wide number of minerals. Often, these chocolate products have a relatively high mineral availability; chocolate milk is even a product for the fortification of nine iron compounds [6,43]. The quantity of minerals present in chocolate is based on the amount of cocoa bean solids [6]. As a result, dark chocolate tends to possess more minerals than milk chocolate [6]. Some significant minerals in chocolate include magnesium, copper, potassium, and calcium [6,11]. These minerals have important vascular roles often aiding in reducing the risk of cardiovascular disease by decreasing blood pressure [11]. Magnesium, especially, is known for its important role in heart health; it has antiarrhythmic and hypotensive properties [11,44]. One study throughout Asia, Australia, and Europe shows that magnesium therapy treatment in patients with acute myocardial infarction reduces the risk of arrhythmic events and increases survival [44]. In addition to this, research shows that a low magnesium intake (<50%) is an important risk factor for stroke, hypertension, and arrhythmias. Chocolate can be a source of magnesium; about 44 g of milk chocolate can provide 8% (24.6 mg), while dark chocolate can provide 15% (51 mg) of the USRDA for magnesium [45]. Furthermore, magnesium is a cofactor that aids in muscle relaxation, energy production, protein synthesis, and bone and teeth absorption [11]. Potassium, calcium, and copper also show effects on heart health. For example, low copper intake in early development correlates to cardiovascular abnormalities [6]. It also is important in the synthesis of neurotransmitters and collagen [6]. Studies also show that calcium and potassium have an inverse relationship with blood pressure, reducing the risk of stroke [11]. Potassium is especially vital in regulating cellular osmolarity and membrane potentials; this is important in maintaining important elements of cardiovascular health such as vascular tone [6]. In addition to these four minerals, chocolate also contains others such as phosphorus, zinc, and sodium [11].

Polyphenols

Polyphenols unlike other micro- and macronutrients are not an essential part of the diet; however, they are often consumed due to its ubiquitous nature in many foods [9]. Chocolate and cocoa, especially, contain many of these plant-derived phenolic compounds that possess cardioprotective qualities. Cocoa powder, specifically, contains about 50 mg of polyphenols per gram; in general, cocoa contains more polyphenols than most foods [11,46]. Cocoa is particularly rich in flavonoids, a subgroup of polyphenols [6,11]. About 12%–18% of cocoa powder's dry weight can be attributed to flavonoids [46]. Among the various type of flavonoids such as isoflavones and flavones, the most common type in cocoa is flavanols, which is also called flavan-3-ols [8,11]. Flavanols form complexes with salivary proteins, resulting in the bitter taste of cacao; often, this bitter taste is masked by the processing of cacao beans and the other ingredients in chocolate [8,11]. Cocoa contains these main types of flavanols: catechin, epicatechin, and procyanidins or tannins [8,11,46]. The epicatechin and catechin occur as monomers, while the procyanidins are oligomers composed of epicatechin and catechin subunits [46]. These oligomers are composed of about 12%–48% of a cocoa bean's dry weight [6].

In terms of chemical structure, a flavonoid is composed of two aromatic rings that are attached by an oxygenated heterocycle; flavonoids have a C6-C3-C6 backbone structure [6,8,11,47]. There are various hydroxyl groups attached to the backbone, depending on the type of flavonoid [43].

Flavanol concentration in the body depends on several factors; the factors involved cocoa either in its manufacturing process or after it is consumed. In terms of manufacturing, processing cocoa often reduces the flavanol content; milk chocolate tends to have less flavanols than dark chocolate (5%–7% compared with 20%–30%) [6,46,48]. Cocoa powder and baking chocolate possess the highest flavanol contents [49]. The manufacturing process decreases the flavanol contents of cocoa by more than 10 times [46]. In addition to manufacturing, environmental factors such as soil type, sun exposure, and rainfall can also influence flavanol content [50].

Along with this, after consuming the cocoa, many physiological factors may influence the concentration of flavanol in the body. For example, the intake of cocoa causes the flavanol content and antioxidant capacity of plasma to rise [8]. Also, epicatechin plasma concentrations are dose-dependent, influenced by the flavanol concentration of the chocolate and the amount of chocolate consumed [6]. Depending on the dose of the cocoa, the highest plasma peaks are measured 2–3 h after consumption [50]. Furthermore, the smaller the molecule's size of the flavanol, the higher the concentration of it in the blood [8]. Other factors unique to the individual can also affect the body's interaction with flavanol. For example, the absorption rate of flavanol in liver and intestinal cells, the binding of flavanol to proteins, the accumulation of flavanol in cell, and the urinary excretion of flavanol are all factors that may influence polyphenol concentration in the body [8,46].

COCOA POLYPHENOLS AND CARDIOVASCULAR DISEASE

Although the positive effects of chocolate on cardiovascular disease have been supported, the many mechanisms driving this are still under heavy research. One concentration focuses on the effect of flavanols in the body. High consumption of flavanols that are found in cocoa and chocolate is related to lower rates of human cardiovascular morbidity and mortality [12,33]. Flavanols' properties including its antihypertensive, antiinflammatory, antioxidant, and antiplatelet activity aid in producing these cardioprotective effects; with this, its effects on nitric oxide availability and the endothelium are significant [6–11].

MECHANISMS OF FLAVANOLS AND COCOA ON THE BODY

Endothelial Function and Nitric Oxide

The vascular endothelium has a pivotal role in heart health. The endothelium is a single layer of cells that are located on the intima-lumen interface of all the blood vessels in the body [12]. It functions in maintaining vascular homeostasis, changing with the direction of fluid flow. The endothelium manufactures and releases various vasoactive substances [8]. Cell-to-cell communication via paracrine and autocrine factors such as nitric oxide and prostacyclin controls the endothelium movements. Vascular homeostasis influences vascular permeability, vascular tone, and platelet aggregation [12]. Cardiovascular disease is often indicated by changes in the vascular homeostasis often involving constriction and inflammation. Endothelial dysfunction, which is known as reduced vasodilation, is often characterized by decreased NO production, increased oxidative stress, and abnormal signaling [11,51]. As a result, being able to measure endolithic function and uncovering therapies that target endothelial function can be very useful in preventing, treating, and measuring cardiovascular disease.

Chocolate consumption is found to improve endothelial function through studies using noninvasive brachial artery flow-mediated dilation (FMD) techniques, which involves a high-resolution ultrasound system that takes images of the brachial artery; this method quantifies endothelial function in humans [52]. Typically, low FMD results in patients are associated with a higher risk of cardiovascular events and disease such as myocardial infarction and stroke [12]. An experimental study shows that as greater doses of chocolate are consumed, a greater increase in FMD occurs; this correlation portrays improvement in endothelial function, supporting that higher chocolate consumption can reduce the risk of cardiovascular disease [12]. In order to observe an increase in FMD, the minimal amount of cocoa consumed ranges from 2 to 5 g [12]. Expanding on this, a metaanalysis displayed an increase in FMD after both acute and chronic consumption of chocolate flavanols [53]. Another study using 21 healthy subjects concluded that dark chocolate consumption also improves endothelial function due to increased epicatechin concentrations [54].

The mechanism that improves endothelial function in humans is unknown, yet studies have found that the increased bioavailability of NO after chocolate consumption is significant [8,12,55]. In humans, NO is released from endothelial cells chiefly in reaction to stress signaled by the blood or receptors such as acetylcholine, bradykinin, or serotonin [8]. NO synthase, eNOs, is used to synthesize NO from L-arginine [53]; this is done based on the availability of the cofactor, tetrahydrobiopterin [8]. Activation of synthesis is speculated to be a result of phosphorylation of eNOS in a P13-kinase/Akt pathway [8]. Once the NO is released, it leads to vasodilation, inducing the relaxation of vascular smooth muscle cells [12,47]. In addition to this vasodilation, the release of NO also prevents leukocyte adhesion, smooth muscle proliferation, and platelet adhesion and aggregation [8,11]. As a result, the flavanols in cocoa are notable as they significantly impact the bioavailability of NO in the system. For example, the chocolate's flavanols increase the endothelial NO synthase activity, aiding the conversion of L-arginine into NO; this increase in the concentration of NO in the blood system heightens the vasodilation effects of NO [8]. Furthermore, cocoa also reduces vascular arginase activity, allowing more arginine substrate to be available to be converted to NO [56]. Currently, the exact mechanism of this is unknown, but it is so far supported that an increase in NO may result in the improvement of endothelial function in patients that consume chocolate. Current research shows that reduced NO bioavailability and reduced eNOS function in the blood system play a role in atherosclerosis and cardiovascular disease due to endothelial dysfunction [53,57]. Endothelial dysfunction is also correlated with the increased risk of cardiovascular disease and coronary events such as stroke and myocardial infarction. [52].

Many additional experiments and studies have been conducted to further support this. For example, in an experiment using rabbit aorta, cocoa is found to induce endothelium-dependent relaxation via the activation of eNOS [57]. With this, cocoa is also found to cause NO-dependent vasodilation in humans; one intervention using 27 participants measured the pulse-wave amplitude in human fingers after 5 days of each participant consuming chocolate. Flavanol-poor chocolate was less associated with vasodilation, while flavanol-rich chocolate influenced vasodilation via the NO activation mechanism much more [7]. Human endothelial vasodilation due to the bioavailability of NO is also found in experiments involving the forearm and in patients with cardiovascular disease risk factors [8]. In patients with cardiovascular disease risk factors such as hypertension and smoking, an intervention using FMD instruments found that drinking cocoa-based beverages high in flavanol concentration (175–186 mg) compared with low-flavanol-concentration (<11 mg) cocoa beverage elevates the bioavailable NO concentration ; this intensifies vasodilation [8,58]. With this, people that consume cocoa consistently are found to have higher levels of NO metabolites in their urine than people who do not; this further supports cocoa's effect on NO availability [8,55].

Antioxidant Effects

Consumption of foods containing antioxidants is associated with lower risk of developing cardiovascular disease [6]. Chocolate's antioxidant properties help contribute to this. In terms of components, chocolate flavanols such as epicatechin and catechin and procyanidins have antioxidant properties due to their tricyclic structures; among these, procyanidins are the most significant [11,59]. More specifically, the delocalization of electrons over the aromatic rings and their ability to donate hydrogens allows them to scavenge reactive oxygen species [6,9]. With this, the aromatic rings present neutralize the free metal radicals, (fe2+ and cu2+); this is important as the free radicals help reactive oxygen species, upregulate antioxidant effects, and hinder enzymes [9,11]. The extent of chocolate's antioxidant capacity is figured out using oxygen radical absorbance capacity (ORAC) assay. It is found that chocolate is a highly antioxidant food as it has the highest oxygen radical absorbance capacity (ORAC) and procyanidin content compared with other antioxidant-rich foods such as blueberries and garlic [60].

The consumption of chocolate products and cocoa shows an increase in antioxidant capacity in the body and a decrease in LDL oxidation [10]. Research has concluded that the purified epicatechin present in chocolate has prevented LDL and liposomes from oxidation in vitro [6]. For example, cocoa phenols at 5 µmol/L inhibited oxidation by only 75%, while pure catechin at 5 µmol/L inhibited LDl oxidation by 85%; both have high antioxidant capacities [10]. With this, several studies have also verified that chocolate products decrease LDL oxidation in vitro and in vivo [6]. This is significant as LDL oxidation plays a large role in the development of atherosclerosis in cardiovascular disease; in a study with 394 individuals, they concluded that LDL oxidation is a large factor in carotid atherosclerosis development over a 5-year period [61]. As a result, research continues on the influence of chocolate on LDL oxidation as a method to reduce this significant risk factor. Several studies have displayed that the intake of chocolate slows LDL oxidation, reducing the risk of developing cardiovascular disease [6].

Antioxidant effects of chocolate and cocoa may also come into play with the bioavailability of NO. The antioxidant effects may help increase NO availability after synthesis; this would occur before reaching the smooth vascular muscles [12]. In addition to this, both the antioxidant properties in chocolate coupled with the elevated levels of NO bioavailability help maximize endothelial function [8].

However, the antioxidant properties of chocolate are still widely debated, and more research is needed; recent research suggests that flavanols are not responsible for the antioxidant properties, but increased uric acid levels after the consumption of flavanol-rich food are instead [8].

ANTIPLATELET EFFECTS

Platelet aggregation and dysfunction is a severe characteristic of cardiovascular disease. In the human body, platelets function to regulate vascular tone [6]. Increased platelet aggregation and reactivity increases the risk of arterial thrombosis and other cardiovascular diseases [6]. This platelet aggregation often occurs when the endothelium is damaged, causing the platelets to become activated as they adhere to the exposed fibrous matrix [11]. Activating the platelets allows them to induce inflammatory responses and thrombus formation [11]. As a result, suppressing this platelet aggregation can be vitally important in increasing thrombolytic activity of blood platelets, reducing cardiovascular damage [11]. Several studies both in vivo and in vitro have demonstrated cocoa's platelet inhibitory properties. This occurs due to the high flavanol content of cocoa; however, it is important to comment that not all flavonoid-containing foods reduce platelet aggregation [6,8]. The exact mechanism of this has yet to be discovered, but it is hypothesized that these flavanols may affect platelets by changing ligand-receptor affinity, signaling pathways, and membrane fluidity [11,62]. It has also been proposed that the flavanols affect the platelet nitric oxide [11]. Despite the lack of known mechanism, chocolate consumption has significant effects on platelets. For instance, research shows that within hours of ingesting 897 mg/ml cocoa beverage, the ADP-/collagen-activated, platelet-related hemostasis decreases [63]. Justified by the reduction in the expression of the glycoprotein IIb/IIIa surface proteins, these effects occur 2 h after consumption [63]. In 2007, a platelet function study of 1535 participants shows that healthy participants who consumed chocolate before being tested has reduced platelet activity compared with those who did not consume chocolate; the study concluded that even small amounts of chocolate caused significant antiplatelet effects [64]. In a small study of about 30 participants, platelet activation decreased 2–6 h after the consumption of a flavonoid-rich cocoa beverage [65].

Current evidence supports the advantageous effect of cocoa on platelet activation; platelet activation plays an important role in inflammation and thrombosis, which can lead to cardiovascular disease [66]. The effects of flavanol are most likely causing the positive effects of cocoa on platelet aggregation. However, minerals such as calcium, which has shown to increase platelet activation, may also influence this [67]. Other minerals such as potassium, magnesium, or stearic acid may also contribute to the antiplatelet effects [11].

ANTIHYPERTENSIVE

The antihypertensive properties of flavanols may influence cocoa's effect on lowering blood pressure. The specific mechanisms of these blood-pressure-lowering effects are not exactly known yet; more research in the area needs to be conducted. Current research suggests that stearic acid, theobromine, increased eNOS activity, improvement of endothelial function may all contribute to this antihypertensive property [8,33,39,68]. The study of theobromine may be a future avenue of study that this compound is associated with antihypertensive properties; it is a significant component of chocolate. [14,68].

SUMMARY

Overall, chocolate consumption is associated with decreased risk in developing cardiovascular disease. Important risk factors such as blood pressure for cardiovascular disease are targeted by chocolate's properties and components. For example, the polyphenols present in chocolate impact the bioavailability of NO, endothelial function, and platelet aggregation. Other components such as the minerals and the lipids also impact the heart health positively. With this, it is important to keep in consideration that chocolate consumption does not solely impact heart health; it can also influence obesity and other health issues. It is important to note that not all components of chocolate and the effects on health are discussed in this review. For example, theobromine, fiber, plant sterols are also significant components in chocolate. With that, effects of chocolate such as its antiinflammatory, insulin resistance, and immune response impact on the human body and heart are not discussed.

REFERENCES

[1] Reddy S.K., Yusuf S. Emerging epidemic of cardiovascular disease in developing countries. Circulation [Internet] 1998;97(6):596–601. https://doi.org/10.1161/01.CIR.97.6.596. Available from http://circ.ahajournals.org/content/97/6/596.

[2] World Health Organization. Cardiovascular Diseases (CVDs) [Internet]. World Health Organization; 2016. Available from http://www.who.int/mediacentre/factsheets/fs317/en/ [updated Sep. 2016; cited 6 Oct. 2016].

[3] Centers for Disease Control and Prevention. Heart Disease [Internet]. Atlanta, GA: Centers for Disease Control and Prevention; 2015. Available from http://www.cdc.gov/heartdisease/facts.htm [updated Aug. 2015; cited 6 Oct. 2016].

[4] Erdman J.W., Carson L., Kwik-Uribe C., Evans E.M., Allen R.R. Effects of cocoa flavanols on risk factors for cardiovascular disease. Asia Pac J Clin Nutr 2008;17(S1):284–7. Available from http://apjcn.nhri.org.tw/server/APJCN/17%20Suppl%201//284.pdf [cited 6 Oct. 2016].

[5] WHO MONICA Project Principal Investigators. The world health organization monica project (monitoring trends and determinants in cardiovascular disease): a major international collaboration. J Clin Epidemiol 2004;41(2):105–14. https://doi.org/10.1016/0895-4356(88)90084-4. Available from http://www.sciencedirect.com/science/article/pii/0895435688900844 [cited 5 Nov. 2016].

[6] Steinberg F.M., Bearden M.M., Keen C.L. Cocoa and chocolate flavonoids: implications for cardiovascular health. J Am Diet Assoc 2003;103(2):https://doi.org/10.1053/jada.2003.50028. Available from http://www.sciencedirect.com/science/article/pii/S0002822302000329 [cited 5 Nov. 2016].

[7] Fisher N.D., Hughes M., Gerhard-Herman M., Hollenberg N.K. Flavanol-rich cocoa induces nitric-oxide-dependent vasodilation in healthy humans. J Hypertens 2003;21(12):2281–6. Available from https://www.ncbi.nlm.nih.gov/pubmed/14654748.

[8] Corti R., Flammer A.J., Hollenberg N.K., Lüscher T.F. Cocoa and cardiovascular health. Circulation 2009;119:1433–41. https://doi.org/10.1161/CIRCULATIONAHA.108.827022. Available from http://circ.ahajournals.org/content/119/10/1433.long [cited 22 Nov. 2016].

[9] Ackar D., Lendić V.K., Valek M., Šubarić D., Miličević B., Babić J., et al. Cocoa polyphenols: can we consider cocoa and chocolate as potential functional food? J Chem 2013;7. https://doi.org/10.1155/2013/289392. Available from https://www.hindawi.com/journals/jchem/2013/289392/cta/ [cited 6 Oct. 2016].

[10] Waterhouse A.L., Shirley J.R., Donovan J.L. Antioxidants in chocolate. Lancet 1996;348(9030):834. https://doi.org/10.1016/S0140-6736(05)65262-2. Available from http://www.sciencedirect.com/science/article/pii/S0140673605652622 [cited 8 Nov. 2016].

[11] Katz D.L., Doughty K., Ali A. Cocoa and chocolate in human health and disease. Antioxid Redox Signal 2011;15(10):2779–811. https://doi.org/10.1089/ars.2010.3697. Available from https://www.ncbi.nlm.nih.gov/pmc/articles/PMC4696435/ [cited 4 Nov. 2016].

[12] Monahan K.D. Effect of cocoa/chocolate ingestion on brachial artery flow-mediated dilation and its relevance to cardiovascular health and disease in humans. Arch Biochem Biophys 2012;527(2):90–4. https://doi.org/10.1016/j.abb.2012.02.021. Available from http://www.sciencedirect.com/science/article/pii/S0003986112000823 [cited 5 Nov. 2016].

[13] Wilson P.K., Hurst J.W. Chocolate as medicine: a quest over the centuries. Cambridge, United Kingdom: The Royal Society of Chemistry; 2012 3–173 p.

[14] Hart C.L., Ksir C. Drugs, society and human behavior. 6th ed. New York, NY: McGraw Hill Education; 2015. 255–257 p.

[15] International Cocoa Organization. Origins of cocoa and its spread around the world [Internet]. London, UK: International Cocoa Organization; 2013. Available from http://www.icco.org/about-cocoa/growing-cocoa.html [updated Mar. 2013; cited 6 Oct. 2016].

[16] KEW: Royal Botanical Gardens. Theobroma Cacao (cocoa tree) [Internet]. KEW: Royal Botanical Gardens. Available from http://www.kew.org/science-conservation/plants-fungi/theobroma-cacao-cocoa-tree [cited 6 Oct. 2016].

[17] Watson R.R., Preedy V.R., Zibadi S. Chocolate in health and nutrition. 7th ed. Humana Press; 2013. 289-297 p. Available from https://link.springer.com/book/10.1007%2F978-1-61779-803-0.

[18] Cadbury Chocolate. Cocoa and Chocolate [Internet]. World Agroforestry. Available from http://www.worldagroforestry.org/treesandmarkets/inaforesta/documents/cocoa%20and%20chocolate/cocoa%20and%20chocolate.pdf [cited 6 Oct. 2016].

[19] Bart-Plangea A., Baryehb E.A. The physical properties of Category B cocoa beans. J Food Eng 2003;60(3):219–27. https://doi.org/10.1016/S0260-8774(02)00452-1. Available from http://www.sciencedirect.com/science/article/pii/S0260877402004521 [cited 4 Nov. 2016].

[20] Lim T.K. Edible medicinal and non-medicinal plants. Netherlands: Springer; 2011. p. 208–51.

[21] World Cocoa Foundation. Cocoa value chain: from farmer to consumer. World Cocoa Foundation; 2016. Available from http://www.worldcocoa-foundation.org/about-cocoa/cocoa-value-chain/ [cited 6 Oct. 2016].

[22] Andres-Lacueva C., Monagas M., Khan N., Izquierdo-Pulido M., Urpi-Sarda M., Permanyer J., et al. Flavanol and flavonol contents of cocoa powder products: influence of the manufacturing process. J Agric Food Chem 2008;56((9):3111–7. https://doi.org/10.1021/jf0728754. Available from http://pubs.acs.org/doi/abs/10.1021/jf0728754 [cited 13 Mar. 2017].

[23] Egan B., Laken M., Donovan J., Woolson R. Does dark chocolate have a role in the prevention and management of hypertension? Hypertension 2010;55:1289–95. https://doi.org/10.1161/HYPERTENSIONAHA.110.151522. Available from http://hyper.ahajournals.org/content/55/6/1289.long [cited 13 Mar. 2017].

[24] Chevaux KA J.L., Villar M.E., et al. Proximate, mineral and procyanidin content of certain foods and beverages consumed by the Kuna Amerinds of Panama. J Food Composit Anal 2001;14:553–63. Available from https://michaellustgarten.wordpress.com/page/10/ [cited 15 Nov. 2016].

[25] Hollenberg N.K., Martinez G., McCullough M., Meinking T., Passan D., Preston M., et al. Aging, acculturation, salt intake, and hypertension in the Kuna of Panama. Hypertension 1997;29(1-2):171–6. Available from https://www.ncbi.nlm.nih.gov/pubmed/9039098 [cited 13 Nov. 2016].

[26] Bayard V., Chamorro F., Motta J., Hollenberg N.K. Does flavanol intake influence mortality from nitric oxide-dependent processes? Ischemic heart disease, stroke, diabetes mellitus, and cancer in Panama. Int J Med Sci 2007;4(1):53–8. https://doi.org/10.7150/ijms.4.53. Available from http://www.medsci.org/v04p0053.htm [cited 14 Nov. 2016].

[27] McCullough M.L., Chevaux K., Jackson L., Preston M., Martinez G., Schmitz H.H., et al. Hypertension, the Kuna, and the epidemiology of flava-nols. J Cardiovasc Pharmacol 2006;47(2):S103–9. Available from https://www.ncbi.nlm.nih.gov/pubmed/16794446 [cited 8 Nov. 2016].

[28] Buijsse B., Feskens E.J., Kok F.J., Kromhout D. Cocoa intake, blood pressure, and cardiovascular mortality: the Zutphen Elderly Study. Arch Intern Med 2006;166(4):411–7. Available from https://www.ncbi.nlm.nih.gov/pubmed/16505260 [cited 4 Dec. 2016].

[29] Larsson S.C., Åkesson A., Gigante B., Wolk A., et al. Chocolate consumption and risk of myocardial infarction: a prospective study and meta-analysis. Heart 2016;102(13):1017–22. https://doi.org/10.1136/heartjnl-2015-309203. Available from https://www.ncbi.nlm.nih.gov/pubmed/26936339 [cited 12 Nov. 2016].

[30] Mink P.J., Scrafford C.G., Barraj L.M., Harnack L., Hong C.P., Nettleton J.A., et al. Flavonoid intake and cardiovascular disease mortality: a prospective study in postmenopausal women. Am J Clin Nutr 2007;85(3):895–909. Available from https://www.ncbi.nlm.nih.gov/pubmed/17344514 [cited 5 Nov. 2016].

[31] Buijsse B., Weikert C., Drogan D., Bergmann M., Boeing H. Chocolate consumption in relation to blood pressure and risk of cardiovascular disease in German adults. Eur Heart J 2010;31(13):1616–23. https://doi.org/10.1093/eurheartj/ehq068. Available from https://www.ncbi.nlm.nih.gov/pubmed/20354055 [cited 28 Oct. 2016].

[32] Djoussé L., Hopkins P.N., Arnett D.K., Pankow J.S., Borecki I., North K.E., et al. Chocolate consumption is inversely associated with calcified atherosclerotic plaque in the coronary arteries: the NHLBI Family Heart Study. Clin Nutr 2011;30(1):38–43. https://doi.org/10.1016/j.clnu.2010.06.011. Available from https://www.ncbi.nlm.nih.gov/pubmed/20655129 [cited 24 Nov. 2016].

[33] Buitrago-Lopez A., Sanderson J., Johnson L., Warnakula S., Wood A., Angelantonio E. Chocolate consumption and cardiometabolic disorders: systematic review and meta-analysis. BMJ 2011;343–4488. https://doi.org/10.1136/bmj.d4488. Available from https://www.ncbi.nlm.nih.gov/pmc/articles/PMC3163382/ [cited 5 Nov. 2016].

[34] Grassi D., Desideri G., Ferri C. Blood pressure and cardiovascular risk: what about cocoa and chocolate? Arch Biochem Biophys 2010;501(1):112–5. https://doi.org/10.1016/j.abb.2010.05.020. Available from http://www.sciencedirect.com/science/article/pii/S0003986110001876 [cited 18 Nov. 2016].

[35] Taubert D., Berkels R., Roesen R., Klaus W. Chocolate and blood pressure in elderly individuals with isolated systolic hypertension. JAMA 2003;290(8):1029–30. https://doi.org/10.1001/jama.290.8.1029. Available from http://jamanetwork.com/journals/jama/article-abstract/197169 [cited 18 Nov. 2016].

[36] Grassi D., Necozione S., Lippi C., Croce G., Valeri L., Pasqualetti P., et al. Cocoa reduces blood pressure and insulin resistance and improves endothelium-dependent vasodilation in hypertensives. Hypertension 2005;46(2):398–405. https://doi.org/10.1161/01.HYP.0000174990.46027.70. Available from https://www.ncbi.nlm.nih.gov/pubmed/16027246 [cited 8 Dec. 2016].

[37] Taubert D., Roesen R., Schömig E. Effect of cocoa and tea intake on blood pressure: a meta-analysis. Arch Intern Med 2007;167(7):626–34. https://doi.org/10.1001/archinte.167.7.626. Available from https://www.ncbi.nlm.nih.gov/pubmed/17420419 [cited 27 Nov. 2016].

[38] Ried K., Sullivan T., Fakler P1., Frank O.R., Stocks N.P. Does chocolate reduce blood pressure? A meta-analysis. BMC Med 2010. https://doi.org/10.1186/1741-7015-8-39. Available from https://www.ncbi.nlm.nih.gov/pmc/articles/PMC2908554/ [cited 18 Nov. 2016].

[39] Taubert D., Roesen R., Lehmann C., Jung N., Schömig E., et al. Effects of low habitual cocoa intake on blood pressure and bioactive nitric oxide. JAMA 2017;298(1):49–60. https://doi.org/10.1001/jama.298.1.49. Available from http://jamanetwork.com/journals/jama/fullarticle/207783 [cited 13 Nov. 2016].

[40] Woodside J.V., McKinley M.C., Young I.S. Saturated and trans fatty acids and coronary heart disease. Curr Atheroscl Rep 2008;10(6):460–6. Available from https://www.ncbi.nlm.nih.gov/pubmed/18937892 [cited 4 Nov. 2016].

[41] Bracco U. Effect of triglyceride structure on fat absorption. Am J Clin Nutr 1994;60(6):1002S–9S. Available from https://www.ncbi.nlm.nih.gov/pubmed/7977140 [cited 8 Dec. 2016].

[42] Grassi D., Desideri G., Necozione S., Lippi C., Casale R., Properzi G., et al. Blood pressure is reduced and insulin sensitivity increased in glucose-intolerant, hypertensive subjects after 15 days of consuming high-polyphenol dark chocolate. J Nutr 2008;138(9):1671–6. Available from https://www.ncbi.nlm.nih.gov/pubmed/18716168 [cited 5 Nov. 2016].

[43] Douglas F.W., Rainey N.H., Wong N.P., Edmondson L.F., LaCroix D.E. Color, flavor, and iron bioavailability in iron-fortified chocolate milk. J Dairy Sci 1981;64(9):1785–93. https://doi.org/10.3168/jds.S0022-0302(81)82767-1. Available from http://www.journalofdairyscience.org/article/S0022-0302(81)82767-1/abstract [cited 8 Nov. 2016].

[44] Delva P. Magnesium and coronary heart disease. Mol Asp Med 2003;24(1–3):63–78. https://doi.org/10.1016/S0098-2997(02)00092-4. Available from http://www.sciencedirect.com/science/article/pii/S0098299702000924 [cited 15 Nov. 2016].

[45] Steinberg F., Bearden M., Keen C. Cocoa and chocolate flavonoids: implications for cardiovascular health. J Am Diet Assoc 2003;103(2):215–23. https://doi.org/10.1053/jada.2003.50028. Available from http://www.andjrnl.org/article/S0002002D8223(02)00032-9/abstract.

[46] Khan N., Khymenets O., Urpí-Sardà M., Tulipani S., Garcia-Aloy M., Monagas M., et al. Cocoa polyphenols and inflammatory markers of cardiovascular disease. Nutrients 2014;6(2):844–80. https://doi.org/10.3390/nu6020844. Available from https://www.ncbi.nlm.nih.gov/pmc/articles/PMC3942736/ [cited 1 Dec. 2016].

[47] Kim J., Kim J., Shim J., Lee C., Lee K., Lee H. Cocoa phytochemicals: recent advances in molecular mechanisms on health. Crit Rev Food Sci Nutr 2011;54(11):1458–72. https://doi.org/10.1080/10408398.2011.641041. Available from http://www.tandfonline.com/doi/citedby/10.1080/10408398.2011.641041?scroll=top&needAccess=true [cited 13 Mar. 2017].

[48] Miller K.B., Hurst W.J., Flannigan N., Ou B., Lee C.Y., Smith N., et al. Survey of commercially available chocolate- and cocoa-containing products in the United States. 2. Comparison of flavan-3-ol content with nonfat cocoa solids, total polyphenols, and percent cacao. Agric Food Chem 2009;57(19):9169–80. https://doi.org/10.1021/jf901821x. Available from https://www.ncbi.nlm.nih.gov/pubmed/19754118 [cited 4 Nov. 2016].

[49] Manach C., Scalbert A., Morand C., Rémésy C., Jiménez L. Polyphenols: food sources and. bioavailability. Am J Clin Nutr 2004;79((5):727–47. Available from http://ajcn.nutrition.org/content/79/5/727.long [cited 23 Nov. 2016].

[50] Rein D., Lotito S., Holt R.R., Keen C.L., Schmitz H.H., Fraga C.G., et al. Epicatechin in human plasma: in vivo determination and effect of chocolate consumption on plasma oxidation status. J Nutr 2000;130(8S):2109S–14S. Available from https://www.ncbi.nlm.nih.gov/pubmed/10917931 [cited 16 Nov. 2016].

[51] Vogel R.A. Measurement of endothelial function by brachial artery flow-mediated vasodilation. Am J Cardiol 2001;88(2A):31E–4E. Available from http://www.sciencedirect.com/science/article/pii/S0002914901017647 [cited 22 Nov. 2016].

[52] Gokce N. Clinical assessment of endothelial function-ready for prime time? Circul Cardiovasc Imag 2011;4(4):348–50. https://doi.org/10.1161/CIRCIMAGING.111.966218. Available from https://www.ncbi.nlm.nih.gov/pmc/articles/PMC3144162/ [cited 8 Nov. 2016].

[53] Hooper L., Kroon P.A., Rimm E.B., Cohn J.S., Harvey I., LeCornu K., et al. Flavonoids, flavonoid-rich foods, and cardiovascular risk: a meta-analysis of randomized controlled trials. Am J Clin Nutr 2008;88(1):38–50 [cited 6 Nov. 2016].

[54] Engler M.B., Engler M.M., Chen C.Y., Malloy M.J., Browne A., Chiu E.Y., et al. Flavonoid-rich dark chocolate improves endothelial function and increases plasma epicatechin concentrations in healthy adults. J Am Coll Nutr 2004;23(3):197–204. Available from https://www.ncbi.nlm.nih.gov/pubmed/15190043 [cited 8 Nov. 2016].

[55] Schroeter H., Heiss C., Balzer J., Kleinbongard P., Keen C.L., Hollenberg N.K., et al. Epicatechin mediates beneficial effects of flavanol-rich cocoa on vascular function in humans. Proc Natl Acad Sci 2006;103(4):1024–9. https://doi.org/10.1016/j.abb.2008.02.040. Available from https://www.ncbi.nlm.nih.gov/pubmed/16418281 [cited 8 Dec. 2016].

[56] Schnorr O., Brossette T., Momma T.Y., Kleinbongard P., Keen C.L., Schroeter H., et al. Cocoa flavanols lower vascular arginase activity in human endothelial cells in vitro and in erythrocytes in vivo. Arch Biochem Phys 2008;476(2):211–5. https://doi.org/10.1016/j.abb.2008.02.040. Available from https://www.ncbi.nlm.nih.gov/pubmed/18348861 [cited 15 Nov. 2016].

[57] Oemar B.S., Tschudi M.R., Godoy N., Brovkovich V., Malinski T., Luscher T.F. Reduced endothelial nitric oxide synthase expression and production in human atherosclerosis. Circulation 1998;97:2494–8. Available from https://www.ncbi.nlm.nih.gov/pubmed/9657467 [cited 19 Nov. 2016].

[58] Heiss C., Kleinbongard P., Dejam A., Perré S., Schroeter H., Sies H., Kelm M. Acute consumption of flavanol-rich cocoa and the reversal of endothelial dysfunction in smokers. J Am Coll Cardiol 2005;46(7):1276–83. Available from https://www.ncbi.nlm.nih.gov/pubmed/16198843 [cited 7 Nov. 2016].

[59] Ramiro-Puig E., Castell M. Cocoa: antioxidant and immunomodulator. Br J Nutr 2009;101(7):931–40. https://doi.org/10.1017/S0007114508169896. Available from https://www.ncbi.nlm.nih.gov/pubmed/19126261 [cited 4 Nov. 2016].

[60] Paquette M. Health and longevity. Persp Psychiatr Care [Internet] 2008;44(2):67–70. https://doi.org/10.1111/j.1744-6163.2008.00154.x. Available from http://onlinelibrary.wiley.com/doi/10.1111/j.1744-6163.2008.00154.x/abstract [cited 13 Mar. 2017].

[61] Aoki T., Abe T., Yamada E., Matsuto T., Okada M. Increased LDL susceptibility to oxidation accelerates future carotid artery atherosclerosis. Lipids Health Dis 2012;11(4):https://doi.org/10.1186/1476-511X-11-4. Available from https://www.ncbi.nlm.nih.gov/pubmed/22230558 [cited 7 Oct. 2016].

[62] Holt R.R., Actis-Goretta L., Momma T.Y., Keen C.L. Dietary flavanols and platelet reactivity. J Cardiovasc Pharmacol 2006;47(2):S187–96. Available from https://www.ncbi.nlm.nih.gov/pubmed/16794457 [cited 1 Dec. 2016].

[63] Pearson D.A., Paglieroni T.G., Rein D., Wun T., Schramm D.D., Wang J.F., et al. The effects of flavanol-rich cocoa and aspirin on ex vivo platelet function. Thromb Res 2002;106(4–5):191–7. Available from https://www.ncbi.nlm.nih.gov/pubmed/12297125 [cited 12 Nov. 2016].

[64] Bordeaux B., Yanek L.R., Moy T.F., White L.W., Becker L.C., Faraday N., et al. Causalchocolate consumption and inhibition of platelet function. Prev Cardiol [Internet] 2007;10((4):175–80. Available from https://www.ncbi.nlm.nih.gov/pubmed/17917513 [cited 3 Nov. 2016].

[65] Rein D., Paglieroni T.G., Wun T., Pearson D.A., Schmitz H.H., Gosselin R., et al. Cocoa inhibits platelet activation and function. J Clin Nutr 2000;72(1):30–5. Available from https://www.ncbi.nlm.nih.gov/pubmed/10871557 [cited 8 Nov. 2016].

[66] Olas B. The multifunctionality of berries toward blood platelets and the role of berry phenolics in cardiovascular disorders. Platelets [Internet] 2016;25:1–10. https://doi.org/10.1080/09537104.2016.1235689. Available from https://www.ncbi.nlm.nih.gov/pubmed/27778523 [cited 9 Nov. 2016].

[67] Pearson D.A., Holt R.R., Rein D., Paglieroni T., Schmitz H.H., Keen C.L. Flavanols and platelet reactivity. Clin Dev Immunol 2005;12(1):1–9. Available from https://www.ncbi.nlm.nih.gov/pubmed/15712593 [cited 8 Nov. 2016].

[68] Kelly C.J. Effects of theobromine should be considered in future studies. Am J Clin Nutr [Internet] 2005;82(2):486–7. Available from http://ajcn.nutrition.org/content/82/2/486.2.full [cited 8 Nov. 2016].

FURTHER READING

[1] Palmer RM, Ashton DS, Moncada S. Vascular endothelial cells synthesize nitric oxide from L-arginine. Nature 1988;333(6174):664–6. Available from https://www.ncbi.nlm.nih.gov/pubmed/3131684 [cited 5 Nov. 2016].

[2] Karim M, McCormick K, Kappagoda CT. Effects of cocoa extracts on endothelium-dependent relaxation. J Nutr 2000;130(8S):2105S–8S. Available from https://www.ncbi.nlm.nih.gov/pubmed/10917930 [cited 20 Nov. 2016].

Chapter 22

Prediabetes: An Emerging Risk Factor for Coronary Artery Disease

Richard B. Stacey, Veronica D'Ambra, Petro Gjini

Department of Internal Medicine, Section on Cardiology, Wake Forest University School of Medicine, Winston-Salem, NC, United States

INTRODUCTION

Prediabetes represents a significant public health problem for much of the world. Classically, most clinicians and investigators considered the risk of mildly elevated blood glucose to be insignificant. The general understanding only assigned true cardiovascular risk to individuals with frank diabetes. However, as more cohort studies began to follow individuals with prediabetes, a different scenario became evident. Clinical studies, both in the general community and the acute-care setting, have found that many of the pathologies associated with diabetes mellitus emerge in the prediabetic state, particularly as it relates to cardiovascular disease.

HISTORY OF PREDIABETES

In one form or another, societies have recognized diabetes and diabetes-like diseases for millennia [1]. In the early 1900s, physicians and scientists started to describe and understand the biochemical derangement associated with diabetes. Toward the midcentury, physicians adopted the glucose tolerance test. The glucose tolerance consisted of measuring a baseline blood glucose, administering a dose of glucose, and monitoring how the blood glucose level changed. While it mainly helped to identify women at highest risk for developing subsequent gestational diabetes, it provided clinicians with the variable nature and function of glucose homeostasis. Through this process, the understanding that an impaired glucose response could be present without overt diabetes mellitus emerged.

Subsequently, in releasing its diabetes classification scheme in 1997, the American Diabetes Association formally recognized impaired fasting glucose as a distinct diagnosis from type 2 diabetes mellitus. The original definition identified impaired fasting glucose from a fasting glucose level of 110–125 mg/dL [2,3]. In later versions, the American Diabetes Association lowered the limit to 100–125 mg/dL [4]. Conversely, the World Health Organization continues to recognize a range of 110–125 mg/dL [5]. Impaired fasting glucose and impaired glucose tolerance from a glucose challenge provide complementary information. A potential downside to a single fasting glucose assessment is misclassification of glucose status, whereas the oral glucose tolerance test directly assesses the body's management of glucose levels [5].

During this time, the hemoglobin A1c emerged as a test to monitor the long-term glucose homeostasis in individuals with diabetes mellitus. With the appealing aspect of a more long-term assessment, researchers established that diabetes mellitus was identified with a hemoglobin A1c higher than 6.5%, impaired fasting glucose with a level of 5.7%–6.5%, and normal glucose metabolism with a level less than 5.7% [4]. In concert with fasting glucose levels, hemoglobin A1c improves identification of those with impaired glucose metabolism. However, there are a minority of individuals who are more susceptible to glycosylation than others as a result of genetic variation. In these individuals, this may make their glucose metabolism appear abnormal when in fact it is normal.

In this same period of time from the late 1990s to the early 2000s, investigators explored abnormal glucose metabolism along with many of its associated cardiovascular risk factors through the concept of the metabolic syndrome. The World Health Organization, the American Heart Association, the US National Cholesterol Education Program Adult Treatment Panel III, the International Diabetes Federation, and the European Group for the Study of Insulin Resistance all included impaired fasting glucose as part of their core definition for the metabolic syndrome [6–9]. In common, they also incorporated other features, such as waist circumference, triglyceride levels, HDL cholesterol levels, and blood pressure.

In summary, the most defining feature of prediabetes remains abnormal glucose metabolism without frank diabetes mellitus. Clinicians establish this most commonly through a fasting glucose level, a marginally elevated hemoglobin A1c or an abnormal glucose tolerance test. Identification of prediabetes represents an important situation where recognition can lead to early intervention and possibly normalization of glucose metabolism. Of individuals with prediabetes, almost half will progress to frank diabetes mellitus within 10 years [10].

THE MAGNITUDE OF PREDIABETES

The growth of impaired fasting glucose as a clinical entity is most easily seen in the United States. Using data from the National Health and Nutrition Examination Survey, investigators demonstrated that prediabetes and diabetes directly impact a significant proportion of individuals in the United States. In 1988–90, almost 8% of the population had diabetes, and an additional 24% were identified as being prediabetic [11]. In the late 1990s, 9% of the population had diabetes, and an additional 26% were diagnosed as prediabetic [11]. In the most recent survey, 14% of the population had diabetes, and an additional 38% were diagnosed as prediabetic [12]. Through these surveys, the most dramatic demographic segment affected by abnormal glucose metabolism was of the older demographic. In the context of an aging society, this may indicate that the public health burden caused by prediabetes may continue to grow substantially. In addition, more children and adolescences are being diagnosed with prediabetes than previously recognized, which may further negatively impact this forecast [13].

In addition to the United States, Europe also suffers with a significant proportion of individuals with prediabetes. In Germany, 16% of the population were found to be either diabetic or prediabetic [14]. As westernized diets and lifestyles proliferate, prediabetes has become a global phenomenon. Through epidemiological studies, China has identified a sizeable proportion of its population with prediabetes. In the Xinjiang region, 10%–15% of the population had diabetes or were prediabetic [15], and in the Three Gorges region, 7% were diabetic, and over 9% were found to be prediabetic [16]. Even in Taiwan, 35% of the population were identified as being prediabetic [17]. In India, prediabetes affects almost 15% of the general population [18]. Africa has also started to see these trends developing [19]. These studies represent a small fraction of the epidemiological work that has established the growing problem of prediabetes.

ESTABLISHED RISKS OF DIABETES MELLITUS

To more fully appreciate the potential risks of prediabetes, one must first recognize the clinical problems associated with diabetes mellitus. Historically, diabetes mellitus causes both macro- and microvascular complications. As for microvascular effects, diabetes mellitus predisposes patients to nephropathy, retinopathy, and neuropathy [20–23]. By damaging individual nephrons, diabetes mellitus causes a more global nephropathy, which, over time, can substantially decrease the kidney function and increase the risk of needing hemodialysis [23–25]. By damaging the retina through microvascular disease, if untreated, diabetes has the potential to cause blindness and remains a significant risk factor for blindness, even with treatment [26,27]. Finally, diabetes mellitus causes a neuropathy, which is associated with pain and sensory deficits [20]. This neuropathy also can limit mobility and mask the pain associated with peripheral skin ulcers, which develop as a result of decreased blood flow from diabetic vascular disease.

As for the macrovascular complications of diabetes mellitus, the most apparent manifestation remains coronary artery disease and peripheral arterial disease [21]. In a landmark study, Haffner and associates found that being diabetic carried the same risk for a future cardiovascular event as having had a prior myocardial infarction [28]. As a result, most risk scoring systems regard diabetes mellitus as a coronary heart disease risk equivalent. Even in individuals with acute coronary syndrome, over 27% were first diagnosed with diabetes [29]. In angina cases without obstructive epicardial coronary artery disease, diabetic microvascular disease was found to cause the symptoms. In reviewing ST-segment elevation myocardial infarctions, nearly 10% were diagnosed with diabetes [30]. Finally, in regard to peripheral vascular disease, diabetes mellitus remains a significant risk factor for severe macrovascular disease requiring amputation [21].

An associated cardiovascular complication of diabetes mellitus is a silent (or asymptomatic) myocardial infarction [31–33]. The occurrence of a silent myocardial infarction results from the convergence of both macro- and microvascular sequelae of diabetes. First, to suppress cardiac pain sensations, autonomic neuropathy must be present. Second, the inflamed atherosclerotic plaques associated with diabetes must result in a plaque rupture that limits coronary blood flow to the extent that myocardium dies. A silent myocardial infarction carries with it almost all the inherent risk of a clinically recognized myocardial infarction [34,35]. While most testing related to coronary artery disease relates to symptomatology, certain guidelines, such as those issued by the European Society of Cardiology [36], provide clinicians with the latitude necessary to identify such myocardial infarctions in high-risk patients, such as diabetics.

EMERGING NONCARDIOVASCULAR RISKS OF PREDIABETES

While the complications and risks of diabetes mellitus remain well described with evidence-based guidelines to assist clinicians, the risks of prediabetes remain less well recognized and appreciated. In assessing the potential relationship between kidney disease and prediabetes, patients with impaired fasting glucose have higher levels of albuminuria, even at normal glomerular filtration rates, which signals potential subclinical kidney damage [37]. Furthermore, in a separate population, those with impaired fasting glucose had a much higher prevalence of chronic kidney disease [38]. Though less established, prediabetics may have a higher prevalence of retinopathy found during funduscopic examination [26]. Finally, in the relationship between prediabetes and peripheral neuropathy, particularly in older individuals, abnormal glucose metabolism relates to peripheral neuropathy and can mask the pain associated with significant osteoarthritis [39–41].

PREDIABETES AND INFLAMMATION

One potential mechanism that links abnormal glucose metabolism to its different pathologies may be an increase in low-grade systemic inflammation, which potentially accelerates the development of known sequela. One of the most common measures of systemic inflammation remains the C-reactive protein (CRP). Diabetics demonstrate an increase in CRP levels, which may forecast future clinical events [42]. In prediabetes, elevated CRP levels have also been found to link with a higher risk of subsequent development of cardiovascular disease [43]. While CRP lacks the specificity to indicate a particular problem, its elevation means an increased risk of clinical events and disease progression in different circumstances [44–46]. While CRP links with many inflammatory processes, in particular, CRP relates to matrix metalloproteinase activity [47]. Matrix metalloproteinases break down the connective tissue in plaque coverings, which further predisposes them to progression and potential plaque ruptures.

ATHEROSCLEROSIS AND INFLAMMATION

At its core, atherosclerosis represents the culmination of both injury and inflammation within the arterial wall. In its beginning, an intimal injury disrupts continuity, which allows for cholesterol deposition within the arterial wall. Initially, this intimal injury may manifest as endothelial dysfunction [45,48,49]. As further injuries take place, more cholesterol deposits in the arterial wall. Recognizing that cholesterol should not freely deposit in the blood vessel wall, macrophages phagocytize the cholesterol particles but unable to move from the area; the macrophages, now called foam cells, remain in the arterial wall [50]. Inflammatory mediators are often released that then recruit other cells, such as smooth muscle cells, to respond. As more injury occurs and inflammatory mediators increase, the plaques grow. Using coronary artery calcification as a surrogate marker for atherosclerosis burden, investigators have demonstrated a consistent association between higher CRP levels and higher coronary artery calcifications [51]. If inflammation remains unchecked, it can erode the covering of the plaque, a process referred to as a plaque rupture [47]. Plaque rupture remains a critical component to both the acute presentation of atherosclerotic disease and its progression.

PREDIABETES AND ATHEROSCLEROSIS

Prediabetics develop more coronary artery calcification than individuals with normal fasting glucose. From a study with over 1000 asymptomatic participants, 21% who had prediabetes, Dr. Lim and colleagues found that those with prediabetes or diabetes had more coronary calcium and were more likely to have a hemodynamically significant coronary stenosis [52]. Furthermore, 19% of diabetics and 11% of prediabetics were found to have coronary plaques compared with only 7% of those with normal glucose metabolism (see Fig. 1). In a separate study involving over 1100 asymptomatic patients, Eun et al. found that 42% of those with prediabetes had coronary calcium as opposed to 25% of those with normal fasting glucose [51]. Using the Multi-Ethnic Study of Atherosclerosis, Kronmal et al. found that prediabetics had a relative risk of 1.35 ($P = .002$) relative to normal fasting glucose participants for the progression of coronary artery calcium [53]. A possible hypothesis is that the systemic inflammation associated with prediabetes accelerates atherogenesis.

PREDIABETES AND ENDOTHELIAL DYSFUNCTION

One of the first manifestations of arterial injury is endothelial dysfunction. Typically, endothelial dysfunction causes arteries to remain more vasoconstricted and less responsive to agents intended to vasodilate. Many complicated pathways exist to connect prediabetes with endothelial dysfunction, but many relate to the concept of oxidative injury, which most likely

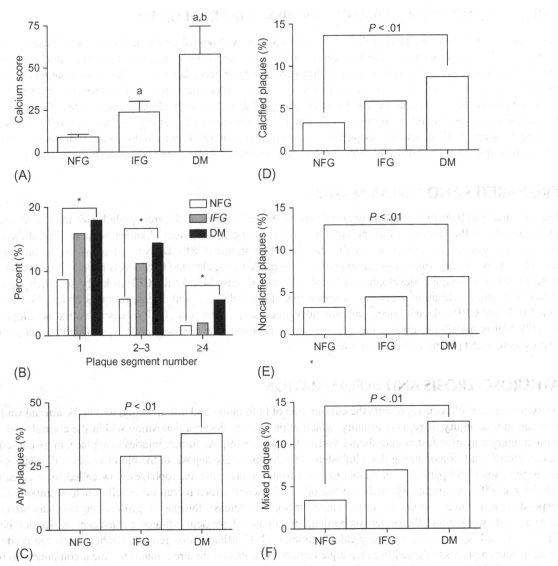

FIG. 1 Coronary calcium score (A), plaque segment number (B), prevalence of any plaques (C), calcified plagues (D), noncalcified plaques (E), and mixed plaques (F), according to the glucose metabolism. (a) *P* < .001 versus NFG; (b) *P* < .001 versus IFG, *P* < .001 versus NFG. *From Atherosclerosis 2009;205(1):156–62.*

results from the associated inflammation. In comparing prediabetics with normal patients, Dr. Su and associates found that prediabetics had reduced vasodilator response [49]. While most studies assessing endothelial dysfunction investigate the brachial artery, endothelial dysfunction can occur in virtually any vascular bed, including the coronary arteries. Using trans-thoracic echocardiography to measure coronary flow reserve, Erdogan et al. found that prediabetics had less coronary flow reserve and more evidence of microvascular disease than their normal fasting glucose counterparts [54]. In some instances, endothelial dysfunction and microvascular disease can result in activity-limiting angina without the presence of obstructive coronary artery disease.

PREDIABETES AND ACUTE CORONARY SYNDROME

The occurrence of acute coronary syndrome indicates that there is an atherosclerotic plaque and that a probable plaque rupture has occurred. Depending on the degree of flow limitation, patients may present with symptoms ranging from angina chest pain to overt infarction of the myocardium. From a study of nearly 700 patients, Kataoka et al. found that in patients with unstable angina, prediabetics had more hemodynamically significant stenosis and longer lesions when present relative to normal patients [55]. Further, at presentation, Sen et al. described that half of all patients who presented with a non-ST-segment elevation myocardial infarction were prediabetic [56]. In a study of over 1000 Scandinavian patients, Kuhl et al. found that over 70% of patients admitted with acute coronary syndrome had abnormal glucose metabolism [57]. Using a glucose tolerance test on patients in South Asia admitted with acute coronary syndrome, Ramachandran et al. confirmed that nearly half of patients were prediabetic [58]. In patients with a non-ST-segment elevation myocardial infarction, prediabetics possessed a higher risk of mortality (25% vs 6%) at 1 year [59]. Further describing this situation, Rioufol et al. found that patients admitted with acute coronary syndrome who were prediabetic not only had an increased 30-day risk of mortality but also had an increased risk of in-hospital mortality as well [60].

PREDIABETES AND ST SEGMENT ELEVATION MYOCARDIAL INFARCTION

An ST-segment elevation myocardial infarction occurs when a thrombus has completely occluded a coronary artery. It represents an urgent medical emergency, the timely intervention of which seeks to salvage myocardium and prevent further loss of cardiac function, the development of congestive heart failure, and to improve mortality. Just as many patients who present with unstable angina and non-ST-segment elevation myocardial infarction are found to be prediabetic, this relationship appears to be even more significant for ST-segment elevation myocardial infarctions. In a cohort of nearly 600 patients, Porter et al. described that nearly a third of patients admitted with ST-segment elevation myocardial infarction were prediabetic [61]. Particularly in lower-risk patients who present with an acute coronary occlusion, prediabetes forecasts 3-month mortality [62]. Finally, prediabetics are also less likely to have myocardial reperfusion postinfarction than patients without prediabetes [63].

PREDIABETES AND SILENT MYOCARDIAL INFARCTIONS

Given that many of the sequela of diabetes begins to emerge during the prediabetic state, particularly neuropathy and atherosclerosis, a natural question would be whether prediabetes is associated with silent myocardial infarctions. With their asymptomatic nature, identification of silent myocardial infarctions remains difficult in the clinical setting where symptoms drive testing. For this reason, cohort studies present the best opportunity to further diagnose and study this elusive entity. Analyzing longitudinal data from over 9000 participants in the Atherosclerosis Risk in Communities Study, Selvin et al. discovered that participants with diabetes and prediabetes had higher troponin levels than those with normal glucose metabolism, a finding that indicates subclinical myocardial damage [33]. With its selection of participants free of overt cardiovascular disease, the Multi-Ethnic Study of Atherosclerosis enrolled over 5000 individuals, most of whom underwent a 12-lead electrocardiogram on inclusion in the study. Using this cohort, Stacey et al. found that prediabetics were more likely to have signs of a prior myocardial infarction on their admission electrocardiogram [64]. There was a higher prevalence of silent myocardial infarctions in those with prediabetes compared with those with normal fasting glucose (3.5% ($n = 72$) vs 1.4% ($n = 30$)). Further, this relationship withstood adjustment for many powerful risk factors specific for silent myocardial infarction, such as age and hypertension. Specifically, after these adjustments, there remained higher odds of a potential silent myocardial infarction in those who were prediabetic compared with those with normal glucose metabolism (OR, 1.60 (95% CI, 1.0–2.5); $P = .048$). Using longitudinal 12-lead electrocardiogram data in an elderly cohort, Stacey et al. found that prediabetes increased the risk of having a silent myocardial infarction during follow-up [65]. Over a mean follow-up of 8.1 years, there were 94 silent myocardial infarctions identified among normal fasting glucose participants and 110 among prediabetic participants. Compared with those with normal glucose metabolism, the crude risk ratio for silent myocardial infarction with prediabetes was 1.53 (95% CI, 1.16–2.01). With adjustment, the hazard ratio for silent myocardial infarction in prediabetics compared with normal fasting glucose was 1.37 (95% CI, 1.03–1.83; $P = .032$).

Of all sequelae, silent myocardial infarctions carry a significant amount of risk for subsequent cardiovascular morbidity and mortality. Though largely unrecognized, a significant proportion of patients with cardiac disease may actually be asymptomatic. Recently, Zhang et al. observed that nearly 50% half of all myocardial infarctions are clinically silent [66]. In the Cardiovascular Health Study, 20% of myocardial infarctions were clinically silent [35], and in the Framingham Study, 25% of all myocardial infarctions were unrecognized [34]. Both of these studies found that the subsequent cardiovascular morbidity and mortality were similar between silent myocardial infarctions and recognized myocardial infarctions. As the number of individuals with prediabetes remains high, the public health burden imposed by potential silent myocardial infarctions may be substantial.

PREDIABETES AND POTENTIAL INTERVENTIONS

With nearly 50% of prediabetics progressing to overt diabetes mellitus within 10 years, preventing and reversing prediabetes remain critical steps in reducing future cardiovascular morbidity and mortality. In evaluating exercise and dietary changes, Pan et al. found that the diet, exercise, and diet-plus-exercise interventions led to a 31 ($P < .03$), a 46 ($P < .0005$), and a 42% ($P < .005$) reduction in progressing to diabetes, respectively [67]. With nearly 10 years of follow-up, Lindstrom et al. also observed a nearly 40% drop in development of diabetes mellitus in those with prediabetes who engaged in both exercise and dietary modifications [68]. Comparing lifestyle modifications to early use of metformin, Knowler et al. discovered that the lifestyle intervention reduced the incidence of diabetes by 58% (95% confidence interval, 48%–66%) and metformin by 31% (95% confidence interval, 17%–43%) [69]. Further lifestyle interventions and potential early pharmacological intervention may help to limit the damage that occurs during the prediabetic state and limit progression to frank diabetes.

CONCLUSIONS

Prediabetes represents a significant threat to public health due to the significant numbers of individuals affected. Many of the consequences of diabetes mellitus begin to manifest during the early stage of prediabetes. Prediabetes increases the risk of cardiovascular disease and merits more focused study and potentially intervention.

REFERENCES

[1] Guaraldi F, Pasquali R. Diabetes: from ancient Egypt to the 18th century. J Assoc Physicians India 2015;63(3):128.

[2] de Vegt F, Dekker JM, Stehouwer CD, Nijpels G, Bouter LM, Heine RJ. The 1997 American Diabetes Association criteria versus the 1985 World Health Organization criteria for the diagnosis of abnormal glucose tolerance: poor agreement in the Hoorn Study. Diabetes Care 1998;21(10):1686–90.

[3] American Diabetes Association. Clinical practice recommendations 1997. Diabetes Care 1997;20(Suppl. 1):S1–S70.

[4] American Diabetes Association. Diagnosis and classification of diabetes mellitus. Diabetes Care 2005;28(Suppl. 1):S37–S42.

[5] Gomez-Perez FJ, Aguilar-Salinas CA, Lopez-Alvarenga JC, Perez-Jauregui J, Guillen-Pineda LE, Rull JA. Lack of agreement between the World Health Organization Category of impaired glucose tolerance and the American Diabetes Association category of impaired fasting glucose. Diabetes Care 1998;21(11):1886–8.

[6] Alberti KZ. Definition, diagnosis, and classification of diabetes mellitus and its complications. World Health Organization; 1999. 32–3.

[7] Expert Panel on Detection, Evaluation, and Treatment of High Blood Cholesterol in Adults. Executive summary of the third report of the National Cholesterol Education Program (NCEP) expert panel on detection, evaluation, and treatment of high blood cholesterol in adults (Adult Treatment Panel III). JAMA 2001;285(19):2486–97.

[8] Grundy SM, Cleeman JI, Daniels SR, Donato KA, Eckel RH, Franklin BA, et al. Diagnosis and management of the metabolic syndrome: an American Heart Association/National Heart, Lung, and Blood Institute Scientific Statement. Circulation 2005;112(17):2735–52.

[9] The IDF Consensus Worldwide Definition of Metabolic Syndrome. International Diabetes Federation; 2006.

[10] Nichols GA, Hillier TA, Brown JB. Progression from newly acquired impaired fasting glucose to type 2 diabetes. Diabetes Care 2007;30(2):228–33.

[11] Cowie CC, Rust KF, Byrd-Holt DD, Eberhardt MS, Flegal KM, Engelgau MM, et al. Prevalence of diabetes and impaired fasting glucose in adults in the U.S. population: National Health and Nutrition Examination Survey 1999–2002. Diabetes Care 2006;29(6):1263–8.

[12] Menke A, Casagrande S, Geiss L, Cowie CC. Prevalence of and trends in diabetes among adults in the United States, 1988–2012. JAMA 2015;314(10):1021–9.

[13] Lee AM, Fermin CR, Filipp SL, Gurka MJ, DeBoer MD. Examining trends in prediabetes and its relationship with the metabolic syndrome in US adolescents, 1999–2014. Acta Diabetol 2017;54(4):373–81. https://doi.org/10.1007/s00592-016-0958-6.

[14] Hauner H, Hanisch J, Bramlage P, Steinhagen-Thiessen E, Schunkert H, Jockel KH, et al. Prevalence of undiagnosed Type-2-diabetes mellitus and impaired fasting glucose in German primary care: data from the German Metabolic and Cardiovascular Risk Project (GEMCAS). Exp Clin Endocrinol Diabetes 2008;116(1):18–25.

[15] Li S, Guo S, He F, Zhang M, He J, Yan Y, et al. Prevalence of diabetes mellitus and impaired fasting glucose, associated with risk factors in rural Kazakh adults in Xinjiang, China. Int J Environ Res Public Health 2015;12(1):554–65.

[16] Qi L, Feng L, Ding X, Mao D, Wang Y, Xiong H. Prevalence of diabetes and impaired fasting glucose among residents in the Three Gorges Reservoir Region, China. BMC Public Health 2014;14:1152.

[17] Chen CM, Yeh MC. The prevalence and determinants of impaired fasting glucose in the population of Taiwan. BMC Public Health 2013;13:1123.

[18] Anjana RM, Pradeepa R, Deepa M, Datta M, Sudha V, Unnikrishnan R, et al. Prevalence of diabetes and prediabetes (impaired fasting glucose and/or impaired glucose tolerance) in urban and rural India: phase I results of the Indian Council of Medical Research-INdia DIABetes (ICMR-INDIAB) study. Diabetologia 2011;54(12):3022–7.

[19] Oguoma VM, Nwose EU, Skinner TC, Digban KA, Onyia IC, Richards RS. Prevalence of cardiovascular disease risk factors among a Nigerian adult population: relationship with income level and accessibility to CVD risks screening. BMC Public Health 2015;15:397.

[20] Asbury AK. Understanding diabetic neuropathy. N Engl J Med 1988;319(9):577–8.

[21] Gregg EW, Li Y, Wang J, Burrows NR, Ali MK, Rolka D, et al. Changes in diabetes-related complications in the United States, 1990–2010. N Engl J Med 2014;370(16):1514–23.

[22] LoGerfo FW, Coffman JD. Current concepts. Vascular and microvascular disease of the foot in diabetes. Implications for foot care. N Engl J Med 1984;311(25):1615–9.

[23] Mogensen CE, Christensen CK. Predicting diabetic nephropathy in insulin-dependent patients. N Engl J Med 1984;311(2):89–93.

[24] Lewis EJ, Hunsicker LG, Bain RP, Rohde RD. The effect of angiotensin-converting-enzyme inhibition on diabetic nephropathy. The Collaborative Study Group. N Engl J Med 1993;329(20):1456–62.

[25] Mogensen CE. Progression of nephropathy in long-term diabetics with proteinuria and effect of initial anti-hypertensive treatment. Scand J Clin Lab Invest 1976;36(4):383–8.

[26] Kawasaki R, Wang JJ, Wong TY, Kayama T, Yamashita H. Impaired glucose tolerance, but not impaired fasting glucose, is associated with retinopathy in Japanese population: the Funagata study. Diabetes Obes Metab 2008;10(6):514–5.

[27] Nijenhuis-Rosien L, Hendriks SH, Kleefstra N, Bilo HJ, Landman GW. Nationwide diabetes-related lower extremity amputation rates in secondary care treated patients with diabetes in the Netherlands (DUDE-7). J Diabetes Complicat 2017;.

[28] Haffner SM, Lehto S, Ronnemaa T, Pyorala K, Laakso M. Mortality from coronary heart disease in subjects with type 2 diabetes and in nondiabetic subjects with and without prior myocardial infarction. N Engl J Med 1998;339(4):229–34.

[29] Conaway DG, O'Keefe JH, Reid KJ, Spertus J. Frequency of undiagnosed diabetes mellitus in patients with acute coronary syndrome. Am J Cardiol 2005;96(3):363–5.

[30] Knudsen EC, Seljeflot I, Abdelnoor M, Eritsland J, Mangschau A, Arnesen H, et al. Abnormal glucose regulation in patients with acute ST-elevation myocardial infarction—a cohort study on 224 patients. Cardiovasc Diabetol 2009;8:6.

[31] Agar JM. Silent myocardial infarction in diabetes mellitus. Med J Aust 1962;49(2):284–5.

[32] Vacek J. Silent myocardial infarction in the diabetic population. Am J Med 1984;76(4):A59. 68.

[33] Selvin E, Lazo M, Chen Y, Shen L, Rubin J, McEvoy JW, et al. Diabetes mellitus, prediabetes, and incidence of subclinical myocardial damage. Circulation 2014;130(16):1374–82.

[34] Kannel WB, Abbott RD. Incidence and prognosis of unrecognized myocardial infarction. An update on the Framingham study. N Engl J Med 1984;311(18):1144–7.

[35] Sheifer SE, Gersh BJ, Yanez 3rd ND, Ades PA, Burke GL, Manolio TA. Prevalence, predisposing factors, and prognosis of clinically unrecognized myocardial infarction in the elderly. J Am Coll Cardiol 2000;35(1):119–26.

[36] Montalescot G, Sechtem U, Achenbach S, Andreotti F, Arden C, Budaj A, et al. 2013 ESC guidelines on the management of stable coronary artery disease: the Task Force on the management of stable coronary artery disease of the European Society of Cardiology. Eur Heart J 2013;34(38):2949–3003.

[37] Won JC, Hong JW, Kim JM, Kim TN, Noh JH, Ko KS, et al. Increased prevalence of albuminuria in individuals with higher range of impaired fasting glucose: the 2011 Korea National Health and Nutrition Examination Survey. J Diabetes Complicat 2015;29(1):50–4.

[38] Plantinga LC, Crews DC, Coresh J, Miller 3rd ER, Saran R, Yee J, et al. Prevalence of chronic kidney disease in US adults with undiagnosed diabetes or prediabetes. Clin J Am Soc Nephrol 2010;5(4):673–82.

[39] Papanas N, Vinik AI, Ziegler D. Neuropathy in prediabetes: does the clock start ticking early? Nat Rev Endocrinol 2011;7(11):682–90.

[40] Ziegler D, Papanas N, Vinik AI, Shaw JE. Epidemiology of polyneuropathy in diabetes and prediabetes. Handb Clin Neurol 2014;126:3–22.

[41] Bongaerts BW, Rathmann W, Heier M, Kowall B, Herder C, Stockl D, et al. Older subjects with diabetes and prediabetes are frequently unaware of having distal sensorimotor polyneuropathy: the KORA F4 study. Diabetes Care 2013;36(5):1141–6.

[42] Bisoendial RJ, Kastelein JJ, Stroes ES. C-reactive protein and atherogenesis: from fatty streak to clinical event. Atherosclerosis 2007;195(2):e10–8.

[43] Aronson D, Bartha P, Zinder O, Kerner A, Shitman E, Markiewicz W, et al. Association between fasting glucose and C-reactive protein in middle-aged subjects. Diabet Med 2004;21(1):39–44.

[44] Costa A, Casamitjana R, Casals E, Alvarez L, Morales J, Masramon X, et al. Effects of atorvastatin on glucose homeostasis, postprandial triglyceride response and C-reactive protein in subjects with impaired fasting glucose. Diabet Med 2003;20(9):743–5.

[45] Panazzolo DG, Sicuro FL, Clapauch R, Maranhao PA, Bouskela E, Kraemer-Aguiar LG. Obesity, metabolic syndrome, impaired fasting glucose, and microvascular dysfunction: a principal component analysis approach. BMC Cardiovasc Disord 2012;12:102.

[46] Simental-Mendia LE, Lazalde B, Zambrano-Galvan G, Simental-Saucedo L, Rabago-Sanchez E, Rodriguez-Moran M, et al. Relation between C-reactive protein and impaired fasting glucose in obese subjects. Inflammation 2012;35(5):1742–6.

[47] Cimmino G, Ragni M, Cirillo P, Petrillo G, Loffredo F, Chiariello M, et al. C-reactive protein induces expression of matrix metalloproteinase-9: a possible link between inflammation and plaque rupture. Int J Cardiol 2013;168(2):981–6.

[48] Eringa EC, Serne EH, Meijer RI, Schalkwijk CG, Houben AJ, Stehouwer CD, et al. Endothelial dysfunction in (pre)diabetes: characteristics, causative mechanisms and pathogenic role in type 2 diabetes. Rev Endocr Metab Disord 2013;14(1):39–48.

[49] Su Y, Liu XM, Sun YM, Jin HB, Fu R, Wang YY, et al. The relationship between endothelial dysfunction and oxidative stress in diabetes and prediabetes. Int J Clin Pract 2008;62(6):877–82.

[50] American Diabetes Association. Diagnosis and classification of diabetes mellitus. Diabetes Care 2004;27(Suppl. 1):S5–S10.

[51] Eun YM, Kang SG, Song SW. Fasting plasma glucose levels and coronary artery calcification in subjects with impaired fasting glucose. Ann Saudi Med 2016;36(5):334–40.

[52] Lim S, Choi SH, Choi EK, Chang SA, Ku YH, Chun EJ, et al. Comprehensive evaluation of coronary arteries by multidetector-row cardiac computed tomography according to the glucose level of asymptomatic individuals. Atherosclerosis 2009;205(1):156–62.

[53] Kronmal RA, McClelland RL, Detrano R, Shea S, Lima JA, Cushman M, et al. Risk factors for the progression of coronary artery calcification in asymptomatic subjects: results from the Multi-Ethnic Study of Atherosclerosis (MESA). Circulation 2007;115(21):2722–30.

[54] Erdogan D, Yucel H, Uysal BA, Ersoy IH, Icli A, Akcay S, et al. Effects of prediabetes and diabetes on left ventricular and coronary microvascular functions. Metabolism 2013;62(8):1123–30.

[55] Kataoka Y, Yasuda S, Morii I, Otsuka Y, Kawamura A, Miyazaki S. Quantitative coronary angiographic studies of patients with angina pectoris and impaired glucose tolerance. Diabetes Care 2005;28(9):2217–22.

[56] Sen K, Mukherjee AK, Dharchowdhury L, Chatterjee A. A study to find out the proportion of prediabetes in patients with acute coronary syndrome in a medical college of Kolkata. J Indian Med Assoc 2008;106(12):776–8.

[57] Kuhl J, Jorneskog G, Wemminger M, Bengtsson M, Lundman P, Kalani M. Long-term clinical outcome in patients with acute coronary syndrome and dysglycaemia. Cardiovasc Diabetol 2015;14:120.

[58] Ramachandran A, Chamukuttan S, Immaneni S, Shanmugam RM, Vishnu N, Viswanathan V, et al. High incidence of glucose intolerance in Asian-Indian subjects with acute coronary syndrome. Diabetes Care 2005;28(10):2492–6.

[59] Facila L, Bertomeu-Gonzalez V, Sanchis J, Bodi V, Nunez J, Llacer A, et al. Glucose levels in non-diabetic patients. Is it a prognostic factor in acute coronary syndrome? Rev Clin Esp 2006;206(6):271–5.

[60] Rioufol G, Zeller M, Oudot A, L'Huillier I, Buffet P, Beer JC, et al. Predictive value of glycemia in acute coronary syndromes. Arch Mal Coeur Vaiss 2004;97(3):47–50.

[61] Porter A, Assali AR, Zahalka A, Iakobishvili Z, Brosh D, Lev EI, et al. Impaired fasting glucose and outcomes of ST-elevation acute coronary syndrome treated with primary percutaneous intervention among patients without previously known diabetes mellitus. Am Heart J 2008;155(2):284–9.

[62] Knudsen EC, Seljeflot I, Abdelnoor M, Eritsland J, Mangschau A, Muller C, et al. Impact of newly diagnosed abnormal glucose regulation on long-term prognosis in low risk patients with ST-elevation myocardial infarction: a follow-up study. BMC Endocr Disord 2011;11:14.

[63] Fefer P, Hod H, Ilany J, Shechter M, Segev A, Novikov I, et al. Comparison of myocardial reperfusion in patients with fasting blood glucose < or =100, 101 to 125, and >125 mg/dl and ST-elevation myocardial infarction with percutaneous coronary intervention. Am J Cardiol 2008;102(11):1457–62.

[64] Stacey RB, Leaverton PE, Schocken DD, Peregoy JA, Bertoni AG. Prediabetes and the association with unrecognized myocardial infarction in the multi-ethnic study of atherosclerosis. Am Heart J 2015;170(5):923–8.

[65] Stacey R, Leaverton P, Schocken D, Peregoy J, Lyles M, Bertoni A, et al. Impaired fasting glucose increases risk for an unrecognized myocardial infarction: the cardiovascular health study. Circulation 2015;132. A16000.

[66] Zhang ZM, Rautaharju PM, Prineas RJ, Rodriguez CJ, Loehr L, Rosamond WD, et al. Race and sex differences in the incidence and prognostic significance of silent myocardial infarction in the atherosclerosis risk in communities (ARIC) study. Circulation 2016;133(22):2141–8.

[67] Pan XR, Li GW, Hu YH, Wang JX, Yang WY, An ZX, et al. Effects of diet and exercise in preventing NIDDM in people with impaired glucose tolerance. The Da Qing IGT and diabetes study. Diabetes Care 1997;20(4):537–44.

[68] Lindstrom J, Peltonen M, Eriksson JG, Ilanne-Parikka P, Aunola S, Keinanen-Kiukaanniemi S, et al. Improved lifestyle and decreased diabetes risk over 13 years: long-term follow-up of the randomised Finnish Diabetes Prevention Study (DPS). Diabetologia 2013;56(2):284–93.

[69] Knowler WC, Barrett-Connor E, Fowler SE, Hamman RF, Lachin JM, Walker EA, et al. Reduction in the incidence of type 2 diabetes with lifestyle intervention or metformin. N Engl J Med 2002;346(6):393–403.

Chapter 23

Mindfulness-Based Therapy and Heart Health

Kathleen P. Ismond*, Cecilia Bukutu†, Sunita Vohra*

*Department of Pediatrics, University of Alberta, Edmonton, AB, Canada, †Institutional Research and Program Development, Concordia University of Edmonton, Edmonton, AB, Canada

INTRODUCTION

Cardiovascular disease (CVD) is the current leading cause of death with an estimated global annual mortality rate of 17 million increasing to 23.6 million by 2030 [1]. In 2013, the leading causes of CVD-related deaths were ischemic heart disease and cerebrovascular disease; ~70% of all CVD deaths occurred in low- to middle-income countries [1]. Mortality rates are influenced by the accessibility and affordability of aspirin, β-blockers, ACEIs, and statins that remain unavailable to 60% of households in low-income countries [2].

CVD is not a new disease as there have been reports describing symptoms and autopsy findings throughout human history that are consistent with CVD. For example, Leonardo da Vinci (1452–1519) provided an early description of arteriosclerosis in older adults, while Hippocrates (400 BCE) was the first to recognize the symptoms and outcomes of stroke [3]. CT imaging of Egyptian mummies c. 1900 BCE revealed calcification of their arterial and heart walls, a key indicator of arteriosclerosis [4].

The Framingham Heart Study was initiated in 1948 to identify risk factors for CVD as it was the general tenet that many factors contributed to disease development. The longitudinal, observational cohort study enrolled 5,209 healthy adults (aged 30–62 years; M/F 5:6) without covert symptoms of CVD [5]. At baseline and every 2 years, study participants underwent an extensive physical examination and lifestyle interview. Over time, the investigators have identified major CVD risk factors associated with lifestyle generating over 1,200 publications in addition to the development of effective treatments and preventive strategies. The modifiable risk factors for CVD are high blood pressure and or blood cholesterol, smoking, obesity, glucose regulation in diabetes (types 1 and 2), and physical inactivity. The nonmodifiable risk factors include genetics, gender, and aging. The Framingham study researchers also identified related characteristics, such as blood triglyceride and high-density lipoprotein (HDL) cholesterol levels, and psychosocial issues.

Despite significant advances in the diagnosis and treatment of CVD, prevention by managing the known risk factors provides the best outcome. Much effort has been invested in developing strategies to minimize CVD risk factors, such as nicotine replacement therapy (NRT) for smoking cessation, weight loss programs for obesity, and a range of exercise therapy programs. However, changing human behavior is challenging. Even the Egyptian members of the pharaoh's court did not forego their privileged diet high in meat in favor of health [4].

The first preliminary results investigating mindfulness-based therapy for the treatment of chronic pain were published in 1982 by Kabat-Zinn [6]. Building upon the mindful meditation practices of Theravada Buddhism, Mahayana Buddhism, and yogic traditions, Kabat-Zinn designed a secular program for nonexperts or meditation novices to achieve a state of "detached self-observation," the opposite of concentration meditation [6]. In this way, pleasant and unpleasant sensations can be explored individually, noted, and dismissed and then move to the next without assigning a hierarchical classification or content value to it. Within the context of chronic pain, detached observation results in a learned recognition of the sensation while reducing the emotional and cognitive feelings associated with the experience, hurt, and suffering [6]. Over Kabat-Zinn's 10-week study, the participants engaged in his "stress reduction and relaxation program." Chronic pain severity was measured at baseline and again at 10 weeks. Upon final analysis, 65% reported a significant decrease in the severity of their chronic pain, and this was directly attributed to the program [6]. Since then, the mindfulness-based stress reduction (MBSR) program or the mindfulness-based cognitive therapy (MBCT, developed specifically for depression) have been investigated to assess their efficacy as either a treatment or preventive strategy to modify human behavior in a range of health conditions, such as cancer, cardiovascular disease, and metabolic syndrome, and addictions for both adults and children [7].

Complementary therapies are widely sought in North America and abroad. In the United States, 40% of the population use complementary therapies, including products (e.g., dietary supplements and herbal extracts), and practices, whether practitioner-based (e.g., chiropractors, massage therapists, and acupuncturists) or through adherence to healing philosophies and self-care (e.g., yoga, music, dance, and mindfulness) [8]. An epidemiological survey of patients with heart disease reported that 36% used complementary therapies in the previous 12 months [8]. Apart from treating CVD, complementary therapies are also used for managing CVD risk factors, such as hypertension, diabetes, and cholesterol levels. Mindfulness has been investigated in patients diagnosed with CVD (see [9,10]) and for prevention of CVD.

In the context of heart disease, most mindfulness-based research has been primarily exploratory characterized by small study sizes, short-term follow-ups after the standard 8-week MBSR training session, and much variation regarding the eligibility criteria for control groups or use of active controls. A 2015 systematic review by Loucks et al. described the gamut of studies assessing MBSR for treatment of CVD risk factors. The authors then suggested a consensus-based mechanism and theoretical framework to explain the relationship between mindfulness and mitigation of CVD risk [11]. A cautionary point is that the review and theory were based on studies that had inherent limitations regarding methodological quality. In retrospect, their efforts to develop a mechanistic explanation and framework to understand the relationship between mindfulness and CVD risk factors may have been premature.

Since the Loucks et al. article, more randomized controlled trials (RCTs) in this field have been published. The current review will focus on assessing RCTs published between March 1, 2012 and March 31, 2017 that assessed the efficacy of mindfulness-based meditation on improving clinical outcomes of the major modifiable CVD risk factors: smoking (tobacco), diabetes mellitus, hypertension, obesity, nutrition, and physical inactivity [1]. To ensure that the findings are clinically relevant, the enrolled participants had to exhibit the CVD risk factor at baseline. This caveat excluded RCTs that recruited only healthy individuals. On a side note, stress is not a CVD risk factor in and of itself but can contribute to the onset or exacerbation of others, such as emotional eating, smoking, and hypertension [1]. The role of mindfulness for the treatment of stress has been reviewed elsewhere (see [12]) and will not be addressed in this chapter.

TOBACCO SMOKING

Tobacco smoking is one of the key risk factors for many conditions, such as cancer, emphysema, and CVD. A total of five RCTs were identified and are summarized in Table 1. Each trial provided the control group with an active intervention unrelated to mindfulness. Due to high attrition rates (ranging from 30% to 50% for the first four RCTs), the studies were underpowered. Three studies used a modified form of MBSR, "Mindfulness Training for Smoking" (MTS) offered over 6 weeks, that was developed to apply mindfulness skills to manage the challenges of smoking cessation, such as coping with smoking triggers and withdrawal symptoms [13–15].

The first trial, by Davis et al. in 2013 [13], recruited college students (M/F, 7:1) who self-reported ≥5 alcohol binges per month. The controls received "Interactive Learning for Smokers" (ILS), a 6-week course combining elements from two national programs: "Freedom from Smoking" (FFS) developed by the American Lung Association and the "Nicotine Dependence Center Program" issued by The Mayo Clinic. Significantly, more days of smoking abstinence was reported for participants in the MTS intervention during the 2 weeks after Quit Day compared with controls according to the intent-to-treat (ITT) population analysis. The difference in the number of days of abstinence is noteworthy considering that the mean baseline smoking rate was 13.75 (standard deviation (SD) 6.36) cigarettes per day. However, no significant difference between the groups was found for the primary outcome of biochemically confirmed, 7-day point prevalence abstinence rate at 2 weeks after the Quit Day (MTS = 20.0% vs control = 4.0%, $P = 0.08$).

In a subsequent trial by Davis et al., the investigators compared MTS with a telephonic tobacco quit line and provided nicotine replacement therapy patches for 4 weeks after the Quit Day [14]. Again, no significant difference in the primary study outcome was found regarding the 7-day point prevalence abstinence rate at 4 or 24 weeks after the Quit Day using the ITT population (M/F 1:1; mean age 41.7 years, SD 13.3). Following randomization, 78 participants dropped out creating significant differences in baseline values between the two study groups. The MTS intervention had a higher percentage of heavier smokers, minorities, and individuals with post-secondary education. Controlling for cigarettes smoked per day and education, initiators in the MTS ($N = 59$) and control ($N = 59$) groups were found to have biochemically confirmed significant differences for the point prevalence abstinence rates at 4 and 24 weeks. The adjusted 4-week continuous abstinence rate after Quit Day favored the MTS group (odds ratio (OR) 3.32 (1.31–8.41), $P = 0.01$). Although the results were promising, participants were not blinded to their randomly assigned study group that was attributed to the high attrition rate occurring between the periods of randomization and program initiation.

TABLE 1 Overview of Tobacco Smoking Randomized Controlled Trials

First Author Year Country	Intervention (I) (N)	Control (C) (N)	Outcomes	Intervention Versus Control	Comments	Attrition I Versus C
Davis [13] 2013 the United States	MTS for 6 weeks (30)	ILS for 6 weeks (25)	Days abstinent for 2 weeks after quit day (ITT)	5.10 (SD 6.00) versus 2.04 days (SD 3.98), $P=0.03$	Age range: 18–29 years	30% or 45%
Davis [14] 2014 the United States	MTS for 4 weeks + NRT for 4 weeks (59)	Telephonic Wisconsin Tobacco Quit Line + NRT for 4 weeks (59)	Abstinence[a] at • 4 weeks • 24 weeks	45.8 versus 25.4%, $P=0.04$ 38.7 versus 20.6%, $P=0.03$	NSD for ITT groups (Inter=105, C=91) at 4 or 24 weeks	78% or 40%
Davis [15] 2014 the United States	MTS for 7 weeks + NRT for 2 weeks (67)	7-week FFS + relaxation programming for 7 weeks + NRT for 2 weeks (68)	Abstinence[a] rate 6 months after quit day (ITT)	25 versus 17.9%, $P=0.35$ (CI: 0.67–0.51)	MTS improved urge intensity, mindfulness, perceived stress, and experiential avoidance	76% or 43%
Schuman-Olivier [16] 2014 the United States	Mindfulness training for 4 weeks; subdivided into high (17) and low (22) based on FFMQ	FFS for 4 weeks; subdivided into high (25) and low (21) based on FFMQ	Average cigarettes per day during 12th-week follow-up comparing high groups	3.9 versus 11.1 cigarettes per day, $P=0.01$	Post hoc analysis of a 2011 study, NSD between low groups	26% or 30%
Ruscio [17] 2015 the United States	Brief mindfulness 20 min/day for 2 weeks (20)	Sham meditation 20 min/day for 2 weeks (17)	Cigarettes per day over time	Group × day interaction, $F(1, 436)=5.50$, $P=0.01$	No emphasis on smoking cessation	7% or 16%

Abbreviations: *CI*, 95% confidence interval; *FFS*, Freedom From Smoking program developed by the American Lung Association; *FFMQ*, Five-Facet Mindfulness Questionnaire; *ILS*, interactive learning for smokers; *ITT*, intent-to-treat; *LMM*, linear mixed model; *MTS*, mindfulness training for smoking; *NRT*, nicotine replacement therapy; *SD*, standard deviation; *SES*, socioeconomic status.

[a]*Biochemically confirmed 7-day point prevalence abstinence.*

In a third study by Davis and colleagues, the team ensured that the participants were blinded throughout, and the controls were provided with a similarly intensive program that also included a relaxation component unrelated to mindfulness [15]. The biochemically confirmed abstinence rate 6 months from quit day did not significantly differ between the two groups according to the ITT analysis ($N=135$; M/F ~1:1; mean age 44.5 years). Importantly, the groups were well matched at baseline. Despite the null outcome, MTS participants reported improved self-management regarding smoking urges, stress, and experiential avoidance.

In 2014, Schuman-Olivier et al. [16] reanalyzed data collected from a 2011 trial [18] that reported that 4 weeks of mindfulness training ($N=39$) versus 4 weeks of FFS ($N=46$) resulted in a greater biochemically confirmed point prevalence abstinent rate at 17 weeks after training (36% vs 6%, $P=0.012$). A key trait of mindfulness is the nonjudgment of personal experiences. To determine if this inherent personal trait influences therapy outcomes, the post hoc analysis equally divided group participants (M/F 5:3; mean age, 46 years) as either high or low according to their dispositional mindfulness determined by the 39-item Five-Facet Mindfulness Questionnaire (FFMQ). The authors concluded that individuals who rarely assessed their personal experience (classified as high) were more likely to respond to mindfulness instead of FFS. Thus, this suggests that there is merit to prescreen smokers and provide those with high personal nonjudgment characteristics with a mindfulness-based smoking cessation program instead of FFS. Follow-on studies exploring this aspect have yet to be reported.

The last study compared brief mindfulness ($N=20$) with sham meditation ($N=17$) each practiced for 20 min/day for 2 weeks [17]. Participants were told to smoke as much or as little as they wished with no emphasis on cessation or abstinence. There was no difference found with respect to either carbon monoxide or cotinine levels over the course of the study. Self-reported daily smoking rates decreased over time in the mindfulness group. The ecological momentary assessment data indicated that the mindfulness group benefited from a reduced overall negative effect of smoking and craving sensations in the time immediate following meditation.

In summary, modifying smoking behavior to the point of abstinence is challenging and requires effective self-control and willingness on the smoker's part. Spurred by early studies such as Brewer et al. [18], recent, better-quality, blinded RCTs have not provided unequivocal evidence that mindfulness significantly increases smoking cessation rates or biochemically confirmed abstinence rates. An initial attempt to identify smokers who respond better to either MTS or FFS using the FFMQ scale indicates that more research is required to identify individuals who may respond preferentially to one program over another. Smoking is a complex behavior requiring a wide array of cessation program options to appeal to the needs and interests of each smoker. As mindfulness is a nonpharmacological intervention that offers additional cognitive benefits to the practitioner and can be used in combination with other therapies (e.g., nicotine replacement), the growing evidence base supports its ongoing use in combination with existing strategies to achieve smoking cessation and abstinence.

DIABETES MELLITUS (TYPES 1 AND 2)

Diabetes has long been associated with increased risk for developing CVD. Poor mental health and depression have been linked with the progression of CVD, while the corollary, improving mental health, is only now being investigated in the hope to improve clinical measures, such as glucose regulation and hemoglobin A1c levels (HbA1c) [19]. Unlike the smoking studies, usual care or wait-listed patients were commonly used as controls, blinding participants was not a priority, and the attrition rates were much lower. The interventions were uniformly 8 weeks in length.

Van Son et al. compared MBCT with usual care in patients with type 1 and 2 diabetes and published two articles describing the short- and long-term outcomes, respectively [20,21]. The cohort had a male-to-female ratio of 1:1, and the mean age was 56.5 years. There was no significant between-group difference with respect to HbA1c levels at either 8th-week [21] or 6th-month [20] posttraining. A study limitation was the inclusion of participants who had slightly elevated HbA1c values (mean of 59 mmol/mol, SD = 13) that falls between the range of the treatment goal (≤53) and more complex diabetes (≤64). However, MBCT led to significant differences in stress reduction, depressive symptoms, and anxiety in the short- and long-term studies.

The two studies by Tovote et al. used wait-listed patients as the controls for comparing the effect of MBCT or CBT in patients with diabetes regarding a range of psychosocial factors, including depression and diabetes-related stress [22,23]. The same cohort used in both studies had a male-to-female ratio of ~1:1 and a mean age of 53.1 years (SD = 11.8). With respect to glucose regulation, HbA1c analyses were restricted to the intervention groups. At the third-month posttraining, there were no significant decreases in HbA1c levels for either intervention even though the mean baseline values were high, 63.4 (SD = 9.6) and 67.1 (SD = 15.2) for the MBCT and CBT groups, respectively [22]. At the 9th-month follow-up, the MBCT group had a mean decrease of 1 mmol/mol HbA1c versus 1.9 mmol/mol for the CBT group. Although a favorable decreasing trend was found, the decreased amounts were not clinically meaningful. The authors concluded that the interventions offered participants' benefits as there were significant within group improvements in depression, well-being, anxiety, and diabetes distress [23].

A 2014 German study enrolled type 2 diabetes patients with early kidney disease to explore the physiological effect of improved depression and stress in response to MBSR therapy [19]. Over the 3-year study, this cohort of 110 very ill patients (M/F 3.6:1, mean age 59 years) experienced 19 events including four deaths. There were no between-group differences for these events and were attributed to events unrelated to the study. The mean urinary albumin-creatinine ratio (ACR) was determined using 24 h urine samples collected on 3 consecutive days. At the first-year follow-up, MBSR patients had decreased urinary ACR values (44–37 mg/g) compared with increased values for the controls (47–55 mg/g, P = 0.05). The relative risk reduction was 59.6% (95% CI; 10.9–81.7) for ACR progression in patients receiving MBSR therapy. As well, a 71.2% (95% CI; 27.7–88.5) relative risk reduction for the carotid intima media thickening (IMT) progression was also associated with MBSR. These changes were not present at follow-up (years 2 and 3) even though there were no changes in the participants' blood pressure management medications. These physical changes correlated with significantly lower stress levels of the MBSR versus the controls after 1 year. Of interest is that the stress levels of the controls were similar to those of MBSR in years 2 and 3 with the authors suggesting that the unblinded controls accessed MBSR materials independently. Without intermittent MBSR booster sessions throughout the 3-year follow-up, those in the MBSR group likely reduced or stopped their daily practices over time. The small sample sizes of the two study groups require that this be repeated on a larger scale to obtain clinically relevant evidence.

An open-label Korean Nursing study compared two interventions, a translated version of MBSR (KMBSR) and recreational walking, with a control group [24]. Each cohort also received education about diabetes and its management. Only patients with type 2 diabetes were recruited (M/F 1:1, mean age 66.3 years). Comparing blood glucose levels from baseline to the end of the 8-week trial revealed no significant differences within or between the study groups. Interestingly, there were no between-group differences in changes with respect to diabetes-related stress response either. Only the walking and control groups experienced improvements in psychological response. Here, KMBSR did not improve cognitive health within a brief time frame of 8 weeks. Meanwhile, the authors noted significant within group changes in the vascular parameters (e.g., serum plasminogen activator inhibitor-1 and blood tissue-type plasminogen activator) in both the KMBSR and walking groups. These parameters are important upstream contributors to the development of arteriosclerosis.

Overall, the RCTs in diabetes have not provided evidence that mindfulness improves glycemic control despite their design inconsistencies. Rather, study findings highlight the variability in cognitive and clinical measures in response to MBSR over time. The studies that assessed vascular factors noted improvements in response to the interventions in this patient population. Although mindfulness did not directly improve glucose regulation, indirect benefits were derived with respect to improving CVD risk factors in diabetics that merit additional, well-designed trials to fully assess the potential. The mindfulness intervention was generally well received by the participants and the studies had attrition rates within the anticipated ranges (Table 2).

TABLE 2 Overview of Diabetes Mellitus Randomized Controlled Trials

First Author Year Country	Intervention(s) (N)	Control (N)	Outcomes	Intervention(s) Versus Control	Comments
Van Son [20] 2013 Netherlands	MBCT (70)	Usual care (69)	HbA_{1c} change from baseline to 8 weeks post training	NSD	Type 1 and 2
Van Son [21] 2014 Netherlands	MBCT (70)	Usual care (69)	HbA_{1c} change from baseline to 6 months post training	NSD	Type 1 and 2
Tovote [22] 2014 Netherlands	MBCT (31) CBT (32)	Wait listed (31)	HbA_{1c} change from baseline to 3 months post training	NSD between MBCT and CBT	Type 1 and 2 plus comorbid depression
Tovote [23] 2014 Netherlands	MBCT (31) CBT (32)	Wait listed (31)	HbA_{1c} change from baseline to 12 months post training	NSD between MBCT and CBT	Type 1 and 2 plus comorbid depression
Kopf [19] 2014 Germany	MBSR (53)	Usual care (57)	Urinary ACR change from baseline to 1 year	−7 versus 8 mg/g, $P=0.05$	Type 2 with early kidney disease, NSD at 2 or 3 years, open label
Jung [24] 2015 Korea	KMBSR+education (21) Walking+education (18)	Education (17)	Blood sugar level (mg/dL) change from baseline to posttraining	KMBSR −7.38 (SD 19.6) Walking −13.11 (SD. 30.4) Controls −7.24 (SD 13.4) NSD between or within groups	Type 2, NSD for psychological or diabetes-related stress

Abbreviations: *ACR*, albumin-creatinine ratio; *CBT*, cognitive-based therapy; *HbA1c*, hemoglobin A1c; *KMBSR*, Korean mindfulness-based stress reduction; *MBCT*, mindfulness-based cognitive therapy; *NSD*, no significant difference; *PAI-1*, plasminogen activator inhibitor-1 (serum); *SD*, significant difference; *t-PA*, tissue-type plasminogen activator.

HYPERTENSION

A key risk factor for stroke, hypertension has been studied in relation to mindfulness interventions in three RCTs (Table 3). As seen in the previous work by Kopf et al., mindfulness was associated with decreases in vascular risk factors in patients with diabetes [19]. Here, we limited the review of recent RCTs to those recruiting individuals diagnosed with hypertension without coronary heart disease or other conditions to minimize potential confounding elements.

The first study compared 8 weeks of MBSR with a control group who participated in a research assistant-led social support group that mirrored the intervention with respect to time commitment and attention received [25]. The participants were recruited from a low-income senior house community and had a mean age of 73 years; no details provided regarding gender. There were no significant differences at baseline with respect to ever-smoking status, education, race, heart rate, systolic blood pressure, and perceived stress. Of the 20 patients, 18 were on hypertensive medications at baseline with 3 being normotensive and 4 prehypertensive. Although participants had comorbid conditions, the most common were gastroesophageal reflux disease or high cholesterol. Multivariate regression results revealed noteworthy changes in both the systolic and diastolic blood pressure values between the groups.

In 2013, Hughes et al. compared MBSR with an active control that consisted of supervised training in progressive muscle relaxation (PMR) [26]. Participants were not receiving medications for hypertension during the study, and all had elevated blood pressure levels at baseline (systolic 120–139 mm Hg or diastolic 80–89 mm Hg) consistent with prehypertension. Otherwise, participants were healthy, nonsmokers without any comorbid conditions treated with medications known to alter blood pressure (M/F 4:6, mean age 50.3 years). Significant between-group differences were seen for both systolic and diastolic blood pressure rates with the intervention group benefiting from greater improvement than controls. The more robust measure of 24 h ambulatory blood pressure monitoring found no significant difference between the groups. The attrition rate was 7 (25%) and 11 (40%) for the intervention and control groups, respectively. This high rate of loss limits the interpretation of the results.

The largest study (N = 87) by Blom et al. compared MBSR with a wait-list control group (M/F 2:3, mean age 56 years) [27]. All participants were diagnosed with stage 1 hypertension according to 24 h ambulatory blood pressure monitoring; none were taking hypertension-related medications. No significant differences were identified either within or between the study groups. Unlike the previous RCTs, 24 h ambulatory monitoring was used to evaluate changes in blood pressure rates, and there was a 4-week gap between completing the intervention and the final blood pressure monitoring. Overall, 14 participants failed to complete the 12-week clinical trial of which 10 were in the control.

TABLE 3 Overview of Hypertension Randomized Controlled Trials

First Author Year Country	Intervention (N)	Control (N)	Outcomes	Intervention Versus Control	Comments
Palta [25] 2012 The United States	MBSR (12)	Social support (8)	Baseline to 8 weeks: Change in systolic BP (clinic) Change in diastolic BP (clinic)	−21.9 (SE 8.3), P = 0.02 −16.7 (SE 4.6), P = 0.003	90% on hypertension-related meds
Hughes [26] 2013 The United States	MBSR (28)	PMR (28)	Baseline to 8 weeks: Change in systolic BP (clinic) Change in diastolic BP (clinic)	−4.8 versus −0.7, P = 0.016 −1.9 versus 1.2, P = 0.008	Prehypertensives, 0% on meds
Blom [27] 2014 Canada	MBSR (46)	Wait-list control (41)	Baseline to fourth-week postintervention: Change in systolic BP ± SD (24 h ambulatory) Change in diastolic BP ± SD (24 h ambulatory)	−0.4 ± 6.7 versus −0.4 ± 7.8, NSD 0.04 ± 4.9 versus −0.4 ± 4.6, NSD	Stage 1 hypertension, 0% on meds

Abbreviations: BP, blood pressure measured in mm Hg; NSD, no significant difference; PMR, progressive muscle relaxation; SE, standard error.

The contradictory results of the three studies reviewed fail to inform if mindfulness training improves blood pressure in the setting of prehypertension or hypertension. In clinic, blood pressure monitoring is known to create an artificial increase in rates due to the "white coat" effect where patients are ill at ease in unfamiliar surroundings [28]. This effect may have contributed to the significant changes recorded in the Palta et al. study, whereas the Hughes et al. team described in detail how they made every effort to minimize this effect [25,26]. Blom and colleagues observed that significant improvements associated with cognitive or mindfulness training were reported in studies where the participants were taking blood pressure medication apart from one study [27]. They theorized that the intervention improved patient compliance with their medication regimens, thereby resulting in improved blood pressure levels. This explanation does not address the significant changes in the unmedicated prehypertensive population trial by Hughes et al. [26].

The variation in trial design, follow-up times, and participant eligibility makes it challenging to identify let alone quantify clinically relevant benefits of mindfulness in hypertension. A review and meta-analysis of meditation and blood pressure concluded that mindfulness affords practitioners with statistically significant systolic and diastolic improvements for hypertensive and normotensive individuals [29]. The authors highlighted the need for large-scale, well-designed trials and the use of 24 h ambulatory blood pressure monitoring. Future research needs to also focus on identifying the mechanism(s) by which meditation effects physiological changes that can alter blood pressure.

OBESITY

Imbalances in food intake regulation are often associated with excess weight and depression. The underlying causes of excess body weight are complex and can involve reward-driven eating behaviors or psychological stress, such as acute isolated or chronic ongoing stressors [30]. Mindfulness-based training has proved promising in helping individuals overcome challenging behaviors and patterned conditioning. As an example, a 2014 RCT reported that a binge eating disorder was resolved in 95% ($N=19$) of individuals receiving a mindfulness-based program at the fourth-month postintervention compared with 48% ($N=21$) of the wait-listed controls [31].

The mainstays of studies in this area have focused on assessing the efficacy of mindfulness, either alone or in combination with other programming, to achieve and maintain weight loss. As reviewed by Katterman and colleagues, both emotional eating and binge eating respond favorably to mindfulness-based meditation practices, although the effect on significant and sustained weight loss is inconclusive [32]. Curiously, several recent RCTs recruited healthy or overweight participants rather than obese persons to assess the effect of a mindfulness intervention on stress eating [33], food intake following food-cue exposure [34], and portion size [35]. The importance of adhering to a program intended to change food behaviors is not compelling when the participant is not at risk. As well, findings from such studies are not clinically relevant if obtained in a population different than the target.

A 2016 study that recruited 194 obese individuals (mean BMI 35.5, M/F 1:4, mean age 47 years) for a 5.5-month diet-exercise program with and without the added component of mindfulness [30]. The primary study outcome was weight loss at 18 months from baseline. The estimated effects favored the intervention ($N=100$) over the control group coupled with an extra weight loss of 1.7 kg (95% CI; −4.7 to 1.2, $P=0.24$). This trend persisted for both fasting glucose measures ($P=0.01$) and changes in the triglyceride/HDL ratio ($P=0.07$). An overall 24% attrition rate was noted; those who were in the intervention group had little interest in mindfulness and lower expected outcomes that led to their dropping out. The authors commented in favor of participant self-assignment to one of the two study groups as this may be a more pragmatic approach reflective of the real world while offering a higher study completion rate for research purposes. The influence of the two mindfulness instructors during the 5.5-month training period affected the mean weight change of the participants after 6 months. At the 12th month, one group continued losing weight, while the other began to regain weight. At the 18th month, these trends continued resulting in a 4.3 kg separation between the groups directly attributed to the instructors and their presentation of the materials.

NUTRITION

Healthy eating is essential for reducing the risk of CVD. As a lifestyle behavior, mindfulness has the potential to improve eating habits especially if driven by negative emotions, such as anxiety or depression. To date, studies have preferentially focused on weight loss trials rather than improving nutritional quality [11]. Evidence of mindfulness on improving dietary patterns in relation to CVD has been primarily obtained from uncontrolled studies that show promising results [11]. However, further study is required preferably using RCTs or well-designed observational trials.

One RCT published in 2014 assessed the effects of mindfulness therapy ($N=27$, M/F 2:3, mean age 53.9 years SD=8.2) versus an active control consisting of diabetes self-management education ($N=25$, M/F 2:3, mean age 54.0 years SD=7.0)

[36]. At the third-month postintervention, each group had significant improvements in depressive symptoms, outcome expectations, nutrition and eating-related self-efficacy, and food-oriented cognitive control. The controls had significantly greater improvements in nutrition knowledge and intake of fruits and vegetables. Meanwhile, the intervention yielded significant improvements in mindfulness. The authors suggested that offering the two programs separately or in combination could provide patients with valuable choices by which they could meet their personal self-care needs.

PHYSICAL INACTIVITY

As a lifestyle behavior, it has been suggested that mindfulness-based interventions could assist individuals overcoming emotional barriers to physical activity. This has not been the subject of many studies yet, and only one recent RCT was identified. The study conducted in Netherland hypothesized that a worksite mindfulness-based multicomponent intervention with e-coaching ("mindful 'vitality in practice'") would increase vigorous physical activity and reduce sedentary behavior at work [37]. This trial randomized 257 workers, mainly highly educated women (mean age, 45.6 years) 1:1 to either the intervention or control groups at two research institutions. At 12 months from baseline, attrition rates were modest at 9.1% for questionnaires and 16% for accelerometers. Using linear mixed effect models and the ITT population, no significant differences were found regarding physical activity at either sixth- or twelfth-month follow-up between the groups. At present, it is inconclusive if mindfulness-based programming improves physical activity in individuals leading sedentary lifestyles.

SUMMARY

Mindfulness-based meditation practice has received much interest of late as the medical research community explores alternative medicines. This review has summarized recent RCTs to better understand the potential of mindfulness in stabilizing or reducing modifiable risks associated with CVD, a leading cause of mortality. The questions being addressed by these studies, for the most part, remain unanswered. However, although the findings overall may be inconclusive, mindfulness as an intervention for smoking reduction or cessation is supported by evidence from well-designed RCTs with blinded participants and active controls. Here, the high attrition rates render the studies underpowered rather than inherent issues with methodological rigor.

For the studies reviewed, no mention has been made regarding adverse events, severe or otherwise attributed to MBSR or MBCT. This is because none of the RCTs reported any. As a nonpharmacological intervention for the treatment of risk factors for CVD, there is the potential for mindfulness to exacerbate existing mental conditions. Conversely, in the studies that measured changes in mental health (e.g., depression, stress, or anxiety) pre- and postintervention, the overall trend was that the participants in the mindfulness intervention improved. This primarily self-reported outcome is consistent with the literature suggesting that mindfulness is beneficial for some and is not detrimental to any.

The next wave of clinical trials will have to be well-designed with well-defined, clinically relevant participant eligibility criteria and active controls to robustly ascertain and quantify if mindfulness-based programs could be offered as therapy to reduce the rates of CVD. A unique aspect of the mindfulness interventions is that there is much similarity between them regardless of the CVD risk factor unlike other therapies that are risk factor-specific. In that sense, the universality and low cost of mindfulness relative to pharmacological treatments make it attractive for future study.

REFERENCES

[1] Benjamin E, Blaha M, Chiuve S, Cushman M, Das S, Deo R, et al. Heart disease and stroke statistics—2017 update: a report from the American Heart Association. Circulation 2017;135:e146–603.

[2] Khatib R, McKee M, Shannon H, Chow C, Rangarajan S, Teo K, et al. Availability and affordability of cardiovascular disease medicines and their effect on use in high-income, middle-income, and low-income countries: an analysis of the PURE study data. Lancet 2016;387(10013):61–9.

[3] World Health Organization. The atlas of heart disease and stroke. Geneva, Switzerland: WHO, Marketing & Dissemination; 2004.

[4] Allam A, Thompson R, Wann S, Miyamoto M, Thomas GS. Computed tomographic assessment of atherosclerosis in ancient Egyptian mummies. JAMA 2009;302(19):2091–4.

[5] Framingham Heart Study. Framingham Heart Study; 2017. https://www.framinghamheartstudy.org.

[6] Kabat-Zinn J. An outpatient program in behavioral medicine for chronic pain patients based on the practice of mindfulness meditation: theoretical considerations and preliminary results. Gen Hosp Psychiatry 1982;4(1):33–47.

[7] Gotink RA, Chu P, Busschbach JJ, Benson H, Fricchione GL, Hunink MG. Standardised mindfulness-based interventions in healthcare: an overview of systematic reviews and meta-analyses of RCTs. PLoS ONE 2015;10(4):e0124344.

[8] Chrysant SG. The clinical significance and costs of herbs and food supplements used by complementary and alternative medicine for the treatment of cardiovascular diseases and hypertension. J Hum Hypertens 2016;30(1):1–6.

[9] Abbott RA, Whear R, Rodgers LR, Bethel A, Thompson Coon J, Kuyken W, et al. Effectiveness of mindfulness-based stress reduction and mindfulness based cognitive therapy in vascular disease: a systematic review and meta-analysis of randomised controlled trials. J Psychosom Res 2014;76(5):341–51.

[10] Younge JO, Gotink RA, Baena CP, Roos-Hesselink JW, Hunink MG. Mind-body practices for patients with cardiac disease: a systematic review and meta-analysis. Eur J Prev Cardiol 2015;22(11):1385–98.

[11] Loucks EB, Schuman-Olivier Z, Britton WB, Fresco DM, Desbordes G, Brewer JA, et al. Mindfulness and cardiovascular disease risk: state of the evidence, plausible mechanisms, and theoretical framework. Curr Cardiol Rep 2015;17(12):112.

[12] Goyal M, Singh S, Sibinga EM, Gould NF, Rowland-Seymour A, Sharma R, et al. Meditation programs for psychological stress and well-being: a systematic review and meta-analysis. JAMA Intern Med 2014;174(3):357–68.

[13] Davış JM, Mills DM, Stankevitz KA, Manley AR, Majeskie MR, Smith SS. Pilot randomized trial on mindfulness training for smokers in young adult binge drinkers. BMC Complement Altern Med 2013;13:215.

[14] Davis JM, Goldberg SB, Anderson MC, Manley AR, Smith SS, Baker TB. Randomized trial on mindfulness training for smokers targeted to a disadvantaged population. Subst Use Misuse 2014;49(5):571–85.

[15] Davis JM, Manley AR, Goldberg SB, Smith SS, Jorenby DE. Randomized trial comparing mindfulness training for smokers to a matched control. J Subst Abuse Treat 2014;47(3):213–21.

[16] Schuman-Olivier Z, Hoeppner BB, Evins AE, Brewer JA. Finding the right match: mindfulness training may potentiate the therapeutic effect of nonjudgment of inner experience on smoking cessation. Subst Use Misuse 2014;49(5):586–94.

[17] Ruscio AC, Muench C, Brede E, Waters AJ. Effect of brief mindfulness practice on self-reported affect, craving, and smoking: a pilot randomized controlled trial using ecological momentary assessment. Nicotine Tob Res 2016;18(1):64–73.

[18] Brewer JA, Mallik S, Babuscio TA, Nich C, Johnson HE, Deleone CM, et al. Mindfulness training for smoking cessation: results from a randomized controlled trial. Drug Alcohol Depend 2011;119(1-2):72–80.

[19] Kopf S, Oikonomou D, Hartmann M, Feier F, Faude-Lang V, Morcos M, et al. Effects of stress reduction on cardiovascular risk factors in type 2 diabetes patients with early kidney disease—results of a randomized controlled trial (HEIDIS). Exp Clin Endocrinol Diabetes 2014;122(6):341–9.

[20] van Son J, Nyklicek I, Pop VJ, Blonk MC, Erdtsieck RJ, Pouwer F. Mindfulness-based cognitive therapy for people with diabetes and emotional problems: long-term follow-up findings from the DiaMind randomized controlled trial. J Psychosom Res 2014;77(1):81–4.

[21] van Son J, Nyklicek I, Pop VJ, Blonk MC, Erdtsieck RJ, Spooren PF, et al. The effects of a mindfulness-based intervention on emotional distress, quality of life, and HbA(1c) in outpatients with diabetes (DiaMind): a randomized controlled trial. Diabetes Care 2013;36(4):823–30.

[22] Tovote KA, Fleer J, Snippe E, Peeters AC, Emmelkamp PM, Sanderman R, et al. Individual mindfulness-based cognitive therapy and cognitive behavior therapy for treating depressive symptoms in patients with diabetes: results of a randomized controlled trial. Diabetes Care 2014;37(9):2427–34.

[23] Tovote KA, Schroevers MJ, Snippe E, Sanderman R, Links TP, Emmelkamp PM, et al. Long-term effects of individual mindfulness-based cognitive therapy and cognitive behavior therapy for depressive symptoms in patients with diabetes: a randomized trial. Psychother Psychosom 2015;84(3):186–7.

[24] Jung HY, Lee H, Park J. Comparison of the effects of Korean mindfulness-based stress reduction, walking, and patient education in diabetes mellitus. Nurs Health Sci 2015;17(4):516–25.

[25] Palta P, Page G, Piferi RL, Gill JM, Hayat MJ, Connolly AB, et al. Evaluation of a mindfulness-based intervention program to decrease blood pressure in low-income African-American older adults. J Urban Health 2012;89(2):308–16.

[26] Hughes JW, Fresco DM, Myerscough R, van Dulmen MH, Carlson LE, Josephson R. Randomized controlled trial of mindfulness-based stress reduction for prehypertension. Psychosom Med 2013;75(8):721–8.

[27] Blom K, Baker B, How M, Dai M, Irvine J, Abbey S, et al. Hypertension analysis of stress reduction using mindfulness meditation and yoga: results from the HARMONY randomized controlled trial. Am J Hypertens 2014;27(1):122–9.

[28] Pickering TG, Shimbo D, Haas D. Ambulatory blood-pressure monitoring. N Engl J Med 2006;354(22):2368–74.

[29] Shi L, Zhang D, Wang L, Zhuang J, Cook R, Chen L. Meditation and blood pressure: a meta-analysis of randomized clinical trials. J Hypertens 2017;35(4):696–706.

[30] Mason AE, Epel ES, Aschbacher K, Lustig RH, Acree M, Kristeller J, et al. Reduced reward-driven eating accounts for the impact of a mindfulness-based diet and exercise intervention on weight loss: data from the SHINE randomized controlled trial. Appetite 2016;100:86–93.

[31] Kristeller J, Wolever RQ, Sheets V. Mindfulness-based eating awareness training (MB-EAT) for binge eating: a randomized clinical trial. Mindfulness 2014;5(3):282–97.

[32] Katterman SN, Kleinman BM, Hood MM, Nackers LM, Corsica JA. Mindfulness meditation as an intervention for binge eating, emotional eating, and weight loss: a systematic review. Eat Behav 2014;15(2):197–204.

[33] Corsica J, Hood MM, Katterman S, Kleinman B, Ivan I. Development of a novel mindfulness and cognitive behavioral intervention for stress-eating: a comparative pilot study. Eat Behav 2014;15(4):694–9.

[34] Fisher N, Lattimore P, Malinowski P. Attention with a mindful attitude attenuates subjective appetitive reactions and food intake following food-cue exposure. Appetite 2016;99(Apr):10–6.

[35] Cavanagh K, Vartanian LR, Herman CP, Polivy J. The effect of portion size on food intake is robust to brief education and mindfulness exercises. J Health Psychol 2014;19(6):730–9.

[36] Miller CK, Kristeller JL, Headings A, Nagaraja H. Comparison of a mindful eating intervention to a diabetes self-management intervention among adults with type 2 diabetes: a randomized controlled trial. Health Educ Behav 2014;41(2):145–54.

[37] van Berkel J, Boot CR, Proper KI, Bongers PM, van der Beek AJ. Effectiveness of a worksite mindfulness-based multi-component intervention on lifestyle behaviors. Int J Behav Nutr Phys Act 2014;11(1):9.

Chapter 24

Lifestyle Impact and Genotype-Phenotype Correlations in Brugada Syndrome

Houria Daimi*[,†], Amel Haj Khelil[†], Khaldoun Ben Hamda[‡], Amelia Aranega*, Jemni B.E. Chibani[†], Diego Franco*

*Department of Experimental Biology, University of Jaén, Jaén, Spain, [†]Biochemistry and Molecular Biology Laboratory, Faculty of Pharmacy, University of Monastir, Monastir, Tunisia, [‡]Cardiology Service, Fattouma Bourguiba Hospital, Monastir, Tunisia

INTRODUCTION

Brugada syndrome (BrS) is a form of cardiac arrhythmia characterized by ST segment elevation with downsloping "coved-type" (type 1) or "saddleback" (type 2) pattern in V1–V3 precordial chest leads [1] and frequently associated with dysfunctional sodium channels. A common presentation of BrS is syncope, which is caused by fast polymorphic ventricular tachycardia. Such syncope typically occurs in the third and fourth decade of life and usually at rest or during sleep. Pediatric cases are thus very rare and generally unmasked after febrile episodes. In some cases, tachycardia does not terminate spontaneously and might degenerate into ventricular fibrillation leading to sudden death [2].

BrS exhibits an autosomal dominant pattern of inheritance, with incomplete penetrance and variable expression. Currently, its global prevalence is estimated at 3–5 per 10000 people, although the incidence is higher in Southeast Asian countries than in the rest of the world. The syndrome is genetically heterogeneous and can arise from pathogenic variants in at least 19 different genes [3]. The major gene associated with BrS is SCN5A, which encodes for the α-subunit of the voltage-gated cardiac sodium channel NaV1.5. Screening for pathogenic variants in this gene uncovers mutations in approximately 20% of BrS patients [4,5]. An additional 15% of patients can be molecularly diagnosed if minor genes described as causing BrS are included in the genetic screening (ABCC9, CACNA1C, CACNA2D1, CACNB2, FGF12, GPD1L, HCN4, KCND3, KCNE1L, KCNE3, KCNH2, KCNJ8, PKP2, RANGRF, SCN1B, SCN2B, SCN3B, SCN10A, SLMAP, and TRPM4) [3,6,7]. Hence, causal genetic variants are not found in a high percentage of patients with BrS.

Genetic testing of BrS patients generally involves sequencing of protein-coding portions and flanking intronic regions of SCN5A, according to recent ESC Guidelines [8]. Although the idea of BrS as a monogenic disease with no or few nongenetic factor interference fascinated the clinicians and geneticists for decades, this idea is slowly vanishing with the revelation of several "external" modulators and precipitating factors related mainly to the day-to-day lifestyle of the diagnosed patients and those who are at a high risk of developing the syndrome. The implication of these factors shaping the disease was further confirmed with the increasing evidences of the lack of genotype/phenotype correlation.

BrS GENETICS AND THE GENOTYPE-PHENOTYPE CORRELATION

SCN5A gene, encoding the α-subunit of the cardiac sodium channel protein, was the first gene to be associated with BrS and still represents the major gene in BrS pathogenesis. However, based on current findings, knowledge of an SCN5A-specific mutation may not provide guidance in formulating a clear BrS diagnosis or determining prognosis [9]. On a metaanalysis performed by Wu et al. [10], it was demonstrated that the risk of cardiac events did not statistically increase in patients with SCN5A gene mutations. Although more than 450 pathogenic variants have been identified in 24 genes encoding sodium, potassium, and calcium channels or associated proteins since BrS was first described, it is yet not clear which of these, if any, are associated with a greater risk of arrhythmic events or sudden death [11]. Furthermore, polymorphisms have been shown to modulate SCN5A mutations, either by aggravation of the channelopathy or by rescuing the pathophysiological effect of the mutation. SCN5A-H558R is a great example of a common polymorphism that has been shown to restore normal sodium channel function in the case of specific mutations and to enhance sodium channel dysfunction in others [12–14]. Regardless of the mechanism by which SCN5A-H558R or any other genetic variant is capable of modulating the BrS pattern, it became evident that the phenotypic expression of SCN5A mutations is very versatile and

Lifestyle in Heart Health and Disease. https://doi.org/10.1016/B978-0-12-811279-3.00024-0

285

controversial. In fact, patients presenting overlapping LQT syndrome (known as an SCN5A gain of function disease) and BrS (SCN5A loss of function disease) have been described [14]. Our own experience includes several families with one or more index case either symptomatic or asymptomatic with spontaneous or induced Brugada ECG give more evidences that the old concept of BrS as a simple Mendelian disease is no longer true suggesting that different variants may contribute to the phenotype, including benign polymorphisms and even nonsense mutations [15]. Moreover, identifying functional SCN5A mutations in healthy individuals among the same family suggest that sodium channel defects may be one of the contributory causes but not necessarily the main one. Our recent study involving a 2-month-old male sporadic BrS patient is one of those cases where the genetic diagnosis of SCN5A in the proband and his family members revealed a set of SCN5A mutations and polymorphisms that are differentially distributed among the family members, however, without any clear genotype-phenotype correlation [16]. Our patient is asymptomatic with no previous family history of the disease in whom the typical type 1 BrS ECG was unmasked upon a postvaccination fever. This case confirms the dynamics of the BrS ECG and the role of fever as the main precipitating factor of arrhythmias in BrS pediatric cases [17] (Fig. 1).

FACTORS PRECIPITATING AND MODULATING THE ECG AND ARRHYTHMIAS IN BrS

The Brugada ECG is often concealed but may be unmasked or modulated. The ST segment elevation in the BrS can be very dynamic and modified by several factors such as fever, diet and raised plasma insulin concentration, autonomic activity, alcohol and cocaine toxicity, and medication [18]. These agents may also induce acquired forms of the BrS.

Hyperthermia

Premature inactivation of the sodium channel in SCN5A mutations associated with BrS is accentuated at high temperatures, suggesting that a febrile state may unmask BrS. Indeed, several case reports have emerged recently demonstrating that febrile illness could reveal Brugada ECG and precipitate VF [16,17]. Cellular electrophysiological studies using patch-clamp methods have demonstrated that SCN5A mutations impair cardiac sodium channel gating more severely at high temperatures, leading to further I_{Na} reduction during fever [19]. Pronounced reduction of I_{Na} during fever is thought to increase arrhythmogenicity in BrS patients. Thus, febrile states in BrS patients, particularly the pediatric ones, should be efficiently managed with the corresponding antipyretic medications. Furthermore, BrS patients and their families should be aware of the prophylactic antipyretic measures that should be taken in order to manage on time any possible febrile states.

Diet and Electrolyte Imbalance

A high glucose concentration and insulin release may be important factors inducing VT in BrS patients. Thus, food intake and traditional gastronomic habits were demonstrated to have a direct impact on patients diagnosed with BrS. Characteristic Brugada ECG changes can be augmented by a large meal, salty or carbohydrate-enriched foods. Recently, Talib and his coworkers [20] have reported a BrS case unmasked in a 53-year-old Muslim man after a large meal intake posterior to a long period of fast during the month of Ramadan. The case presented no family history of sudden cardiac death or any kind of cardiac arrhythmias. Consistent with this observation, a recent study by Nogami et al. [21] found that glucose and insulin could unmask the Brugada ECG that is also in line with a study performed by the Thai Ministry of Public Health (1990) in which an association between the risk of sudden death and a large meal of glutinous rice ("sticky rice") or carbohydrates ingested on the night was established [9].

Although electrolyte disturbances such as hypokalemia, hypocalcemia, and hypomagnesemia, especially hypokalemia, were extensively described as the major precipitating factors for several cardiac arrhythmic syndromes such as LQT [18], knowledge about their role unmasking the Brugada ECG pattern is still limited. Hypokalemia is known to augment ST segment elevation and is proarrhythmogenic in BrS [22], whereas hyponatremia causing BrS is unknown in medical literature [23]. Furthermore, hypokalemia has been implicated as a contributing cause for the high prevalence of sudden cardiac death in regions such as the northeast of Thailand where potassium deficiency is endemic [9].

Leisure Activities and Sports

Differentiating right precordial early repolarization, which may be present in about 4% of athletes, from BrS can be difficult, but downsloping of the ST segment and a slightly prolonged QRS duration (i.e., >100ms) are indicators for the Brugada ECG [24]. Although sudden death typically occurs at rest, the conduction defects explain why some may develop ventricular arrhythmias during exercise [25]. Athletes are prone to metabolic disturbances (electrolyte changes and fluid

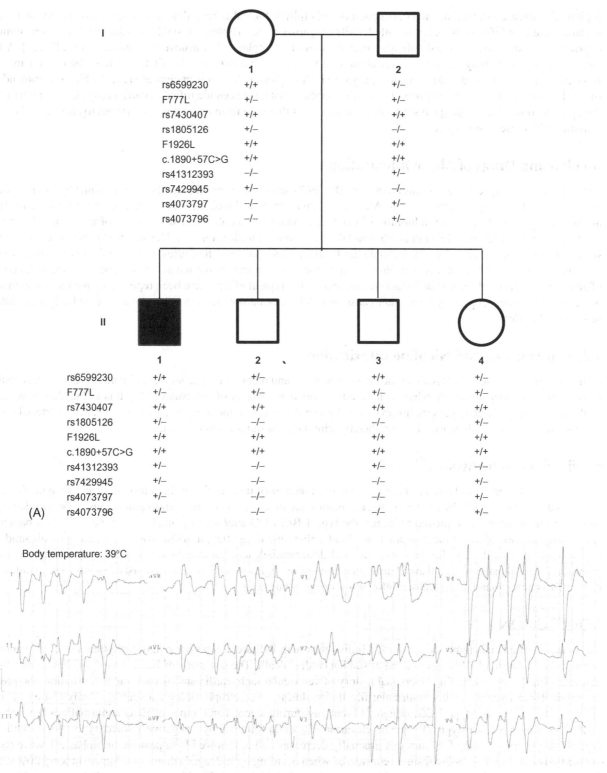

FIG. 1 (A) Pedigree of a Tunisian family with a sporadic pediatric case (2 months old) of Brugada syndrome. The distribution of the different SCN5A polymorphisms/mutations identified among the family members was represented. −/− indicates the absence of the SCN5A variant in the corresponding individual, while +/− and +/+ indicate the presence of the corresponding heterozygous or the homozygous SCN5A variant, respectively. (B) A 12-lead ECG recorded from the Brugada syndrome patient (II-1) after a febrile state (39°C). Note the typical aspect of a "coved-type" of the ST segment elevation in the precordial leads V1, V2, and V3 corresponding to Brugada syndrome type 1 ECG pattern.

depletion), increased vagal tone, and conduction defects (physiological at rest) that may enhance propensity to die at rest or during exercise [26]. Thus, patients with definite diagnosis of BrS (history of sudden cardiac death) are recommended to practice low-intensity sport and only after implantation of an implantable cardioverter defibrillator (ICD) [27]. All non-competitive sports activity can be however allowed to patients with a Brugada-like ECG in whom the risk for malignant ventricular arrhythmias and sudden death is judged to be low. Nonetheless, it is unclear at present which recommendations should be made for phenotypically negative family members that have been identified as carriers (by class 1 antiarrhythmic drug provocation or genotyping). It seems prudent to restrict them too from competitive sports but to allow all leisure-time activities with close follow-up [27].

Alcohol and Drugs of Abuse Intoxication

Few reports have suggested association between BrS ECG manifestations and agents such as dimenhydrinate, cocaine/heroin, alcohol, and opioid intoxication. Although sodium channel blockade is, in general, an established method to uncover Brugada ECG changes, inhibition of calcium channels by alcohol and some drugs of abuse may be the main mechanism of Brugada-type ECG in alcohol and drugs of abuse intoxication [28]. Heroin intoxication is associated with sudden death syndrome [29]. Fatal arrhythmia has been reported as one of the causes of sudden death in heroin users irrespective of the dose and serum concentration. Despite the scarce reports about the association between the BrS and drugs of abuse intoxication, clinical manifestations mimicking that typical of BrS have been reported in association with heroin overdose. Such phenocopy may be another cause of ECG abnormalities seen in heroin overdose and might be linked to sudden cardiac death.

Tobacco Smoking and Nicotine Intoxication

Although there are scarce reports about the effect of tobacco and nicotine triggering the BrS pattern, such effects could not be neglected. Nicotine intoxication has been already reported as a trigger of BrS pattern [30]. It is still unclear how nicotine could induce Brugada ECG pattern; however, it is known that nicotine induces changes in heart rate and arterial pressure through stimulating catecholamine release, thereby stimulating sodium channels [31].

Medications to be Avoided in BrS

The presence of type 1 BrS ECG has been particularly linked to an increased risk for ventricular tachyarrhythmia, cardiac arrest, and sudden death in BrS patients. Recent reports identified a set of drugs and medications to be avoided by BrS patients as these drugs were proved to induce the type 1 BrS ECG and/or arrhythmias in BrS patients. BrS patients are particularly advised to avoid the consumption of antiarrhythmic drugs (i.e., ajmaline, flecainide, and procainamide), the psychotropic treatments (i.e., lithium, lozapine, and desipramine), and the anesthetic medications (i.e., bupivacaine and propofol) [32]. All patients should avoid drugs that unmask or aggravate BrS. A definitive list of agents to be avoided in the case of BrS can be found in the website http://www.brugadadrugs.org.

CONCLUSION

BrS is an inherited arrhythmogenic disorder. Individuals with BrS are at a higher risk of arrhythmia, such as ventricular tachycardia or ventricular fibrillation, and of sudden cardiac death. The diagnosis of BrS is based on clinical and electrocardiographic features including a personal history of ventricular tachycardia and/or ventricular fibrillation, the presence of ventricular tachycardia and/or ventricular fibrillation during an electrophysiological study, a family history of sudden cardiac death or a coved-type ECG, or agonal breathing during sleep. Three types of ECG abnormalities have been described. However, a coved-type ST segment elevation of ≥2 mm followed by a negative T wave (type 1) or a saddleback-type ST segment elevation of ≥2 mm that gradually decreases and is followed by a positive or biphasic T wave (type 2) are the two electrocardiographic features considered when deciding for a Brugada diagnosis. Approximately 20%–25% of BrS patients are genetically diagnosed with pathogenic variations in SCN5A. However, known BrS-susceptibility genes can only explain 30%–35% of clinically diagnosed cases, indicating that 65%–70% of BrS patients remain genetically unsolved. Unlike other cardiac channelopathies such as LQT incomplete penetrance of BrS mutations, dynamic ECG manifestations confer difficulties in establishing a clear genotype-phenotype correlation. Our own experience, among others, gives evidences on the challenges that clinicians and geneticists are facing trying to explain the diagnosis of some BrS cases in the absence of family history and genotype-phenotype correlation. In the BrS, it is not always possible to determine the phenotype from genetic data, as the identification of a genetic defect in the patient is not equivalent to provide a clear diagnosis. Although the functional analysis somehow could confirm the pathogenicity of a given mutation, the effect of the external precipitating/modulating factors could not be neglected, conferring thus more complexity to the BrS diagnosis and to the management of decision making. As already recommended in other cardiovascular diseases, we believe that it is necessary to define a "Brugada lifestyle" excluding main triggers and precipitating factors and providing the Brugada patients with a clear panel of factors to be avoided during their day-to-day life.

REFERENCES

[1] Brugada P, Brugada J. Right bundle branch block, persistent ST segment elevation and sudden cardiac death: a distinct clinical and electrocardiographic syndrome. A multicenter report. J Am Coll Cardiol 1992;20:1391–6.

[2] Napolitano C, Priori SG. Brugada syndrome. Orphanet J Rare Dis 2006;1:35.

[3] Sarquella-Brugada G, Campuzano O, Arbelo E, Brugada J, Brugada R. Brugada syndrome: clinical and genetic findings. Genet Med 2016;18:3–12.

[4] Priori SG, Napolitano C, Gasparini M, Pappone C, Della Bella P, Giordano U, et al. Natural history of Brugada syndrome: insights for risk stratification and management. Circulation 2002;105:1342–7.

[5] Kapplinger JD, Tester DJ, Alders M, Benito B, Berthet M, Brugada J, et al. An international compendium of mutations in the SCN5A-encoded cardiac sodium channel in patients referred for Brugada syndrome genetic testing. Heart Rhythm 2010;7:33–46.

[6] Hennessey JA, Marcou CA, Wang C, Wei EQ, Wang C, Tester DJ, et al. FGF12 is a candidate Brugada syndrome locus. Heart Rhythm 2013;10:1886–94.

[7] Wang Q, Ohno S, Ding WG, Fukuyama M, Miyamoto A, Itoh H, et al. Gain-of-function KCNH2 mutations in patients with Brugada syndrome. J Cardiovasc Electrophysiol 2014;25:522–30.

[8] Priori SG, Blomström-Lundqvist C, Mazzanti A, Blom N, Borggrefe M, Camm J, et al. 2015 ESC Guidelines for the management of patients with ventricular arrhythmias and the prevention of sudden cardiac death: The Task Force for the Management of Patients with Ventricular Arrhythmias and the Prevention of Sudden Cardiac Death of the European Society of Cardiology (ESC). Endorsed by: Association for European Paediatric and Congenital Cardiology (AEPC). Eur Heart J 2015;36:2793–867.

[9] Antzelevitch C, Brugada P, Brugada J, Brugada R. Brugada syndrome: from cell to bedside. Curr Probl Cardiol 2005;30(1):9–54.

[10] Wu W, Tian L, Ke J, Sun Y, Wu R, Zhu J, Ke Q. Risk factors for cardiac events in patients with Brugada syndrome. A PRISMA-compliant meta-analysis and systematic review. Medicine 2016;30(e4214). https://doi.org/10.1097/MD.0000000000004214.

[11] Poelzing S, Forleo C, Samodell M, Dudash L, Sorrentino S, Anaclerio M, et al. SCN5A polymorphism restores trafficking of a Brugada syndrome mutation on a separate gene. Circulation 2006;114:368–76.

[12] Viswanathan PC, Benson DW, Balser JR. A common SCN5A polymorphism modulates the biophysical effects of an SCN5A mutation. J Clin Invest 2003;111:341–6.

[13] Lizotte E, Junttila MJ, Dube MP, Hong K, Benito B, De Zutter M, et al. Genetic modulation of Brugada syndrome by a common polymorphism. J Cardiovasc Electrophysiol 2009;20:1137–41.

[14] Veltmann C, Barajas-Martinez H, Wolpert C, Borggrefe M, Schimpf R, Pfeiffer R, et al. Further insights in the most common SCN5A mutation causing overlapping phenotype of long QT syndrome, Brugada syndrome, and conduction defect. J Am Heart Assoc 2016;5(7):e003379. https://doi.org/10.1161/JAHA.116.003379.

[15] Daimi H, Haj Khelil A, Neji A, Ben Hamda K, Maaoui S, Aranega A, et al. Additive contribution of SCN5A coding and non-coding sequences to the occurrence of Brugada syndrome: a molecular study of two Tunisian families. Gene (in press).

[16] Daimi H, Haj Khelil A, Ben Hamda K, Aranega A, Chibani JBE, Franco D. Absence of family history and phenotype–genotype correlation in pediatric Brugada syndrome: more burden to bear in clinical and genetic diagnosis. Pediatr Cardiol 2015;36:1090–6.

[17] Manohar S, Dahal BR, Gitler B. Fever-induced Brugada syndrome. J Investig Med High Impact Case Rep 2015;3(1).

[18] Nakajima T, Kaneko Y, Kurabayashi M. Unveiling specific triggers and precipitating factors for fatal cardiac events in inherited arrhythmia syndromes. Circ J 2015;79.

[19] Amin AS, Meregalli PG, Bardai A, Wilde AA, Tan HL. Fever increases the risk for cardiac arrest in the Brugada syndrome. Ann Intern Med 2008;149:216–8.

[20] Talib S, van de Poll SWE. Brugada syndrome diagnosed after Ramadan. Lancet 2013;382:100.

[21] Nogami A, Nakao M, Kubota S, Sugiyasu A, Doi H, Yokoyama K, et al. Enhancement of J-ST-segment elevation by the glucose and insulin test in Brugada syndrome. PACE 2003;26.

[22] Araki T, Konno T, Itoh H, Ino H, Shimizu M. Brugada syndrome with ventricular tachycardia and fibrillation related to hypokalemia. Circ J 2003;67:93–5.

[23] Mok NS, Tong CK, Yuen HC. Concomitant-acquired long QT and Brugada syndromes associated with indapamide-induced hypokalemia and hyponatremia. PACE 2008;31:772–5.

[24] Zorzi A, Leoni L, Di Paolo FM, Rigato I, Migliore F, Bauce B, et al. Differential diagnosis between early repolarization of athlete's heart and coved-type Brugada electrocardiogram. Am J Cardiol 2015;115:529e532.

[25] Rossenbacker T, Carroll SJ, Liu H, Kuiperi C, de Ravel TJ, Devriendt K, et al. Novel pore mutation in SCN5A manifests as a spectrum of phenotypes ranging from atrial flutter, conduction disease, and Brugada syndrome to sudden cardiac death. Heart Rhythm 2004;1:610–5.

[26] Heidbuchel H, Corrado D, Biffi A, Hoffmann E, Panhuyzen-Goedkoope N, Hoogsteenf J, et al. Recommendations for participation in leisure-time physical activity and competitive sports of patients with arrhythmias and potentially arrhythmogenic conditions. Part II: Ventricular arrhythmias, channelopathies and implantable defibrillators. Eur J Cardiovasc Prevent Rehabil 2006;13:676–86.

[27] Delise P, Marras E, Bocchino M. Brugada-like electrocardiogram pattern: how to stratify the risk for sudden cardiac death. Is sports activity contra-indicated? J Cardiovasc Med 2006;7:239–45.

[28] van Stigt AH, Overduin RJ, Staats LC, Loen V, van der Heyden MAG. A heart too drunk to drive; AV block following acute alcohol intoxication. Chin J Physiol 2016;59(1):1–8.

[29] Calcaterra S, Glanz J, Binswanger IA. National trends in pharmaceutical opioid related overdose deaths compared to other substance related overdose deaths: 1999–2009. Drug Alcohol Depend 2013;131(3):263–70.

[30] Ondrejka J, Giorgio G. Type 1 Brugada pattern associated with nicotine toxicity. J Emerg Med 2015;49(6):83–186.

[31] Haass M, Kubler W. Nicotine and sympathetic neurotransmission. Cardiovasc Drugs Ther 1997;10:657–65.

[32] Turker I, Ai T, Itoh H, Horie M. Drug-induced fatal arrhythmias: acquired long QT and Brugada syndromes. Pharmacol Ther 2017. https://doi.org/10.1016/j.pharmthera.2017.05.001.

Chapter 25

Social Relationships and Cardiovascular Health: Underlying Mechanisms, Life Course Processes, and Future Directions

Courtney Boen*, Yang C. Yang[†]

*University of Pennsylvania, Philadelphia, PA, United States, [†]Lineberger Comprehensive Cancer Center, University of North Carolina at Chapel Hill, Chapel Hill, NC, United States

INTRODUCTION

Humans are inherently social beings, embedded in a variety of social relationships; we are parents, siblings, friends, co-workers, partners, children, and neighbors. This interweaving of individual lives in the context of social relationships has profound consequences for health and well-being, particularly cardiovascular health. Individuals within one's social network can act as sources of social and emotional support, social strain, and purveyors of health-related information. A wide and growing body of research has examined the role of social relationships in affecting health and disease risk, with studies across disciplines providing overwhelming evidence of the critical importance of good relationships for maintaining and improving health from early to late life [1–5]. In addition to the health-promoting benefits of social connectedness, research also finds that social isolation, or a lack of social ties, poses as a grave health risk, with a lack of social integration being linked to higher rates of cardiovascular disease and associated mortality [6,7] and higher rates of general and cause-specific mortality [8,9]. Studies document that the mortality risk associated with social isolation is comparable with that of unhealthy behaviors such as cigarette smoking [10] and physical inactivity [3].

Given the documented associations between social relationships and health, in general, and cardiovascular health, in particular, it is essential that scholars, researchers, and practitioners across disciplines consider how social ties and corresponding levels of social support and strain affect health and disease risk across the human life course. This chapter will proceed as follows. First, we provide an overview of scientific evidence to date linking social relationships to health, paying particular attention to how social ties may impact heart health and cardiovascular disease risk across the life course. We next explore some of the biological, behavioral, and psychoemotional mechanisms underlying the link between social relationships and cardiovascular health from birth through old age. Finally, we close by highlighting some of the critical gaps in the knowledge about the linkage between social relationships and cardiovascular health and offer recommendations for areas of future research.

SOCIAL RELATIONSHIPS AND CARDIOVASCULAR HEALTH

Overview

Research dating back to the 1970s documents a link between social relationships and health [9,11,12]. Studies offer particularly compelling evidence of the impact of social relationships on cardiovascular health, with a growing body of literature documenting robust associations between social relationships and hypertension [2,13], coronary artery disease [6,14], myocardial infarction [15], and cardiovascular disease [16].

Fig. 1 proposes a conceptual framework for understanding the linkages between social relationships and health and longevity. Research on the health effects of social relationships typically proposes two primary dimensions of social relationships (Box 1): social integration and social support. *Social integration* refers to the structural characteristics of individuals' social relationships and indicates the size, scope, and connectedness of an individual's social network. To capture individual levels of social integration, researchers typically use measures of relationships status [17,18], network range or

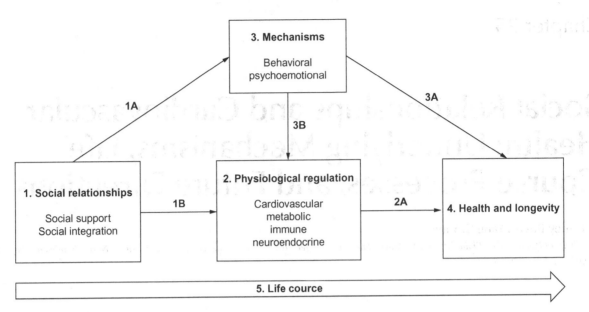

FIG. 1 Conceptual model.

diversity [19], and frequency of contact with social ties [14]. Together, measures of social integration generally describe the patterning and organization of social ties. In contrast, measures of *social support* reflect the functional dimension of social relationships by indicating the level of support an individual receives or perceives from his or her relationships. Measures of social support can include perceived levels of social support [2,20] and loneliness [21,22]. Cohen [23] provides a particularly clear contrast between these two dimensions of social relationships and their relations to health. Cohen defines social support as "a social network's provision of psychological and material resources intended to benefit an individual's ability to cope with stress" (p. 676); by contrast, Cohen argues that "social connectedness is beneficial irrespective of whether one is under stress" (p. 678).

Physiological Pathways

There are a number of direct and indirect mechanisms through which social relationships may impact health, generally, and cardiovascular well-being, specifically, which are visually depicted in Fig. 1. While a number of studies have linked social relationships to health and disease risk, broadly, a smaller body of literature has begun to examine the physiological pathways relating social integration, social support, and health. As described by Berkman and Syme [9], "Adequate tests of the hypothesis that social circumstances alter general susceptibility to disease in humans will not be possible...until data are available on the physiological mechanisms capable of mediating the relationship between social events and disease outcomes." Research on the physiological consequences of social relationship exposures, then, not only helps to answer questions related to biological plausibility but also helps to refine our existing theories about how social relationships "get under the skin" to affect health and mortality and design appropriate intervention strategies to mitigate corresponding disease risk [24].

Studies suggest that social relationships may protect health through conferring social support to individuals, which may buffer against the harmful effects of social stressors on physical health [25]. In response to perceived stressors or threats, activity in the sympathetic nervous system (SNS) and hypothalamic-pituitary-adrenocortical (HPA) axes increases, resulting in the release of stress hormones that upregulate cardiovascular, metabolic, immune, inflammatory, and other physiological functions. In addition to releasing stress hormones, the SNS and HPA axis also play a critical role in regulating several cardiovascular functions, including blood pressure [21,26]. While short-term activation of the SNS and HPA axis is essential to protecting health from acute stressors or infections, long-term increases in activity of these systems in response to chronic or ongoing stressors—such as social isolation or social relationship strain—have been linked to physiological dysregulation, allostatic load, and increased disease and mortality risk (Path 2A). Exposure to chronic stressors and strains has been linked to the development and exacerbation of a number of cardiovascular conditions and risk factors, including hypertension [2], inflammation [27,28], metabolic syndrome [1], body mass index (BMI) and waist circumference [3], and elevated cholesterol [29].

As suggested by Path 1B in Fig. 1, social relationships may buffer individuals from the harmful health effects of stress by conferring emotional support—including demonstrations of love, caring, and encouragement—and instrumental guidance and assistance—including offering assistance to deal with problems —in the face of stressors and strains. Together, the emotional and instrumental social support that individuals perceive or receive from their relationships may aid in maintaining physiological regulation in the face of life challenges and demands. For example, studies have found that older individuals who report higher levels of social support are less likely to experience increases in systolic blood pressure as they age, compared with individuals reporting lower levels of support [2].

Behavioral Mechanisms

In addition to their stress-buffering effects, social relationships may also impact cardiovascular disease risk though a number of behavioral mechanisms, whereby individuals receive behavioral guidance from individuals in their social networks [30–32], as suggested by Paths 1A, 3A, and 3B in Fig. 1. Social network members may monitor, encourage, remind, and/or pressure individuals to adopt certain health behaviors [33], such as healthy eating, physical activity, or health-care utilization. Additionally, individuals develop their own beliefs about the appropriateness of health behaviors through their interactions and comparisons with others in their social groups. For example, individuals develop norms about the acceptability of smoking, exercising, drug use, mental health counseling, and seeking medical treatment by observing and comparing with others in their social networks [33]. Given documented links between health behaviors and markers of physiological regulation (Path 3B) [34,35] and health and mortality risk (Path 3A) [36,37], the link between social relationships and health and disease outcomes is partially explained by health behaviors.

A number of studies find that these behavioral processes undergird the associations between social relationships and cardiovascular health outcomes. For example, using nationally representative data on older adults, Yang et al. [2] found that individuals reporting higher levels of social integration were less likely to develop hypertension over time, and their findings suggested that health behaviors may help to explain this link. The authors found that socially integrated individuals were engaged in more physical activity, consumed fewer alcoholic beverages, and had lower BMIs than less socially integrated individuals, which in turn lowered their hypertension risk. Similarly, Cornwell and Waite [13] examined the factors associated with blood pressure control. They found that individuals with large social networks who reported discussing health issues with their network members had the lowest risk of undiagnosed and uncontrolled hypertension. Their findings also suggested that the link between social relationships and hypertension was partially explained by health-care utilization, such that network members encouraged individuals to seek medical care and treatment.

Psychoemotional Processes

In addition to behavioral guidance and support, individuals may also derive a variety of psychoemotional benefits from their social relationships, which has consequences for cardiovascular health and mortality risk, as indicated in Fig. 1. Research finds that individuals develop a sense of purpose and belonging through their interactions with others, which can prevent anxiety; improve self-esteem; fend off feelings of sadness, hopelessness, and despair; and ultimately improve and protect health [33]. In the context of a social relationship, such as a friendship or marriage, individuals can develop a sense of mattering, where they come to believe that they are important to that person and that he or she depends on them [33]. This sense of purpose and mattering in the context of social relationships has been linked to better health behaviors and improved psychological well-being [4,23,38].

Similarly, the psychoemotional benefits one obtains from their participation in social relationships may also have positive consequences for perceptions of control and mastery and self-esteem. When individuals feel that they matter and are supported, they may feel more in control of their life and confident in their abilities to cope with new challenges or stressors. In turn, individuals who are socially integrated and supported have lower anxiety and depression [39,40] and reduced physiological reactivity in the face of social stressors [3,41], as indicated in Path 3B of Fig. 1. Together, these psychoemotional benefits associated with social relationships reduce disease and mortality risk.

Life Course Processes

Finally, as indicated in Box 5 of Fig. 1, the processes by which social relationships impact physiological well-being and health and mortality risk unfold across the life course, from the time an individual is in utero through old age. The life course perspective has provided key insights into how the social conditions one experiences over the life span affect physiological functioning, biological well-being, and health and disease risk with age. The life course perspective is guided by overarching principles that have important consequences for the study of social relationships and cardiovascular health. For one, this perspective requires an acknowledgment and understanding that human development and aging-related disease such as heart disease are lifelong processes and that the antecedents and consequences of life course changes and occurrences vary according to their timing in a person's life [42]. Further, the life course perspective also acknowledges that social relationships reflect the linkages, dependencies, and sociohistorical influences of individual lives [42]. In this way, then, any attempt to understand how social relationships affect health and disease risk, broadly, and cardiovascular well-being, specifically, must examine how the transitioning in and out of social roles and the forming and dissolution of social ties play out across the life span to affect biology as individuals age.

Research suggests that social relationships affect birth outcomes, with pregnant women reporting higher levels of social support having better birth outcomes and lower risk of low birth weight than less socially supported women [43,44]. Given documented links between birth weight and cardiovascular disease [45], social relationships may serve as an intergenerational process through which social conditions affect heart health and disease risk. Social connectedness in adolescence has been further linked to critical biological markers of cardiovascular disease risk, including BMI, waist circumference, systolic blood pressure, and C-reactive protein [3]. In fact, Yang et al. [3] found that the risk of elevated C-reactive protein associated with social isolation in adolescence was comparable to physical inactivity, a well-established driver of cardiovascular risk. In this way, one's social relationships at younger ages may play a key role in establishing future trajectories of health and well-being.

A number of studies also find that social relationships are critical to improving and sustaining life in late life [1–3,13], as individuals transition out of some roles (e.g., employment) and into new ones (e.g., grandparenthood and widowhood). Using data from two nationally representative, older-adult samples, Yang et al. [3] found that older adults reporting higher levels of social integration had lower cardiovascular disease risk, as indicated by lower risk of hypertension, inflammation, and obesity. Similarly, the authors also found that individuals reporting more social support were at reduced health risk. With the aging of the population in the United States—which puts a greater number of individuals at risk for heart-related health problems—developing policies and programs that encourage the development and maintenance of supportive social relationships will be critical to improving population health [2].

GAPS IN THE LITERATURE AND DIRECTIONS FOR FUTURE RESEARCH

Despite a large and growing body of research focused on the role of social relationships in affecting cardiovascular disease and mortality risk, critical gaps in literature remain. First, most studies utilize single measures of social relationships, such as perceived isolation or marital status, which limits understanding of the roles of different dimensions of social relations in population health in several key ways. For one, the reliance on single indicators of social relationships makes it difficult to determine the unique influences of particular social network or social support characteristics on health [13]. For example, while a count of one's social ties may be indicative of the structure of an individual's social network, this type of measure does not provide any information about the functional support the individual receives from those ties. In fact, it is possible for individuals with vast numbers of social ties to still perceive low levels of support and high levels of loneliness; on the other hand, having one or two close social relationships may lead individuals to perceive high levels of support [46]. Further, the use of single indicators limits understanding of how different dimensions of social relationship quantity, quality, and function may jointly and uniquely affect health and disease risk [2]. As such, it remains relatively unexplored whether social integration and social support are independent determinants of cardiovascular health or whether they represent interrelated dimensions of social relations. Finally, by including just one or two indicators of social relationships in empirical examinations of population health disparities, studies risk underestimating the role of social relationships in affecting health. Research documents that multidimensional measures of social relationships that include both the structural and functional dimensions of social ties better predict mortality than single, one-dimensional measures [5]. For this reason, restricting analyses to just one or two single indicators of social relationships risks underestimating contribution of social ties and corresponding levels of social support to population health differences. As such, more research utilizing diverse, multidimensional, and nuanced measures of social relationships is needed in order to more comprehensively capture the role of social relationships in affecting cardiovascular health and mortality risk.

Second, most studies on the associations between social relationships and health, generally, and cardiovascular disease and well-being, specifically, rely on cross-sectional data. As such, concerns about possible selection biases and reverse causality plague many of these studies. For example, research suggests that individuals with stress-related illnesses, such as hypertension, may withdraw from social relationships, limit social activities, and thus perceive lower levels of social support [47]. In this way, there is concern that any documented associations between social relationships and health may be a result of reverse causality, whereby individuals in poor health retreat from their social ties and perceive high levels of social isolation. Further, evidence also suggests that cross-sectional and longitudinal analyses offer varying evidence of the links between different dimensions of social relationships and health. For example, Yang et al. [2] used a combination of cross-sectional and longitudinal modeling strategies to document differing associations between social integration, social support, and blood pressure across study designs. While their cross-sectional models indicated that social integration was more predictive of systolic blood pressure and hypertension risk than social support, results from the longitudinal models showed that both dimensions of social relationships were predictive of change in the outcomes over time. Not only will future studies utilizing longitudinal data temper concerns about selection and reverse causality, but these studies will also improve understanding of how the structure and function of social relationships affect changes in cardiovascular risk over time and across the life course, which is essential to improving policy, programmatic, and intervention efforts in an aging population.

Finally, future research should continue to examine the biological, psychological, and behavioral mechanisms linking social relationships to cardiovascular outcomes. Most research on the associations between social relationships and health focuses on disease outcomes, so less is known about the biological pathways undergirding these associations, leaving questions about biological plausibility unanswered. Further, research utilizing predisease biomarkers of cardiovascular risk can also improve our ability to design policies and programs that intervene prior to the onset of disease. By examining how social relationships relate to predisease markers of cardiovascular well-being, research can elucidate the role of social ties in delaying the onset or slowing the progression of cardiovascular diseases. Similarly, by including measures of possible behavioral and psychosocial mechanism in empirical examinations, studies can document how social relationships affect cardiovascular risk by improving or worsening health-related behaviors and affecting individual psychological states. Improving scientific understanding of the mechanisms linking social relationships to cardiovascular health will further contribute to the design and implementation of effective disease prevention and intervention efforts.

CONCLUSION

Taken together, the vast and diverse literature on social relationships and cardiovascular health consistently documents a strong and robust association between social integration, social support, and a range of cardiovascular outcomes. Across the human life span, social relationships have been linked to a variety of cardiovascular risk factors and outcomes, with more integrated and supported individuals having consistently lower disease and mortality risk than others. Given that one in four deaths in the United States is attributable to heart disease [48], it is essential that research and intervention efforts aimed at understanding and improving cardiovascular outcomes consider how the development, maintenance, and erosion of social relationships affect population health and longevity.

REFERENCES

[1] Yang YC, Li T, Ji Y. Impact of social integration on metabolic functions: evidence from a nationally representative longitudinal study of US older adults. BMC Public Health 2013;13(1):1210.

[2] Yang YC, Boen C, Mullan Harris K. Social relationships and hypertension in late life: evidence from a nationally representative longitudinal study of older adults. J Aging Health 2015;27(3):403–31. https://doi.org/10.1177/0898264314551172.

[3] Yang YC, Boen C, Gerken K, Li T, Schorpp K, Harris KM. Social relationships and physiological determinants of longevity across the human life span. Proc Natl Acad Sci U S A 2016;113(3):578–83. https://doi.org/10.1073/pnas.1511085112.

[4] Berkman LF, Glass T, Brissette I, Seeman TE. From social integration to health: Durkheim in the new millennium. Soc Sci Med 2000;51(6):843–57.

[5] Holt-Lunstad J, Smith TB, Layton JB. Social relationships and mortality risk: a meta-analytic review. PLoS Med 2010;7(7):e1000316.

[6] Orth-Gomér K, Rosengren A, Wilhelmsen L. Lack of social support and incidence of coronary heart disease in middle-aged Swedish men. Psychosom Med 1993;55(1):37–43.

[7] Eng PM, Rimm EB, Fitzmaurice G, Kawachi I. Social ties and change in social ties in relation to subsequent total and cause-specific mortality and coronary heart disease incidence in men. Am J Epidemiol 2002;155(8):700–9.

[8] Steptoe A, Shankar A, Demakakos P, Wardle J. Social isolation, loneliness, and all-cause mortality in older men and women. Proc Natl Acad Sci U S A 2013;110(15):5797–801.

[9] Berkman LF, Syme SL. Social networks, host resistance, and mortality: a nine-year follow-up study of alameda county residents. Am J Epidemiol 1979;109(2):186–204.

[10] House JS, Landis KR, Umberson D. Social relationships and health. Science 1988;241(4865):540–5.

[11] Cassel J. The contribution of the social environment to host resistance. Am J Epidemiol 1976;104(2):107–23.

[12] Cobb S. Presidential address—1976. Social support as a moderator of life stress. Psychosom Med 1976;38(5):300–14.

[13] Cornwell EY, Waite LJ. Social network resources and management of hypertension. J Health Soc Behav 2012;53(2):215–31. https://doi.org/10.1177/0022146512446832.

[14] Brummett BH, Barefoot JC, Siegler IC, et al. Characteristics of socially isolated patients with coronary artery disease who are at elevated risk for mortality. Psychosom Med 2001;63(2):267–72.

[15] Berkman LF, Leo-Summers L, Horwitz RI. Emotional support and survival after myocardial infarction. Ann Intern Med 1992;117(12):1003–9.

[16] Stringhini S, Berkman L, Dugravot A, Ferrie JE, Marmot M, Kivimaki M. Socioeconomic status, structural and functional measures of social support, and mortality the British Whitehall II cohort study, 1985–2009. Am J Epidemiol 2012;175(12):1275–83.

[17] Litwak E, Messeri P, Wolfe S, Gorman S, Silverstein M, Guilarte M. Organizational theory, social supports, and mortality rates: a theoretical convergence. Am Sociol Rev 1989;54(1):49–66.

[18] Hu Y, Goldman N. Mortality differentials by marital status: an international comparison. Demography 1990;27(2):233–50.

[19] Haines VA, Hurlbert JS. Network range and health. J Health Soc Behav 1992;33(3):254–66.

[20] Wethington E, Kessler RC. Perceived support, received support, and adjustment to stressful life events. J Health Soc Behav 1986;27(1):78–89.

[21] Hawkley LC, Masi CM, Berry JD, Cacioppo JT. Loneliness is a unique predictor of age-related differences in systolic blood pressure. Psychol Aging 2006;21(1):152.

[22] Cacioppo JT, Hawkley LC. Social isolation and health, with an emphasis on underlying mechanisms. Perspect Biol Med 2003;46(3):S39–52.

[23] Cohen S. Social relationships and health. Am Psychol 2004;59(8):676.

[24] Uchino BN. Social support and health: a review of physiological processes potentially underlying links to disease outcomes. J Behav Med 2006;29(4):377–87.

[25] Cohen S, Wills TA. Stress, social support, and the buffering hypothesis. Psychol Bull 1985;98(2):310.

[26] Christou DD, Jones PP, Jordan J, Diedrich A, Robertson D, Seals DR. Women have lower tonic autonomic support of arterial blood pressure and less effective baroreflex buffering than men. Circulation 2005;111(4):494–8. https://doi.org/10.1161/01.CIR.0000153864.24034.A6.

[27] Yang YC, Schorpp K, Harris KM. Social support, social strain and inflammation: evidence from a national longitudinal study of US adults. Soc Sci Med 2014;107:124–35.

[28] Yang YC, McClintock MK, Kozloski M, Li T. Social isolation and adult mortality: the role of chronic inflammation and sex differences. J Health Soc Behav 2013;54(2):183–203. https://doi.org/10.1177/0022146513485244.

[29] Thomas PD, Goodwin JM, Goodwin JS. Effect of social support on stress-related changes in cholesterol level, uric acid level, and immune function in an elderly sample. Am J Psychiatry 1985;142(6):735–7.

[30] Marsden PV, Friedkin NE. Network studies of social influence. Sociol Methods Res 1993;22(1):127–51.

[31] Smith KP, Christakis NA. Social networks and health. Annu Rev Sociol 2008;34:405–29.

[32] Umberson D, Crosnoe R, Reczek C. Social relationships and health behavior across life course. Annu Rev Sociol 2010;36:139.

[33] Thoits PA. Mechanisms linking social ties and support to physical and mental health. J Health Soc Behav 2011;52(2):145–61. https://doi.org/10.1177/0022146510395592.

[34] McDade TW, Hawkley LC, Cacioppo JT. Psychosocial and behavioral predictors of inflammation in middle-aged and older adults: the Chicago health, aging, and social relations study. Psychosom Med 2006;68(3):376–81.

[35] McDade TW, Lindau ST, Wroblewski K. Predictors of C-reactive protein in the national social life, health, and aging project. J Gerontol B Psychol Sci Soc Sci 2011;66(1):129–36. https://doi.org/10.1093/geronb/gbq008.

[36] Davis MA, Neuhaus JM, Moritz DJ, Lein D, Barclay JD, Murphy SP. Health behaviors and survival among middle aged and older men and women in the NHANES I epidemiologic follow-up study. Prev Med 1994;23(3):369–76.

[37] Krueger PM, Chang VW. Being poor and coping with stress: health behaviors and the risk of death. Am J Public Health 2008;98(5):889–96.

[38] Uchino BN. Social support and physical health: understanding the health consequences of relationships. New Haven, CT: Yale University Press; 2004.

[39] Mirowsky J, Ross CE. Social causes of psychological distress. New Brunswick, NJ: Transaction Publishers; 2003.

[40] Turner RJ, Lloyd DA. The stress process and the social distribution of depression. J Health Soc Behav 1999;40(4):374–404.

[41] Taylor SE, Klein LC, Lewis BP, Gruenewald TL, Gurung RA, Updegraff JA. Biobehavioral responses to stress in females: tend-and-befriend, not fight-or-flight. Psychol Rev 2000;107(3):411.

[42] Pavalko EK, Willson AE. Life course approaches to health, illness and healing. In: Handbook of the sociology of health, illness, and healing. New York: Springer; 2011. p. 449–64.

[43] Collins NL, Dunkel-Schetter C, Lobel M, Scrimshaw SC. Social support in pregnancy: psychosocial correlates of birth outcomes and postpartum depression. J Pers Soc Psychol 1993;65(6):1243.

[44] Feldman PJ, Dunkel-Schetter C, Sandman CA, Wadhwa PD. Maternal social support predicts birth weight and fetal growth in human pregnancy. Psychosom Med 2000;62(5):715–25.

[45] Rich-Edwards JW, Stampfer MJ, Manson JE, et al. Birth weight and risk of cardiovascular disease in a cohort of women followed up since 1976. BMJ 1997;315(7105):396–400.

[46] Kiecolt-Glaser JK, Gouin J, Hantsoo L. Close relationships, inflammation, and health. Neurosci Biobehav Rev 2010;35(1):33–8.

[47] Maier SF, Watkins LR. Cytokines for psychologists: implications of bidirectional immune-to-brain communication for understanding behavior, mood, and cognition. Psychol Rev 1998;105(1):83.

[48] Centers for Disease Control. Underlying cause of death 1999–2013 on CDC WONDER online database. Data are from the multiple cause of death files, 1999–2013, as compiled from data provided by the 57 vital statistics jurisdictions through the vital statistics cooperative program; Updated 2015 [Accessed 3 February 2015].

Chapter 26

Trace Elements and Coronary Artery Disease

Ayşegül Bayir

Faculty of Medicine, Emergency Department, Selçuk University, Konya, Turkey

INTRODUCTION

Coronary artery disease is the most common cause of mortality and morbidity in developed and developing countries. Risk factors such as hypertension, hypercholesterolemia, and smoking dependence are present in less than 50% of the patients diagnosed with coronary artery disease. Free oxygen radicals are responsible for degenerative diseases such as coronary artery disease. Peroxidation of low-density lipoproteins (LDL) increases free radicals. This is an important factor for increased foam cell formation and atherosclerosis. Most trace elements have antioxidant properties that inhibit or reduce the formation of free oxygen radicals. It has been known for many years that trace element levels are important factors in the development and course of many diseases such as cardiovascular diseases, especially coronary artery disease. Numerous studies have reported that there is a relationship between low serum zinc (Zn) level, low copper (Cu) level, low selenium (Se) level, and high iron (Fe) levels and coronary artery disease development. In this section, the relationship between trace elements and coronary artery diseases will be explained.

THE IMPACT OF TRACE ELEMENTS ON CORONARY ARTERY DISEASES

Arterial thickness is an independent predictor of cardiovascular mortality in general population. Wall elasticity of the large elastic arteries depends on extracellular matrix proteins such as elastin and collagen. Depending on aging and some diseases, the elastic fibrils degrade and break up. Most trace elements have been found to be important in the pathogenesis of atherosclerosis. Elements such as magnesium (Mg), cobalt (Co), lithium (Li), and manganese (Mn) are considered to have a positive effect on cardiovascular diseases, whereas cadmium (Cd), lead (Pb), and silver (Ag) have a negative effect on atherosclerosis [1].

It has been reported that serum trace elements can be used as diagnostic and prognostic markers in coronary artery diseases. Serum zinc (Zn) and copper (Cu) levels in patients with myocardial infarction have been reported to vary with time. In addition, there is a relation between serum Cu and Fe levels and death from coronary artery disease. Cu and Fe play an important role in oxidative stress that is an important factor in the pathogenesis of ischemic heart diseases. Zn, Cu, and Fe, on the other hand, act as cofactors for superoxide dismutase (SOD) and glutathione peroxidase (GPx), which are major antioxidant enzymes in the body [2].

Supporting this information, plasma Zn levels in female patients diagnosed with acute coronary syndromes (ACS) were found to be significantly lower than in the healthy control group. It has also been reported that plasma malondialdehyde (MDA) levels, which are indicative of lipid peroxidation level in female patients with ACS, are significantly higher than in the healthy control group. In conclusion, it is emphasized that lipid peroxidation and Zn levels are important in patients with ACS and should be monitored during diagnose and treatment period [3].

Jain found that the serum level of Cu increased in the postinfarction from 24th hour and the Zn level decreased in the patients who had acute myocardial infarction [4]. It was determined in the study that both Cu and Zn levels returned to normal in 14 days after acute myocardial infarction. Altekin et al. [2] investigated the relationship between serum Cu, Fe, Se, and Zn levels and cardiac troponins and CK-MB levels in patients with ACS. As a result of the study, there was a positive correlation between cardiac troponins and serum Cu and Fe levels and negative correlation between serum Zn and Se levels. Serum trace element levels were correlated with myocardial injury grade.

Lifestyle in Heart Health and Disease. https://doi.org/10.1016/B978-0-12-811279-3.00026-4

Cu homeostasis is tightly balanced. The excess of Cu in the body is also associated with severe diseases as much as Cu inadequacy. The relationship between Cu metabolism and cardiovascular diseases is contradictory. Sidhu reported that chelation therapy in patients with stable coronary artery disease and a history of acute myocardial infarction prevented poor cardiovascular outcome [5]. Grammer investigated the relationship between serum ceruloplasmin and Cu levels in all patients with angiographic coronary artery disease and deaths due to cardiovascular causes and all causes in a LURIC study (the Ludwigshafen Risk and Cardiovascular Health study) conducted with a total of 3253 participants. Studies have shown that high ceruloplasmin and Cu levels are associated with both death risk due to all causes and cardiovascular death risk [6]. In another study with 2233 participants, there was a positive correlation between serum Cu level and 10-year coronary risk. In the same study, it was also reported that there was a strong and inverse correlation between serum Cu/Zn ratio and 10-year coronary risk [7].

In a study investigating the relationship between serum Zn and Cu levels and development of ischemic cardiomyopathy, serum Cu levels of patients diagnosed with ischemic cardiomyopathy were found to be significantly higher than in the healthy control group. According to the New York Heart Association (NYHA) classification, patients with NYHA III had higher serum Cu levels, lower Zn levels, and lower Zn/Cu levels than NYHA II patients. It has been stated that serum Cu levels play a role in the development of ischemic cardiomyopathy, the symptoms of ischemic cardiomyopathy can be reversed, and the progress can be stopped by using a chelator, which requires more extensive clinical studies [8].

Besides these studies, there are also experimental and clinical studies investigating the effects of trace element support on oxidative stress and cardiac dysfunction after acute myocardial infarction. Experimental myocardial infarction was performed with isoproterenol in rats pretreated with 0.1 mg/kg Se, 400 µg/kg chromium (Cr), and 30 mg/kg Zn for 28 days. It was reported in an experimental myocardial infarction model performed with subcutaneous isoproterenol (85 mg/day) for 2 days in rats given trace element support [9] that the levels of enzymes such as SOD, catalase (CAT), and GPx decreased; cardiac necrosis factor-α (TNF-α) level increased; and vascular endothelial growth factor (VEGF) decreased. Al-Rasheed reported that cardiac enzymes after myocardial infarction in the Se-supplemented rats, lipid peroxidation, SOD, CAT, GPx, TNF-α, and VEGF levels were within normal limits but dyslipidemia did not improve. In the rats given Cr supplementation, significant improvement was detected in the levels of markers except VEGF together with the lipid profile. Zn supplementation was found to have more positive effects on dyslipidemia angiogenesis [9]. As a result, trace elements given daily trace element support have been reported to have positive effects on myocardial performance by preventing oxidative damage, providing angiogenesis, antiinflammatory, and antihyperlipidemic effects.

The lack of Fe is has been reported to be a risk factor for cardiovascular diseases in recent years. Fe deficiency independent of anemia is an effective factor for prognosis in patients with heart failure. Regardless of the increased hemoglobin level in patients with heart failure, intravenous (IV) iron treatment has been found to increase functional capacity. There are few studies investigating the prevalence and clinical significance of Fe deficiency in patients with ischemic coronary syndrome. In a study conducted by Meroño, the relationship between iron deficiency, 30-day mortality, readmission to hospital, functional capacity, and quality of life was investigated in patients diagnosed with ACS [10]. It was found in the study that the lack of Fe was fairly frequent in patients with ACSs, it had no effect on cardiovascular mortality and morbidity in the medium term (30 days), but it was associated with poorer functional recovery, worse functional capacity, and quality of life.

Since iron is an important element in the formation of reactive oxygen species, the relationship between high iron levels and atherosclerotic heart diseases has also been investigated. Free Fe or catalytic Fe, free of transferrin or ferritin in circulation, has the potential to produce high levels of reactive oxygen products. In a study of 1701 patients diagnosed with ACS, the relationship between catalytic iron levels and all-cause mortality, recurrent acute myocardial infarction, recurrent ischemia, heart failure, and bleeding was investigated for a mean follow-up of 10 months [11]. Catalytic Fe levels were found to be significantly higher in dying patients than in survivors. It was concluded that high catalytic Fe level was correlated with increased mortality rates due to all causes. However, in this study, there was no significant relationship between catalytic Fe level and recurrent myocardial infarction, recurrent ischemia, heart failure, and bleeding. It has also been reported that catalytic Fe is not useful as an adjunct to biochemical markers for risk stratification of ischemic events.

In a study conducted by Huang [12], serum levels of Fe and inflammatory markers, interleukin 6 (IL) levels were examined independent of the anemia at the eighth hour before and after intervention in patients having acute myocardial infarction and undergoing percutaneous coronary intervention (PCI). In addition, patients were assessed by echocardiography within the first 2 days and 6 months after PCI. In the study, it was found that when the serum Fe level of the patients decreased, the TIMI risk score increased—there was a significant negative correlation between them—and when the serum Fe level was lower, the serum IL 6 level increased. Lower serum Fe levels before PCI have been reported to have adverse effects on the improvement of left ventricular systolic function after 6 months of PCI. At the end of the study, it was reported that higher serum Fe levels were cardioprotective and serum Fe levels could be used as a biomarker for post-ST elevated acute myocardial infarction complications and could be used for follow-up after a therapeutic approach.

Experimental studies have shown that reactive oxygen species produced during ischemia/reperfusion injury can be reduced with Se support. Se is a trace element that is responsible for the synthesis activity of certain antioxidant enzymes such as GPx and selenoprotein synthesis known to be protective against oxidative stress. Inadequate GPx activity is associated with atherosclerosis and poor prognosis in patients with coronary artery disease. In studies investigating the effects of Se deficiency on long-term prognosis in patients with ACS and stable angina pectoris (SAP), the hypothesis is based on increased plaque rupture and/or ischemia/reperfusion injury rates. In a study involving a total of 1731 ACS and SAP patients [13], participants were followed for cardiac death for 6 years. During the study period, a total of 190 patients died of cardiac causes. Mean serum Se levels of the patients with ACS who died were significantly lower than the mean serum Se levels of survivors. There was no significant correlation between serum Se levels and cardiovascular outcomes in SAP patients.

There are some studies reporting that there is a negative relationship between dietary intake of low Se or the level of Se and the risk of developing cardiovascular disease. However, studies of Se support in preventing cardiovascular disease do not support this hypothesis [14].

Some studies carried out in patients with STEMI and NSTEMI have reported that there is a positive correlation between peak troponin I (TnI) levels and serum Ser levels, indicating the degree or extent of myocardial damage [15]. In contrast, a similar study of patients with ACS diagnosed by Bayır et al. found different results [16]. In this study, it was reported that there was a negative correlation between serum TnI levels and Se, Zn, and Cu levels in patients diagnosed with ACS and low Se, Zn, and Cu levels could be a risk factor for ACS.

In another study conducted by Islamoğlu et al., serum Zn and Cu levels of 67 patients with atherosclerosis were compared with the healthy control group. As a result of this study, serum Zn and Cu levels of patients with atherosclerosis were significantly lower than those of healthy control group. However, it was stated that serum Zn and Cu levels did not determine the degree of atherosclerosis [17].

Magnesium (Mg) is another element that has been investigated for the prognosis in patients with ACS. In a study of patients who underwent stenting after ACS, serum Mg levels and major cardiovascular events were investigated for a mean of 24 months after stenting. There was no significant relationship between serum Mg levels and major cardiovascular events in patients with UAP in the study. In contrast, low serum Mg levels have been reported to be a strong predictor of major cardiovascular events in patients with acute myocardial infarction [18].

There are many studies in different parts of the world reporting that there is a relationship between drinking water Mg concentration and coronary artery disease. In the Atherosclerosis Risk in Communities (ARIC) study [19], 13,922 healthy participants without coronary artery disease were followed for 4–7 years, and participants with lower serum Mg levels were found to have a higher risk of coronary artery disease.

Similarly, the study of the National Health and Nutrition Examination Survey Epidemiologic Follow-up has demonstrated a significant inverse association between serum Mg levels and coronary heart disease and all-cause mortality [20].

The fact that Mg may have a preventive effect of cardiovascular diseases is partly explained with the reduction of inflammatory response. The clinical inflammatory syndrome characterized by leukocyte and macrophage activation, the release of inflammatory cytokines and proteins, and the release of dense free oxygen radicals were observed in experimental animals developed hypomagnesemia. While increase in Mg level in extracellular fluid reduces inflammation, decrease in Mg level causes the activation of phagocytes and endothelial cells. Similarly, in experimental hypomagnesemia studies, inflammation has been shown to cause proatherogenic changes in the hypertriglyceridemia and lipoprotein profile [21].

Magnesium has negative effects on myocardial infarction, infarct size, cardiac arrhythmia, endothelial platelet aggregation, coagulation system, vascular tone, and lipid metabolism. Experimental studies have shown that Mg has antiplatelet effects, inhibits thrombus formation and reocclusion in coronary vessels, and facilitates fibrinolysis-related recanalization [22,23].

In placebo-controlled randomized double-blind studies with Mg, it was determined that infarct size was reduced in patients receiving IV Mg supplementation compared with placebo group. It was reported in Killip class I patients who underwent acute myocardial infarction that infarct area was reduced by 20% and inhospital mortality was reduced with Mg supplementation [24]. In a large, placebo-controlled study, congestive heart failure, all-cause mortality in 28 days, and deaths due to ischemic heart disease during a 4.5-year follow-up were found to be reduced by 25%, 24%, and 20%, respectively, in patients receiving IV magnesium [25]. Inhospital mortality was reduced by 50% in patients given 22 g IV Mg for 48 h after infarction and the patients with an acute myocardial infarction and who were not suitable for the treatment of reperfusion [26].

In another study conducted by Zhang et al. [27], another study conducted with 58,615 healthy participants aged 40–79 years, Mg intake in the diet was questioned, and patients were followed for an average of 14.7 years in terms of cardiovascular disease mortality. At the end of follow-up period, 2690 deaths due to cardiovascular diseases were detected.

As a result, magnesium intake in the diet was found to be inversely related to cardiovascular disease-related mortality, especially in female participants.

As can be seen, keeping the trace elements at a certain limit in the body is very important for the prevention of cardiovascular diseases, especially coronary artery diseases and death rates due to coronary artery diseases. There are also some studies reporting that the frequency of coronary artery disease decreases with the supplement of these elements.

REFERENCES

[1] Subrahmanyam G, Pathapati RM, Ramalingam K, Indira SA, Kantha K, Soren B. Arterial stiffness and trace elements in apparently healthy population – a cross-sectional study. J Clin Diagn Res 2016;10(9):LC12–5.

[2] Altekin E, Coker C, Sişman AR, Onvural B, Kuralay F, Kirimli O. The relationship between trace elements and cardiac markers in acute coronary syndromes. J Trace Elem Med Biol 2005;18(3):235–42.

[3] Cikim G, Canatan H, Gursu MF, Gulcu F, Baydas G, Kilicoglu AE. Levels of zinc and lipid peroxidation in acute coronary syndrome. Biol Trace Elem Res 2003;96(1–3):61–9.

[4] Jain VK, Mohan G. Serum zinc and copper in myocardial infarction with particular reference to prognosis. Biol Trace Elem Res 1991;31(3):317–22.

[5] Sidhu MS, Saour BM, Boden WE. A TACTful reappraisal of chelation therapy in cardiovascular disease. Nat Rev Cardiol 2014;11(3):180–3.

[6] Grammer TB, Kleber ME, Silbernagel G, Pilz S, Scharnagl H, Lerchbaum E, et al. Copper, ceruloplasmin, and long-term cardiovascular and total mortality (the Ludwigshafen Risk and Cardiovascular Health Study). Free Radic Res 2014;48(6):706–15.

[7] Ghayour-Mobarhan M, Shapouri-Moghaddam A, Azimi-Nezhad M, Esmaeili H, Parizadeh SM, Safarian M, et al. The relationship between established coronary risk factors and serum copper and zinc concentrations in a large Persian cohort. J Trace Elem Med Biol 2009;23(3):167–75.

[8] Shokrzadeh M, Ghaemian A, Salehifar E, Aliakbari S, Saravi SS, Ebrahimi P. Serum zinc and copper levels in ischemic cardiomyopathy. Biol Trace Elem Res 2009;127(2):116–23.

[9] Al-Rasheed NM, Attia HA, Mohamed RA, Al-Rasheed NM, Al-Amin MA. Preventive effects of selenium yeast, chromium picolinate, zinc sulfate and their combination on oxidative stress, inflammation, impaired angiogenesis and atherogenesis in myocardial infarction in rats. J Pharm Pharm Sci 2013;16(5):848–67.

[10] Meroño O, Cladellas M, Ribas-Barquet N, Poveda P, Recasens L, Bazán V, et al. Iron deficiency is a determinant of functional capacity and health-related quality of life 30 days after an acute coronary syndrome. Rev Esp Cardiol (Engl Ed) 2016;70(5):363–70.

[11] Steen DL, Cannon CP, Lele SS, Rajapurkar MM, Mukhopadhyay B, Scirica BM, et al. Prognostic evaluation of catalytic iron in patients with acute coronary syndromes. Clin Cardiol 2013;36(3):139–45.

[12] Huang CH, Chang CC, Kuo CL, Huang CS, Chiu TW, Lin CS, et al. Serum iron concentration, but not hemoglobin, correlates with TIMI risk score and 6-month left ventricular performance after primary angioplasty for acute myocardial infarction. PLoS One 2014;9(8):e104495.

[13] Lubos E, Sinning CR, Schnabel RB, Wild PS, Zeller T, Rupprecht HJ, et al. Serum selenium and prognosis in cardiovascular disease: results from the AtheroGene study. Atherosclerosis 2010;209(1):271–7.

[14] Stranges S, Navas-Acien A, Rayman MP, Guallar E. Selenium status and cardiometabolic health: state of the evidence. Nutr Metab Cardiovasc Dis 2010;20(10):754–60.

[15] Kutil B, Ostadal P, Vejvoda J, Kukacka J, Cepova J, Alan D, et al. Alterations in serum selenium levels and their relation to troponin I in acute myocardial infarction. Mol Cell Biochem 2010;345(1–2):23–7.

[16] Bayır A, Kara H, Kıyıcı A, Oztürk B, Akyürek F. Levels of selenium, zinc, copper, and cardiac troponin I in serum of patients with acute coronary syndrome. Biol Trace Elem Res 2013;154(3):352–6.

[17] Islamoglu Y, Evliyaoglu O, Tekbas E, Cil H, Elbey MA, Atilgan Z, et al. The relationship between serum levels of Zn and Cu and severity of coronary atherosclerosis. Biol Trace Elem Res 2011;144(1–3):436–44.

[18] An G, Du Z, Meng X, Guo T, Shang R, Li J, et al. Association between low serum magnesium level and major adverse cardiac events in patients treated with drug-eluting stents for acute myocardial infarction. PLoS One 2014;9(6):e98971.

[19] Liao F, Folsom AR, Brancati FL. Is low magnesium concentration a risk factor for coronary heart disease? The Atherosclerosis Risk in Communities (ARIC) Study. Am Heart J 1998;136:480–90.

[20] Ford ES. Serum magnesium and ischemic heart disease: findings from national sample of US adults. Int J Epidemiol 1999;28:645–51.

[21] Shechter M. Magnesium and cardiovascular system. Magnes Res 2010;23(2):60–72.

[22] Bonetti PO, Lerman LO, Lerman A. Endothelial dysfunction. Endothelial dysfunction. A marker of atherosclerotic risk. Arterioscl Thromb Vasc Biol 2003;23:168–75.

[23] Rukshin V, Santos R, Gheorghiu M, Shah PK, Kar S, Padmanabhan S, et al. A prospective, nonrandomized, open-labeled pilot study investigating the use of magnesium in patients undergoing nonacute percutaneous coronary intervention with stent implantation. J Cardiovasc Pharmacol Ther 2003;8:193–200.

[24] Morton BC, Nair RC, Smith FM, McKibbon TG, Poznanski WJ. Magnesium therapy in acute myocardial infarction. Magnesium 1984;3:346–52.

[25] Woods KL, Fletcher S. Long term outcome after intravenous magnesium sulphate in suspected acute myocardial infarction: the second Leicester Intravenous Magnesium Intervention Trial (LIMIT-2). Lancet 1994;343:816–9.

[26] Shechter M, Hod H, Chouraqui P, Kaplinsky E, Rabinowitz B. Magnesium therapy in acute myocardial infarction when patients are not candidates for thrombolytic therapy. Am J Cardiol 1995;75:321–3.

[27] Zhang W, Iso H, Ohira T, Date C, Tamakoshi A. JACC Study Group. Associations of dietary magnesium intake with mortality from cardiovascular disease: the JACC study. Atherosclerosis 2012;221(2):587–95.

Decompressive Hemicraniectomy for Severe Stroke: An Updated Review of the Literature

Theresa L. Green*,†, Janine Pampellonne*, Lee Moylan†

*Queensland University of Technology, Brisbane, QLD, Australia, †Royal Brisbane and Women's Hospital, Brisbane, QLD, Australia

BACKGROUND

Decompressive hemicraniectomy (DHC) is a therapeutic surgical option for treatment of massive (i.e., CT—large volume with edema) middle cerebral artery infarction (MCA), lobar intracerebral hemorrhage (lobar ICH), and severe aneurysmal subarachnoid hemorrhage (aSAH) [1–5]. The mechanism of neurological deterioration relates to brain edema with midline shift and pressure on vital brain structures producing coma and death [6,7]. Best medical therapy currently consists of close monitoring, hyperventilation, osmotherapy (the use of hypertonic solutions to produce dehydration for the treatment of cerebral edema), hypothermia, or barbiturate-induced coma [8]. Modern surgical techniques and careful patient selection enhance the survival and clinical outcomes postoperatively in patients with massive MCA infarction, large lobar ICH, or severe SAH [2,8–10]. Medical advances in the treatment of acute stroke conditions, specifically potentially life-saving treatments such as DHC, have created a situation whereby individuals and families are faced with making difficult decisions during the very acute stage of stroke; full understanding of the potential outcome is limited. Recent evidence suggests that outcomes based on the modified Rankin score (most common functional outcome measure used in stroke research) may be insufficient to determine acceptability of the procedure for patients and clinicians [11,12].

DHC has been shown to be effective in reducing mortality and improving functional outcomes in these three life-threatening conditions [8,13]. However, assessing clinical benefit also needs to include features such as morbidity, quality of life (QoL), and caregiver outcomes [14–17]. Indeed, clinicians continue to debate recommending an aggressive surgical intervention without sufficient evidence to support long-term outcomes such as illness-related morbidity and health-related QoL [18]. QoL is defined as persons' perception of their life that can be affected by such things as their physical and psychological state and level of independence. Health-related quality of life (HRQoL) is a broad, multidimensional concept that extends beyond physical health. While it is considered a reflection of an individual's subjective perception of their health and mental, emotional, and social function, there is no single definitive definition of HRQoL [19].

THE REVIEW PROCESS

Methods

This chapter provides an update of a narrative review conducted in 2011, examining the state of the literature on patient and caregiver outcomes following DHC, specifically QoL and caregiver burden. The review is situated within our theoretical assumption that severe stroke and life-saving interventions influence long-term outcomes beyond functional recovery and mortality. We used the JBI review framework [20] to guide this comprehensive synthesis of the literature related to psychosocial outcomes following DHC for severe stroke. For the purposes of this review, we will use the term QoL as this was the term used in our previous study although many of the articles retained for this review did use the term HRQoL.

Search Methods

First, we searched the online databases Academic Search Premier, Cochrane Library, Cumulative Index to Nursing and Allied Health Literature (CINAHL), Embase, Joanna Briggs Institute (JBI), Medline, ProQuest, PubMed, and PsycINFO (Fig. 1). Dissertations or conference proceedings were not included in the final search. Key terms for this phase of the search included ("decompressive hemicraniectomy" OR craniectomy) AND ("severe stroke" OR stroke) AND "quality of life" ("caregiver burden" OR caregiver*). We reviewed papers published in English language journals that examined the recovery of patients with severe stroke, including acute ischemic stroke, lobar ICH, and SAH and those with a focus

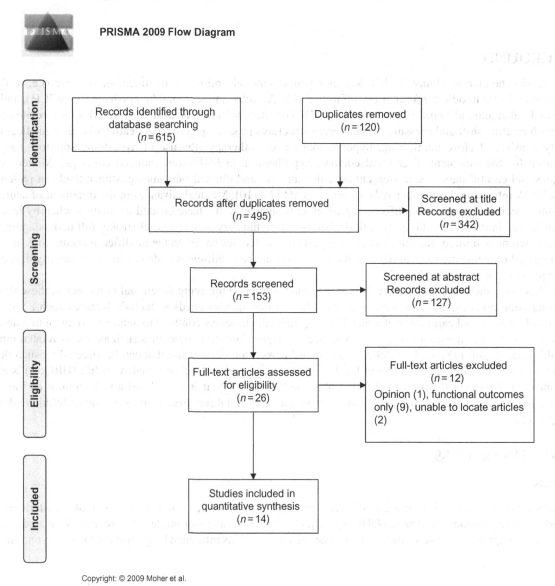

PRISMA 2009 Flow Diagram

FIG. 1 Impact of decompressive hemicraniectomy for severe stroke on quality of life and caregivers: an updated review of the literature. *From Moher D, Liberati A, Tetzlaff J, Altman DG, The PRISMA Group. Preferred reporting items for systematic reviews and meta-analyses: the PRISMA statement. PLoS Med 2009;6(7):e1000097. http://dx.doi.org/10.1371/journal.pmed1000097.*

on QoL and caregiver burden. Preliminary criteria for inclusion in this review were papers (a) in which stroke recovery following DHC for severe stroke was described, (b) that showed relevance to the concept of QoL or included a focus on the caregiver, (c) that pertained to adult populations (i.e., ≥18 years), and (d) that were published between Jan. 2009 and Mar. 2017. This initial search strategy yielded 615 titles and abstracts. Dissertations and duplicates were then removed. The first author then assessed the titles and available abstracts electronically using the preliminary inclusion criteria. The nonapplicable papers resulting from the broad search terms related primarily to the functional recovery of patients with severe stroke and radiological poststroke outcomes (Fig. 1).

Search Outcome

Combining the terms severe stroke, DHC, lobar ICH, and subarachnoid hemorrhage with each of the terms "quality of life" and "caregivers" revealed a broad set of journal papers with a specific focus on quantitative and qualitative studies describing QoL or health-related QoL and caregiver outcomes following severe stroke ($n = 153$). Reference lists from original papers were scanned for additional relevant papers. Many papers included in this review were drawn from research conducted in North America and Europe. Using the preliminary inclusion criteria, 26 papers relevant to the broad topic of QoL and caregiver outcomes following severe stroke were retained. All papers were successfully retrieved.

Like the original review, inclusion in the final sample for this review was guided by three criteria: (1) the report of original quantitative or qualitative research, discussion paper, or literature review; (2) a focus on patients with DHC for severe stroke due to massive MCA infarction, lobar ICH, or aSAH discharged home postrehabilitation; and (3) a conceptualization of stroke recovery that included a focus on QoL or caregiver outcomes [15]. A final sample of 14 papers in which QoL of the patient and/or caregiver and caregiver outcomes, including burden or strain, were specifically addressed were retained for full review.

Quality Appraisal

Twelve of the 14 articles retained for the review were screened for quality using the Pace et al. *mixed-method appraisal tool* (MMAT) [21]. This tool was developed to appraise qualitative, quantitative, and mixed-method research studies using a scoring matrix. It has undergone initial efficiency and reliability testing and content validation [22–24]. The other two articles were narrative reviews of the literature. The quality appraisal criteria included screening questions: research questions/objectives and data collection processes; quantitative sampling strategy, sample representativeness, measurement strategies, and response rate questions; and mixed-method research design, integration, and limitation questions. Data were collected by two reviewers in duplicate (TG and JP) on QoL and caregiver outcomes (Appendix, Review Matrix). Based on the appraisal tool results, all quantitative and mixed-method studies demonstrated good-quality design and methods (Table 1).

RESULTS

Primary Study Characteristics

From the 153 unique quantitative or mixed-method studies screened, 14 studies contained data on post-DHC quality of life and/or caregiver outcomes (329 participants—316 patients and 18 caregivers; mean age 49.4 years, range 19–72 years). The majority of the studies were retrospective cohort studies examining DHC outcomes spanning the years 1995–2013 ($n = 8$), with four prospective cohort studies (two mixed methods) and two literature reviews.

The results of the review will be categorized below as either QoL (including psychosocial measures) for patients and caregivers or specific caregiver outcomes.

Quality of Life

Overall, DHC was found to reduce mortality and improve functional outcomes [14,25–28] in the review cohort. However, many patients continued to live with significant physical impairment that limited their activities of daily living and influenced perceptions of QoL [15,25–27,29–35]. In most cases, the physical aspects of QoL were rated significantly worse than psychosocial domains, but overall, QoL was deemed acceptable. Some caregivers and patients reported that despite the patient's significant disability and high incidence of depression, they were satisfied with their life post DHC [34–36]. Geurts et al. found that improved QoL noted at 1 year post DHC was sustained at third-year follow-up in the surgical intervention group versus the nonsurgical cohort [26].

TABLE 1 Characteristics of Included Studies

First Author	Country	Method Used	Sample	Mean Age	MMAT Quality
Benejam	Spain	Prospective cohort	29 Pt	49	d
Geurts	The Netherlands	Observational cohort	64 Pt	50	d
Green	Canada	Literature review	355 Pt	54	n/a
Green	Canada	Mixed-methods convergent parallel	27 Pt 13 CG	54	d
Kelly	The United States	Retrospective cohort	32 Pt	55	c
McKenna	Ireland	Case series (patients)	5 Pt	36	d
McKenna	Ireland	Mixed methods convergent parallel (caregivers)	5 CG	n/r	d
Pfefferkorn	Germany	Retrospective cohort	57 Pt	59	d
Rahme	Ohio	Literature review	382 Pt	50	n/a
Schmidt	Germany	Retrospective cohort	20 Pt	52	d
Sundseth	Norway	Retrospective cohort	26 Pt	48	d
van Middelaar	The Netherlands	Retrospective cohort	25 Pt	48	d
von Sarnowski	Germany	Retrospective cohort	11 Pt	48	d
Weil	Canada	Retrospective cohort	14 Pt	44	d

Cg, caregivers/spouse; *Pt*, patients; *MMAT*, mixed-method appraisal tool, scoring.
[a] 25% of criteria met.
[b] 50% of criteria met.
[c] 75% of criteria met.
[d] 100% of criteria met.

In this review, it was apparent that QoL, particularly in the physical domain, was reduced for many stroke survivors. However, as identified in some of the research, it may well be that patient and family assessments of QoL following DHC can be tempered by time and the subjective experience of living with stroke. This might be reflected as well in the responses that most caregivers would make the decision again and many patients would, in retrospect, also support the same decision again [14,15,31,33–35].

Caregiver Outcomes

Only two studies were found for this review in which caregiver outcomes following DHC were specifically addressed [14,30]. In the study by Green et al., caregivers, while generally satisfied with caregiving, demonstrated increasing dissatisfaction with their family function (distressed range) and were moderately burdened by caregiving needs of their spouse/family member following DHC [14]. McKenna et al. noted that while overall carer scores were within average normative scores, over time, negative indicators appeared, in the areas of satisfaction with performance as a caregiver and diminishing personal life satisfaction [30].

DISCUSSION

It was apparent from the current review of the literature that the parameters used to define "good functional outcomes" and the scales used to measure functional and psychosocial outcomes still vary considerably. Time frames for assessments were also inconsistent across the studies (ranging from 4 months to several years post-DHC), making comparison of outcomes over time difficult.

In several studies, the mRS (score 0–6, lower better) was reported to be >3, indicating moderately severe physical and functional disability [25,27,31]. The mRS is a very easy to administer functional outcome scale and is the most commonly used scale in stroke research. However, there is a risk of variability in ratings [37] as the mRS has very broad categories, which may result in a patient being attributed a lower (or higher) score due to assessor perceptions of disability. Even with an mRS score of "3," however, patients often do have significant disability and require assistance in many activities of daily living [37–39]. Full independence post-DHC was also reported as the exception, not the norm.

Stroke can be a devastating disease, leaving patients and families to cope with severe poststroke functional, cognitive, and emotional disabilities. For some patients, the severity of their stroke is life-threatening, and DHC offers some hope of survival. However, the full consequences of this intervention are still not fully understood in terms of the long-term impact on patients and caregivers/families. Indeed, considering the age range for participants included in studies in this review, patients undergoing DHC are generally under 60 years of age [15]; this potentially can mean DHC patients may survive another 20–30 years, which has long-term implications for families, caregivers, and health-care system. Despite this, most patients and families interviewed in the studies in this review did not regret their decision to undergo surgical intervention and would do so again if circumstances necessitated such a decision.

It is possible to extrapolate from the data presented here an impression of the amount of care relegated to informal care-givers following hospital and rehabilitation discharge of patients surviving DHC. Indeed, it can be understood that many DHC survivors will require considerable physical, psychological, and emotional support for extended periods of time. As most informal care is provided by family members, these elements of care may place considerable stress and burden on the care providers. Caregiver burden, while not measured in most studies in this review, was noted to be high in the two that did assess caregiver outcomes. The scales used to assess burden typically reflect elements of health such as emotional well-being, stress, relationships, and coping ability [40]. It is also worth noting that while DHC may save lives, it does place a burden on informal caregivers that eventually may precipitate or manifest in poor physical and mental/emotional health. However, as noted previously [15], it would be speculative to make assumptions about perceptions of caregiver burden based on these limited data. This highlights the need to provide descriptions, from a quantitative and qualitative perspective, of patient and caregiver outcomes following DHC for severe stroke.

CONCLUSIONS

While there is a growing body of literature describing functional outcomes and morbidity/mortality post-DHC, and studies examining QoL and caregiver outcomes following stroke in general abound, there continues to be a significant gap in the literature in relation to QoL and caregiver outcomes following DHC for acute stroke. As only a small number of patients undergo DHC each year, this does create difficulty in obtaining a critical mass of patients for study recruitment. As well, many of these stroke survivors remain moderately to severely disabled postdischarge; thus, the ability to complete standard assessment measures may be compromised.

Despite these difficulties, this review highlights the need for ongoing investigation of long-term neurobehavioral and psychosocial outcomes following hemicraniectomy to aid health-care professionals in discussions with families prior to surgical intervention. As well, understanding patient and caregiver HRQoL post DHC will generate opportunities to improve the health-care experience of patients, caregivers, and health-care providers by implementing appropriately contextualized interventions.

APPENDIX REVIEW MATRIX FOR LITERATURE REVIEW

First Author, Year	Focus	No. of Subjects	Geographic Location	Subject Characteristics	Sample, Design	Salient Findings, Comments
Benejam et al., 2009	Quality of life (QoL) and neurobehavioral changes in survivors of malignant middle cerebral artery (MCA) infarction	29 patients	Barcelona, Spain	Male and female aged ≤65 years, infarction >50% of MCA territory on CT or MRI	Convenience sample, prospective cohort—disability rated using the Barthel index (BI) with a score >60 deemed functionally independent (7 pts >60). QoL assessed using the sickness impact profile with mean of 38.2 (range 29–72)	Decompressive hemicraniectomy (DHC) reduces mortality and improves functional outcome (5 pts mRS 0–3, 14 pts mRS 4) and younger patients fared better, QoL scores higher (worse) in physical than psychosocial domains. Depression scores indicative of mild-severe depression in 13 of the 29 participants
Geurts et al., 2013	Surgical decompression for space-occupying cerebral infarction: outcomes at 3 years in the randomized HAMLET trial	64 patients randomized to decompression (n=32) versus best medical treatment (n=32)	The Netherlands	Male and female patients ≤60 years of age with hemispheric infarction randomly assigned to intervention (DHC) versus control	Observational cohort. Primary outcome mRS at 1 year (good 0–3, poor 4,5 or death). Functional outcome at 3 years, case fatality, dependence (BI), QoL (SF-36), depression, caregiver strain index (CSI)	Outcomes obtained at 1 year sustained at 3 years post DHC (HAMLET): − No effect on poor functional outcome − Reduced case fatality − Improved QoL in surgical group − 16 surgical patients and 8 controls lived at home
Green et al., 2010	QoL and caregiver outcomes following DHC for severe stroke	—		MCA infarction. Intracerebral hemorrhage (ICH). Lobar intracerebral hemorrhage (ICH)	Narrative review	Good functional outcomes associated with younger age and fewer comorbid conditions; QoL reduced for many stroke survivors although not assessed routinely in studies included in the review; most patients and caregivers would make same decisions
Green et al., 2015	Aggressive surgical interventions for severe stroke: impact on QoL, caregiver burden and family outcomes	27 patients (19 female and 8 male) and 13 spouses	Calgary, Canada	Retrospective cohort. Hemicraniectomy; surgical clot removal. Male and female. 1–7 years post intervention	Mixed-method convergent parallel design. Primary outcomes QoL and caregiver burden, secondary outcomes family function, depression, morbidity and mortality, cognitive status. MOCA, EQ-5D (0–100, higher better), RNLI, PHQ-9 (depression), BAKAS caregiving outcome scale (BCOS), Family Assessment Device (FAD, 0 (better), 4 (worse))	Significant improvements in stroke severity and patient outcome scores over time, no statistically significant differences between patients and caregiver groups, family assessment scores scored at high end of normal (1.95, worse) with 7 patients and 4 CGs rating FF in distressed range (>2); BCOS: 15–105 point scale, higher scores better, mean 52.6, range 43.6–57.7. Patients valued ability to interact meaningfully with family unit. Spouses valued patient independence in activities of daily living (ADLs) and appropriate behavior in social situations. All patients and 12 spouses would make the same decision to consent for DHC

Study	Aim	N	Location	Population	Design / Outcome measures	Results
Kelly et al., 2014	Comparing clinician- and patient-reported outcome measures after hemicraniectomy for ischemic stroke	32 patients	Cleveland, the United States	Predominantly female ischemic stroke patients who had undergone hemicraniectomy between Jan 2009 and May 2013 at Cleveland Clinic, the United States	Outpatient follow-up at 3±2 months (early) and 9±3 months (late) NIHSS, mRS, EQ-5D, PHQ-9, SIS-16	11 of 32 patients completed all HSMs during both follow-up periods. Both HSM types (clinician- and patient-reported outcome measures) improved over time. Clinician-reported median NIHSS scores improved from 12 to 7 ($P=.003$). Median mRS scores little improvement from 4 to 3 ($P=.2$). Patient-reported median EQ-5D scores improved from 0.33 to 0.69 ($P=.03$). Median PHQ-9 scores improved from 9 to 1 ($P=.06$). SIS-16 scores improved from 23 to 57 ($P=.01$). EQ-5D and mRS score differences between periods were correlated ($r=-.65$, $P=.03$), only the EQ-5D showed significant improvement over time
McKenna et al., 2012	Long-term neuropsychological and psychosocial outcomes of survivors of malignant middle cerebral artery infarction treated via decompressive hemicraniectomy	5 patients	Belfast, Ireland	Stroke survivors who had sustained MCA infarction and received DH between 6 months and 10 years prior to time of study	Case series design [4]. Assessed—neuropsychological domains (IQ, attention, visual/auditory memory, executive function, and visual-spatial ability) Psychosocial domains assessed—self-rated depression and anxiety (HADS), QoL (SAQOL-39)	All patients experienced neuropsychological impairments Social support and preserved abilities may help to protect against depression and unacceptable QoL. DH following stroke does not necessarily lead to unacceptable long-term neuropsychological and psychosocial outcomes. All 4 stroke survivors and caregivers retrospectively considered surgery as a favorable course of action

APPENDIX REVIEW MATRIX FOR LITERATURE REVIEW—CONT'D

First Author, Year	Focus	No. of Subjects	Geographic Location	Subject Characteristics	Sample, Design	Salient Findings, Comments
McKenna et al., 2013 (Subanalysis of McKenna et al., 2012)	Experience and level of burden of caregivers of survivors of malignant middle cerebral artery infarctions who had undergone decompressive hemicraniectomy	6 caregivers	Belfast Ireland	Family caregivers nominated by stroke survivors who had sustained malignant middle cerebral artery infarction and undergone subsequent decompressive hemicraniectomy within the preceding 10 years	Mixed-method convergent parallel design Caregiver experience, burden, levels of depression, and anxiety Semistructured interviews with carer, thematic content analysis of carer interview data Sense of competence and HADS results compared with normative data	Caregivers overall SCQ scores were within average limits (1 SD of the mean of normative sample, 42 ± 13.9) (sample range 42–54) Subscale 1—satisfaction with stroke survivor as a recipient of care ≤50% of the max possible score (range 25–50%) (increasing score = decreasing levels of satisfaction) Subscale 2—satisfaction with one's own performance as a caregiver ≤50% of the max possible score (range 37.5%–52.1%) (increasing score = decreasing levels of satisfaction) Subscale 3—consequences of involvement in care for the personal life of the caregiver ≥50% (range 50%–78.1%) (increasing score = more negative consequences) Caregivers did not experience clinically significant levels of depression All caregivers identified methods of coping and, with the exception of one carer, areas of posttraumatic growth
Pfefferkorn et al., 2009	Long-term outcome after suboccipital decompressive craniectomy (SDC) for malignant cerebellar infarction	57 patients	Munich, Germany	Male and female, acute cerebellum infarction that were treated by SDC between 1995 and 2006 at an institution in Germany	Retrospective cohort study Outcomes—mortality, long-term functional outcome, QoL, prognostic factors Structured interview, mRS, SF-36	SDC is safe procedure for patients with malignant cerebellar infarction Substantial early mortality related to infarct, not procedure Acceptable long-term outcome especially in the absence of brain stem infarction however QoL moderately lower than in healthy controls
Rahme et al., 2012	Decompressive hemicraniectomy for malignant middle cerebral artery infarction: is life worth living?	382 patients from 16 studies	Cleveland, Ohio	Male and female, mean age 50.1 years	Literature review	Despite significant physical disability and high incidence of depression, majority of patients are satisfied with their life following DHC for malignant MCA infarction and do not regret having undergone the procedure

| Schmidt et al., 2011 | Cognition after DHC for malignant media infarction of nonspeech-dominant hemisphere | 20 patients | Germany | Male and female, malignant MCA infarction of nonspeech-dominant hemisphere and treated using DHC with duraplasty at a university medical center in Germany between 1999 and 2003. DHC ≥1 year prior to reexamination

Control group of persons matched by age, gender and level of education | Retrospective observational study

Functional outcome (BI, NEADL, or NIHSS), full neuropsychological assessment, health-related QoL | All patients but one were neurologically handicapped, half of them severely. Age was significantly correlated with poorer values on the Rankin scale and Barthel index All cognitive domain z values were significantly lower than in the control group. Upon reexamination, 18 of 20 patients were found to be cognitively impaired to a degree that fulfilled the formal DSM IV criteria for dementia

Patients with nonspeech-dominant hemispheric infarctions treated with DHC are at high risk of depression and severe cognitive impairment; however, DHC should not be rejected as a treatment option |
| Sundseth et al., 2014 | Long-term outcome and QoL after DHC in speech-dominant swollen middle cerebral artery infarction versus nondominant side infarction | 26 patients | Norway | Male and female treated with DHC due to malignant middle cerebral artery infarction at Oslo University Hospital Rikshospitalet, Norway between May 1998 and Oct. 2010 | Retrospective cohort study

Primary outcome—global functioning (mRS)

Secondary outcomes—neurological impairment (NIHSS), dependency (BI), cognitive impairment (MMSE), outcome after severe brain injury (GOS), anxiety and depression (HADS), QoL (short form-36) | There was no statistical or clinical difference in functional outcome and QoL in patients with speech-dominant compared with nondominant side infarction |

APPENDIX REVIEW MATRIX FOR LITERATURE REVIEW—CONT'D

First Author, Year	Focus	No. of Subjects	Geographic Location	Subject Characteristics	Sample, Design	Salient Findings, Comments
van Middelaar et al., 2015	QoL, symptoms of depression, and caregiver burden after surgical decompression for a space-occupying middle cerebral artery infarct	25 patients	Amsterdam, The Netherlands	Male and female underwent surgical decompression for space-occupying MCA infarct between Oct. 2007 and Mar. 2012, ≥6 months post DHC	Retrospective cohort study QoL (VAS, SF-36, 0–100 where higher score = better QoL), depression (HADS, ≥8 = depression), caregiver burden (CSI, ≥7 = considerable level of stress for caregiver), functional outcome (mRS), cognition (MMSE)	In chronic phase after DHC, QoL is comparable with that in the general population, whereas physical QoL is worse. Depression is relatively uncommon, but the majority of caregivers feel substantially burdened. The majority of patients and caregivers would in retrospect again decide for the surgery
von Sarnowski et al., 2011	Long-term health-related QoL after DHC in stroke patients with life-threatening space-occupying brain edema	11 patients	Germany	Male and female underwent hemicraniectomy due to space-occupying anterior circulation cerebral infarction between 2000 and 2007, 9–51 months after DHC	Retrospective cohort study Outcome—long-term health-related QoL, NIH stroke scale, BI, mRS, neuropsychological Tests, and HRQoL scales(SF-36, NHP), Questions on Life Satisfaction, HADS, and EQ-5D	Physical components of health-related QoL are highly impaired; however, patients achieved a satisfying level of psychological well-being. The majority of patients and caregivers would in retrospect again agree to undergo life-saving DHC
Weil et al., 2011	Quality of Life following hemicraniectomy for malignant MCA territory infarction	14 patients		Male and female <60 years, good previous QoL, CT-documented extensive MCA territory infarction, underwent DHC between 2001 and 2009	Functional Outcome—GOS, mRS, BI QoL, SIS	Most patients report satisfactory QoL despite significant disability

REFERENCES

[1] Honeybul S, Ho KM. The current role of decompressive craniectomy in the management of neurological emergencies. Brain Inj 2013;27(9):979–91.

[2] Cruz-Flores S, Berge E, Whittle IR. Surgical decompression for cerebral oedema in acute ischaemic stroke. Cochrane Database Syst Rev 2012;1:CD003435. https://doi.org/10.1002/14651858.CD003435.pub2.

[3] Hofmeijer J, Kappelle L, Algra A, Amelink G, Gijn J, Worp H. Surgical decompression for space-occupying cerebral infarction (the Hemicraniectomy After Middle Cerebral Artery infarction with Life-threatening Edema Trial [HAMLET]): a multicentre, open, randomised trial. Lancet Neurol 2009;8(4):326–33. Available from http://onlinelibrary.wiley.com/o/cochrane/clcentral/articles/001/CN-00686001/frame.html.

[4] Rajwani K, Crocker M. Decompressive craniectomy for the treatment of malignant middle cerebral artery infarction: an up-to-date systematic review. Br J Neurosurg 2016;30(2):180. Available from http://onlinelibrary.wiley.com/o/cochrane/clcentral/articles/371/CN-01160371/frame.html.

[5] Fung C, Murek M, Z'Graggen WJ, Krähenbühl AK, Gautschi OP, Schucht P, et al. Decompressive hemicraniectomy in patients with supratentorial intracerebral hemorrhage. Stroke 2012;43(12):3207–11.

[6] Subramaniam S, Hill MD. Decompressive hemicraniectomy for malignant middle cerebral artery infarction: an update. Neurologist 2009;15(4):178–84.

[7] Amorim RL, de Andrade AF, Gattás GS, Paiva WS, Menezes M, Teixeira MJ, et al. Improved hemodynamic parameters in middle cerebral artery infarction after decompressive craniectomy. Stroke 2014;45(5):1375–80.

[8] Back L, Nagaraja V, Kapur A, Eslick GD. Role of decompressive hemicraniectomy in extensive middle cerebral artery strokes: a meta-analysis of randomised trials. Int Med J 2015;45(7):711–7.

[9] Streib CD, Hartman LM, Molyneaux BJ. Early decompressive craniectomy for malignant cerebral infarction: meta-analysis and clinical decision algorithm. Neurol Clin Pract 2016;6(5):433–43.

[10] Dawson J, Hopkins P, Ling J, Walsh D, Tolias C. Retrospective observation of 6-month survival following decompressive craniectomy in a London major trauma and stroke centre. Crit Care 2012;16:S112.

[11] Honeybul S, Ho KM, Blacker DW. ORACLE stroke study: opinion regarding acceptable outcome following decompressive hemicraniectomy for ischemic stroke. Neurosurgery 2016;79(2):231–6.

[12] Honeybul S, Ho K. Decompressive craniectomy – a narrative review and discussion. Aust Crit Care 2014;27(2):85–91.

[13] Stephen H, Kwok H. Decompressive craniectomy for neurological emergencies: a systematic review. Connect World Crit Care Nurs 2015;9(4):148.

[14] Green T, Demchuk A, Newcommon N. Aggressive surgical interventions for severe stroke: impact on quality of life, caregiver burden and family outcomes. Can J Neurosci Nurs 2015;37(2):15–25.

[15] Green TL, Newcommon N, Demchuk A. Quality of life and caregiver outcomes following decompressive hemicraniectomy for severe stroke: a narrative literature review. Can J Neurosci Nurs 2010;32(2):24–33.

[16] Silva SM, Corrêa FI, Pereira GS, Faria CDCM, Corrêa JCF. Construct validity of the items on the Stroke Specific Quality of Life (SS-QOL) questionnaire that evaluate the participation component of the International Classification of Functioning, Disability and Health. Disabil Rehabil 2016;1–7. https://doi.org/10.1080/09638288.2016.1250117.

[17] Silva SM, Corrêa FI, Faria CDCM, Corrêa JCF. Psychometric properties of the stroke specific quality of life scale for the assessment of participation in stroke survivors using the rasch model: a preliminary study. J Phys Ther Sci 2015;27(2):389–92.

[18] Honeybul S, Gillett GR, Ho KM, Janzen C, Kruger K. Is life worth living? Decompressive craniectomy and the disability paradox. J Neurosurg 2016;125(3):775–8.

[19] Salter KL, Moses MB, Foley NC, Teasell RW. Health-related quality of life after stroke: what are we measuring? Int J Rehabil Res 2008;31(2):111–7.

[20] Jordan Z, Lockwood C, Aromataris E, Munn Z. The updated JBI model for evidence-based healthcare. Adelaide: The Joanna Briggs Institute; 2016.

[21] Pace R, Pluye P, Bartlett G, Macaulay AC, Salsberg J, Jagosh J, et al. Testing the reliability and efficiency of the pilot Mixed Methods Appraisal Tool (MMAT) for systematic mixed studies review. Int J Nurs Stud 2012;49(1):47–53.

[22] Pace R, Pluye P, Bartlett G, Macaulay A, Salsberg J, Jagosh J, et al, editors. Reliability of a tool for concomitantly appraising the methodological quality of qualitative, quantitative and mixed methods research: a pilot study. In: 38th Annual meeting of the North American Primary Care Research Group (NAPCRG), Seattle, USA; 2010.

[23] Pluye P, Gagnon M-P, Griffiths F, Johnson-Lafleur J. A scoring system for appraising mixed methods research, and concomitantly appraising qualitative, quantitative and mixed methods primary studies in Mixed Studies Reviews. Int J Nurs Stud 2009;46(4):529–46.

[24] Souto RQ, Khanassov V, Hong QN, Bush PL, Vedel I, Pluye P. Systematic mixed studies reviews: updating results on the reliability and efficiency of the mixed methods appraisal tool. Int J Nurs Stud 2015;52(1):500–1.

[25] Benejam B, Sahuquillo J, Poca MA, Frascheri L, Solana E, Delgado P, et al. Quality of life and neurobehavioral changes in survivors of malignant middle cerebral artery infarction. J Neurol 2009;256(7):1126–33.

[26] Geurts M, Van Der Worp HB, Kappelle LJ, Amelink GJ, Algra A, Hofmeijer J. Surgical decompression for space-occupying cerebral infarction: outcomes at 3 years in the randomized HAMLET trial. Stroke 2013;44(9):2506–8.

[27] Kelly ML, Rosenbaum BP, Kshettry VR, Weil RJ. Comparing clinician- and patient-reported outcome measures after hemicraniectomy for ischemic stroke. Clin Neurol Neurosurg 2014;126:24–9.

[28] Pfefferkorn T, Eppinger U, Linn J, Birnbaum T, Herzog J, Straube A, et al. Long-term outcome after suboccipital decompressive craniectomy for malignant cerebellar infarction. Stroke 2009;40(9):3045–50.

[29] McKenna A, Wilson FC, Caldwell S, Curran D, Nagaria J, Convery F. Long-term neuropsychological and psychosocial outcomes of decompressive hemicraniectomy following malignant middle cerebral artery infarctions. Disabil Rehabil 2012;34(17):1444–55.

[30] McKenna A, Wilson FC, Caldwell S, Curran D, Nagaria J, Convery F, et al. Decompressive hemicraniectomy following malignant middle cerebral artery infarctions: a mixed methods exploration of carer experience and level of burden. Disabil Rehabil 2013;35(12):995–1005.

[31] Schmidt H, Heinemann T, Elster J, Djukic M, Harscher S, Neubieser K, et al. Cognition after malignant media infarction and decompressive hemicraniectomy – a retrospective observational study. BMC Neurol 2011;11:77.

[32] Sundseth J, Sundseth A, Thommessen B, Johnsen L, Altmann M, Sorteberg W, et al. Long-term outcome and quality of life after craniectomy in speech-dominant swollen middle cerebral artery infarction. Neurocrit Care 2015;22(1):6–14.

[33] van Middelaar T, Richard E, van der Worp HB, van den Munckhof P, Nieuwkerk PT, Visser MC, et al. Quality of life after surgical decompression for a space-occupying middle cerebral artery infarct: a cohort study. BMC Neurol 2015;15(1).

[34] von Sarnowski B, Kleist-Welch Guerra W, Kohlmann T, Moock J, Khaw AV, Kessler C, et al. Long-term health-related quality of life after decompressive hemicraniectomy in stroke patients with life-threatening space-occupying brain edema. Clin Neurol Neurosurg 2012;114(6):627–33.

[35] Weil AG, Rahme R, Moumdjian R, Bouthillier A, Bojanowski MW. Quality of life following hemicraniectomy for malignant MCA territory infarction. Can J Neurol Sci 2011;38(3):434–8.

[36] Rahme R, Zuccarello M, Kleindorfer D, Adeoye OM, Ringer AJ. Decompressive hemicraniectomy for malignant middle cerebral artery territory infarction: is life worth living? J Neurosurg 2012;117(4):749–54.

[37] Quinn TJ, Dawson J, Walters MR, Lees KR. Exploring the reliability of the modified Rankin Scale. Stroke 2009;40(3):762–6.

[38] Quinn TJ, Dawson J, Walters MR, Lees KR. Reliability of the modified Rankin Scale. Stroke 2009;40(10):3393–5.

[39] Banks JL, Marotta CA. Outcomes validity and reliability of the modified Rankin scale: implications for stroke clinical trials. Stroke 2007;38(3):1091–6.

[40] Bakas T, Champion V, Perkins SM, Farran CJ, Williams LS. Psychometric testing of the revised 15-item Bakas Caregiving Outcomes Scale. Nurs Res 2006;55(5):346–55.

Index

Note: Page numbers followed by *f* indicate figures and *t* indicate tables.

Edwards Brothers Inc.
Ann Arbor MI. USA
January 22, 2018